Springer Handbook
of Crystal Growth

Govindhan Dhanaraj, Kullaiah Byrappa,
Vishwanath Prasad, Michael Dudley (Eds.)

Springer Handbook of Crystal Growth
Organization of the Handbook

Part A Fundamentals of Crystal Growth and Defect Formation
1 Crystal Growth Techniques and Characterization: An Overview
2 Nucleation at Surfaces
3 Morphology of Crystals Grown from Solutions
4 Generation and Propagation of Defects During Crystal Growth
5 Single Crystals Grown Under Unconstrained Conditions
6 Defect Formation During Crystal Growth from the Melt

Part B Crystal Growth from Melt Techniques
7 Indium Phosphide: Crystal Growth and Defect Control by Applying Steady Magnetic Fields
8 Czochralski Silicon Single Crystals for Semiconductor and Solar Cell Applications
9 Czochralski Growth of Oxide Photorefractive Crystals
10 Bulk Crystal Growth of Ternary III–V Semiconductors
11 Growth and Characterization of Antimony-Based Narrow-Bandgap III–V Semiconductor Crystals for Infrared Detector Applications
12 Crystal Growth of Oxides by Optical Floating Zone Technique
13 Laser-Heated Pedestal Growth of Oxide Fibers
14 Synthesis of Refractory Materials by Skull Melting Technique
15 Crystal Growth of Laser Host Fluorides and Oxides
16 Shaped Crystal Growth

Part C Solution Growth of Crystals
17 Bulk Single Crystals Grown from Solution on Earth and in Microgravity
18 Hydrothermal Growth of Polyscale Crystals
19 Hydrothermal and Ammonothermal Growth of ZnO and GaN
20 Stoichiometry and Domain Structure of KTP-Type Nonlinear Optical Crystals
21 High-Temperature Solution Growth: Application to Laser and Nonlinear Optical Crystals
22 Growth and Characterization of KDP and Its Analogs

Part D Crystal Growth from Vapor
23 Growth and Characterization of Silicon Carbide Crystals
24 AlN Bulk Crystal Growth by Physical Vapor Transport
25 Growth of Single-Crystal Organic Semiconductors
26 Growth of III–Nitrides with Halide Vapor Phase Epitaxy (HVPE)
27 Growth of Semiconductor Single Crystals from Vapor Phase

Part E Epitaxial Growth and Thin Films

28 Epitaxial Growth of Silicon Carbide by Chemical Vapor Deposition
29 Liquid-Phase Electroepitaxy of Semiconductors
30 Epitaxial Lateral Overgrowth of Semiconductors
31 Liquid-Phase Epitaxy of Advanced Materials
32 Molecular-Beam Epitaxial Growth of HgCdTe
33 Metalorganic Vapor-Phase Epitaxy of Diluted Nitrides and Arsenide Quantum Dots
34 Formation of SiGe Heterostructures and Their Properties
35 Plasma Energetics in Pulsed Laser and Pulsed Electron Deposition

Part F Modeling in Crystal Growth and Defects

36 Convection and Control in Melt Growth of Bulk Crystals
37 Vapor Growth of III Nitrides
38 Continuum-Scale Quantitative Defect Dynamics in Growing Czochralski Silicon Crystals
39 Models for Stress and Dislocation Generation in Melt Based Compound Crystal Growth
40 Mass and Heat Transport in BS and EFG Systems

Part G Defects Characterization and Techniques

41 Crystalline Layer Structures with X-Ray Diffractometry
42 X-Ray Topography Techniques for Defect Characterization of Crystals
43 Defect-Selective Etching of Semiconductors
44 Transmission Electron Microscopy Characterization of Crystals
45 Electron Paramagnetic Resonance Characterization of Point Defects
46 Defect Characterization in Semiconductors with Positron Annihilation Spectroscopy

Part H Special Topics in Crystal Growth

47 Protein Crystal Growth Methods
48 Crystallization from Gels
49 Crystal Growth and Ion Exchange in Titanium Silicates
50 Single-Crystal Scintillation Materials
51 Silicon Solar Cells: Materials, Devices, and Manufacturing
52 Wafer Manufacturing and Slicing Using Wiresaw

Subject Index

使 用 说 明

1.《晶体生长手册》原版为一册，分为A～H部分。考虑到使用方便以及内容一致，影印版分为6册：第1册—Part A，第2册—Part B，第3册—Part C，第4册—Part D、E，第5册—Part F、G，第6册—Part H。

2.各册在页脚重新编排页码，该页码对应中文目录。保留了原书页眉及页码，其页码对应原书目录及主题索引。

3.各册均给出完整6册书的章目录。

4.作者及其联系方式、缩略语表各册均完整呈现。

5.主题索引安排在第6册。

6.文前介绍基本采用中英文对照形式，方便读者快速浏览。

材料科学与工程图书工作室

联系电话　0451-86412421
　　　　　0451-86414559
邮　　箱　yh_bj@yahoo.com.cn
　　　　　xuyaying81823@gmail.com
　　　　　zhxh6414559@yahoo.com.cn

Springer 手册精选系列

晶体生长手册

晶体生长模型及缺陷表征

【第5册】

Springer
Handbook of

Crystal

Growth

〔美〕Govindhan Dhanaraj 等主编

（影印版）

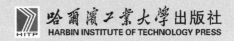

黑版贸审字08-2012-047号

Reprint from English language edition:
Springer Handbook of Crystal Growth
by Govindhan Dhanaraj, Kullaiah Byrappa, Vishwanath Prasad
and Michael Dudley
Copyright © 2010 Springer Berlin Heidelberg
Springer Berlin Heidelberg is a part of Springer Science+Business Media
All Rights Reserved

This reprint has been authorized by Springer Science & Business Media for distribution in China Mainland only and not for export there from.

图书在版编目（CIP）数据

晶体生长手册. 5, 晶体生长模型及缺陷表征 =Handbook of Crystal Growth. 5, Modeling in Crystal Growth and Defects Characterization : 英文 / (美)德哈纳拉(Dhanaraj,G.) 等主编. —影印本. —哈尔滨：哈尔滨工业大学出版社, 2013.1
（Springer手册精选系列）
ISBN 978-7-5603-3870-5

Ⅰ.①晶… Ⅱ.①德… Ⅲ.①晶体生长 – 手册 – 英文②晶体缺陷 – 手册 – 英文 Ⅳ.①O78-62②O77-62

中国版本图书馆CIP数据核字(2012)第292329号

材料科学与工程
图书工作室

责任编辑	杨 桦　许雅莹　张秀华
出版发行	哈尔滨工业大学出版社
社　　址	哈尔滨市南岗区复华四道街10号 邮编 150006
传　　真	0451-86414749
网　　址	http://hitpress.hit.edu.cn
印　　刷	哈尔滨市石桥印务有限公司
开　　本	787mm×960mm 1/16 印张 25.25
版　　次	2013年1月第1版 2013年1月第1次印刷
书　　号	ISBN 978-7-5603-3870-5
定　　价	76.00元

（如因印刷质量问题影响阅读，我社负责调换）

序 言

多年以来，有很多探索研究已经成功地描述了晶体生长的生长工艺和科学，有许多文章、专著、会议文集和手册对这一领域的前沿成果做了综合评述。这些出版物反映了人们对体材料晶体和薄膜晶体的兴趣日益增长，这是由于它们的电子、光学、机械、微结构以及不同的科学和技术应用引起的。实际上，大部分半导体和光器件的现代成果，如果没有基本的、二元的、三元的及其他不同特性和大尺寸的化合物晶体的发展则是不可能的。这些文章致力于生长机制的基本理解、缺陷形成、生长工艺和生长系统的设计，因此数量是庞大的。

本手册针对目前备受关注的体材料晶体和薄膜晶体的生长技术水平进行阐述。我们的目的是使读者了解经常使用的生长工艺、材料生产和缺陷产生的基本知识。为完成这一任务，我们精选了50多位顶尖科学家、学者和工程师，他们的合作者来自于22个不同国家。这些作者根据他们的专业所长，编写了关于晶体生长和缺陷形成共计52章内容：从熔体、溶液到气相体材料生长；外延生长；生长工艺和缺陷的模型；缺陷特性的技术以及一些现代的特别课题。

本手册分为七部分。Part A介绍基础理论：生长和表征技术综述，表面成核工艺，溶液生长晶体的形态，生长过程中成核的层错，缺陷形成的形态。

Part B介绍体材料晶体的熔体生长，一种生长大尺寸晶体的关键方法。这一部分阐述了直拉单晶工艺、泡生法、布里兹曼法、浮区熔融等工艺，以及这些方法的最新进展，例如应用磁场的晶体生长、生长轴的取向、增加底基和形状控制。本部分涉及材料从硅和Ⅲ-Ⅴ族化合物到氧化物和氟化物的广泛内容。

第三部分，本书的Part C关注了溶液生长法。在前两章里讨论了水热生长法的不同方面，随后的三章介绍了非线性和激光晶体、KTP和KDP。通过在地球上和微重力环境下生长的比较给出了重力对溶液生长法的影响的知识。

Part D的主题是气相生长。这一部分提供了碳化硅、氮化镓、氮化铝和有机半导体的气相生长的内容。随后的Part E是关于外延生长和薄膜的，主要包括从液相的化学气相淀积到脉冲激光和脉冲电子淀积。

Part F介绍了生长工艺和缺陷形成的模型。这些章节验证了工艺参数和产生晶体质量问题包括缺陷形成的直接相互作用关系。随后的Part G展示了结晶材料特性和分析的发展。Part F和G说明了预测工具和分析技术在帮助高质量的大尺寸晶体生长工艺的设计和控制方面是非常好用的。

最后的Part H致力于精选这一领域的部分现代课题，例如蛋白质晶体生长、凝胶结晶、原位结构、单晶闪烁材料的生长、光电材料和线切割大晶体薄膜。

我们希望这本施普林格手册对那些学习晶体生长的研究生，那些从事或即将从事这一领域研究的来自学术界和工业领域的研究人员、科学家和工程师以及那些制备晶体的人是有帮助的。

我们对施普林格的Dr. Claus Acheron，Dr. Werner Skolaut和le-tex的Ms Anne Strobach的特别努力表示真诚的感谢，没有他们本书将无法呈现。

我们感谢我们的作者编写了详尽的章节内容和在本书出版期间对我们的耐心。一位编者（GD）感谢他的家庭成员和Dr. Kedar Gupta(ARC Energy 的CEO)，感谢他们在本书编写期间的大力支持和鼓励。还对Peter Rudolf, David Bliss, Ishwara Bhat和Partha Dutta在A、B、E部分的编写中所给予的帮助表示感谢。

Nashua, New Hampshire, April 2010	G. Dhanaraj
Mysore, India	K. Byrappa
Denton, Texas	V. Prasad
Stony Brook, New York	M. Dudley

Preface

Over the years, many successful attempts have been made to describe the art and science of crystal growth, and many review articles, monographs, symposium volumes, and handbooks have been published to present comprehensive reviews of the advances made in this field. These publications are testament to the growing interest in both bulk and thin-film crystals because of their electronic, optical, mechanical, microstructural, and other properties, and their diverse scientific and technological applications. Indeed, most modern advances in semiconductor and optical devices would not have been possible without the development of many elemental, binary, ternary, and other compound crystals of varying properties and large sizes. The literature devoted to basic understanding of growth mechanisms, defect formation, and growth processes as well as the design of growth systems is therefore vast.

The objective of this Springer Handbook is to present the state of the art of selected topical areas of both bulk and thin-film crystal growth. Our goal is to make readers understand the basics of the commonly employed growth processes, materials produced, and defects generated. To accomplish this, we have selected more than 50 leading scientists, researchers, and engineers, and their many collaborators from 22 different countries, to write chapters on the topics of their expertise. These authors have written 52 chapters on the fundamentals of crystal growth and defect formation; bulk growth from the melt, solution, and vapor; epitaxial growth; modeling of growth processes and defects; and techniques of defect characterization, as well as some contemporary special topics.

This Springer Handbook is divided into seven parts. Part A presents the fundamentals: an overview of the growth and characterization techniques, followed by the state of the art of nucleation at surfaces, morphology of crystals grown from solutions, nucleation of dislocation during growth, and defect formation and morphology.

Part B is devoted to bulk growth from the melt, a method critical to producing large-size crystals. The chapters in this part describe the well-known processes such as Czochralski, Kyropoulos, Bridgman, and floating zone, and focus specifically on recent advances in improving these methodologies such as application of magnetic fields, orientation of the growth axis, introduction of a pedestal, and shaped growth. They also cover a wide range of materials from silicon and III–V compounds to oxides and fluorides.

The third part, Part C of the book, focuses on solution growth. The various aspects of hydrothermal growth are discussed in two chapters, while three other chapters present an overview of the nonlinear and laser crystals, KTP and KDP. The knowledge on the effect of gravity on solution growth is presented through a comparison of growth on Earth versus in a microgravity environment.

The topic of Part D is vapor growth. In addition to presenting an overview of vapor growth, this part also provides details on vapor growth of silicon carbide, gallium nitride, aluminum nitride, and organic semiconductors. This is followed by chapters on epitaxial growth and thin films in Part E. The topics range from chemical vapor deposition to liquid-phase epitaxy to pulsed laser and pulsed electron deposition.

Modeling of both growth processes and defect formation is presented in Part F. These chapters demonstrate the direct correlation between the process parameters and quality of the crystal produced, including the formation of defects. The subsequent Part G presents the techniques that have been developed for crystalline material characterization and analysis. The chapters in Parts F and G demonstrate how well predictive tools and analytical techniques have helped the design and control of growth processes for better-quality crystals of large sizes.

The final Part H is devoted to some selected contemporary topics in this field, such as protein crystal growth, crystallization from gels, in situ structural studies, growth of single-crystal scintillation materials, photovoltaic materials, and wire-saw slicing of large crystals to produce wafers.

We hope this Springer Handbook will be useful to graduate students studying crystal growth and to researchers, scientists, and engineers from academia and industry who are conducting or intend to conduct research in this field as well as those who grow crystals.

We would like to express our sincere thanks to Dr. Claus Acheron and Dr. Werner Skolaut of Springer and Ms Anne Strohbach of le-tex for their extraordinary efforts without which this handbook would not have taken its final shape.

We thank our authors for writing comprehensive chapters and having patience with us during the publication of this Handbook. One of the editors (GD) would like to thank his family members and Dr. Kedar Gupta (CEO of ARC Energy) for their generous support and encouragement during the entire course of editing this handbook. Acknowledgements are also due to Peter Rudolf, David Bliss, Ishwara Bhat, and Partha Dutta for their help in editing Parts A, B, E, and H, respectively.

Nashua, New Hampshire, April 2010	G. Dhanaraj
Mysore, India	K. Byrappa
Denton, Texas	V. Prasad
Stony Brook, New York	M. Dudley

About the Editors

Govindhan Dhanaraj is the Manager of Crystal Growth Technologies at Advanced Renewable Energy Company (ARC Energy) at Nashua, New Hampshire (USA) focusing on the growth of large size sapphire crystals for LED lighting applications, characterization and related crystal growth furnace development. He received his PhD from the Indian Institute of Science, Bangalore and his Master of Science from Anna University (India). Immediately after his doctoral degree, Dr. Dhanaraj joined a National Laboratory, presently known as Rajaramanna Center for Advanced Technology in India, where he established an advanced Crystal Growth Laboratory for the growth of optical and laser crystals. Prior to joining ARC Energy, Dr. Dhanaraj served as a Research Professor at the Department of Materials Science and Engineering, Stony Brook University, NY, and also held a position of Research Assistant Professor at Hampton University, VA. During his 25 years of focused expertise in crystal growth research, he has developed optical, laser and semiconductor bulk crystals and SiC epitaxial films using solution, flux, Czochralski, Bridgeman, gel and vapor methods, and characterized them using x-ray topography, synchrotron topography, chemical etching and optical and atomic force microscopic techniques. He co-organized a symposium on Industrial Crystal Growth under the 17th American Conference on Crystal Growth and Epitaxy in conjunction with the 14th US Biennial Workshop on Organometallic Vapor Phase Epitaxy held at Lake Geneva, WI in 2009. Dr. Dhanaraj has delivered invited lectures and also served as session chairman in many crystal growth and materials science meetings. He has published over 100 papers and his research articles have attracted over 250 rich citations.

Kullaiah Byrappa received his Doctor's degree in Crystal Growth from the Moscow State University, Moscow in 1981. He is Professor of Materials Science, Head of the Crystal Growth Laboratory, and Director of the Internal Quality Assurance Cell of the University of Mysore, India. His current research is in crystal engineering of polyscale materials through novel solution processing routes, particularly covering hydrothermal, solvothermal and supercritical methods. Professor Byrappa has co-authored the Handbook of Hydrothermal Technology, and edited 4 books as well as two special editions of Journal of Materials Science, and published 180 research papers including 26 invited reviews and book chapters on various aspects of novel routes of solution processing. Professor Byrappa has delivered over 60 keynote and invited lectures at International Conferences, and several hundreds of colloquia and seminars at various institutions around the world. He has also served as chair and co-chair for numerous international conferences. He is a Fellow of the World Academy of Ceramics. Professor Byrappa is serving in several international committees and commissions related to crystallography, crystal growth, and materials science. He is the Founder Secretary of the International Solvothermal and Hydrothermal Association. Professor Byrappa is a recipient of several awards such as the Sir C.V. Raman Award, Materials Research Society of India Medal, and the Golden Jubilee Award of the University of Mysore.

Vishwanath "Vish" Prasad is the Vice President for Research and Economic Development and Professor of Mechanical and Energy Engineering at the University of North Texas (UNT), one of the largest university in the state of Texas. He received his PhD from the University of Delaware (USA), his Masters of Technology from the Indian Institute of Technology, Kanpur, and his bachelor's from Patna University in India all in Mechanical Engineering. Prior to joining UNT in 2007, Dr. Prasad served as the Dean at Florida International University (FIU) in Miami, where he also held the position of Distinguished Professor of Engineering. Previously, he has served as a Leading Professor of Mechanical Engineering at Stony Brook University, New York, as an Associate Professor and Assistant Professor at Columbia University. He has received many special recognitions for his contributions to engineering education. Dr. Prasad's research interests include thermo-fluid sciences, energy systems, electronic materials, and computational materials processing. He has published over 200 articles, edited/co-edited several books and organized numerous conferences, symposia, and workshops. He serves as the lead editor of the Annual Review of Heat Transfer. In the past, he has served as an Associate Editor of the ASME Journal of Heat. Dr. Prasad is an elected Fellow of the American Society of Mechanical Engineers (ASME), and has served as a member of the USRA Microgravity Research Council. Dr. Prasad's research has focused on bulk growth of silicon, III-V compounds, and silicon carbide; growth of large diameter Si tube; design of crystal growth systems; and sputtering and chemical vapor deposition of thin films. He is also credited to initiate research on wire saw cutting of large crystals to produce wafers with much reduced material loss. Dr. Prasad's research has been well funded by US National Science Foundation (NSF), US Department of Defense, US Department of Energy, and industry.

Michael Dudley received his Doctoral Degree in Engineering from Warwick University, UK, in 1982. He is Professor and Chair of the Materials Science and Engineering Department at Stony Brook University, New York, USA. He is director of the Stony Brook Synchrotron Topography Facility at the National Synchrotron Light Source at Brookhaven National Laboratory, Upton New York. His current research focuses on crystal growth and characterization of defect structures in single crystals with a view to determining their origins. The primary technique used is synchrotron topography which enables analysis of defects and generalized strain fields in single crystals in general, with particular emphasis on semiconductor, optoelectronic, and optical crystals. Establishing the relationship between crystal growth conditions and resulting defect distributions is a particular thrust area of interest to Dudley, as is the correlation between electronic/optoelectronic device performance and defect distribution. Other techniques routinely used in such analysis include transmission electron microscopy, high resolution triple-axis x-ray diffraction, atomic force microscopy, scanning electron microscopy, Nomarski optical microscopy, conventional optical microscopy, IR microscopy and fluorescent laser scanning confocal microscopy. Dudley's group has played a prominent role in the development of SiC and AlN growth, characterizing crystals grown by many of the academic and commercial entities involved enabling optimization of crystal quality. He has co-authored some 315 refereed articles and 12 book chapters, and has edited 5 books. He is currently a member of the Editorial Board of Journal of Applied Physics and Applied Physics Letters and has served as Chair or Co-Chair for numerous international conferences.

List of Authors

Francesco Abbona
Università degli Studi di Torino
Dipartimento di Scienze Mineralogiche
e Petrologiche
via Valperga Caluso 35
10125 Torino, Italy
e-mail: *francesco.abbona@unito.it*

Mohan D. Aggarwal
Alabama A&M University
Department of Physics
Normal, AL 35762, USA
e-mail: *mohan.aggarwal@aamu.edu*

Marcello R.B. Andreeta
University of São Paulo
Crystal Growth and Ceramic Materials Laboratory,
Institute of Physics of São Carlos
Av. Trabalhador Sãocarlense, 400
São Carlos, SP 13560-970, Brazil
e-mail: *marcello@if.sc.usp.br*

Dino Aquilano
Università degli Studi di Torino
Facoltà di Scienze Matematiche, Fisiche e Naturali
via P. Giuria, 15
Torino, 10126, Italy
e-mail: *dino.aquilano@unito.it*

Roberto Arreguín-Espinosa
Universidad Nacional Autónoma de México
Instituto de Química
Circuito Exterior, C.U. s/n
Mexico City, 04510, Mexico
e-mail: *arrespin@unam.mx*

Jie Bai
Intel Corporation
RA3-402, 5200 NE Elam Young Parkway
Hillsboro, OR 97124-6497, USA
e-mail: *jie.bai@intel.com*

Stefan Balint
West University of Timisoara
Department of Computer Science
Blvd. V. Parvan 4
Timisoara, 300223, Romania
e-mail: *balint@math.uvt.ro*

Ashok K. Batra
Alabama A&M University
Department of Physics
4900 Meridian Street
Normal, AL 35762, USA
e-mail: *ashok.batra@aamu.edu*

Handady L. Bhat
Indian Institute of Science
Department of Physics
CV Raman Avenue
Bangalore, 560012, India
e-mail: *hlbhat@physics.iisc.ernet.in*

Ishwara B. Bhat
Rensselaer Polytechnic Institute
Electrical Computer
and Systems Engineering Department
110 8th Street, JEC 6031
Troy, NY 12180, USA
e-mail: *bhati@rpi.edu*

David F. Bliss
US Air Force Research Laboratory
Sensors Directorate Optoelectronic Technology
Branch
80 Scott Drive
Hanscom AFB, MA 01731, USA
e-mail: *david.bliss@hanscom.af.mil*

Mikhail A. Borik
Russian Academy of Sciences
Laser Materials and Technology Research Center,
A.M. Prokhorov General Physics Institute
Vavilov 38
Moscow, 119991, Russia
e-mail: *borik@lst.gpi.ru*

Liliana Braescu
West University of Timisoara
Department of Computer Science
Blvd. V. Parvan 4
Timisoara, 300223, Romania
e-mail: lilianabraescu@balint1.math.uvt.ro

Kullaiah Byrappa
University of Mysore
Department of Geology
Manasagangotri
Mysore, 570 006, India
e-mail: kbyrappa@gmail.com

Dang Cai
CVD Equipment Corporation
1860 Smithtown Ave.
Ronkonkoma, NY 11779, USA
e-mail: dcai@cvdequipment.com

Michael J. Callahan
GreenTech Solutions
92 Old Pine Drive
Hanson, MA 02341, USA
e-mail: mjcal37@yahoo.com

Joan J. Carvajal
Universitat Rovira i Virgili (URV)
Department of Physics and Crystallography
of Materials and Nanomaterials (FiCMA-FiCNA)
Campus Sescelades, C/ Marcel·lí Domingo, s/n
Tarragona 43007, Spain
e-mail: joanjosep.carvajal@urv.cat

Aaron J. Celestian
Western Kentucky University
Department of Geography and Geology
1906 College Heights Blvd.
Bowling Green, KY 42101, USA
e-mail: aaron.celestian@wku.edu

Qi-Sheng Chen
Chinese Academy of Sciences
Institute of Mechanics
15 Bei Si Huan Xi Road
Beijing, 100190, China
e-mail: qschen@imech.ac.cn

Chunhui Chung
Stony Brook University
Department of Mechanical Engineering
Stony Brook, NY 11794-2300, USA
e-mail: chuchung@ic.sunysb.edu

Ted Ciszek
Geolite/Siliconsultant
31843 Miwok Trl.
Evergreen, CO 80437, USA
e-mail: ted_ciszek@siliconsultant.com

Abraham Clearfield
Texas A&M University
Distinguished Professor of Chemistry
College Station, TX 77843-3255, USA
e-mail: clearfield@chem.tamu.edu

Hanna A. Dabkowska
Brockhouse Institute for Materials Research
Department of Physics and Astronomy
1280 Main Str W.
Hamilton, Ontario L8S 4M1, Canada
e-mail: dabkoh@mcmaster.ca

Antoni B. Dabkowski
McMaster University, BIMR
Brockhouse Institute for Materials Research,
Department of Physics and Astronomy
1280 Main Str W.
Hamilton, Ontario L8S 4M1, Canada
e-mail: dabko@mcmaster.ca

Rafael Dalmau
HexaTech Inc.
991 Aviation Pkwy Ste 800
Morrisville, NC 27560, USA
e-mail: rdalmau@hexatechinc.com

Govindhan Dhanaraj
ARC Energy
18 Celina Avenue, Unit 77
Nashua, NH 03063, USA
e-mail: dhanaraj@arc-energy.com

Ramasamy Dhanasekaran
Anna University Chennai
Crystal Growth Centre
Chennai, 600 025, India
e-mail: rdhanasekaran@annauniv.edu;
rdcgc@yahoo.com

Ernesto Diéguez
Universidad Autónoma de Madrid
Department Física de Materiales
Madrid 28049, Spain
e-mail: *ernesto.dieguez@uam.es*

Vijay K. Dixit
Raja Ramanna Center for Advance Technology
Semiconductor Laser Section,
Solid State Laser Division
Rajendra Nagar, RRCAT.
Indore, 452013, India
e-mail: *dixit@rrcat.gov.in*

Sadik Dost
University of Victoria
Crystal Growth Laboratory
Victoria, BC V8W 3P6, Canada
e-mail: *sdost@me.uvic.ca*

Michael Dudley
Stony Brook University
Department of Materials Science and Engineering
Stony Brook, NY 11794-2275, USA
e-mail: *mdudley@notes.cc.sunysb.edu*

Partha S. Dutta
Rensselaer Polytechnic Institute
Department of Electrical, Computer
and Systems Engineering
110 Eighth Street
Troy, NY 12180, USA
e-mail: *duttap@rpi.edu*

Francesc Díaz
Universitat Rovira i Virgili (URV)
Department of Physics and Crystallography
of Materials and Nanomaterials (FiCMA-FiCNA)
Campus Sescelades, C/ Marcel·lí Domingo, s/n
Tarragona 43007, Spain
e-mail: *f.diaz@urv.cat*

Paul F. Fewster
PANalytical Research Centre,
The Sussex Innovation Centre
Research Department
Falmer
Brighton, BN1 9SB, UK
e-mail: *paul.fewster@panalytical.com*

Donald O. Frazier
NASA Marshall Space Flight Center
Engineering Technology Management Office
Huntsville, AL 35812, USA
e-mail: *donald.o.frazier@nasa.gov*

James W. Garland
EPIR Technologies, Inc.
509 Territorial Drive, Ste. B
Bolingbrook, IL 60440, USA
e-mail: *jgarland@epir.com*

Thomas F. George
University of Missouri-St. Louis
Center for Nanoscience,
Department of Chemistry and Biochemistry,
Department of Physics and Astronomy
One University Boulevard
St. Louis, MO 63121, USA
e-mail: *tfgeorge@umsl.edu*

Andrea E. Gutiérrez-Quezada
Universidad Nacional Autónoma de México
Instituto de Química
Circuito Exterior, C.U. s/n
Mexico City, 04510, Mexico
e-mail: *30111390@escolar.unam.mx*

Carl Hemmingsson
Linköping University
Department of Physics, Chemistry
and Biology (IFM)
581 83 Linköping, Sweden
e-mail: *cah@ifm.liu.se*

Antonio Carlos Hernandes
University of São Paulo
Crystal Growth and Ceramic Materials Laboratory,
Institute of Physics of São Carlos
Av. Trabalhador Sãocarlense
São Carlos, SP 13560-970, Brazil
e-mail: *hernandes@if.sc.usp.br*

Koichi Kakimoto
Kyushu University
Research Institute for Applied Mechanics
6-1 Kasuga-kouen, Kasuga
816-8580 Fukuoka, Japan
e-mail: *kakimoto@riam.kyushu-u.ac.jp*

Imin Kao
State University of New York at Stony Brook
Department of Mechanical Engineering
Stony Brook, NY 11794-2300, USA
e-mail: *imin.kao@stonybrook.edu*

John J. Kelly
Utrecht University,
Debye Institute for Nanomaterials Science
Department of Chemistry
Princetonplein 5
3584 CC, Utrecht, The Netherlands
e-mail: *j.j.kelly@uu.nl*

Jeonggoo Kim
Neocera, LLC
10000 Virginia Manor Road #300
Beltsville, MD, USA
e-mail: *kim@neocera.com*

Helmut Klapper
Institut für Kristallographie
RWTH Aachen University
Aachen, Germany
e-mail: *klapper@xtal.rwth-aachen.de;*
helmut-klapper@web.de

Christine F. Klemenz Rivenbark
Krystal Engineering LLC
General Manager and Technical Director
1429 Chaffee Drive
Titusville, FL 32780, USA
e-mail: *ckr@krystalengineering.com*

Christian Kloc
Nanyang Technological University
School of Materials Science and Engineering
50 Nanyang Avenue
639798 Singapore
e-mail: *ckloc@ntu.edu.sg*

Solomon H. Kolagani
Neocera LLC
10000 Virginia Manor Road
Beltsville, MD 20705, USA
e-mail: *harsh@neocera.com*

Akinori Koukitu
Tokyo University of Agriculture and Technology
(TUAT)
Department of Applied Chemistry
2-24-16 Naka-cho, Koganei
184-8588 Tokyo, Japan
e-mail: *koukitu@cc.tuat.ac.jp*

Milind S. Kulkarni
MEMC Electronic Materials
Polysilicon and Quantitative Silicon Research
501 Pearl Drive
St. Peters, MO 63376, USA
e-mail: *mkulkarni@memc.com*

Yoshinao Kumagai
Tokyo University of Agriculture and Technology
Department of Applied Chemistry
2-24-16 Naka-cho, Koganei
184-8588 Tokyo, Japan
e-mail: *4470kuma@cc.tuat.ac.jp*

Valentin V. Laguta
Institute of Physics of the ASCR
Department of Optical Materials
Cukrovarnicka 10
Prague, 162 53, Czech Republic
e-mail: *laguta@fzu.cz*

Ravindra B. Lal
Alabama Agricultural and Mechanical University
Physics Department
4900 Meridian Street
Normal, AL 35763, USA
e-mail: *rblal@comcast.net*

Chung-Wen Lan
National Taiwan University
Department of Chemical Engineering
No. 1, Sec. 4, Roosevelt Rd.
Taipei, 106, Taiwan
e-mail: *cwlan@ntu.edu.tw*

Hongjun Li
Chinese Academy of Sciences
R & D Center of Synthetic Crystals,
Shanghai Institute of Ceramics
215 Chengbei Rd., Jiading District
Shanghai, 201800, China
e-mail: *lh_li@mail.sic.ac.cn*

Elena E. Lomonova
Russian Academy of Sciences
Laser Materials and Technology Research Center,
A.M. Prokhorov General Physics Institute
Vavilov 38
Moscow, 119991, Russia
e-mail: *lomonova@lst.gpi.ru*

Ivan V. Markov
Bulgarian Academy of Sciences
Institute of Physical Chemistry
Sofia, 1113, Bulgaria
e-mail: *imarkov@ipc.bas.bg*

Bo Monemar
Linköping University
Department of Physics, Chemistry and Biology
58183 Linköping, Sweden
e-mail: *bom@ifm.liu.se*

Abel Moreno
Universidad Nacional Autónoma de México
Instituto de Química
Circuito Exterior, C.U. s/n
Mexico City, 04510, Mexico
e-mail: *carcamo@unam.mx*

Roosevelt Moreno Rodriguez
State University of New York at Stony Brook
Department of Mechanical Engineering
Stony Brook, NY 11794-2300, USA
e-mail: *roosevelt@dove.eng.sunysb.edu*

S. Narayana Kalkura
Anna University Chennai
Crystal Growth Centre
Sardar Patel Road
Chennai, 600025, India
e-mail: *kalkura@annauniv.edu*

Mohan Narayanan
Reliance Industries Limited
1, Rich Branch court
Gaithersburg, MD 20878, USA
e-mail: *mohan.narayanan@ril.com*

Subramanian Natarajan
Madurai Kamaraj University
School of Physics
Palkalai Nagar
Madurai, India
e-mail: *s_natarajan50@yahoo.com*

Martin Nikl
Academy of Sciences of the Czech Republic (ASCR)
Department of Optical Crystals, Institute of Physics
Cukrovarnicka 10
Prague, 162 53, Czech Republic
e-mail: *nikl@fzu.cz*

Vyacheslav V. Osiko
Russian Academy of Sciences
Laser Materials and Technology Research Center,
A.M. Prokhorov General Physics Institute
Vavilov 38
Moscow, 119991, Russia
e-mail: *osiko@lst.gpi.ru*

John B. Parise
Stony Brook University
Chemistry Department
and Department of Geosciences
ESS Building
Stony Brook, NY 11794-2100, USA
e-mail: *john.parise@stonybrook.edu*

Srinivas Pendurti
ASE Technologies Inc.
11499, Chester Road
Cincinnati, OH 45246, USA
e-mail: *spendurti@asetech.com*

Benjamin G. Penn
NASA/George C. Marshall Space Flight Center
ISHM and Sensors Branch
Huntsville, AL 35812, USA
e-mail: *benjamin.g.penndr@nasa.gov*

Jens Pflaum
Julius-Maximilians Universität Würzburg
Institute of Experimental Physics VI
Am Hubland
97078 Würzburg, Germany
e-mail: *jpflaum@physik.uni-wuerzburg.de*

Jose Luis Plaza
Universidad Autónoma de Madrid
Facultad de Ciencias,
Departamento de Física de Materiales
Madrid 28049, Spain
e-mail: *joseluis.plaza@uam.es*

Udo W. Pohl
Technische Universität Berlin
Institut für Festkörperphysik EW5-1
Hardenbergstr. 36
10623 Berlin, Germany
e-mail: *pohl@physik.tu-berlin.de*

Vishwanath (Vish) Prasad
University of North Texas
1155 Union Circle
Denton, TX 76203-5017, USA
e-mail: *vish.prasad@unt.edu*

Maria Cinta Pujol
Universitat Rovira i Virgili
Department of Physics and Crystallography
of Materials and Nanomaterials (FiCMA-FiCNA)
Campus Sescelades, C/ Marcel·lí Domingo
Tarragona 43007, Spain
e-mail: *mariacinta.pujol@urv.cat*

Balaji Raghothamachar
Stony Brook University
Department of Materials Science and Engineering
310 Engineering Building
Stony Brook, NY 11794-2275, USA
e-mail: *braghoth@notes.cc.sunysb.edu*

Michael Roth
The Hebrew University of Jerusalem
Department of Applied Physics
Bergman Bld., Rm 206, Givat Ram Campus
Jerusalem 91904, Israel
e-mail: *mroth@vms.huji.ac.il*

Peter Rudolph
Leibniz Institute for Crystal Growth
Technology Development
Max-Born-Str. 2
Berlin, 12489, Germany
e-mail: *rudolph@ikz-berlin.de*

Akira Sakai
Osaka University
Department of Systems Innovation
1-3 Machikaneyama-cho, Toyonaka-shi
560-8531 Osaka, Japan
e-mail: *sakai@ee.es.osaka-u.ac.jp*

Yasuhiro Shiraki
Tokyo City University
Advanced Research Laboratories,
Musashi Institute of Technology
8-15-1 Todoroki, Setagaya-ku
158-0082 Tokyo, Japan
e-mail: *yshiraki@tcu.ac.jp*

Theo Siegrist
Florida State University
Department of Chemical
and Biomedical Engineering
2525 Pottsdamer Street
Tallahassee, FL 32310, USA
e-mail: *siegrist@eng.fsu.edu*

Zlatko Sitar
North Carolina State University
Materials Science and Engineering
1001 Capability Dr.
Raleigh, NC 27695, USA
e-mail: *sitar@ncsu.edu*

Sivalingam Sivananthan
University of Illinois at Chicago
Department of Physics
845 W. Taylor St. M/C 273
Chicago, IL 60607-7059, USA
e-mail: *siva@uic.edu; siva@epir.com*

Mikhail D. Strikovski
Neocera LLC
10000 Virginia Manor Road, suite 300
Beltsville, MD 20705, USA
e-mail: *strikovski@neocera.com*

Xun Sun
Shandong University
Institute of Crystal Materials
Shanda Road
Jinan, 250100, China
e-mail: *sunxun@icm.sdu.edu.cn*

Ichiro Sunagawa
University Tohoku University (Emeritus)
Kashiwa-cho 3-54-2, Tachikawa
Tokyo, 190-0004, Japan
e-mail: *i.sunagawa@nifty.com*

Xu-Tang Tao
Shandong University
State Key Laboratory of Crystal Materials
Shanda Nanlu 27, 250100
Jinan, China
e-mail: *txt@sdu.edu.cn*

Vitali A. Tatartchenko
Saint – Gobain, 23 Rue Louis Pouey
92800 Puteaux, France
e-mail: *vitali.tatartchenko@orange.fr*

Filip Tuomisto
Helsinki University of Technology
Department of Applied Physics
Otakaari 1 M
Espoo TKK 02015, Finland
e-mail: *filip.tuomisto@tkk.fi*

Anna Vedda
University of Milano-Bicocca
Department of Materials Science
Via Cozzi 53
20125 Milano, Italy
e-mail: *anna.vedda@unimib.it*

Lu-Min Wang
University of Michigan
Department of Nuclear Engineering
and Radiological Sciences
2355 Bonisteel Blvd.
Ann Arbor, MI 48109-2104, USA
e-mail: *lmwang@umich.edu*

Sheng-Lai Wang
Shandong University
Institute of Crystal Materials,
State Key Laboratory of Crystal Materials
Shanda Road No. 27
Jinan, Shandong, 250100, China
e-mail: *slwang@icm.sdu.edu.cn*

Shixin Wang
Micron Technology Inc.
TEM Laboratory
8000 S. Federal Way
Boise, ID 83707, USA
e-mail: *shixinwang@micron.com*

Jan L. Weyher
Polish Academy of Sciences Warsaw
Institute of High Pressure Physics
ul. Sokolowska 29/37
01/142 Warsaw, Poland
e-mail: *weyher@unipress.waw.pl*

Jun Xu
Chinese Academy of Sciences
Shanghai Institute of Ceramics
Shanghai, 201800, China
e-mail: *xujun@mail.shcnc.ac.cn*

Hui Zhang
Tsinghua University
Department of Engineering Physics
Beijing, 100084, China
e-mail: *zhhui@tsinghua.edu.cn*

Lili Zheng
Tsinghua University
School of Aerospace
Beijing, 100084, China
e-mail: *zhenglili@tsinghua.edu.cn*

Mary E. Zvanut
University of Alabama at Birmingham
Department of Physics
1530 3rd Ave S
Birmingham, AL 35294-1170, USA
e-mail: *mezvanut@uab.edu*

Zbigniew R. Zytkiewicz
Polish Academy of Sciences
Institute of Physics
Al. Lotnikow 32/46
02668 Warszawa, Poland
e-mail: *zytkie@ifpan.edu.pl*

Acknowledgements

F.37 Vapor Growth of III Nitrides
by Dang Cai, Lili Zheng, Hui Zhang

This work was supported by the DOD Multidisciplinary University Research Initiative (MURI) program administered by the Office of Naval Research under Grant N00014-01-1-1-0716 monitored by Dr. Colin E. Wood. We would like to express our gratitude to Drs. Williams Mecouch and Zlatko Sitar from North Carolina State University for providing experimental data.

目 录

缩略语

Part F 晶体生长及缺陷模型

36 熔体生长晶体体材料的传导和控制 ……………………………………………… 3
  36.1 运输过程的物理定律 ………………………………………………… 5
  36.2 熔体的流动结构 ……………………………………………………… 7
  36.3 外力对流动的控制 …………………………………………………… 16
  36.4 前 景 …………………………………………………………………… 26
  参考文献 …………………………………………………………………… 26

37 Ⅲ族氮化物的气相生长 ………………………………………………………… 31
  37.1 Ⅲ族氮化物的气相生长概述 ………………………………………… 32
  37.2 AlN/GaN气相淀积的数学模型 ……………………………………… 36
  37.3 气相淀积AlN/GaN的表征 …………………………………………… 39
  37.4 GaN的IVPE生长模型——个案研究 ………………………………… 46
  37.5 气相GaN/AlN膜生长的表面形成 …………………………………… 62
  37.6 结 语 …………………………………………………………………… 63
  参考文献 …………………………………………………………………… 64

38 生长直拉硅晶体中连续尺寸量子缺陷动力学 ………………………………… 69
  38.1 微缺陷的发现 ………………………………………………………… 71
  38.2 无杂质时的缺陷动力学 ……………………………………………… 72
  38.3 有氧时的直拉缺陷动力学 …………………………………………… 92
  38.4 有氮时的直拉缺陷动力学 …………………………………………… 101
  38.5 直拉硅单晶中空位的横向合并 ……………………………………… 109
  38.6 结 论 …………………………………………………………………… 116
  参考文献 …………………………………………………………………… 120

39 熔体基底化合物晶体生长中应力和位错产生的模型 ………………………… 123
  39.1 综 述 …………………………………………………………………… 123
  39.2 晶体生长过程 ………………………………………………………… 124
  39.3 半导体材料的位错分布 ……………………………………………… 125
  39.4 位错产生的模型 ……………………………………………………… 127

  39.5 晶体的金刚石结构 ᠁ 131
  39.6 半导体的变形特性 ᠁ 134
  39.7 Haasen模型对晶体生长的应用 ᠁ 138
  39.8 替代模式 ᠁ 139
  39.9 模型概述和数值实现 ᠁ 148
  39.10 数值结果 ᠁ 150
  39.11 总 结 ᠁ 162
  参考文献 ᠁ 163

40 BS和EFG系统中的质量和热量传输 ᠁ 167
  40.1 杂质分布的基预测模型——垂直BS系统 ᠁ 168
  40.2 杂质分布的基预测模型——EFG系统 ᠁ 177
  参考文献 ᠁ 188

Part G 缺陷表征及技术

41 晶体层结构的X射线衍射表征 ᠁ 193
  41.1 X射线衍射 ᠁ 194
  41.2 层结构的基本直接X射线衍射分析 ᠁ 195
  41.3 设备和理论思考 ᠁ 200
  41.4 从低到高的复杂性分析实例 ᠁ 201
  41.5 快速分析 ᠁ 207
  41.6 薄膜微映射 ᠁ 208
  41.7 展 望 ᠁ 209
  参考文献 ᠁ 210

42 晶体缺陷表征的X射线形貌技术 ᠁ 213
  42.1 X射线形貌的基本原则 ᠁ 214
  42.2 X射线形貌技术的发展历史 ᠁ 216
  42.3 X射线形貌技术和几何学 ᠁ 218
  42.4 X射线形貌技术理论背景 ᠁ 223
  42.5 X射线形貌上缺陷的对比原理 ᠁ 228
  42.6 X射线形貌上的缺陷分析 ᠁ 233
  42.7 目前的应用状况和发展 ᠁ 237
  参考文献 ᠁ 238

43 半导体的缺陷选择性刻蚀 ... 241
43.1 半导体的湿法刻蚀：机制 ... 242
43.2 半导体的湿法刻蚀：结构和缺陷选择性 ... 247
43.3 缺陷选择性刻蚀方法 ... 249
参考文献 ... 261

44 晶体的透射电子显微镜表征 ... 265
44.1 缺陷的TEM表征的理论基础 ... 265
44.2 半导体系统TEM应用的典型实例 ... 281
44.3 结语：目前的应用状况和发展 ... 302
参考文献 ... 303

45 点缺陷的电子自旋共振表征 ... 309
45.1 电子自旋共振 ... 310
45.2 EPR分析 ... 312
45.3 ERP技术范围 ... 322
45.4 辅助仪器和支持技术 ... 326
45.5 总结与最终思考 ... 333
参考文献 ... 334

46 半导体缺陷特性的正电子湮没光谱表征 ... 339
46.1 正电子湮没光谱 ... 340
46.2 点缺陷的识别及其电荷状态 ... 348
46.3 缺陷、掺杂和电子补偿 ... 353
46.4 点缺陷和生长条件 ... 357
46.5 总结 ... 364
参考文献 ... 364

Contents

List of Abbreviations

Part F Modeling in Crystal Growth and Defects

36 Convection and Control in Melt Growth of Bulk Crystals
Chung-Wen Lan 1215
- 36.1 Physical Laws for Transport Processes 1217
- 36.2 Flow Structures in the Melt 1219
- 36.3 Flow Control by External Forces 1228
- 36.4 Outlook 1238
- **References** 1238

37 Vapor Growth of III Nitrides
Dang Cai, Lili Zheng, Hui Zhang 1243
- 37.1 Overview of Vapor Growth of III Nitrides 1244
- 37.2 Mathematical Models for AlN/GaN Vapor Deposition 1248
- 37.3 Characteristics of AlN/GaN Vapor Deposition 1251
- 37.4 Modeling of GaN IVPE Growth – A Case Study 1258
- 37.5 Surface Evolution of GaN/AlN Film Growth from Vapor 1274
- 37.6 Concluding Remarks 1275
- **References** 1276

38 Continuum-Scale Quantitative Defect Dynamics in Growing Czochralski Silicon Crystals
Milind S. Kulkarni 1281
- 38.1 The Discovery of Microdefects 1283
- 38.2 Defect Dynamics in the Absence of Impurities 1284
- 38.3 Czochralski Defect Dynamics in the Presence of Oxygen ... 1304
- 38.4 Czochralski Defect Dynamics in the Presence of Nitrogen . 1313
- 38.5 The Lateral Incorporation of Vacancies in Czochralski Silicon Crystals 1321
- 38.6 Conclusions 1328
- **References** 1332

39 Models for Stress and Dislocation Generation in Melt Based Compound Crystal Growth
Vishwanath (Vish) Prasad, Srinivas Pendurti 1335
- 39.1 Overview 1335
- 39.2 Crystal Growth Processes 1336
- 39.3 Dislocations in Semiconductors Materials 1337

	39.4	Models for Dislocation Generation	1339
	39.5	Diamond Structure of the Crystal	1343
	39.6	Deformation Behavior of Semiconductors	1346
	39.7	Application of the Haasen Model to Crystal Growth	1350
	39.8	An Alternative Model	1351
	39.9	Model Summary and Numerical Implementation	1360
	39.10	Numerical Results	1362
	39.11	Summary	1374
	References		1375

40 Mass and Heat Transport in BS and EFG Systems
Thomas F. George, Stefan Balint, Liliana Braescu 1379

	40.1	Model-Based Prediction of the Impurity Distribution – Vertical BS System	1380
	40.2	Model-Based Prediction of the Impurity Distribution – EFG System	1389
	References		1400

Part G Defects Characterization and Techniques

41 Crystalline Layer Structures with X-Ray Diffractometry
Paul F. Fewster 1405

	41.1	X-Ray Diffractometry	1406
	41.2	Basic Direct X-Ray Diffraction Analysis from Layered Structures	1407
	41.3	Instrumental and Theoretical Considerations	1412
	41.4	Examples of Analysis from Low to High Complexity	1413
	41.5	Rapid Analysis	1419
	41.6	Wafer Micromapping	1420
	41.7	The Future	1421
	References		1422

42 X-Ray Topography Techniques for Defect Characterization of Crystals
Balaji Raghothamachar, Michael Dudley, Govindhan Dhanaraj 1425

	42.1	Basic Principles of X-Ray Topography	1426
	42.2	Historical Development of the X-Ray Topography Technique	1428
	42.3	X-Ray Topography Techniques and Geometry	1430
	42.4	Theoretical Background for X-Ray Topography	1435
	42.5	Mechanisms for Contrast on X-Ray Topographs	1440
	42.6	Analysis of Defects on X-Ray Topographs	1445
	42.7	Current Application Status and Development	1449
	References		1450

43 Defect-Selective Etching of Semiconductors
Jan L. Weyher, John J. Kelly 1453
43.1 Wet Etching of Semiconductors: Mechanisms 1454
43.2 Wet Etching of Semiconductors: Morphology and Defect Selectivity .. 1459
43.3 Defect-Selective Etching Methods 1461
References 1473

44 Transmission Electron Microscopy Characterization of Crystals
Jie Bai, Shixin Wang, Lu-Min Wang, Michael Dudley 1477
44.1 Theoretical Basis of TEM Characterization of Defects 1477
44.2 Selected Examples of Application of TEM to Semiconductor Systems . 1493
44.3 Concluding Remarks: Current Application Status and Development .. 1514
References 1515

45 Electron Paramagnetic Resonance Characterization of Point Defects
Mary E. Zvanut 1521
45.1 Electronic Paramagnetic Resonance 1522
45.2 EPR Analysis 1524
45.3 Scope of EPR Technique 1534
45.4 Supplementary Instrumentation and Supportive Techniques 1538
45.5 Summary and Final Thoughts 1545
References 1546

46 Defect Characterization in Semiconductors with Positron Annihilation Spectroscopy
Filip Tuomisto 1551
46.1 Positron Annihilation Spectroscopy 1552
46.2 Identification of Point Defects and Their Charge States 1560
46.3 Defects, Doping, and Electrical Compensation 1565
46.4 Point Defects and Growth Conditions 1569
46.5 Summary 1576
References 1576

List of Abbreviations

μ-PD	micro-pulling-down
1S-ELO	one-step ELO structure
2-D	two-dimensional
2-DNG	two-dimensional nucleation growth
2S-ELO	double layer ELO
3-D	three-dimensional
4T	quaterthiophene
6T	sexithienyl
8MR	eight-membered ring
8T	hexathiophene

A

a-Si	amorphous silicon
A/D	analogue-to-digital
AA	additional absorption
AANP	2-adamantylamino-5-nitropyridine
AAS	atomic absorption spectroscopy
AB	Abrahams and Burocchi
ABES	absorption-edge spectroscopy
AC	alternate current
ACC	annular capillary channel
ACRT	accelerated crucible rotation technique
ADC	analog-to-digital converter
ADC	automatic diameter control
ADF	annular dark field
ADP	ammonium dihydrogen phosphate
AES	Auger electron spectroscopy
AFM	atomic force microscopy
ALE	arbitrary Lagrangian Eulerian
ALE	atomic layer epitaxy
ALUM	aluminum potassium sulfate
ANN	artificial neural network
AO	acoustooptic
AP	atmospheric pressure
APB	antiphase boundaries
APCF	advanced protein crystallization facility
APD	avalanche photodiode
APPLN	aperiodic poled LN
APS	Advanced Photon Source
AR	antireflection
AR	aspect ratio
ART	aspect ratio trapping
ATGSP	alanine doped triglycine sulfo-phosphate
AVT	angular vibration technique

B

BA	Born approximation
BAC	band anticrossing
BBO	BaB_2O_4
BCF	Burton–Cabrera–Frank
BCT	$Ba_{0.77}Ca_{0.23}TiO_3$
BCTi	$Ba_{1-x}Ca_xTiO_3$
BE	bound exciton
BF	bright field
BFDH	Bravais–Friedel–Donnay–Harker
BGO	$Bi_{12}GeO_{20}$
BIBO	BiB_3O_6
BLIP	background-limited performance
BMO	$Bi_{12}MO_{20}$
BN	boron nitride
BOE	buffered oxide etch
BPD	basal-plane dislocation
BPS	Burton–Prim–Slichter
BPT	bipolar transistor
BS	Bridgman–Stockbarger
BSCCO	Bi–Sr–Ca–Cu–O
BSF	bounding stacking fault
BSO	$Bi_{20}SiO_{20}$
BTO	$Bi_{12}TiO_{20}$
BU	building unit
BaREF	barium rare-earth fluoride
BiSCCO	$Bi_2Sr_2CaCu_2O_n$

C

C–V	capacitance–voltage
CALPHAD	calculation of phase diagram
CBED	convergent-beam electron diffraction
CC	cold crucible
CCC	central capillary channel
CCD	charge-coupled device
CCVT	contactless chemical vapor transport
CD	convection diffusion
CE	counterelectrode
CFD	computational fluid dynamics
CFD	cumulative failure distribution
CFMO	Ca_2FeMoO_6
CFS	continuous filtration system
CGG	calcium gallium germanate
CIS	copper indium diselenide
CL	cathode-ray luminescence
CL	cathodoluminescence
CMM	coordinate measuring machine
CMO	$CaMoO_4$
CMOS	complementary metal–oxide–semiconductor
CMP	chemical–mechanical polishing
CMP	chemomechanical polishing

COD	calcium oxalate dihydrate		DS	directional solidification
COM	calcium oxalate-monohydrate		DSC	differential scanning calorimetry
COP	crystal-originated particle		DSE	defect-selective etching
CP	critical point		DSL	diluted Sirtl with light
CPU	central processing unit		DTA	differential thermal analysis
CRSS	critical-resolved shear stress		DTGS	deuterated triglycine sulfate
CSMO	$Ca_{1-x}Sr_xMoO_3$		DVD	digital versatile disk
CST	capillary shaping technique		DWBA	distorted-wave Born approximation
CST	crystalline silico titanate		DWELL	dot-in-a-well
CT	computer tomography			
CTA	$CsTiOAsO_4$			
CTE	coefficient of thermal expansion			
CTF	contrast transfer function			

E

CTR	crystal truncation rod		EADM	extended atomic distance mismatch
CV	Cabrera–Vermilyea		EALFZ	electrical-assisted laser floating zone
CVD	chemical vapor deposition		EB	electron beam
CVT	chemical vapor transport		EBIC	electron-beam-induced current
CW	continuous wave		ECE	end chain energy
CZ	Czochralski		ECR	electron cyclotron resonance
CZT	Czochralski technique		EDAX	energy-dispersive x-ray analysis
			EDMR	electrically detected magnetic resonance
			EDS	energy-dispersive x-ray spectroscopy
			EDT	ethylene dithiotetrathiafulvalene

D

			EDTA	ethylene diamine tetraacetic acid
D/A	digital to analog		EELS	electron energy-loss spectroscopy
DBR	distributed Bragg reflector		EFG	edge-defined film-fed growth
DC	direct current		EFTEM	energy-filtered transmission electron microscopy
DCAM	diffusion-controlled crystallization apparatus for microgravity		ELNES	energy-loss near-edge structure
DCCZ	double crucible CZ		ELO	epitaxial lateral overgrowth
DCPD	dicalcium-phosphate dihydrate		EM	electromagnetic
DCT	dichlorotetracene		EMA	effective medium theory
DD	dislocation dynamics		EMC	electromagnetic casting
DESY	Deutsches Elektronen Synchrotron		EMCZ	electromagnetic Czochralski
DF	dark field		EMF	electromotive force
DFT	density function theory		ENDOR	electron nuclear double resonance
DFW	defect free width		EO	electrooptic
DGS	diglycine sulfate		EP	EaglePicher
DI	deionized		EPD	etch pit density
DIA	diamond growth		EPMA	electron microprobe analysis
DIC	differential interference contrast		EPR	electron paramagnetic resonance
DICM	differential interference contrast microscopy		erfc	error function
DKDP	deuterated potassium dihydrogen phosphate		ES	equilibrium shape
			ESP	edge-supported pulling
DLATGS	deuterated L-alanine-doped triglycine sulfate		ESR	electron spin resonance
			EVA	ethyl vinyl acetate
DLTS	deep-level transient spectroscopy			
DMS	discharge mass spectroscopy			
DNA	deoxyribonucleic acid			

F

DOE	Department of Energy		F	flat
DOS	density of states		FAM	free abrasive machining
DPH-BDS	2,6-diphenylbenzo[1,2-b:4,5-b']diselenophene		FAP	$Ca_5(PO_4)_3F$
			FCA	free carrier absorption
DPPH	2,2-diphenyl-1-picrylhydrazyl		fcc	face-centered cubic
DRS	dynamic reflectance spectroscopy		FEC	full encapsulation Czochralski

FEM	finite element method	HIV-AIDS	human immunodeficiency virus–acquired immunodeficiency syndrome
FES	fluid experiment system		
FET	field-effect transistor	HK	high potassium content
FFT	fast Fourier transform	HLA	half-loop array
FIB	focused ion beam	HLW	high-level waste
FOM	figure of merit	HMDS	hexamethyldisilane
FPA	focal-plane array	HMT	hexamethylene tetramine
FPE	Fokker–Planck equation	HNP	high nitrogen pressure
FSLI	femtosecond laser irradiation	HOE	holographic optical element
FT	flux technique	HOLZ	higher-order Laue zone
FTIR	Fourier-transform infrared	HOMO	highest occupied molecular orbital
FWHM	full width at half-maximum	HOPG	highly oriented pyrolytic graphite
FZ	floating zone	HOT	high operating temperature
FZT	floating zone technique	HP	Hartman–Perdok
		HPAT	high-pressure ammonothermal technique
		HPHT	high-pressure high-temperature
		HRTEM	high-resolution transmission electron microscopy

G

GAME	gel acupuncture method
GDMS	glow-discharge mass spectrometry
GE	General Electric
GGG	gadolinium gallium garnet
GNB	geometrically necessary boundary
GPIB	general purpose interface bus
GPMD	geometric partial misfit dislocation
GRI	growth interruption
GRIIRA	green-radiation-induced infrared absorption
GS	growth sector
GSAS	general structure analysis software
GSGG	$Gd_3Sc_2Ga_3O_{12}$
GSMBE	gas-source molecular-beam epitaxy
GSO	Gd_2SiO_5
GU	growth unit

HRXRD	high-resolution x-ray diffraction
HSXPD	hemispherically scanned x-ray photoelectron diffraction
HT	hydrothermal
HTS	high-temperature solution
HTSC	high-temperature superconductor
HVPE	halide vapor-phase epitaxy
HVPE	hydride vapor-phase epitaxy
HWC	hot-wall Czochralski
HZM	horizontal ZM

I

IBAD	ion-beam-assisted deposition
IBE	ion beam etching
IC	integrated circuit
IC	ion chamber
ICF	inertial confinement fusion
ID	inner diameter
ID	inversion domain
IDB	incidental dislocation boundary
IDB	inversion domain boundary
IF	identification flat
IG	inert gas
IK	intermediate potassium content
ILHPG	indirect laser-heated pedestal growth
IML-1	International Microgravity Laboratory
IMPATT	impact ionization avalanche transit-time
IP	image plate
IPA	isopropyl alcohol
IR	infrared
IRFPA	infrared focal plane array
IS	interfacial structure
ISS	ion-scattering spectroscopy
ITO	indium-tin oxide
ITTFA	iterative target transform factor analysis
IVPE	iodine vapor-phase epitaxy

H

HA	hydroxyapatite
HAADF	high-angle annular dark field
HAADF-STEM	high-angle annular dark field in scanning transmission electron microscope
HAP	hydroxyapatite
HB	horizontal Bridgman
HBM	Hottinger Baldwin Messtechnik GmbH
HBT	heterostructure bipolar transistor
HBT	horizontal Bridgman technique
HDPCG	high-density protein crystal growth
HE	high energy
HEM	heat-exchanger method
HEMT	high-electron-mobility transistor
HF	hydrofluoric acid
HGF	horizontal gradient freezing
HH	heavy-hole
HH-PCAM	handheld protein crystallization apparatus for microgravity
HIV	human immunodeficiency virus

J

JDS	joint density of states
JFET	junction FET

K

K	kinked
KAP	potassium hydrogen phthalate
KDP	potassium dihydrogen phosphate
KGW	$KY(WO_4)_2$
KGdP	$KGd(PO_3)_4$
KLYF	$KLiYF_5$
KM	Kubota–Mullin
KMC	kinetic Monte Carlo
KN	$KNbO_3$
KNP	$KNd(PO_3)_4$
KPZ	Kardar–Parisi–Zhang
KREW	$KRE(WO_4)_2$
KTA	potassium titanyl arsenate
KTN	potassium niobium tantalate
KTP	potassium titanyl phosphate
KTa	$KTaO_3$
KTaN	$KTa_{1-x}Nb_xO_3$
KYF	KYF_4
KYW	$KY(WO_4)_2$

L

LACBED	large-angle convergent-beam diffraction
LAFB	L-arginine tetrafluoroborate
LAGB	low-angle grain boundary
LAO	$LiAlO_2$
LAP	L-arginine phosphate
LBIC	light-beam induced current
LBIV	light-beam induced voltage
LBO	LiB_3O_5
LBO	$LiBO_3$
LBS	laser-beam scanning
LBSM	laser-beam scanning microscope
LBT	laser-beam tomography
LCD	liquid-crystal display
LD	laser diode
LDT	laser-induced damage threshold
LEC	liquid encapsulation Czochralski
LED	light-emitting diode
LEEBI	low-energy electron-beam irradiation
LEM	laser emission microanalysis
LEO	lateral epitaxial overgrowth
LES	large-eddy simulation
LG	$LiGaO_2$
LGN	$La_3Ga_{5.5}Nb_{0.5}O_{14}$
LGO	$LaGaO_3$
LGS	$La_3Ga_5SiO_{14}$
LGT	$La_3Ga_{5.5}Ta_{0.5}O_{14}$
LH	light hole
LHFB	L-histidine tetrafluoroborate
LHPG	laser-heated pedestal growth
LID	laser-induced damage
LK	low potassium content
LLNL	Lawrence Livermore National Laboratory
LLO	laser lift-off
LLW	low-level waste
LN	$LiNbO_3$
LP	low pressure
LPD	liquid-phase diffusion
LPE	liquid-phase epitaxy
LPEE	liquid-phase electroepitaxy
LPS	$Lu_2Si_2O_7$
LSO	Lu_2SiO_5
LST	laser scattering tomography
LST	local shaping technique
LT	low-temperature
LTa	$LiTaO_3$
LUMO	lowest unoccupied molecular orbital
LVM	local vibrational mode
LWIR	long-wavelength IR
LY	light yield
LiCAF	$LiCaAlF_6$
LiSAF	lithium strontium aluminum fluoride

M

M–S	melt–solid
MAP	magnesium ammonium phosphate
MASTRAPP	multizone adaptive scheme for transport and phase change processes
MBE	molecular-beam epitaxy
MBI	multiple-beam interferometry
MC	multicrystalline
MCD	magnetic circular dichroism
MCT	HgCdTe
MCZ	magnetic Czochralski
MD	misfit dislocation
MD	molecular dynamics
ME	melt epitaxy
ME	microelectronics
MEMS	microelectromechanical system
MESFET	metal-semiconductor field effect transistor
MHP	magnesium hydrogen phosphate-trihydrate
MI	morphological importance
MIT	Massachusetts Institute of Technology
ML	monolayer
MLEC	magnetic liquid-encapsulated Czochralski

MLEK	magnetically stabilized liquid-encapsulated Kyropoulos	NTRS	National Technology Roadmap for Semiconductors
MMIC	monolithic microwave integrated circuit	NdBCO	$NdBa_2Cu_3O_{7-x}$
MNA	2-methyl-4-nitroaniline		
MNSM	modified nonstationary model		

O

MOCVD	metalorganic chemical vapor deposition
MOCVD	molecular chemical vapor deposition
MODFET	modulation-doped field-effect transistor
MOMBE	metalorganic MBE
MOS	metal–oxide–semiconductor
MOSFET	metal–oxide–semiconductor field-effect transistor
MOVPE	metalorganic vapor-phase epitaxy
mp	melting point
MPMS	mold-pushing melt-supplying
MQSSM	modified quasi-steady-state model
MQW	multiple quantum well
MR	melt replenishment
MRAM	magnetoresistive random-access memory
MRM	melt replenishment model
MSUM	monosodium urate monohydrate
MTDATA	metallurgical thermochemistry database
MTS	methyltrichlorosilane
MUX	multiplexor
MWIR	mid-wavelength infrared
MWRM	melt without replenishment model
MXRF	micro-area x-ray fluorescence

OCP	octacalcium phosphate
ODE	ordinary differential equation
ODLN	opposite domain LN
ODMR	optically detected magnetic resonance
OEIC	optoelectronic integrated circuit
OF	orientation flat
OFZ	optical floating zone
OLED	organic light-emitting diode
OMVPE	organometallic vapor-phase epitaxy
OPO	optical parametric oscillation
OSF	oxidation-induced stacking fault

P

PAMBE	photo-assisted MBE
PB	proportional band
PBC	periodic bond chain
pBN	pyrolytic boron nitride
PC	photoconductivity
PCAM	protein crystallization apparatus for microgravity
PCF	primary crystallization field
PCF	protein crystal growth facility
PCM	phase-contrast microscopy
PD	Peltier interface demarcation
PD	photodiode
PDE	partial differential equation
PDP	programmed data processor
PDS	periodic domain structure
PE	pendeo-epitaxy
PEBS	pulsed electron beam source
PEC	polyimide environmental cell
PECVD	plasma-enhanced chemical vapor deposition
PED	pulsed electron deposition
PEO	polyethylene oxide
PET	positron emission tomography
PID	proportional–integral–differential
PIN	positive intrinsic negative diode
PL	photoluminescence
PLD	pulsed laser deposition
PMNT	$Pb(Mg, Nb)_{1-x}Ti_xO_3$
PPKTP	periodically poled KTP
PPLN	periodic poled LN
PPLN	periodic poling lithium niobate
ppy	polypyrrole
PR	photorefractive
PSD	position-sensitive detector
PSF	prismatic stacking fault

N

N	nucleus
N	nutrient
NASA	National Aeronautics and Space Administration
NBE	near-band-edge
NBE	near-bandgap emission
NCPM	noncritically phase matched
NCS	neighboring confinement structure
NGO	$NdGaO_3$
NIF	National Ignition Facility
NIR	near-infrared
NIST	National Institute of Standards and Technology
NLO	nonlinear optic
NMR	nuclear magnetic resonance
NP	no-phonon
NPL	National Physical Laboratory
NREL	National Renewable Energy Laboratory
NS	Navier–Stokes
NSF	National Science Foundation
nSLN	nearly stoichiometric lithium niobate
NSLS	National Synchrotron Light Source
NSM	nonstationary model

PSI	phase-shifting interferometry	RTV	room temperature vulcanizing
PSM	phase-shifting microscopy	R&D	research and development
PSP	pancreatic stone protein		
PSSM	pseudo-steady-state model		
PSZ	partly stabilized zirconium dioxide		

S

PT	pressure–temperature		
PV	photovoltaic	S	stepped
PVA	polyvinyl alcohol	SAD	selected area diffraction
PVD	physical vapor deposition	SAM	scanning Auger microprobe
PVE	photovoltaic efficiency	SAW	surface acoustical wave
PVT	physical vapor transport	SBN	strontium barium niobate
PWO	$PbWO_4$	SC	slow cooling
PZNT	$Pb(Zn, Nb)_{1-x}Ti_xO_3$	SCBG	slow-cooling bottom growth
PZT	lead zirconium titanate	SCC	source-current-controlled
		SCF	single-crystal fiber
		SCF	supercritical fluid technology

Q

		SCN	succinonitrile
		SCW	supercritical water
QD	quantum dot	SD	screw dislocation
QDT	quantum dielectric theory	SE	spectroscopic ellipsometry
QE	quantum efficiency	SECeRTS	small environmental cell for real-time studies
QPM	quasi-phase-matched		
QPMSHG	quasi-phase-matched second-harmonic generation	SEG	selective epitaxial growth
		SEM	scanning electron microscope
QSSM	quasi-steady-state model	SEM	scanning electron microscopy
QW	quantum well	SEMATECH	Semiconductor Manufacturing Technology
QWIP	quantum-well infrared photodetector		
		SF	stacking fault

R

		SFM	scanning force microscopy
		SGOI	SiGe-on-insulator
RAE	rotating analyzer ellipsometer	SH	second harmonic
RBM	rotatory Bridgman method	SHG	second-harmonic generation
RC	reverse current	SHM	submerged heater method
RCE	rotating compensator ellipsometer	SI	semi-insulating
RE	rare earth	SIA	Semiconductor Industry Association
RE	reference electrode	SIMS	secondary-ion mass spectrometry
REDG	recombination enhanced dislocation glide	SIOM	Shanghai Institute of Optics and Fine Mechanics
RELF	rare-earth lithium fluoride		
RF	radiofrequency	SL	superlattice
RGS	ribbon growth on substrate	SL-3	Spacelab-3
RHEED	reflection high-energy electron diffraction	SLI	solid–liquid interface
		SLN	stoichiometric LN
RI	refractive index	SM	skull melting
RIE	reactive ion etching	SMB	stacking mismatch boundary
RMS	root-mean-square	SMG	surfactant-mediated growth
RNA	ribonucleic acid	SMT	surface-mount technology
ROIC	readout integrated circuit	SNR	signal-to-noise ratio
RP	reduced pressure	SNT	sodium nonatitanate
RPI	Rensselaer Polytechnic Institute	SOI	silicon-on-insulator
RSM	reciprocal space map	SP	sputtering
RSS	resolved shear stress	sPC	scanning photocurrent
RT	room temperature	SPC	Scientific Production Company
RTA	$RbTiOAsO_4$	SPC	statistical process control
RTA	rapid thermal annealing	SR	spreading resistance
RTCVD	rapid-thermal chemical vapor deposition	SRH	Shockley–Read–Hall
RTP	$RbTiOPO_4$	SRL	strain-reducing layer
RTPL	room-temperature photoluminescence	SRS	stimulated Raman scattering
RTR	ribbon-to-ribbon		

SRXRD	spatially resolved XRD	TTV	total thickness variation
SS	solution-stirring	TV	television
SSL	solid-state laser	TVM	three-vessel solution circulating method
SSM	sublimation sandwich method	TVTP	time-varying temperature profile
ST	synchrotron topography	TWF	transmitted wavefront
STC	standard testing condition	TZM	titanium zirconium molybdenum
STE	self-trapped exciton	TZP	tetragonal phase
STEM	scanning transmission electron microscopy		
STM	scanning tunneling microscopy		

U

UC	universal compliant
UDLM	uniform-diffusion-layer model
UHPHT	ultrahigh-pressure high-temperature
UHV	ultrahigh-vacuum
ULSI	ultralarge-scale integrated circuit
UV	ultraviolet
UV-vis	ultraviolet–visible
UVB	ultraviolet B

STOS	sodium titanium oxide silicate
STP	stationary temperature profile
STS	space transportation system
SWBXT	synchrotron white beam x-ray topography
SWIR	short-wavelength IR
SXRT	synchrotron x-ray topography

T

TCE	trichloroethylene
TCNQ	tetracyanoquinodimethane
TCO	thin-film conducting oxide
TCP	tricalcium phosphate
TD	Tokyo Denpa
TD	threading dislocation
TDD	threading dislocation density
TDH	temperature-dependent Hall
TDMA	tridiagonal matrix algorithm
TED	threading edge dislocation
TEM	transmission electron microscopy
TFT-LCD	thin-film transistor liquid-crystal display
TGS	triglycine sulfate
TGT	temperature gradient technique
TGW	Thomson–Gibbs–Wulff
TGZM	temperature gradient zone melting
THM	traveling heater method
TMCZ	transverse magnetic-field-applied Czochralski
TMOS	tetramethoxysilane
TO	transverse optic
TPB	three-phase boundary
TPRE	twin-plane reentrant-edge effect
TPS	technique of pulling from shaper
TQM	total quality management
TRAPATT	trapped plasma avalanche-triggered transit
TRM	temperature-reduction method
TS	titanium silicate
TSC	thermally stimulated conductivity
TSD	threading screw dislocation
TSET	two shaping elements technique
TSFZ	traveling solvent floating zone
TSL	thermally stimulated luminescence
TSSG	top-seeded solution growth
TSSM	Tatarchenko steady-state model
TSZ	traveling solvent zone

V

VAS	void-assisted separation
VB	valence band
VB	vertical Bridgman
VBT	vertical Bridgman technique
VCA	virtual-crystal approximation
VCSEL	vertical-cavity surface-emitting laser
VCZ	vapor pressure controlled Czochralski
VDA	vapor diffusion apparatus
VGF	vertical gradient freeze
VLS	vapor–liquid–solid
VLSI	very large-scale integrated circuit
VLWIR	very long-wavelength infrared
VMCZ	vertical magnetic-field-applied Czochralski
VP	vapor phase
VPE	vapor-phase epitaxy
VST	variable shaping technique
VT	Verneuil technique
VTGT	vertical temperature gradient technique
VUV	vacuum ultraviolet

W

WBDF	weak-beam dark-field
WE	working electrode

X

XP	x-ray photoemission
XPS	x-ray photoelectron spectroscopy
XPS	x-ray photoemission spectroscopy
XRD	x-ray diffraction
XRPD	x-ray powder diffraction
XRT	x-ray topography

Y

YAB	$YAl_3(BO_3)_4$
YAG	yttrium aluminum garnet
YAP	yttrium aluminum perovskite
YBCO	$YBa_2Cu_3O_{7-x}$
YIG	yttrium iron garnet
YL	yellow luminescence
YLF	$LiYF_4$
YOF	yttrium oxyfluoride
YPS	$(Y_2)Si_2O_7$
YSO	Y_2SiO_5

Z

ZA	Al_2O_3-$ZrO_2(Y_2O_3)$
ZLP	zero-loss peak
ZM	zone-melting
ZNT	ZN-Technologies
ZOLZ	zero-order Laue zone

Part F Modeling in Crystal Growth and Defects

36 Convection and Control in Melt Growth of Bulk Crystals
Chung-Wen Lan, Taipei, Taiwan

37 Vapor Growth of III Nitrides
Dang Cai, Ronkonkoma, USA
Lili Zheng, Beijing, China
Hui Zhang, Beijing, China

38 Continuum-Scale Quantitative Defect Dynamics in Growing Czochralski Silicon Crystals
Milind S. Kulkarni, St. Peters, USA

39 Models for Stress and Dislocation Generation in Melt Based Compound Crystal Growth
Vishwanath (Vish) Prasad, Denton, USA
Srinivas Pendurti, Cincinnati, USA

40 Mass and Heat Transport in BS and EFG Systems
Thomas F. George, St. Louis, USA
Stefan Balint, Timisoara, Romania
Liliana Braescu, Timisoara, Romania

1214

36. Convection and Control in Melt Growth of Bulk Crystals

Chung-Wen Lan

During melt growth of bulk crystals, convection in the melt plays a critical role in the quality of the grown crystal. Convection in the melt can be induced by buoyancy force, rotation, surface tension gradients, etc., and these usually coexist and interact with one another. The dominant convection mode is also different for different growth configurations and operation conditions. Due to the complexity of the hydrodynamics, the control of melt convection is nontrivial and requires a better understanding of the melt flow structures. Finding a proper growth condition for optimum melt flow is difficult and the operation window is often narrow. Therefore, to control the convection effectively, external forces, such as magnetic fields and accelerated rotation, are used in practice. In this chapter, we will first discuss the convections and their effects on the interface morphology and segregation for some melt growth configurations. The control of the flows by external forces will also be discussed through some experimental and simulation results.

36.1	Physical Laws for Transport Processes 1217
	36.1.1 Conservation Equations 1217
	36.1.2 Boundary Conditions 1218
36.2	Flow Structures in the Melt..................... 1219
	36.2.1 ZM Configuration........................... 1219
	36.2.2 Bridgman Configuration................ 1225
36.3	Flow Control by External Forces.............. 1228
	36.3.1 Steady Magnetic Field 1229
	36.3.2 Rotation...................................... 1233
	36.3.3 Vibration..................................... 1237
36.4	Outlook 1238
References 1238	

Bulk crystals used in electronic and optoelectronic devices require low defects and good composition uniformity, and melt convection during growth of bulk crystals plays a crucial role. The growth interface, which is critical to defect formation, is significantly affected by heat flow. Dopant incorporation and distribution on both macroscopic and microscopic scales are affected by melt convection as well. Therefore, the control of melt flow is necessary for crystal growth in practice. This chapter aims to provide a basic understanding of melt convection and its control through a few heuristic experimental and simulation examples, focusing on the effects of melt convection on the growth interface and composition uniformity.

Various melt growth techniques have been used for the growth of semiconductor and oxide bulk crystals. They can be grouped into three categories: namely, the Czochralski, Bridgman, and zone-melting configurations, as illustrated in Fig. 36.1. The Czochralski (CZ) configuration shown in Fig. 36.1a (*left*) is the most popular growth process, especially for silicon and compound semiconductors. Dislocation-free single crystals up to 16 inch in diameter have been grown, while 8–12 inch-diameter silicon has been produced routinely for integrated-circuit applications. Due to its practical importance, research into melt convections and their control has been very extensive [36.1–4]. The melt flow in the CZ melt is dauntingly complicated. The rising buoyancy flow from the heated crucial wall turns inwards radially and joins the thermocapillary (Marangoni) flow at the melt surface. On the other hand, the rotating crystal sucks up the melt axially at the center of the melt, spinning it up in a thin Ekman layer. The radially outward flow meets the buoyancy and Marangoni flows, leading to a complicated flow structure [36.5]. The rotation of the crucible further

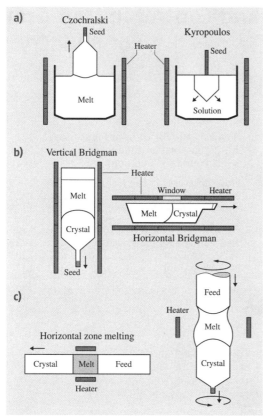

Fig. 36.1a–c Schematic of some configurations for bulk crystal growth from the melt: (**a**) Czochralski (*left*) and Kyropolous configurations; (**b**) vertical (*left*) and horizontal Bridgman configurations; (**c**) zone melting (*left*) and floating zone configurations

complicates the flow, and a Taylor–Proudman column appears under the crystal, while outside this column the melt rotates with the crucible. Such a flow easily becomes nonaxisymmetric [36.6]. Furthermore, in commercial large-scale silicon growth, the melt flow is usually turbulent. With the advance of computing power, by using the large-eddy simulation (LES), *Lukanin* et al. [36.7] successfully simulated 300 mm-diameter CZ silicon growth. The calculated thermal fields and the growth front shape are in good agreement with the measured ones. The Kyropolous method is a variant of the CZ configuration, but its crystal growth is carried out by slow cooling, usually without crystal pulling. The Bridgman and zone-melting (ZM) configurations, as depicted in Fig. 36.1b,c, respectively, are also

important in applications, but the crystal size that can be grown, as compared with that from the CZ configuration, is much smaller. Nevertheless, 6 inch-diameter GaAs ingots grown by the vertical Bridgman (VB) configuration are now commercially available. Beside the general impression, their applications in compound semiconductors and oxides have increased rapidly in recent years because of the ease of operation and the lower cost in mass production. Furthermore, due to the use of the ampoule, especially for VB, very low thermal gradients are allowed and thus the thermal stress in the grown crystal is greatly reduced. Due to the thermally stable configuration, the buoyancy convection in the melt for VB growth is much weaker and simpler. The buoyancy convection is usually generated by the radial thermal gradients caused by the interface deformation or the radial heating from the furnace. Some reviews of earlier research can be found elsewhere [36.8–12]. In particular, the convections in the melt are discussed in detail by [36.10].

For the Bridgman and ZM configurations, the growth orientation can be vertical or horizontal. For the horizontal configuration [on the right of Fig. 36.1b and on the left of Fig. 36.1c for the horizontal Bridgman (HB) and horizontal ZM (HZM) configurations, respectively], an open boat is usually used due to the ease of observation. As we will discuss shortly, due to gravity being perpendicular to the thermal gradient, buoyancy convection, as well as the Marangoni convection due to the free surface, is much stronger as compared with that in the vertical configuration [36.13, 14]. Similarly, due to the radial heating, the buoyancy convection in the ZM configuration (Fig. 36.1c) is also strong. For both horizontal configurations, the growth interface is often significantly affected by the melt convection [36.14, 15]. Another variation for the ZM configuration is the so-called floating-zone (FZ) process, as illustrated on the right of Fig. 36.1c. In the FZ configuration, because the molten zone is supported by its surface tension, contamination from the ampoule material can be avoided. However, due to the limitation of the zone stability, the grown crystal size is usually small, being less than 1 cm for most materials. Nevertheless, FZ growth has been a popular process for research [36.16–21]. Moreover, visualization is easy, and power consumption and material usage are small. Meanwhile, the heat flow phenomena in FZ growth are complicated and interesting. The interplay of the heat flow and the interfaces, as well as the crystal quality, can be investigated effectively. Therefore, much attention has been paid to the flow instability, especially the Marangoni flow, for the past 20 years [36.17, 20]. For the growth of silicon using induction-heated FZ method, the needle-eye configuration is used. Due to the stabilization by the magnetic force, the grown silicon crystal diameter can be up to 8 inch. The convection in the molten zone has also been investigated in detail [36.22, 23].

In general, to grow crystal with good quality, careful control of convections is necessary, so that heat and mass transport are favored for dopant segregation and the shape of the growth interface. For example, a flat or slightly convex growth front toward the melt is desired to minimize parasitic nucleation. Dopant uniformity, both radial and axial, is also a major concern to ensure product quality. Therefore, the basic understanding of the role of the convections in the growth interface and the segregation is important for crystal growth. In addition, these flows affect the growth in many different ways. For the VB configuration, the convection in the melt is usually quite weak and stable. Therefore, the interface shape is mainly determined by global heat transfer and solidification [36.9, 12]. Although the convection is weak, due to the low diffusivity of dopants added to the melt, dopant uniformity is influenced dramatically by the flow. Therefore, manipulating the melt flow to obtain a desired dopant distribution has become a key issue in VB growth. On the contrary, for the ZM (both vertical and horizontal) or HB configurations, the buoyancy convection or the Marangoni flow is usually strong, so that dopant can be well mixed, but the interface shape is strongly affected.

Besides the understanding of melt convections, finding an effective strategy for better growth control is also important. Tuning the temperature profile through furnace design to achieve a desired condition is typical, but this approach is laborious, rather inflexible, and not always very successful. Thus, the use of external forces to assist in controlling crystal growth is becoming popular. Examples include magnetic fields [36.24–33], accelerated crucible rotation (ACRT) [36.34–38], vibration [36.39–47], centrifugation [36.48–58], and reduced gravity [36.59]. Again, we will briefly discuss the results for achieving better growth control, in terms of both interface and dopant distribution, through the use of external forces. Although the introduction is not extensive, we shall illustrate some important concepts that help improve understanding and control of melt convections. Particularly, we shall discuss the origin of the convections through the basic conservation equations and boundary conditions for mass, momentum, energy, and species first. From the basic equations, the mechanisms for flow control by external forces can be easily understood. As discussed by *Müller* and *Ostrogorsky* [36.10], for buoyancy convection, the growth configurations can be characterized by the orientation of the body force with respect to the thermal gradients, which mainly depend on the way of heating. The crystal or crucible rotation and the Marangoni flow are imposed from the boundaries. Therefore, the governing equations and boundary conditions are indeed necessary in understanding and further controlling the convections.

In the next section, the basic governing equations and some boundary conditions are summarized. Since these equations govern the physical laws, the mechanisms and driving forces for melt convections can be easily realized; the control of melt flows by external forces can be better realized as well. In Sect. 36.2, we discuss the basic flow structures of the crystal growth processes through flow visualization experiments using a transparent material and simulation results. In Sect. 36.3, how to control the melt convections through external forces is discussed. The discussion will be focused on the Bridgman and ZM configurations. Due to space limitation, the CZ configuration is not discussed. An outlook is given in Sect. 36.4, where some recent research in melt convection and its control is briefly discussed.

36.1 Physical Laws for Transport Processes

36.1.1 Conservation Equations

The heat flow and dopant transport in crystal growth from the melt are governed by the conservation of momentum, energy, and mass. In the melt, one can consider in general an incompressible heat problem in a rotational field having a constant angular speed Ω and a uniform magnetic field B; the crystal is rotating at a speed Ω_c. In a dimensionless form, the conservation laws for mass, momentum, energy, and dopant based on the Boussinesq approximation for laminar flow can be expressed by the following equations in a rotating frame [36.51, 56]:

$$\nabla v = 0 , \qquad (36.1)$$

$$\frac{\partial v}{\partial \tau} + v \nabla v = -\nabla P + \mathrm{Pr} \nabla^2 v + F , \qquad (36.2)$$

$$\frac{\partial T}{\partial \tau} + v\nabla T = \nabla^2 T + q , \qquad (36.3)$$

$$\frac{\partial C}{\partial \tau} + v\nabla C = \frac{\text{Pr}}{\text{Sc}}\nabla^2 C , \qquad (36.4)$$

where the dimensionless body force $\boldsymbol{F}$ is given as

$$\begin{aligned}\boldsymbol{F} = &-\text{Pr}\left[\text{Ra}_T\left(T - T_{\text{ref}}\right) + \text{Ra}_S\left(C - C_{\text{ref}}\right)\right] \\ &\times \left[\boldsymbol{e}_g - \text{Fr}\boldsymbol{e}_c\right] + \text{Ha}^2\text{Pr}\left(-\nabla\Phi + v\boldsymbol{e}_B\right) \\ &\times \boldsymbol{e}_B - \text{Ta}^{1/2}\text{Pr}\boldsymbol{e}_\Omega \times v ,\end{aligned}$$

where v, τ, P, T, and C are dimensionless velocity, time, pressure, temperature, and dopant concentration, respectively; $\boldsymbol{e}_g$, $\boldsymbol{e}_c$, $\boldsymbol{e}_\Omega$, and $\boldsymbol{e}_B$ are unit vectors in the gravitational, centrifugal, angular rotational, and magnetic field directions, respectively; and T_{ref} and C_{ref} are the dimensionless reference temperature and dopant concentrations, respectively. The dimensionless thermal Rayleigh Ra_T, Hartmann Ha, Taylor Ta, and Froude Fr numbers and their physical meaning are summarized as

$$\text{Ra}_T = \frac{\beta_T g \Delta T L^3}{\nu_m \alpha_m} = \frac{\text{Buoyancy force}}{\text{Viscous force}} ,$$

$$\text{Ta} = 4\Omega^2 L^4/\nu_m^2 = \frac{\text{Coriolis force}}{\text{Viscous force}} ,$$

$$\text{Ha} = |\boldsymbol{B}|L(\sigma/\mu_m)^{1/2} = \frac{\text{Lorentz force}}{\text{Viscous force}} ,$$

$$\text{Fr} = \Omega^2 L/g = \frac{\text{Centrifugal force}}{\text{Gravitatonal force}} ,$$

where L is the characteristic length, σ is the electrical conductivity, μ_m is the melt viscosity, and g is the gravitational acceleration. Additional dimensionless parameters are the Prandtl number $\text{Pr} = \nu_m/\alpha_m$ and the Schmidt number $\text{Sc} = \nu_m/D$, where ν_m is the kinematic viscosity and α_m and D are the thermal and dopant diffusivities, respectively. Also, Ra_S is the solutal Rayleigh number, similar to the thermal one but where the driving force is based on $\beta_S \Delta C$, where β_S is the solutal expansion coefficient, and ΔC the concentration difference. For molten oxides, Pr is about unity and the thermal field can be affected easily by convection. However, for molten semiconductors or metals, $\text{Pr} \ll 1$ and the effect of flow on the heat transfer and the interface shape is much less. On the other hand, for most dopants in the melts, $\text{Sc} \gg 1$, indicating that the dopant field and thus segregation are dominated by convection rather than by molecular diffusion.

36.1.2 Boundary Conditions

To solve the above equations, a set of proper boundary conditions is necessary [36.55]. For crystal growth, the energy and solute balances at the interfaces are particularly important and cannot be ignored. Taking the Bridgman configuration as an example, if the dimensionless ampoule moving speed is v_{amp}, the balances for energy (ignoring internal radiation) and dopant (ignoring solid-state diffusion) at the growth front are

$$\begin{aligned}0 = &Q_c - Q_m + \gamma \text{St} \\ &\times \left[\left(v_{\text{amp}} \frac{dh_c}{d\tau}\right)\boldsymbol{n}\boldsymbol{e}_z + r\Omega_c \boldsymbol{n}\boldsymbol{e}_\phi\right] ,\end{aligned} \qquad (36.5)$$

$$\begin{aligned}0 = &\boldsymbol{n}\nabla C|_m + \frac{\text{Sc}}{\text{Pr}}(1-K)C \\ &\times \left[\left(v_{\text{amp}} - \frac{dh_c}{d\tau}\right)(\boldsymbol{n}\boldsymbol{e}_z) + r\Omega_c^*(\boldsymbol{n}\boldsymbol{e}_\phi)\right] ,\end{aligned} \qquad (36.6)$$

respectively, where Q_c and Q_m are the normal heat fluxes in the crystal and melt sides, respectively, γ_c is the density ratio of the crystal and the melt, St is the Stefan number scaling the heat of fusion by the sensible heat of the melt ($\text{St} = \Delta H/Cp_m\Delta T$), h_c is the dimensionless interface position, $\boldsymbol{e}_z$ is the unit vector in the axial direction, K is the segregation coefficient of the dopant, $\boldsymbol{e}_\phi$ is the unit vector in the azimuthal direction, and $\boldsymbol{n}$ is the unit normal vector. If the interface shape is not axisymmetric ($\boldsymbol{n}\boldsymbol{e}_\phi \neq 0$), the freezing rate may oscillate and melt-back may even occur ($\boldsymbol{n}\boldsymbol{e}_\phi > 0$) during rotation if the growth rate is very small [36.15].

For the FZ and HB configurations, the free surface requires normal and tangential shear stress balances for the momentum equations. At the free surface, the shear stress balance is imposed ([36.60]

$$\tau : \boldsymbol{n}\boldsymbol{s} = \text{Ma}\,\partial T/\partial s , \qquad (36.7)$$

where $\tau : \boldsymbol{n}\boldsymbol{s}$ is the shear stress at the n–s plane of the free surface; $\boldsymbol{n}$ and $\boldsymbol{s}$ are the unit normal and tangential vectors at the free surface, respectively. Also, Ma is the Marangoni number, which is defined as

$$\text{Ma} = \frac{|\partial \gamma/\partial T|T_m L}{\rho_m \nu_m \alpha_m} ,$$

where $\partial \gamma/\partial T$ is the surface tension–temperature coefficient of the melt. Two tangential directions need to be considered for the stress balance. In addition, the kinematic condition ($\boldsymbol{n}v = 0$) at the free surface and the normal stress balance (the Young–Laplace equation) are also satisfied, i.e.,

$$\tau : \boldsymbol{n}\boldsymbol{n} = (2H)\text{Bo} + \lambda_0 , \qquad (36.8)$$

where $2H$ is the mean curvature scaled by $1/L$ and $\text{Bo} = \gamma/(\rho_m g_0 L^2)$ is the static Bond number, where γ is the surface tension of the melt. The detailed procedure for calculating the mean curvature can be found elsewhere [36.61]. Also, λ_0 is a reference pressure head that needs to be determined to satisfy the growth angle constraint for a steady growth.

The governing equations and the associated boundary conditions can be solved by numerical methods. For example, *Lan* and *Liang* [36.62] have developed a multiblock three-dimensional (3-D) finite-volume method with multigrid acceleration to solve the above equations. In each block, a structure mesh is adopted. During iterations, the interface is found by locating the melting temperature. The iterations continue until all the variables converge.

In the following sections, we will discuss first the flow structures and their effect on crystal growth through a few examples, focusing on the Bridgman and ZM configurations. These examples provide a heuristic introduction of melt flow structure and its effects on interface shape and segregation. Much more complicated cases exist for large-scale Czochralski growth of silicon, where the melt flow is usually turbulent [36.1, 7, 63]. However, due to space limitation, we do not attempt to touch on this topic. This introduction is not extensive, and the interested reader can find further discussion in the related references [36.5, 9, 10].

36.2 Flow Structures in the Melt

Flow structures in the melt have significant influence on interface morphology and dopant segregation. Oscillatory melt flows can also cause growth striations, as a result of microscopic composition nonuniformity [36.5]. Therefore, understanding the basic flow structures in the melt is important for growth control. Furthermore, beside the flow structures, the heat of fusion released during crystal growth (the Stefan effect) could significantly affect the interface shape. Therefore, the interaction of the heat flow and solidification is the key factor for the interface shape. In this section, typical convection structures in the ZM and Bridgman configurations will be discussed. The buoyancy, Marangoni, and forced convections are illustrated. Solutal effects on the melt convection will be discussed as well.

36.2.1 ZM Configuration

For the ZM configuration, if the molten zone generated by a resistance heater is confined in the ampoule, the buoyancy convection is caused by radial heating. The typical buoyancy flow is like that in Fig. 36.2a, observed during the ZM growth of a 11 mm-diameter sodium nitrate (NaNO$_3$) crystal [36.64]. As shown, due to radial heating, the hotter and lighter melt near the ampoule floats upward, while the cooler and heavier melt near the centerline sinks, thus producing the flow loop. Because NaNO$_3$ is a high-Prandtl-number material (Pr = 9.2), the heat transfer in the molten zone is significantly enhanced. As a result, the growth interface is also affected, having a gull-wing shape.

On the other hand, with the presence of the free surface for the FZ configuration, the flow structure is dramatically changed due to the Marangoni flow. The Marangoni flow is induced by the nonuniform surface tension due to temperature variation at the free surface. For small-scale growth, the Marangoni flow could be dominant in the molten zone. Figure 36.2b shows a visualized Marangoni flow in a floating molten zone of a 4 mm-diameter NaNO$_3$ rod [36.65]. As shown, there are four flow cells near the free surface, induced by the surface-tension difference along the free surface; the surface tension is higher near the solid side with a lower temperature. Due to the flow, the heat absorbed from the heater is brought toward to the melt–solid interfaces, leading to very convex interfaces. Compared with that in Fig. 36.2a, the interface convexity is significantly larger than the buoyancy-flow induced one. The effects of buoyancy and Marangoni convections can be better understood from the simulation results shown

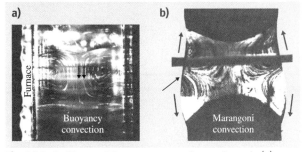

Fig. 36.2a,b Observed flow structures and zone shapes in (**a**) zone-melting growth of NaNO$_3$ (11 mm diameter, growth rate of 1 cm/h); (**b**) observed flow patterns in a stationary floating zone of NaNO$_3$ (4 mm diameter)

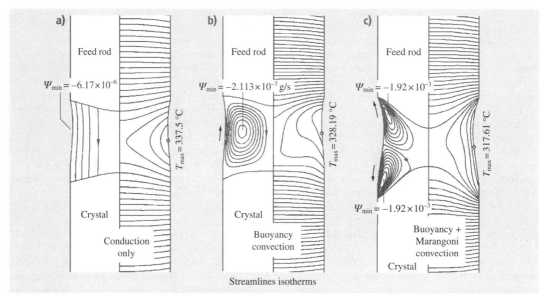

Fig. 36.3a–c Effect of convection modes on the interface shapes for FZ growth of a 4 mm-diameter NaNO$_3$ crystal: (a) conduction mode; (b) buoyancy convection; (c) buoyancy and Marangoni convections; the *left-hand side* shbows the flow patterns and the *right-hand side* the isotherms

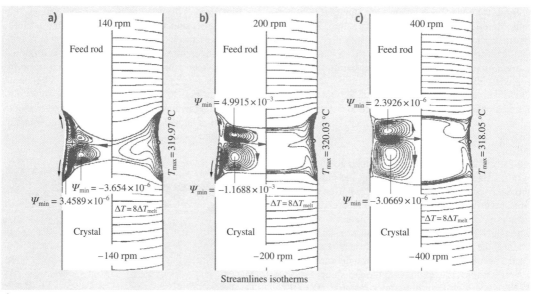

Fig. 36.4a–c Effect of counter-rotation on the flow structures and interface shapes for FZ growth of a 6 mm-diameter NaNO$_3$ crystal: (a) 140 rpm; (b) 200 rpm; (c) 400 rpm

in Fig. 36.3 [36.66]. In Fig. 36.3a, the streamlines are due to the rod feeding and crystal pulling (downwards). With the buoyancy flow in Fig. 36.3b, the flow structure and the resulted interface shape are similar to the ones in Fig. 36.2a. As the Marangoni effect is considered, the interface shape becomes very convex, consistent with that observed in Fig. 36.2b.

Forced convection can also be introduced for FZ growth through rotation. In general, counter-rotating the feed and crystal rods is often used, but usually below 30 rpm. *Lan* and *Kou* [36.66] proposed an extreme situation by using high-speed rotation, say 200 rpm, for zone shape control. Figure 36.4 shows the effects of counter-rotation for a FZ growth of a 6 mm-diameter NaNO$_3$ crystal. As shown, the flow structures due to the forced convection are also very different from the previous ones. The counter-rotating rods act as a centrifugal fan, sucking the melt axially toward the solid and spinning outward radially in the thin Ekman boundary layer near the solid. Thus, the induced convection cells are near the center core of the molten zone. Because their flow directions are different from the Marangoni cells, the Marangoni flow is suppressed with increasing rotation speed. The photograph of the visualized flow at 200 rpm is shown in Fig. 36.5a. Due to the lens effect of the melt as a result of its different reflective index from the air, the Marangoni flow cannot be seen. From this side view, it can be imagined that the heat absorbed from the heater can be brought into the zone center effectively. As a result, the interfaces become flat and the zone can be kept very short by using a smaller heating power. This idea has been applied to the growth of 1 cm-diameter NaNO$_3$ crystal with a very short and stable zone [36.67]. A photograph of the growth is shown in Fig. 36.5b. Computer simulation has also been carried out, and good agreement with experimental observation has been obtained.

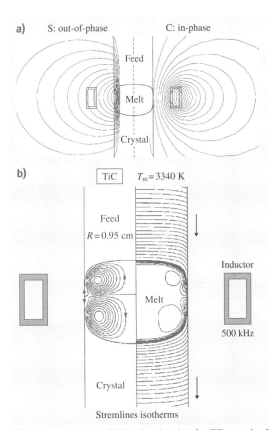

Fig. 36.6a,b Effect of induction heating for FZ growth of 2 cm-diameter TiC: (**a**) calculated in-phase (*right*) and out-of-phase (*left*) magnetic components; (**b**) calculated flow patterns (*left*) and isotherms (*right*) and the zone shape

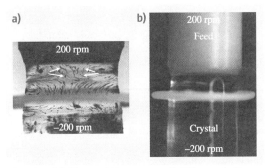

Fig. 36.5 (**a**) Observed flow patterns in FZ growth (6 mm-diameter rod) under 200 rpm counter-rotation; (**b**) FZ growth of a 1 cm-diameter NaNO$_3$ crystal under 200 rpm counter-rotation

Furthermore, for electrically conductive melt, the molten zone can also be produced by an induction coil; the heat is generated by the Joule heating of the induced eddy current. In addition, the Lorentz force can induce significant magnetic stirring and this leads to a severe distortion of the molten zone. Especially, if no insulation is used for the zone, the significant radiative heat loss from the molten zone causes a very concave growth front. Figure 36.6a shows calculated magnetic fields for FZ growth of a 2 cm-diameter TiC rod, and Fig. 36.6b shows the calculated flow (*right*) and thermal fields (*left*) as well as the zone shape. From the distortion of the isotherms, the magnetic stirring inside the molten zone is quite vigorous. The Marangoni flow in this case is weak as compared with the flow generated by the magnetic stirring. Increasing the coil frequency reduces the skin depth and magnetic stirring. The interface concavity can be reduced as well. However, a more effective way for interface control is to reduce the radiation heat loss, for which the use of insulation or a secondary heater is helpful [36.68]. In addition, the power (induction current) required and also the magnetic stirring can be significantly reduced.

The flow structures in the previous examples are axisymmetric. However, in reality, several factors can cause symmetry breaking. Imperfect heating or alignment can cause asymmetric zone as well. Inherent nonlinear bifurcation is also typical. *Lan* and *Liang* [36.69] have demonstrated through numerical simulation the nonlinear symmetry breaking leading to three-dimensional (3-D) flows for a ZM growth for a 2 cm-diameter GaAs crystal in a quartz ampoule (wall thickness 2 mm). The bifurcation diagrams for the zone length and surface temperature differences using the heater temperature (T_p) as the parameter are illustrated in Fig. 36.7a and b, respectively. Some plots for thermal and flow fields indicated in Fig. 36.7 are further illustrated in Fig. 36.8. First, we start the solution at 1500 °C, as shown in Figs. 36.7 and 36.8a, the solution is axisymmetric and stable. On further increasing the heater temperature, we can observe a subcritical bifurcation at about 1513 °C. However, the new mode that bifurcates from the axisymmetric ($m0$) mode is an $m2$ mode instead of an $m1$ mode. Two stable solutions b and c at 1510 °C are shown in Fig. 36.8b,b′, respectively. The $m2$ mode c has a twofold symmetry, which can be clearly seen from the cut of the isotherms at the middle section of the molten zone ($z = L/2$). As shown at the bottom of Fig. 36.8b′, although they are both symmetric with respect to the centerline, the results are quite different. By comparing the interface shapes from the two cuts, one may also get a 3-D view of the interface shapes. Further increasing the heater temperature, as shown in Fig. 36.7, we encounter a secondary bifurcation, and the $m2$ mode becomes unstable. The new mode branching from $m2$ is the $m1$ mode again. This bifurcation is subcritical as well. Two stable solutions c and c′ are further illustrated in Fig. 36.8c,c′, respectively.

It is believed that the existence of the $m2$ mode at lower T_p is due to the much smaller aspect ratio of the molten zone. As the zone length is increased by increasing the heating power, the $m1$ mode becomes dominant. Furthermore, the subcriticality is due to the existence of the deformable interfaces. The bifurcation becomes supercritical if we fix the interfaces. Such an observation is similar to that in the two-phase Rayleigh–Bernard

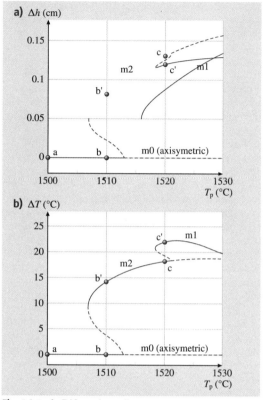

Fig. 36.7a,b Bifurcation diagrams of ZM growth of 1 inch-diameter GaAs using the heater temperature as the parameter of: **(a)** the zone length deviation and **(b)** the surface temperature deviation

problem discussed by *Lan* and *Wang* [36.70]. Again, although Ra_w [$Ra_w = (\beta_T g_0 \Delta T L^3)/(\nu_m \alpha_m)$; $\Delta T = T_w - T_m$, where T_w is the wall temperature] is up to 5.78×10^5 ($\Delta T = 63.33$ K is the maximum temperature in the molten zone) for the $m1$ mode at $T_p = 1530$ °C, the solution is still stable. Similar observations were found by *Baumgartl* et al. [36.71].

Besides the symmetry breaking, the bifurcation to unsteady state is often encountered, and this time-dependent flow is often related to growth striation; for example, in ZM growth of GaAs single crystals at $Ra_w \approx 4 \times 10^5$, 3-D striation patterns were found [36.71]. In the previous experiment, a steady axisymmetric mode bifurcated to a 3-D toroidal mode first at $Ra_w \approx 3 \times 10^4$ and then to a time-dependent mode at a higher Ra_w ($\approx 4 \times 10^4$) for a parabolic thermal profile at the melt surface. A steady one-roll

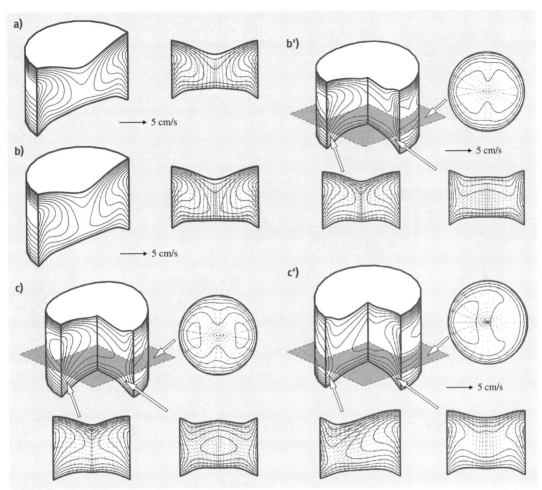

Fig. 36.8a–c Some solutions of Fig. 36.7: (**a**) 2-D solution at $T_p = 1500\,°\text{C}$; (**b**) 2-D solution at $T_p = 1510\,°\text{C}$; (**b'**) 3-D $m0$ solution at $T_p = 1510\,°\text{C}$; (**c**) 3-D $m2$ solution at $T_p = 1520\,°\text{C}$; (**c'**) 3-D $m1$ solution at $T_p = 1520\,°\text{C}$

($m1$) mode was found for 8×10^4 ($\text{Ra}_w \approx 1.85 \times 10^5$). Beyond $\text{Ra}_w \approx 1.85 \times 10^5$, the $m1$ flow became time dependent.

Similar to the nonlinear bifurcations in the ZM configuration in an ampoule, the FZ configuration also has a similar behavior. Indeed, the flow bifurcation is caused by Marangoni flow, and extensive studies have been made of this subject [36.17, 20, 21, 56].

In addition to the vertical configuration, the horizontal ZM (HZM) configuration is also used in bulk crystal growth and material purification [36.72]. The observed flow structure for a horizontal molten zone of succinonitrile (SCN) is shown in Fig. 36.9a. Due to the buoyancy force, there are two flow cells in the melt. The melt flows upwards at the center, and downwards near the interfaces. The upward melt (lighter) is also heated at the top wall and then flows to the two sides to melt back the interfaces. As a result, the upper part of the zone becomes wider. The downward melt (heavier) from the interfaces is warmed up again from the bottom wall at the center and then flows upwards. Although the picture shown in Fig. 36.9a appears to be two dimensional, the flow structure is fully 3-D. However, the flow patterns at other light-cut planes are much more difficult to observe. The trapezoidal zone shape in Fig. 36.9a is caused by the buoyancy flow and is consistent with

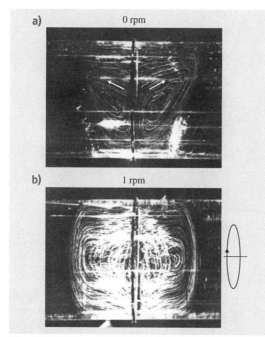

Fig. 36.9a,b Flow patterns in horizontal zone-melting growth of SCN: (**a**) without rotation; (**b**) with 1 rpm rotation

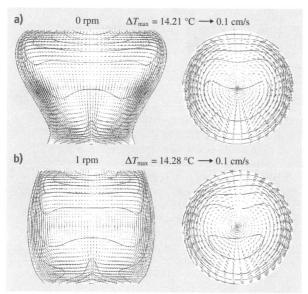

Fig. 36.10a,b Simulated flow and thermal fields: (**a**) 0 rpm; (**b**) 1 rpm rotation

that reported [36.72]. Interestingly, if we impose only 1 rpm ampoule rotation, a significant change of the zone shape is observed, as shown in Fig. 36.9b. As shown, the zone shape becomes much more symmetric, and the upper and lower zone lengths are about the same. As compared with Fig. 36.9a, the rotation reduces the upper zone length, while widens the lower part; the zone length seems to be averaged by the rotation. The flow structure is also changed quite significantly. In fact, due to rotation, the flow visualization became much more difficult. The tracer particles (aluminum powder) are heavier than the melt. As a result, larger particles tended to settle down, as shown by the bright area at the melt bottom in Fig. 36.10a.

Computer simulation has also been conducted for the observed flows and interface shape [36.73], as shown in Fig. 36.10. Indeed, the simulation captures the key features of the experiments shown in Fig. 36.9: both the flow structures and the interface shapes. For the case without rotation, as shown in Fig. 36.10a, the flow patterns and interface shapes are in good agreement with Fig. 36.9a. Due to the buoyancy force, the hotter (lighter) melt at the bottom floats upwards. It is warmed up again at the top wall, and flows to the two sides, leading to a significant back melting of the interfaces at the top. The downward melt near the interfaces is cooler and heavier, but it is heated up again by the heater at the bottom wall and floats up at the center. One can also get a better picture from the distorted isotherms, which are consistent with the flow directions. The right-hand side view of Fig. 36.10a also illustrates the buoyancy convection; the melt near the wall is heated and flows upward. One cannot see a downward flow from this plane, because the major downward flows appear near the interfaces. With 1 rpm rotation, as shown in Fig. 36.10b, the zone shape becomes much more axisymmetric, which agrees well with the observation in Fig. 36.9b. The flow structure shown in Fig. 36.10b is also similar to that in Fig. 36.9b as well. One shall pay more attention on the flow shapes and vortices. The right figure of Fig. 36.10b also indicates that, although the effect of rotation becomes more obvious, the change of the flow and isotherms is small.

The success of the full model calculation in Fig. 36.10b is due to consideration of the interface energy balance. As shown in (36.5), for steady rotation without ampoule translation, the term $\gamma_c \mathrm{St}(r\Omega_c n e_\phi)$ plays a crucial role. During rotation, solidification at the interface occurs at $\gamma_c \mathrm{St}(r\Omega_c n e_\phi) < 0$, while melting occurs at $\gamma_c \mathrm{St}(r\Omega_c n e_\phi) > 0$; for example, for the left interface, at the front side, $\gamma_c \mathrm{St}(r\Omega_c n e_\phi) < 0$ and solidification dominates. The heat of fusion is released to the melt and it distorts the interface toward the rotational direction. Meanwhile, the melting occurs at the back side, and the heat of fusion is extracted from the melt, making the solid interface intrude toward the melt in the rotational direction there. As a result, the zone length is averaged out with the period of rotation. This mechanism (the Stefan effect) is dominant for the interface shapes here and cannot be ignored in the simulation.

36.2.2 Bridgman Configuration

The Bridgman configuration is also widely adopted for melt growth of bulk crystals. Unlike the ZM configuration, in which significant superheating is necessary to form a stable zone, the Bridgman configuration is suitable for growth that requires low superheating or thermal gradients. For the HB configuration using an open boat, due to the free surface, the Marangoni convection is also important. Nevertheless, due to the much larger melt volume, the buoyancy flow is also strong. To understand its basic flow structure, we take the HB growth of $NaNO_3$ as an example. The top view of the growth in a transparent furnace is shown in Fig. 36.11a [36.53]. During growth, the interface is always concave, due to the release of heat of fusion and the effect of melt convection, although the latter factor is much more significant. As will be discussed shortly, the Marangoni flow is responsible for the sharp contact angle between the interface and the crucible [36.13]. Such an acute angle often induces parasitic nucleation from the crucible wall, leading to a polycrystalline growth. The visualized flow is shown in Fig. 36.11b. As shown, the buoyancy and Marangoni flows are in the same direction, and the hot melt near the surface flows from the left to the right, pushing the interface to the crystal side. Meanwhile, because of the no-slip boundary condition at the solid wall, the isotherms at the solid wall are nearly pinned by the energy balance there with the conduction in the ampoule. Accordingly, the interface becomes very concave. The calculated flow patterns and isotherms, shown in Fig. 36.11b and c, respectively, further explain the situation.

Further understanding of the flow structures can be obtained from the numerical simulation of HB growth of GaAs by *Lan* and *Liang* [36.61]. Figure 36.12 shows the effects of convection modes on the thermal and velocity fields (on the planes at $z = 0$ and $y = 0$) and the growth front (crucible puling speed is 5 mm/h). As shown in Fig. 36.12a, without any driving forces for convection, the growth front is only slightly concave;

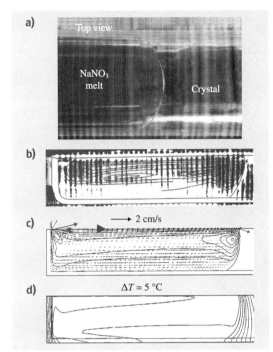

Fig. 36.11a–d Effect of buoyancy and Marangoni flows on the HB growth of $NaNO_3$ crystals: (**a**) observed interface seen from the top; (**b**) observed flow patterns; (**c**) calculated flow patterns; (**d**) calculated isotherms and interface shape

it is quit flat viewed from the top. When buoyancy force is considered in Fig. 36.12b, as shown in the side view of the flow fields, a clockwise natural convection is induced. Although the maximum melt velocity is less than 1.5 cm/s, the isotherms in the melt are highly distorted by the convection. As a result, the interface shape becomes highly distorted and very concave. The interface position at the bottom is not affected much, but the upper interface is melted backward significantly. A 3-D view of the growth front shape is illustrated on the right-hand side for comparison. Clearly, the effect of buoyancy convection on the growth front shape is significant.

When the Marangoni effect is included, as shown in Fig. 36.12c, the flow structures are not changed much; the flow in the side view is still clockwise in direction. However, the melt velocity near the free surface and the growth interface becomes much higher; the maximum velocity is about 5 cm/s. The isotherms near the free surface are thus further affected, even though the overall view of the isotherms is not changed much. As

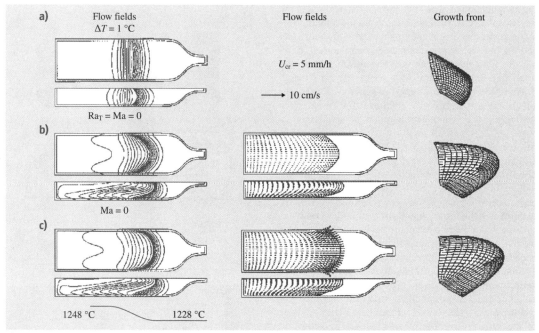

Fig. 36.12a–c Effects of buoyancy and Marangoni flows on the flow, thermal fields, and the interface shape for the HB growth of GaAs single crystal: (**a**) conduction mode; (**b**) buoyancy convection; (**c**) buoyancy and Marangoni convection

a result, the growth front near the top surface is melted back more. As shown in the side view, the growth interface also becomes sharper. Because the driving force for the Marangoni convection is proportional to local thermal gradients at the free surface, the flow near the interface is thus stronger, where the thermal gradients are higher. In addition, the flow direction there becomes more perpendicular to the isotherms, leading to a diverged flow near the interface, as shown in the top view of the flow fields. This diverged flow also melts back the interface toward the crucible and results in a rounder interface shape at the free surface when viewed from the top. More importantly, the interface–crucible contact angle becomes much smaller as well, which from the crystal growth point of view, is more likely to induce parasitic nucleation and the formation of polycrystals. This is consistent with the observation in Fig. 36.11a.

In practice, control of the growth front shape is important for crystal quality. The results in Fig. 36.12 clearly illustrate the significance of the buoyancy and Marangoni convections for the growth interface. However, since the direction of driving forces for both convection modes is not likely to be changed, the room for modulating the convection and thus controlling the interface is not great. The convections can be reduced with smaller thermal gradients, but the smaller thermal gradients also cause other problems. With a fixed growth rate, the heat of fusion will cause a more de-

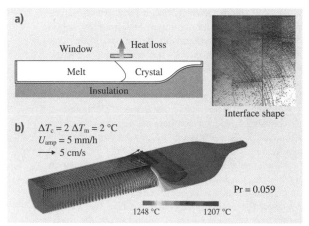

Fig. 36.13 (**a**) Schematic of HB growth of GaAs with a viewing window for local radiation cooling; on the right is the observed interface shape (cross section); (**b**) simulated flow and thermal fields

formed interface shape for lower thermal gradients. Constitutional supercooling [36.73] is a concern as well.

For this configuration, an effective approach is to provide local cooling at the top interface. This concept has been adopted for the growth of GaAs using a view window above the interface, as shown on the left of Fig. 36.13a [36.61]; the observed interface shape [36.74] is shown on the right of Fig. 36.13a. This window provides local radiative cooling and thus controls the interface shape. In addition to the top radiative cooling, booster heating (through the center zone) at the bottom of the interface is also useful, and this further reduces the interface concavity of the lower interface. Computer simulation based on these ideas has been conducted [36.61] and the result is shown in Fig. 36.13b. As shown, the temperature is greatly reduced near the window and the temperature gradients are also increased significantly there; the isotherm spacing in the crystal is increased to 2 °C, which is twice that in the melt. Because the quartz crucible has a higher emissivity than GaAs, the lowest temperature beneath the window is located at the crucible surface. Interestingly, although the interface is still slightly concave, its shape is reversed; the upper interface becomes closer to the hot zone. The growth interface is now more perpendicular to the crucible wall. A similar interface shape was also observed in crystal growth experiments [36.75]. A flatter interface is possible by reducing the heat loss through the window. Local cooling using an air jet was applied to the HB growth of NaNO$_3$, and a convex interface was successfully obtained [36.14].

Unlike the previous examples with large radial thermal gradients, the vertical Bridgman (VB) configuration is thermally stable and the radial thermal gradients are small. Accordingly, the thermally driven buoyancy convection is significantly weaker; for example, if the furnace thermal profile is linear, the buoyancy flow is only induced by the interface deformation. When stationary, the interface deflection is mainly due to the difference in the thermal conductivity between the melt and the crystal. If the thermal conductivity of the melt is larger than that of the crystal, the interface is concave. The concave interface induces radial thermal gradients and thus buoyancy flow. The concavity increases with the crystal growth speed due to the release of the heat of fusion. The buoyancy convection due to the interface concavity appears near the growth interface and affects the radial segregation significantly [36.76–78].

Figure 36.14a shows the simulated buoyancy flow and solute fields in VB growth of gallium-doped germa-

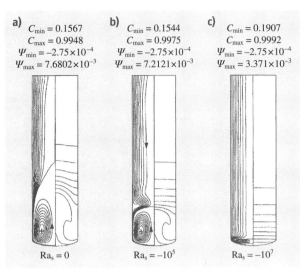

Fig. 36.14a–c Solutal effect on the buoyancy convection for VG growth of gallium-doped germanium: (a) $Ra_S = 0$; (b) $Ra_S = -10^5$; (c) $Ra_S = -10^7$; in each figure on the *left-hand side* of the simulated results is the flow fields and on the *right-hand side* the solutal fields

nium in a small graphite ampoule at $Ra_T = 10^7$ [36.78]; $Ra_T = 2.489 \times 10^8$ for normal gravity at a thermal gradient of $50\,K/cm$. Because the ampoule is pulled downward, the streamlines in the bulk melt indicate the material flow due to the ampoule translation. Also, the interface concavity is here is due to the larger thermal conductivity of the melt than that of the crystal. In addition to the thermal convection, the solutal effect can also affect the flow; for example, if a solute is added to germanium, solute segregation causes density variation in the melt, leading to solutal convection. Figure 36.14b,c show the effect of solute (silicon), in terms of solutal Rayleigh number Ra_S, on the flow. Because silicon dissolves more in the germanium solid than in the melt, silicon is depleted near the interface during crystal growth. Due to the less and lighter silicon near the interface, the melt density near the interface increases and this suppresses the thermal convection. As shown in Fig. 36.14c, the buoyancy convection is almost suppressed due to the solutal effect.

Through the visualization experiments of SCN containing acetone (lighter) or salol (heavier), the thermal–solutal convection can be better understood. Figure 36.15a,b shows the interface shape evolution during Bridgman growth of SCN containing 0.064 wt % acetone and 0.15 wt % salol, respectively; the pulling speed is 1.6 μm/s [36.47]. The simulated results are

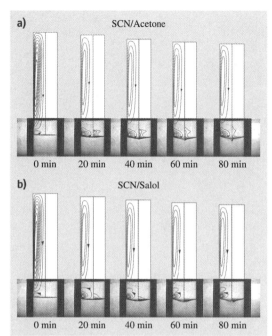

Fig. 36.15a,b Double diffusive convections in VB crystal growth of SCN: (**a**) SCN/acetone; (**b**) SCN/salol; in each figure on the *left-hand side* of the simulated results is the flow fields and on the *right-hand side* the solutal fields

also shown for comparison. In each simulated result, the left-hand side shows the streamlines, and the right-hand side the acetone fields. As shown in Fig. 36.15a for SCN/acetone stationary, the interface is flat and the convection near the interface is extremely weak. The upper cell is caused by the radial heating from the hot zone. As the solidification starts, the interface becomes concave and the flow cell near the interface is enhanced. As a result, the acetone rejected during solidification is redistributed by the flow, with increasing acetone concentration toward the center of the interface. Because acetone is lighter than SCN, the radial acetone gradients also enhance the flow, leading to a highly localized solute distribution at the center of the interface. The local acetone accumulation further causes a depression of the interface there, which becomes obvious at about 40 min. As the supercooling is built up, morphological breakdown can occur when the supercooling overcomes the interfacial energy. The simulated results at 60 and 80 min show a deep depression (pit) at the center of the interface, where high constitutional supercooling exists. The convection in the bulk melt remains about the same during crystal growth.

The evolution of interface morphology for SCN/Salol shows a similar behavior. However, as shown in Fig. 36.15b, the depression shape is wider. The morphological breakdown occurs at about the same time, i.e., about 40 min. The bottom of the breakdown area is much flatter than that for SCN/acetone. The convection near the interface is much weaker than that for SCN/acetone, as shown in Fig. 36.15b. Such a weaker flow is due to the heavier solute, which suppresses the flow. In other words, the radial density due to thermal gradients is counterbalanced by the solutal gradients; one can observe this contribution from the source term of (36.1). In addition, the flow cell near the interface is closer to the ampoule wall as compared with that in Fig. 36.15a. More importantly, the concentration profile is quite uniform near the interface, as shown by the much flatter isoconcentration lines. This also indicates that the convective effect on the solute transport is much weaker.

Because the convection in the VB configuration in the previous examples is rather weak, nonlinear flow bifurcation to symmetry breaking or oscillatory flows is not likely. Nevertheless, in reality, a perfect axisymmetric growth condition is hard to achieve. An asymmetric furnace thermal profile or slight ampoule tilting can lead to significantly 3-D convections [36.79]. As a result, the segregation behavior predicted by an axisymmetric numerical model is often erroneous. Nevertheless, since the convection is rather weak, flow suppression by using rotation or magnetic field could be quite effective. We will thus briefly discuss its flow control by using external forces.

36.3 Flow Control by External Forces

So far, we have illustrated the convections in the ZM and Bridgman configurations. As just discussed, for the FZ configuration, crystal rotation is an effective way to introduce forced convection. However, in many cases, the suppression of unstable flows to avoid growth striations and the manipulation of local flow to reduce segregation are necessary. The use of external forces is particularly effective for such purposes. Various external forces have been considered for use in the melt growth of bulk crystals. The static magnetic field is one of the most popular ones, and has been used widely for flow suppression [36.24, 27, 31], but it is restricted to electrically conductive melt. Rotation, steady or unsteady, can also be used, and its applications have no special limitations. In the following sections, we will discuss flow control through a few examples.

36.3.1 Steady Magnetic Field

A steady magnetic field is an effective way to suppress the flow if the melt is electrically conductive. Two examples for FZ and VB configurations are given here.

As discussed previously, in the FZ configuration, the free surface of the molten zone often induces significant Marangoni flow, and it often lead to unstable heat flows, striations, and distorted interfaces, even in microgravity environment [36.21]. The control of the unsteady Marangoni flow has been an important topic in crystal growth, and has attracted extensive research over the years [36.19]. For silicon growth, magnetic fields have been known useful for suppressing unsteady flows, so that steady growth can be obtained and the grown crystal is striation free [36.18, 21]. Both axial [36.18, 21] and transversal magnetic fields [36.28, 80] have been investigated. In the FZ Si growth experiments by *Dold* et al. [36.18], a striation-free core in the grown crystal was found, and the core size increased with increasing magnetic field. It was believed that the melt inside the core was significantly suppressed by the magnetic field. *Lan* [36.81] made the first attempt to simulate the growth using an axisymmetric model. The results on the core size and dopant (phosphorous) distribution are shown in Fig. 36.16a,b, respectively. As shown in Fig. 36.16a, the streamlines are stretched along the direction of the applied magnetic field. Also, the radius of the suppressed core increases with increasing magnetic field strength, and the calculated core sizes at different axial magnetic field strengths are consistent with the measured ones. In addition, the calculated phosphorous distributions in Fig. 36.16b are consistent with the measured ones as well. Although the results based on the axisymmetric model are consistent with the experimental observations, the oscillatory and 3-D behavior cannot be seen in the simulation.

Further 3-D numerical simulations have been carried out recently [36.60]. Figure 36.17 shows the calculated results for an axial magnetic field of 0.5 T. At this condition, a steady-state result is obtained. However, as shown in Fig. 36.17, the result is not exactly axisymmetric but has a fourfold symmetry, which can be seen from the thermal and velocity fields on the z–y plane (growth direction is in the x-axis). The results on the x–y and x–z planes also show that the flow in the core region is greatly suppressed, but the thermocapillary flow is still quite strong near the melt surface. This is consistent with previous 2-D simulations shown in Fig. 36.16 [36.81]. In Fig. 36.17, the flow and thermal fields on the x–z and x–y planes happen

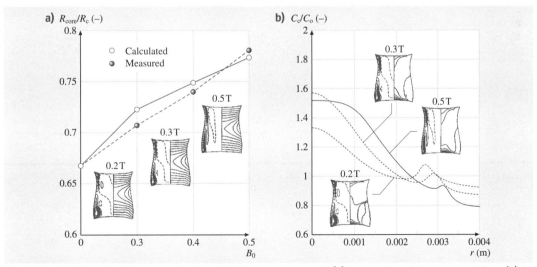

Fig. 36.16a,b Effect of axial magnetic fields on FZ silicon crystal growth: (**a**) the core size of the suppressed flow; (**b**) the radial phosphorous distribution; the simulated flow and solutal fields are also shown in (**a**) for comparison

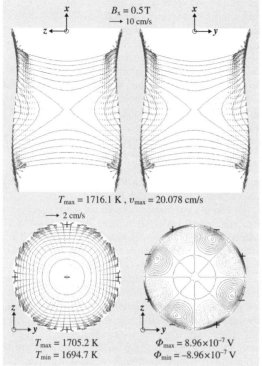

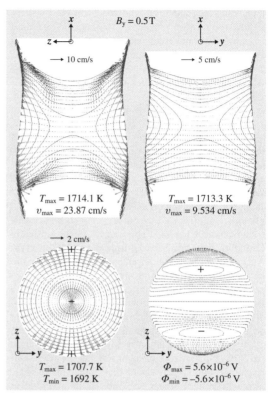

Fig. 36.17 Three-dimensional simulated flow, thermal, and potential fields in a FZ silicon growth under an axial magnetic field of 0.5 T

to be identical due to symmetry. However, if we examine the result at $x = 4.1$ cm (the middle of the molten zone) on the y–z plane, the temperature is lower at the

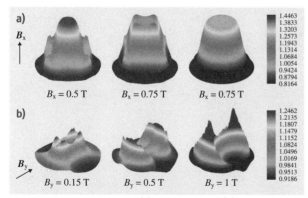

Fig. 36.19a,b Effect of axial (**a**) and transversal (**b**) magnetic fields on the dopant distribution in the growth crystal

Fig. 36.18 Three-dimensional simulated flow, thermal, and potential fields in a FZ silicon growth under a transversal magnetic field of 0.5 T

four corners. Due to such nonuniform thermal gradients, there are eight flow vortices caused by the thermocapillary force. As a result, the induced electrical potential distribution on the same plane also has an eight-cell structure.

When the magnetic field strength is less than 0.46 T, the symmetry disappears and the result becomes 3-D and time dependent. In the experiments by [36.18], a magnetic field strength greater than 0.24 T was found to be necessary to obtain nearly striation-free crystals. In fact, even at 0.5 T, irregular striation patterns were visible near the crystal surface. Therefore, our results are still in reasonable consistency with the observations.

With a transversal field, the steady-state result is obtained at a much lower magnetic field strength of 0.15 T. The calculated result for 0.5 T is shown in Fig. 36.18; the velocity scales at different planes are different for clarity. As shown, the results are asymmetric, but have a twofold symmetry when viewed from the y–z plane.

The surface zone length on the plane (x–z plane) perpendicular to the magnetic direction is much longer than that on the parallel plane. If we examine the flow fields, clearly, as mentioned previously, the melt flow on the plane perpendicular to the magnetic field is not as suppressed as that on the parallel plane, leading to a longer zone length there. If we view the results at the y–z plane (at $x = 4.1$ cm), the isotherms have an ellipsoid shape, while the maximum temperature appears at the surface of the x–z plane.

The dopant distribution is also significantly affected by the flow. Figure 36.19 summarizes the dopant distributions obtained from previous results with more results added. As shown, under axial fields, the segregation increases with increasing magnetic field strength due to flow damping. At 0.5 and 0.75 T of the axial field, the fourfold distribution is caused by flow such as that shown in Fig. 36.17. Under transversal fields, two concentration peaks align parallel to the field direction at 0.5 and 1 T. This is simply due to the poorer mixing

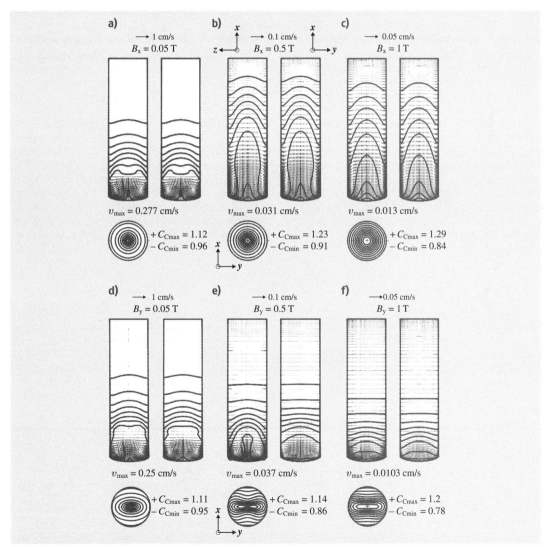

Fig. 36.20a–f Effect of axial and transversal magnetic fields on the flow and dopant fields

in the x–y plane, where the flow is suppressed more. On the other plane, the convection remains strong, so that the segregation is less. If we compare the segregation under both magnetic configurations, it is clear that the radial segregation is much less in the transversal field, even though it is highly asymmetric. Besides, as mentioned earlier, the minimum magnetic field strength to suppress the unsteady thermocapillary flow is also lower for the transversal field. Therefore, in real applications, it is believed that the transversal field may still be useful. If rotation is applied, the asymmetry may be reduced. The effect of rotation for the VB configuration has been investigated by *Lan* and *Yen* [36.60]. The effect of centrifugal pumping has also been discussed, where the conductivity of the crystal is important.

Flow suppression by static magnetic fields in the Bridgman configuration is even more effective, as could be clearly illustrated through numerical simulation. Again the gallium-doped germanium growth in a graphite ampoule discussed previously is used for illustration, with $Ra_T = 2.489 \times 10^8$. Figure 36.20 shows some flow structures and dopant fields for several axial and horizontal fields. For the case of axial magnetic fields (Fig. 36.20a–c), two lower cells induced by radial thermal gradients are stretched in the axial direction by the axial magnetic field. Because the flow tending to across the magnetic lines will be suppressed, the flow cells are elongated by the axial magnetic field. Interestingly, as the cell is stretched axially, the solute penetrates more into the bulk melt. As will be shown shortly, for the case of $B_x = 0.5\,\text{T}$, the bulk mixing is much enhanced. With the same magnetic strength, the horizontal field (Fig. 36.20d–f) is slightly more effective in suppressing the flow in terms of the maximum melt velocity. This is because the axial melt motion on the x–y plane is greatly suppressed; the axial melt motion is induced by radial thermal gradients. Although the flow on the plane (x–z plane) perpendicular to the magnetic direction is not suppressed effectively, the overall flow penetration into the upper bulk melt is significantly reduced. One can compare the dopant distributions in the x–z plane and the x–y plane to get a better idea of this. In addition, it is also clear that the solute mixing decreases monotonically with increasing magnetic field. Nevertheless, the flow and solute fields become asymmetric. One can further examine the solute concentration in the crystal ($C_c = KC$) at the interface; the solute field is stretched in the applied magnetic direction.

To illustrate the bulk solute mixing and the radial segregation, we have also calculated the effective segregation coefficient K_{eff} from the pseudo-steady-state results for various magnetic fields, $K_{\text{eff}} = \langle C \rangle / C_0$ [36.82, 83], where $\langle C \rangle$ is the average solute concentration in the melt. Figure 36.21a shows the effect of field strength (in terms of Ha number) on K_{eff}. As shown, the diffusion growth can be reached more quickly by applying a transversal field. However, for the axial field, there is a decrease in K_{eff}, i.e., better global dopant mixing, at

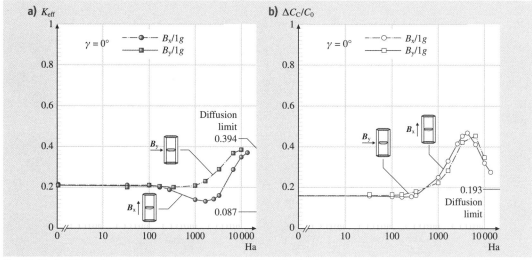

Fig. 36.21a,b Effect of axial and transversal magnetic fields: (**a**) effective segregation coefficient K_{eff}; (**b**) radial dopant segregation

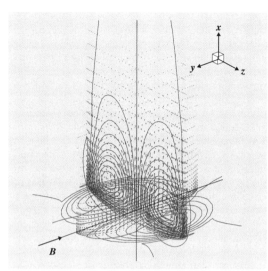

Fig. 36.22 Induced electrical potential and flow fields under a transversal magnetic field (1 T)

Apparently, even though the magnetic field is quite effective in terms of flow damping, it is clear that to suppress the flow completely is also difficult. It should be noticed that the damping effect is effective only for the flow in the plane parallel to the magnetic field. The flow damping in the plane perpendicular to the magnetic field is much less effective. This could be explained by a buildup of the electric potential that reduces the induced electric current for the Lorentz force. Fig. 36.22 illustrates the flow and the induced electric potential fields due to the transversal field at 0.5 T. As shown, the potential gradients build up due to the thermal toroidal cells and the transversal fields. The potential contour surfaces look like a pair of kidneys (with one positive and one negative potential value), with a symmetry plane at $z = 0$. In other words, due to the potential gradients, the net currents, $j = \sigma(-\nabla \Phi + v \times e_B)$, for the Lorentz force ($j \times e_B$) become smaller. Accordingly, the flow suppression becomes less effective. It can also be seen from the previous case for the FZ configuration that the flow perpendicular to the magnetic field cannot be suppressed effectively.

Ha ≈ 2000. Again, as discussed regarding Fig. 36.20b, the flow cells are stretched axially by the axial field and penetrate deeply into the bulk melt, leading to better bulk dopant mixing. The amount of radial segregation by both types of fields in Fig. 36.21b remains similar; $\Delta C = C_{c\,max} - C_{c\,min}$ is the maximum dopant concentration difference at the interface.

36.3.2 Rotation

Beside a static magnetic field, the use of rotation has been popular for crystal growth. There are a few ways of using rotation. The use of a centrifuge [36.51, 84], i.e., so-called centrifugal or high-gravity processing,

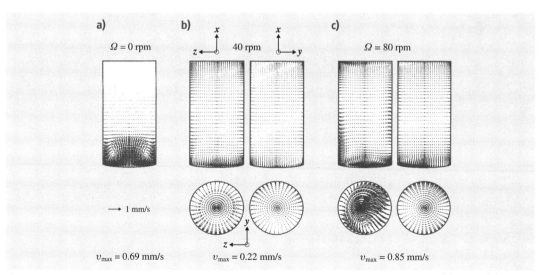

Fig. 36.23a–c Effect of free-swing rotation on the flow fields for a gradient freeze growth of GaAs: (**a**) 0 rpm; (**b**) 40 rpm; (**c**) 80 rpm

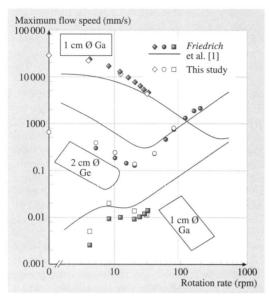

Fig. 36.24 Effect of rotation rate on the maximum melt velocity for various growth orientations

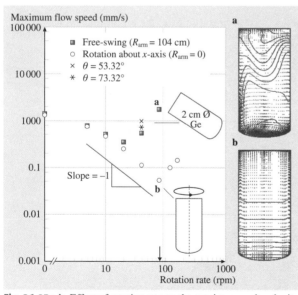

Fig. 36.25a,b Effect of rotation rate on the maximum melt velocity for rotation about the growth axis and the free-swing configuration. The flow and dopant fields for the free-swing (**a**) and rotation about growth axis (**b**) configurations are shown on the right

is particularly interesting. ACRT has also been widely used [36.34, 36–38]. The way of using the centrifuge in crystal growth has always been in the free-swinging configuration, in which the sample is placed at the end of a rotating arm. In such a case, the resultant acceleration is antiparallel to the axial thermal gradient, which is thus intuitively thermally stable. Accordingly, the convection can be suppressed at a certain rotation rate, or the so-called magic-g level [36.84], where the Coriolis force balances the gravitational acceleration. Beyond the magic-g level, the centrifugal acceleration becomes important and the centrifugal thermal convection increases. Figure 36.23 shows simulated flow patterns of Bridgman growth of germanium in a free-swing centrifuge at different rotation speed [36.51, 56]. For 0 rpm (Fig. 36.23a) the flow is axisymmetric and its structure is typical for the VB configuration, with a concave interface; the heating temperature profile is linear. At 40 rpm the flow near the growth interface is significantly suppressed by the Coriolis force and the flow structure is also changed dramatically. Although the averaged resultant gravity direction is still antiparallel to the growth axis, the centrifugal acceleration and the Coriolis force in the melt are asymmetric, leading to the 3-D flow. Due to the nonuniform forces, the global convection increases slightly away from the interface. As the rotation speed is further increased to 80 rpm, the centrifugal force becomes dominant and the convection increases, as can be seen from the larger velocity vectors. The flows in the x–y plane are also shown, but are in general featureless except for the flow near the growth interface.

We also present two flow patterns in the y–z plane, shown at the bottom of the figures for 40 and 80 rpm, respectively. One is at $x = 4.2$ cm and the other at 4.9 cm. The edge of the interface is at about $x = 4.0$ cm. As shown, near the interface the flow is mostly counterclockwise, but at some places the flow may be in the opposite direction. Interestingly, at 80 rpm, the flow pattern at $x = 4.9$ cm shows two cells with different flow directions. Therefore, the flow seen from the top does not have a well-defined structure. Closer to the growth interface, the counterclockwise flow seems to be clearer. Nevertheless, as the interface becomes flat, the flow pattern is also changed.

The maximum melt velocity as a function of rotation speed is further illustrated in Fig. 36.24 (open symbols), where the results of *Friedrich* et al. [36.51] (filled symbols and solid lines) are also included for comparison. We also performed calculations for gallium melt (in both the free-swing and horizontal configurations). The comparison with the previous study is also shown in Fig. 36.24. As shown, they are all in good agreement. The solid lines in Fig. 36.24 are from the scaling analysis of *Friedrich* et al. [36.51]. As shown for the free-swing case of germanium, there is a minimum of convection at about 20 rpm. This is supposed to be the so-called magic-g level, having the least axial dopant segregation. At this critical rotation rate, the Coriolis force balances the two gravitational forces. Beyond this value, the centrifugal force becomes dominant and enhances the convection.

Interestingly, in a recent numerical study by *Lan* and *Tu* [36.85], rotation about the growth axis could give a much better result. As shown in Fig. 36.25, the melt flow can be suppressed more effectively by this configuration, and its flow and dopant fields are still axisymmetric. On the contrary, the free-swing, or near, configuration at the magic-g level generates 3-D flows and severe dopant nonuniformity; a side view of the flow and dopant fields at 80 rpm is shown in Fig. 36.25. *Lan* [36.55] also performed a numerical simulation for a similar system and found that the flow direction (thermal convection) near the solidification front could be reversed at high speeds. As a result, an inversion of radial dopant distribution was found, clearly due to the centrifugal acceleration. In fact, in some of the earlier numerical studies of Bridgman crystal growth in a rotating ampoule [36.53, 54, 86], the centrifugal acceleration was ignored. As a result, the convection decreases monotonically with increasing rotation speed. This is correct only at low rotation rate (small Fr number). When the rotation speed is higher than the magic-g level, the convection is enhanced by centrifugal acceleration. Clearly, the centrifugal force is perpendicular to the axial gradient, and the buoyancy flow due to the centrifugal force is generated. This flow direction happens to be in the opposite direction to that near the interface due to the concave interface.

To validate the idea proposed by *Lan* and *Tu* [36.85], visualization experiments on a rotating table using SCN doped with a small amount of ethanol or acetone were performed by *Lan* et al. [36.57, 58]. Figure 36.26 shows the effect of rotation on the interface morphology after 3 h of growth (the growth rate was 2.5 μm/s with 0.07 wt% of acetone). Before the critical rotation speed was reached (Fig. 36.26a–c), the breakdown location from a planar to a cellular interface was a good indication of the acetone accumulation at the center of the interface; the initial stage of the growth also showed pit formation, as illustrated in Fig. 36.15a (60 min). Again,

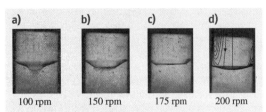

Fig. 36.26a–d Effect of rotation speed (about the growth axis) on interface morphology for the VB growth containing 0.007 wt% (after 3 h of crystal growth); the ampoule translation speed is 2.5 μm/s; (a) 100 rpm; (b) 150 rpm; (c) 175 rpm; (d) 200 rpm

the morphological breakdown of the interface was due to constitutional supercooling. With a high enough ro-

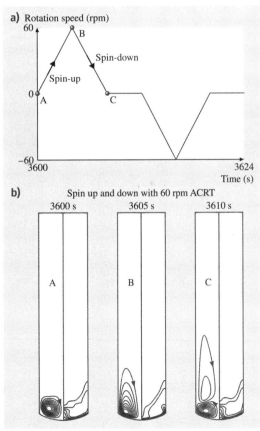

Fig. 36.27a,b Effect of ACRT on the flow and solute fields: (a) ACRT rotation cycle pattern; (b) instantaneous flow and solute fields at the moments indicated in (a)

tation speed, as shown in Fig. 36.26d at 200 rpm, the acetone was pushed outward to the edge of the interface, while the growth interface remained smooth. The calculated flow and solute fields near the growth interface are also illustrated in Fig. 36.26d. Therefore, by using an appropriate rotation rate, rotation about the growth axis is believed to be a useful approach for growth control, and this is not restricted to electrically conductive materials.

Beside steady rotation, ACRT is also a useful technique. Both the rotation cycle pattern and the period are critical to the flow control. Again, we take SCN as an example. The ACRT cycle pattern is shown in Fig. 36.27a, and some instantaneous flow patterns and acetone concentrations corresponding to the stages A, B, and C indicated in Fig. 36.27a are shown in Fig. 36.27b. The spin-up flows at 3600 and 3605 s near the interface are counterclockwise in direction. The spin-down flow at 3605 s is in the opposite flow direction. The acetone fields are significantly affected by the instantaneous flow as well. Similarly, the isotherms (not shown here) near the interface are found to be distorted toward the center of the interface. As a result,

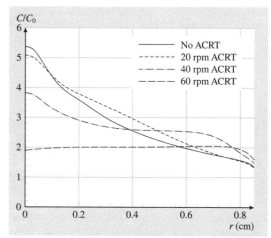

Fig. 36.29 Effect of the rotation amplitude in ACRT on the averaged radial acetone distribution

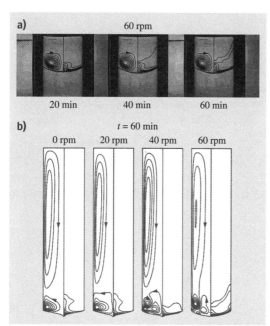

Fig. 36.28 (a) Flow and solute fields and the interface shapes at different growth periods for 60 rpm ACRT. (b) Average flow and solute fields and interface shapes for different maximum rotation magnitudes in ACRT

the interface becomes significantly concave at 60 rpm. The comparison of the calculated and observed interface shapes at different growth period for 60 rpm ACRT is shown in Fig. 36.28a. As shown, they are in good agreement. In addition, the interface when stationary is flat and it takes about 20 min to reach a steady shape.

The averaged flow and acetone fields for different ACRT amplitudes at 60 min after solidification are shown in Fig. 36.28b, from which the acetone concentration profiles at the interface are plotted in Fig. 36.29. As shown in Fig. 36.28b, it is clear that the original upper cell is not affected much by ACRT, except at 60 rpm. In other words, the solute mixing is confined to the region near the interface. However, at 60 rpm, the global mixing is slightly enhanced due to the connection of the lower and upper flow cells. Therefore, the maximum acetone concentration is significantly lower for 60 rpm. On the other hand, the lower cell is significantly affected by ACRT and this significantly affects the radial acetone segregation. As shown in Fig. 36.28, in general, as the rotation amplitude increases, except for 20 rpm, the acetone accumulation at the interface center decreases and the radial uniformity increases for the acetone concentration at the interface. From Fig. 36.29 it is thus clear that ACRT improves radial acetone uniformity and lowers its concentration due to improved mixing by ACRT. Particularly, at 60 rpm, the much slower acetone accumulation, as a result of better global mixing, reduces supercooling and enhances morphological stability.

36.3.3 Vibration

The ACRT mentioned previously usually requires a rotation cycle having a period long enough to develop Ekman flows. The Ekman time scale can be estimated by $R_c/\sqrt{\Omega \nu}$ [36.87], where R_c is the crystal radius, Ω is the rotation speed, and ν is the kinematic viscosity of the melt. For growth of SCN in small to medium-sized vertical Bridgman systems, the Ekman time scale is up to a few seconds. An alternative approach to applying ACRT is to use a cycle time that is much shorter than the Ekman time. This method is known as the angular vibration technique (AVT) [36.45]. In this technique the ampoule is vibrated at a frequency greater than 1 Hz in the rotational direction to generate a radial outward Schlichting flow near the growth front [36.45, 47].

Figure 36.30a,b shows some simulated results for AVT with different frequencies for SCN/acetone and SCN/salol, respectively. The observed interface shapes after 1 h growth are put together for comparison. As shown, the simulated interface concavity for both cases agrees quite well with the experiments. It should be noticed that the interface is at the upper boundary of the breakdown area. More importantly, from 0 to 5 Hz, the interface concavity decreases with frequency, while from 5 to 20 Hz the concavity increases with frequency. The reason is quite clear from the simulation. From 0 to 5 Hz, the flow above the interface is weakened by vibration because of the radial outward streaming flow induced by the angular vibration. As a result, the solute distribution becomes more uniform and this reduces the interface concavity cased by the local solute accumulation. On the other hand, from 5 to 20 Hz, the Schlichting flow becomes dominant. Since the flow is in the clockwise direction and the isotherms are distorted with the

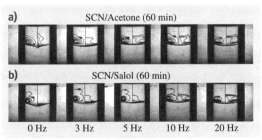

Fig. 36.30a,b Effect of angular vibration frequency on the flow and solute fields and the interface shape for VB growth of SCN: (**a**) containing 0.064 wt % acetone; (**b**) 0.15 wt % salol

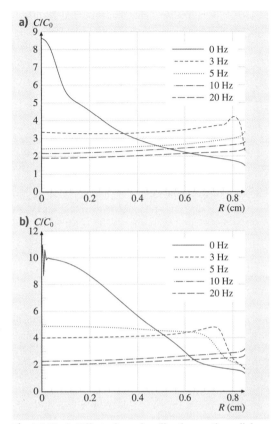

Fig. 36.31a,b Effect of angular vibration on the radial solute distributions for VB growth of SCN: (**a**) containing 0.064 wt % acetone; (**b**) 0.15 wt % salol

flow, the interface concavity increases with the vibration intensity (frequency).

The radial solute concentrations extracted from the simulated results in Fig. 36.30 are plotted in Fig. 36.31 for SCN/acetone and SCN/salol, respectively. As shown, for both cases, the radial segregation reverses from 0 to 3 Hz. This indicates that the Schlichting streaming flow is strong enough to overcome the buoyancy force and is able to push the solute from the interface center to the rim. From both Figs. 36.30 and 36.31, it is clear that, when the frequency is greater than 10 Hz, the Schlichting flow dominates and the solutal effect becomes insignificant, which can be seen from the solutal fields as well as the radial solute segregation profiles.

Besides angular vibration, vibration can also be applied axially. However, as discussed by *Lan* [36.68], the thermovibration requires radial gradients, and the effect is not as significant as that by angular vibration. Further discussion on flow control by vibration for crystal growth has been given by *Lyubimov* et al. [36.40].

36.4 Outlook

So far, we have discussed melt convection and its control for ZM and Bridgman bulk crystal growth through a few examples. As illustrated, the interface shape, which is a key factor for crystal quality, and the composition uniformity are significantly affected by the convective heat and mass transports due to the flow. We have introduced the basic flow structures through flow visualization experiments and numerical simulations. From these, one can better understand the interplay of the transport processes and the interface in crystal growth processes. The melt flow is affected by body forces, which can be the buoyancy (gravitational and centrifugal), Coriolis, and Lorentz forces. With a free surface, the thermocapillary force can also drive the melt flow. Thermal and solutal gradients are the sources for the buoyancy convection. Therefore, if these gradients are antiparallel to the gravitational or centrifugal acceleration, the convection can be minimized, and this is typically the case for vertical Bridgman growth in normal gravity or growth in a centrifuge under a free-swing configuration. In such a configuration, the flow is much more stable. Nevertheless, the residual flow can induce significant composition nonuniformity. The use of microgravity, magnetic fields, rotation or vibration is useful in manipulating the flow and thus improving crystal uniformity. Furthermore, suppressing the flow also helps improve axial composition uniformity, and a static magnetic field is particularly effective if the melt is electrically conductive.

Although we have not been able to discuss the flow and its control for Czochralski crystal growth, the concepts learned from previous examples are still useful. For example, similar to that in the FZ growth, in Czochralski growth of oxide crystals, the interface inversion can be easily controlled by crystal rotation [36.88]. However, the convection in the CZ configuration is much more complicated and hard to elucidate. For example, the buoyancy instability leading to spoke patterns at the melt surface [36.6, 89] and its coupling with rotation leading to baroclinic instability [36.90–92] are also typical in CZ oxide growth. The baroclinic instability has also been reported for CZ silicon growth [36.93]. Moreover, in large-scale CZ silicon growth, the flow is often turbulent. To suppress the turbulent flow, magnetic fields have been widely used in practice [36.27].

In contrast to flow damping, generating a favorable flow is also useful in bulk crystal growth. This includes the use of rotating and traveling magnetic fields [36.35, 94], as well as dynamic and alternating magnetic fields [36.4]. Recently, *Watanabe* et al. [36.30] proposed the electromagnetic Czochralski (EMCZ) method, in which the Lorentz force was used to generate a controlled melt flow. To generate this controlled melt flow, they created a Lorentz force by combining static magnetic fields with an electric current passing through the melt from a growing crystal. The controlled melt flow in the EMCZ method makes it easy to control the temperature distribution around the crystal, so that the growth interface can be better controlled [36.95]. Since the rotating flow washing over the crucible's inner surface can be adjusted, the oxygen content in the grown crystal can be controlled as well [36.32].

Over the years, extensive research has been applied to the investigation of melt flows, and many techniques have been proposed for their control. Still, with the increasing demands for crystal quality for advanced electronic and optoelectronic applications, this research will continue. Especially, their connection to as-grown defects and crystal quality still requires much more research to obtain deeper understanding. On the other hand, computer modeling has become an indispensable tool in the analysis and design of bulk crystal growth systems. Convection structures and their control in melt growth have been well stimulated and better understood [36.60, 96].

References

36.1 K. Kakimoto, M. Watanabe, M. Eguchi, T. Hibiya: Ordered structure in non-axisymmetric flow of silicon melt convection, J. Cryst. Growth **126**, 435–440 (1993)

36.2 K. Kakimoto, H. Ozoe: Oxygen distribution at a solid–liquid interface of silicon under transverse magnetic fields, J. Cryst. Growth **212**, 429–437 (2000)

36.3 I.Y. Evstratov, V.V. Kalaev, A.I. Zhmakin, Y.N. Makarov, A.G. Abramov, N.G. Ivanov, A.B. Korsakov, E.M. Smirnov, E. Dornberger, J. Virbulis, E. Tomzig, W. von Ammon: Numerical study of 3D unsteady melt convection during industrial-scale CZ Si-crystal growth, J. Cryst. Growth **237-239**, 1757–1761 (2002)

36.4 E. Tomzig, J. Virbulis, W. von Ammon, Y. Gelfgat, L. Gorbunov: Application of dynamic and combined magnetic fields in the 300 mm silicon single-crystal growth, Mater. Sci. Semicond. Process. **5**, 347–351 (2003)

36.5 D.T.J. Hurle: *Crystal Pulling from the Melt* (Springer, Berlin Heidelberg 1993)

36.6 A.D.W. Jones: An experimental model of the flow in Czochralski growth, J. Cryst. Growth **61**, 235–244 (1983)

36.7 D.P. Lukanin, V.V. Kalaev, Y.N. Makarov, T. Wetzel, J. Virbulis, W. von Ammon: Advances in the simulation of heat transfer and prediction of the melt-crystal interface shape in silicon CZ growth, J. Cryst. Growth **266**, 20–27 (2004)

36.8 R.A. Brown: Theory of transport processes in single crystal growth from the melt, AIChE J. **34**, 881–911 (1989)

36.9 F. Dupret, N. van den Bogaert: Modeling Bridgman and Czochralski growth. In: *Handbook of Crystal Growth: Growth Mechanisms and Dynamics*, Vol. 2b, ed. by D.T.J. Hurle (North-Holland, Amsterdam 1994)

36.10 G. Müller, A.G. Ostrogorsky: Convection in melt growth. In: *Handbook of Crystal Growth Growth Mechanisms and Dynamics*, Vol. 2b, ed. by D.T.J. Hurle (North-Holland, Amsterdam 1994)

36.11 Y.F. Zou, G.-X. Wang, H. Zhang, V. Prasad: Mechanisms of thermo-solutal transport and segregation in high-pressure liquid-encapsulated Czochralski crystal growth, J. Heat Transf. **121**, 148–159 (1999)

36.12 C.W. Lan: Recent progress of crystal growth modeling and growth control, Chem. Eng. Sci. **59**, 1437–1457 (2004)

36.13 M.C. Liang, C.W. Lan: Three-dimensional thermocapillary and buoyancy convections and interface shape in horizontal Bridgman crystal growth, J. Cryst. Growth **180**, 587–596 (1997)

36.14 C.W. Lan, M.C. Su, M.C. Liang: A visualization and computational study of horizontal Bridgman crystal growth, J. Cryst. Growth **208**, 717–725 (1999)

36.15 C.W. Lan, J.H. Chian, T.Y. Wang: Interface control mechanisms in horizontal zone-melting with slow rotation, J. Cryst. Growth **218**, 115–124 (2000)

36.16 C.W. Lan: Heat Transfer, Fluid Flow, and Interface Shapes in Floating-Zone Crystal Growth. Ph.D. Thesis (University of Wisconsin, Madison 1991)

36.17 M. Levenstam, G. Amberg: Hydrodynamical instabilities of thermocapillary flow in a half-zone, J. Fluid Mech. **297**, 357–372 (1995)

36.18 P. Dold, A. Cröll, K.W. Benz: Floating-zone growth of silicon in magnetic fields. I. Weak static axial fields, J. Cryst. Growth **183**, 545–553 (1998)

36.19 S. Nakamura, T. Hibiya, K. Kakimoto, N. Imaishi, S. Nishizawa, A. Hirata, K. Mukai, S. Yoda, T.S. Morita: Temperature fluctuations of the Marangoni flow in a liquid bridge of molten silicon under microgravity on board the TR-IA-4 rocket, J. Cryst. Growth **186**, 85–94 (1998)

36.20 M. Prange, M. Wanschura, H.C. Kuhlmann, H.J. Rath: Linear stability of thermocapillary convection in cylindrical liquid bridges under axial magnetic fields, J. Fluid Mech. **394**, 281–302 (1999)

36.21 M. Schweizer, A. Croll, P. Dold, T. Kaiser, M. Lichtensteiger, K.W. Benz: Measurement of temperature fluctuations and microscopic growth rates in a silicon floating zone under microgravity, J. Cryst. Growth **203**, 500–510 (1999)

36.22 A. Mühlbauer, A. Muiznieks, J. Virbulis, A. Lüdge, H. Riemann: Interface shape, heat transfer and fluid flow in the floating zone growth of large silicon crystals with the needle-eye technique, J. Cryst. Growth **151**, 66–79 (1995)

36.23 G. Ratnieks, A. Muiznieks, A. Mühlbauer: Modelling of phase boundaries for large industrial FZ silicon crystal growth with the needle-eye technique, J. Cryst. Growth **255**, 227–240 (2003)

36.24 H.F. Utech, M.C. Flemming: Elimination of solute banding in indium antimonide crystals by growth in a magnetic Field, J. Appl. Phys. **37**, 2021–2023 (1966)

36.25 K.M. Kim, P. Smetana: Oxygen segregation in CZ silicon crystal-growth on applying a high axial magnetic-field, J. Electrochem. Soc. **133**, 1682–1686 (1986)

36.26 K.M. Kim: Suppression of thermal convection by transverse magnetic field, J. Electrochem. Soc. **129**, 427–429 (1982)

36.27 D.T.J. Hurle, R.W. Series: Use of a magnetic field in melt growth. In: *Handbook of Crystal Growth*, Vol. 2a, ed. by D.T.J. Hurle (North-Holland, Amsterdam 1994)

36.28 F.M. Herrmann, G. Müller: Growth of 20 mm diameter GaAs crystals by the floating-zone technique with controlled As-vapour pressure under microgravity, J. Cryst. Growth **156**, 350–360 (1995)

36.29 P. Dold, K.W. Benz: Rotating magnetic fields: fluid flow and crystal growth applications, Prog. Cryst. Growth Charact. Mater. **38**, 7–38 (1999)

36.30 M. Watanabe, M. Eguchi, T. Hibiya: Silicon crystal growth by the electromagnetic Czochralski (EMCZ) method, Jpn. J. Appl. Phys. **38**, L10–L13 (1999)

36.31 M. Watanabe, K.W. Yi, T. Hibiya, K. Kakimoto: Direct observation and numerical simulation of molten silicon flow during crystal growth under magnetic fields by x-ray radiography and large-scale computation, Progr. Crystal Growth Charact. Mater. **38**, 215–238 (1999)

36.32 M. Watanabe, M. Eguchi, W. Wang, T. Hibiya, S. Kuragaki: Controlling oxygen concentration and distribution in 200 mm diameter Si crystals using the electromagnetic Czochralski (EMCZ) method, J. Cryst. Growth **237-239**, 1657–1662 (2002)

36.33 A. Mitric, T. Duffar, C. Diaz-Guerra, V. Corregidor, L.C. Alves, C. Garnier, G. Vian: Growth of $Ga_{1-x}In_xSb$ alloys by vertical Bridgman technique under alternating magnetic field, J. Cryst. Growth **287**, 224–229 (2006)

36.34 H.J. Scheel: Accelerated crucible rotation: a novel stirring technique in high-temperature solution growth, J. Cryst. Growth **13/14**, 560–565 (1971)

36.35 P. Capper, J.J. Gosney: Method of growing crystalline cadmium mercury telluride grown by method, UK Patent 2098879 (1982)

36.36 P. Capper, J.J.G. Gosney, C.L. Jones: Application of the accelerated crucible rotation technique to the Bridgman growth of $Cd_xHg_{1-x}Te$: simulations and crystal growth, J. Cryst. Growth **70**, 356–364 (1984)

36.37 P. Capper, J.C. Brice, C.L. Jones, W.G. Coates, J.J.G. Gosney, C.K. Ard., I. Kenworthy: Interfaces and flow regimes in ACRT grown $Cd_xHg_{1-x}Te$ crystals, J. Cryst. Growth **89**, 171–176 (1988)

36.38 W.G. Coates, P. Capper, C.L. Jones, J.J.G. Gosney, C.K. Ard, I. Kenworthy, A. Clark: Effect of ACRT rotation parameters on Bridgman grown $Cd_xHg_{1-x}Te$ crystals, J. Cryst. Growth **94**, 959–966 (1989)

36.39 A.V. Anilkumar, R.N. Grugel, R.N. Shen, T.G. Wang: Control of thermocapillary convection in a liquid bridge by vibration, J. Appl. Phys. **73**, 4165–4170 (1993)

36.40 D.V. Lyubimov, T.P. Lyubimova, S. Meradji, B. Roux: Vibrational control of crystal growth from liquid phase, J. Cryst. Growth **180**, 648–659 (1997)

36.41 W.S. Liu, M.F. Wolf, D. Elwell, R.S. Feigelson: Low frequency vibrational stirring: a new method for radial mixing solutions and melts during growth, J. Cryst. Growth **82**, 589–597 (1987)

36.42 W. Yuan, M. Banan, L.L. Regel, W.R. Wilcox: The effect of vertical vibration of the ampoule on the direction solidification of InSb-GaSb alloy, J. Cryst. Growth **151**, 235–242 (1995)

36.43 V. Uspenski, J.J. Favier: High frequency vibration and natural convection in Bridgman-scheme crystal growth, Int. J. Heat Mass Transf. **37**, 691–698 (1994)

36.44 C.W. Lan: Effect of axisymmetric magnetic fields on heat flow and interface in floating-zone silicon crystal growth, Model. Simul. Mater. Sci. Eng. **6**, 423–445 (1998)

36.45 W.C. Yu, Z.B. Chen, W.T. Hsu, B. Roux, T.P. Lyubimova, C.W. Lan: Reversing radial segregation and suppressing morphological instability during Bridgman crystal growth by angular vibration, J. Cryst. Growth **271**, 474–480 (2004)

36.46 W.C. Yu, Z.B. Chen, W.T. Hsu: Effects of angular vibration on the flow, segregation, and interface morphology in vertical Bridgman crystal growth, Int. J. Heat Mass Transf. **50**, 58–66 (2007)

36.47 Y.C. Liu, W.C. Yu, B. Roux, T.P. Lyubimova, C.W. Lan: Thermal-solutal flows and segregation and their control by angular vibration in vertical Bridgman crystal growth, Chem. Eng. Sci. **61**, 7766–7773 (2006)

36.48 W.A. Arnold, W.R. Wilcox, F. Carlson, A. Chait, L.L. Regel: Transport modes during crystal growth in a centrifuge, J. Cryst. Growth **119**, 24–40 (1992)

36.49 G. Müller, G. Neumann, W. Weber: The growth of homogeneous semiconductor crystals in a centrifuge by the stabilizing influence of the Coriolis force, J. Cryst. Growth **119**, 8–23 (1992)

36.50 W.R. Wilcox, L.L. Regel: Influence of centrifugation on transport phenomena, 46th Int. Astronaut. Congr. (Oslo 1995)

36.51 J. Friedrich, J. Baumgartl, H.J. Leister, G. Müller: Experimental and theoretical analysis of convection and segregation in vertical Bridgman growth under high gravity on a centrifuge, J. Cryst. Growth **167**, 45–55 (1996)

36.52 W.R. Wilcox, L.L. Regel, W.A. Arnold: Convection and segregation during vertical Bridgman growth with centrifugation, J. Cryst. Growth **187**, 543–558 (1998)

36.53 C.W. Lan: Effects of ampoule rotation on flows and segregation in vertical Bridgman crystal growth, J. Cryst. Growth **197**, 983–991 (1999)

36.54 A. Yeckel, F.P. Doty, J.J. Derby: Effect of steady ampoule rotation on segregation in high-pressure vertical Bridgman growth of cadmium zinc telluride, J. Cryst. Growth **203**, 87–102 (1999)

36.55 C.W. Lan: Effects of centrifugal acceleration on flows and segregation in vertical Bridgman crystal growth, J. Cryst. Growth **229**, 595–600 (2001)

36.56 C.W. Lan, C.H. Chian: Three-dimensional simulation of Marangoni convection in floating-zone crystal growth, J. Cryst. Growth **230**, 172–180 (2001)

36.57 C.W. Lan, Y.W. Yang, C.Y. Tu: Reversing radial segregation and suppression morphological instability in directional solidification by rotation, J. Cryst. Growth **235**, 619–625 (2002)

36.58 C.W. Lan, Y.W. Yang, H.Z. Chen, I.F. Lee: Segregation and morphological instability due to double diffusive convection in rotational directional solidification, Metal. Mater. Trans. A **33**, 3011–3017 (2002)

36.59 A.F. Witt, H.C. Gatos, M. Lichtensteiger, M.C. Lavine, C.J. Herman: Crystal growth and steady state segregation under zero gravity, J. Electrochem. Soc. **122**, 276–283 (1975)

36.60 C.W. Lan, B.C. Yeh: Three-dimensional analysis of flow and segregation in vertical Bridgman crystal growth under a transversal magnetic field with ampoule rotation, J. Cryst. Growth **266**, 200–206 (2004)

36.61 M.C. Lan, M.C. Liang: A three-dimensional finite-volume/Newton method for thermal-capillary

36.62 C.W. Lan, M.C. Liang: Multigrid methods for incompressible heat flow problems with an unknown interface, J. Comput. Phys. **152**, 55–77 (1999)

36.63 A. Lipchin, R.A. Brown: Hybrid finite-volume/finite-element simulation of heat transfer and melt turbulence in Czochralski crystal growth of silicon, J. Cryst. Growth **216**, 192–203 (2000)

36.64 C.W. Lan, D.T. Yang: A numerical study on heat flow and interface of the vertical zone-melting crystal growth, Numer. Heat Transf., Part A **129**, 131–145 (1996)

36.65 C.W. Lan: Newton's method for solving heat transfer, fluid flow and interface shapes in a floating molten zone, Int. J. Numer. Method Fluids **19**, 41–65 (1994)

36.66 C.W. Lan, S. Kou: Heat-transfer, fluid-flow and interface shapes in floating-zone crystal-growth, J. Cryst. Growth **108**, 351–366 (1991)

36.67 S. Kou, C.W. Lan: Contactless heater floating zone refining and crystal growth, US Patent 5217565 (1993)

36.68 C.W. Lan: Heat transfer, fluid flow, and interface shapes in zone melting processing with induction heating, J. Electrochem. Soc. **145**, 3926–3935 (1998)

36.69 C.W. Lan, M.C. Liang: Three-dimensional simulation of vertical zone-melting crystal growth: Symmetry breaking to multiple states, J. Cryst. Growth **208**, 327–340 (2000)

36.70 C.W. Lan, C.H. Wang: Three-dimensional bifurcations of a two-phase Rayleigh–Benard problem in a cylinder, Int. J. Heat Mass Transf. **44**, 1823–1838 (2001)

36.71 J. Baumgartl, W. Budweiser, G. Müller, G. Neumann: Studies of buoyancy driven convection in a vertical cylinder with parabolic temperature profile, J. Cryst. Growth **97**, 9–17 (1989)

36.72 W.G. Pfann: *Zone Melting* (Wiley, New York 1958)

36.73 C.W. Lan: Effects of axial vibration on vertical zone-melting processing, Int. J. Heat Mass Transf. **43**, 1987–1997 (2000)

36.74 W.W. Mullins, R.F. Sekerka: The stability of a planar interface during solidification of a dilute binary alloy, J. Appl. Phys. **35**, 444–451 (1964)

36.75 K.H. Lie, J.T. Hsu, Y.D. Guo, T.P. Chen: Influence of through-window radiation on the horizontal Bridgman process for rectangular shaped GaAs crystals, J. Cryst. Growth **109**, 205–211 (1991)

36.76 P.M. Adornato, R.A. Brown: Convection and segregation in directional solidification of dilute and non-dilute binary alloy: effects of ampoule and furnace design, J. Cryst. Growth **80**, 155–190 (1987)

36.77 D. Hofmann, T. Jung, G. Müller: Growth of 2 inch Ge:Ga crystals by the dynamic verical gradient freeze process and its numerical modeling including transient segregation, J. Cryst. Growth **128**, 213–218 (1992)

36.78 C.W. Lan, F.C. Chen: A finite-volume method for solute segregation in directional solidification and comparison with a finite-element method, Comput. Methods Appl. Mech. Eng. **31**, 191–207 (1996)

36.79 M.C. Liang, C.W. Lan: Three-dimensional convection and solute segregation in vertical Bridgman crystal growth, J. Cryst. Growth **167**, 320–332 (1996)

36.80 G.D. Robertson, D.J. O'Connor: Magnetic field effects on float-zone Si crystal growth: strong axial fields, J. Cryst. Growth **76**, 111–122 (1986)

36.81 C.W. Lan: Effect of axisymmetric magnetic fields on radial dopant segregation of floating-zone silicon growth in a mirror furnace, J. Cryst. Growth **169**, 269–278 (1996)

36.82 J.A. Burton, R.C. Prim, W.P. Slichter: The distribution of solute in crystals grown from the melt. Part I. Theoretical, J. Chem. Phys. **21**, 1987–1991 (1953)

36.83 D.H. Kim, P.M. Adornato, R.A. Brown: Effect of vertical magnetic field on convection and segregation in vertical Bridgman crystal growth, J. Cryst. Growth **89**, 339–356 (1988)

36.84 L.L. Rodot, M. Rodot, W.R. Wilcox: Material processing in high gravity-proceedings of the 1st international workshop on material processing in high gravity, J. Cryst. Growth **119**, R8 (1992)

36.85 C.W. Lan, C.Y. Tu: Three-dimensional analysis of heat flow, segregation, and interface shape of gradient-freeze growth in a centrifuge, J. Cryst. Growth **226**, 406–418 (2001)

36.86 M.R. Foster: The effect of rotation on vertical Bridgman growth at large Rayleigh number, J. Fluid Mech. **409**, 185–221 (2000)

36.87 A. Yeckel, J.J. Derby: Effect of accelerated crucible rotation on melt composition in high-pressure vertical Bridgman growth of cadmium zinc telluride, J. Cryst. Growth **209**, 734–750 (2000)

36.88 S.H. Lee, Y.J. Kim, S.H. Cho, E.P. Yoon: The influence of the Czochralski growth parameters on the growth of lithium niobate single crystals, J. Cryst. Growth **125**, 175–180 (1992)

36.89 Q. Xiao, J.J. Derby: Three-dimensional melt flows in Czochralski oxide growth: High-resolution, massively parallel, finite element computations, J. Cryst. Growth **152**, 169–181 (1995)

36.90 M.P. Gates, B. Cockayne: Purification of sodium tungstate, Nature **207**, 855 (1965)

36.91 C.D. Brandle: Flow transitions in Czochralski oxide melts, J. Cryst. Growth **57**, 65–70 (1982)

36.92 C.J. Jing, N. Imaishi, T. Sato, Y. Miyazawa: Three-dimensional numerical simulation of oxide melt flow in Czochralski configuration, J. Cryst. Growth **216**, 372–388 (2000)

36.93 M. Watanabe, M. Eguchi, K. Kakimoto: The baroclnic flow instability in rotating silicon melt, J. Cryst. Growth **128**, 288–292 (1993)

36.94 P. Dold, K.W. Benz: Modification of fluid flow and heat transport in vertical Bridgman configurations

36.95 M. Watanabe, D. Vizman, J. Friedrich: Large modification of crystal-melt interface shape during Si crystal growth by using electromagnetic Czochralski method (EMCZ), J. Cryst. Growth **292**, 252–256 (2006)

36.96 G. Müller, J. Friedrich: Challenges in modeling of bulk crystal growth, J. Cryst. Growth **266**, 1–19 (2004)

37. Vapor Growth of III Nitrides

Dang Cai, Lili Zheng, Hui Zhang

Good understanding of transport phenomena in vapor deposition systems is critical to fast and effective crystal growth system design. Transport phenomena are complicated and are related to operating conditions, such as temperature, velocity, pressure, and species concentration, and geometrical conditions, such as reactor geometry and source–substrate distance. Due to the limited in situ experimental monitoring, design and optimization of growth is mainly performed through semi-empirical and trial-and-error methods. Such an approach is only able to achieve improvement in the deposition sequence and cannot fulfill the increasingly stringent specifications required in industry. Numerical simulation has become a powerful alternative, as it is fast and easy to obtain critical information for the design and optimization of the growth system. The key challenge in vapor deposition modeling lies in developing an accurate simulation model of gas-phase and surface reactions, since very limited kinetic information is available in the literature. In this chapter, GaN thin-film growth by iodine vapor-phase epitaxy (IVPE) is used as an example to present important steps for system design and optimization by the numerical modeling approach. The advanced deposition model will be presented for multicomponent fluid flow, homogeneous gas-phase reaction inside the reactor, heterogeneous surface reaction on the substrate surface, heat transfer, and species transport. Thermodynamic and kinetic analysis will be presented for gas-phase and surface reactions, together with a proposal for the reaction mechanism based on experiments. The prediction of deposition rates is presented. Finally, the surface evolution of film growth from vapor is analyzed for the case in which surface diffusion determines crystal grain size and morphology. Key control parameters for film instability are identified for quality improvement.

37.1 Overview of Vapor Growth of III Nitrides 1244
    37.1.1 Various GaN/AlN Vapor-Growth Systems 1244
    37.1.2 Modeling of AlN/GaN Vapor Deposition 1246

37.2 Mathematical Models for AlN/GaN Vapor Deposition 1248
    37.2.1 Transport Equations 1248
    37.2.2 Growth Kinetics 1249
    37.2.3 Numerical Solution 1251

37.3 Characteristics of AlN/GaN Vapor Deposition 1251
    37.3.1 Theoretical Analysis of Heat and Mass Transfer 1251
    37.3.2 Thermodynamic and Kinetic Analysis of Chemical Reactions 1254

37.4 Modeling of GaN IVPE Growth – A Case Study 1258
    37.4.1 Scaling Analysis 1258
    37.4.2 Computational Issues 1258
    37.4.3 Gas-Phase and Surface Reactions Analysis 1259
    37.4.4 Geometrical and Operational Conditions Optimization.............. 1264
    37.4.5 Effect of Total Gas Flow Rate on Substrate Temperature 1264
    37.4.6 Effect of Substrate Rotation on Deposition Rate and Deposition Uniformity 1269
    37.4.7 Quasi-equilibrium Model for Deposition Rate Prediction 1270
    37.4.8 Kinetic Deposition Model 1271

37.5 Surface Evolution of GaN/AlN Film Growth from Vapor 1274

37.6 Concluding Remarks 1275

References 1276

37.1 Overview of Vapor Growth of III Nitrides

Chemical vapor deposition (CVD) systems have been widely used to grow thin-film and bulk GaN/AlN crystals, which have a broad range of industrial applications, especially in the field of optoelectronics. For example, they have been used to manufacture optoelectronic devices such as light-emitting diodes (LEDs), laser diodes (LDs), and detectors [37.46, 47]. Due to their wide bandgap and high breakdown field, they have become important materials for high-temperature/high-power electronics [37.48–51].

In this section, GaN/AlN vapor-growth systems are briefly introduced first. An iodine vapor deposition system for GaN bulk growth is taken as an example to explain the modeling of comprehensive transport phenomena and chemical reactions in crystal growth from vapor. Different numerical models, their limitations, and future treads are then reviewed. GaN thin-film growth by the iodine vapor-phase epitaxy (IVPE) technique is presented in this chapter as an example for discussion.

37.1.1 Various GaN/AlN Vapor-Growth Systems

For conventional semiconductors such as silicon, the Bridgman or Czochralski methods are usually used to grow ingots from a melt. However, the high vapor pressure of nitrogen at the melting temperature

Table 37.1 Research groups of GaN/AlN growth

Researchers	Growth techniques	Substrate	Growth conditions	Size	Maximum growth rate
Speck et al. (1997–2006)	MBE [37.1–8]	MOCVD GaN/ sapphire template (0001)	GaN: 10^{-11} Torr, 650–800 °C	–	750 nm/h
Karpov et al. (1998–2004)	HVPE [37.9, 10] Sublimation growth [37.11–15]	– GaN:6H-SiC AlN:SiC	GaN: 760 Torr 900–1100 °C GaN: 1100–1250 °C AlN: 760 Torr; > 1700 °C	– GaN: $15 \times 15 \times 0.5$ mm^3 AlN: $10 \times 10 \times 0.08$ mm^3	GaN: 40 μm/h GaN: 1 mm/h AlN: 1 mm/h
Edgar et al. (2000–2006)	MOCVD [37.16–18] Sublimation growth [37.19–24]	GaN:3C-SiC/ Si(100) AlN:6H-SiC	GaN: 76 Torr; 950 °C AlN: 100–800 Torr; 1700–1900 °C	– –	GaN: 1.5 μm/h AlN: 1 mm/h
Bliss et al. (1999–2005)	CVRP [37.25] IVPG [37.26, 27] HVTE [37.28, 29]	GaN:quartz GaN:sapphire GaN:sapphire	GaN: 1 atm; 900 °C GaN: 75–750 Torr; 910–1025 °C AlN: 3.5–760 Torr, 1100–1300 °C	GaN: 9×2 mm$^2 \times 100$ μm GaN: 32 μm thickness 0.05–1 mm diameter AlN: 75 μm thickness 2.5–5 mm diameter	GaN: 25–100 μm/h GaN: 10–11 μm/h AlN: 40 μm/h
Sitar (2001–2006)	IVPE [37.30] Sublimation growth [37.31–37]	GaN/AlN:sapphire AlN/SiC	GaN: 400–800 Torr, 1000–1200 °C AlN: 500–900 Torr; 2200–2300 °C	GaN: 50 mm in diameter AlN: 18 mm in diameter	GaN: 75 μm/h AlN: 5 mm/h
Slack et al. (2000–2005)	Sublimation growth [37.38–40]	–	AlN: ≈ 1800 °C (Rojo et al., 2001)	AlN: 150 mm^2	AlN: 0.9 mm/h
Spencer et al. (1993–2004)	MOCVD [37.41–45]	4H-SiC or 6H-SiC	AlN: 10 Torr, 1160–1190 °C		AlN: 1 μm/h

of GaN/AlN hampers melt growth of the materials. Other growth techniques were developed to fabricate GaN/AlN thick films at lower temperature. Metalorganic chemical vapor deposition (MOCVD) [37.52–55] and molecular-beam epitaxy (MBE) [37.56–59] technologies have been extensively used to grow high-quality GaN/AlN films for device applications. Both methods, however, have the drawback of high cost. Owing to its cost effectiveness, the halide vapor-phase epitaxy (HVPE) technique has gained more attention for depositing thick AlN/GaN layers [37.25, 26, 60–63]. In HVPE, the growth rate is mainly determined by the mass flow rate of the reactants, since surface reactions inside the reactor are close to equilibrium due to high temperature on the substrate. High growth rates ($> 20\,\mu\text{m/h}$) can easily be achieved in HVPE. Besides HVPE, sublimation methods [37.11, 31, 64–69] have also been used for AlN/GaN growth. In sublimation growth, high growth rate and good quality crystal are achieved [37.70] as the results of using high growth temperature ($\approx 2000\,°\text{C}$) and employing the repeated seeding method [37.65]. Different AlN/GaN vapor-growth systems are summarized in Table 37.1.

HVPE method has been used to produce GaN thick films or ingots with growth rates of up to $50\,\mu\text{m/h}$ and acceptable thickness uniformity [37.73, 74]. In the process, GaN is grown from the reactions of $NH_3/GaCl$ or $NH_3/GaCl_3$. The use of HCl gas to obtain $GaCl_x$ in the reactor creates contamination due to its corrosive effects. To prevent this problem, *Bliss* et al. [37.26, 27] developed an iodine vapor-phase epitaxy (IVPE) growth system for the first time, in which solid iodine is vaporized and reacts with gallium source to form GaI_x. Using a horizontal IVPE system, GaN film at growth rate on the order of $10\,\mu\text{m/h}$ has been obtained, and a low level of yellow luminescence of thus-produced GaN film has been observed, indicating the beneficial effect of reduced contamination in the process gases. To investigate the feasibility of higher GaN growth rate and better deposition quality, a vertical up-flow IVPE system for GaN has been designed and built at North Carolina State University to grow GaN thick film on a squared SiC substrate [37.30, 71, 72].

A schematic of the vertical reactor is presented in Fig. 37.1a. Iodine vapor is used to transport Ga from source to substrate to grow GaN. Ammonia gas is introduced to provide the nitrogen source for GaN growth. The reactor consists of four concentric tubes: furnace tube, reactor tube, outer silica tube, and inner silica tube. Iodine is carried by N_2 and H_2 gases flowing upwards through the inner silica. At the top of the inner silica tube is the gallium source within a BN holder. The iodine reacts with Ga melt to form GaI_x. A silica nozzle is placed above the gallium source to enhance Ga replacement. The gas flowing out of the silica nozzle is, therefore, a mixture of nitrogen, iodine, hydrogen, GaI_x, and gallium. N_2 gas flows between the inner and outer silica tubes to shield the Ga from the ammonia in the gas phase and enhance reactions on the substrate. Finally, the ammonia and nitrogen mixture flows through the space between the reactor tube and the outer silica tube. The system is heated by a re-

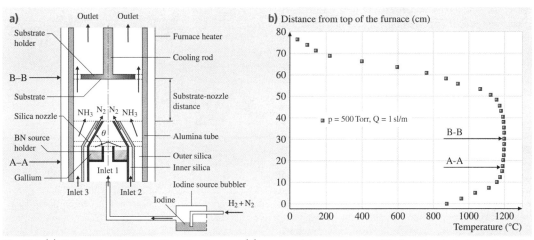

Fig. 37.1 (a) Schematic of the reaction chamber and (b) measured temperature profile from top to bottom along the furnace wall (after [37.30, 71, 72])

Table 37.2 List of simulation models for CVD process

Model	Length scale	Function
Ab initio methods, density function theory (DFT) [37.75, 76]	Electronic scale: 10^{-13}–10^{-9} m	Calculate transition state structures, surface reaction pathways, etc.
Molecular dynamics [37.75, 77, 78]	Atomic scale: 10^{-10}–10^{-6} m	Simulate the real atomic motion
Kinetic Monte Carlo (KMC) [37.79–84]	Microstructure scale: 10^{-9}–10^{-5} m	Bridge atomic scale and microscopic scale in dynamic simulations
Level-sets or hybrid models [37.77, 85, 86]	Film scale: 10^{-7}–10^{-3} m	Simulate grain growth, recrystallization, etc.
Continuum models [37.87, 88]	Reactor scale: 10^{-5}–10 m	Simulate heat and mass transport in the CVD reactor

sistant heater. Calibration of the furnace temperature has been conducted under reactor pressure of 500 Torr and N_2 flow rate of 1 slm. Figure 37.1b shows the temperature profile measured by a thermocouple along the furnace wall. By setting the growth temperature at 1200 °C, the uniform-temperature zone achieved is about 37.1 cm. The gallium source (position A–A) and the substrate (position B–B) are both positioned in the uniform-temperature zone. The gallium source is positioned close to the bottom of the zone to ensure that the entire growth area is located in the uniform-temperature zone.

37.1.2 Modeling of AlN/GaN Vapor Deposition

Physics-based theoretical modeling and simulation are widely used for better understanding of growth mechanisms and identifying the important issues related to growth processes, material characteristics, and dopant incorporation. Advanced models integrating the key aspects in the growth process can help identify the material stability limits, define the ideal parametric window(s), and improve crystal perfection. In addition, the physical mechanisms that limit the growth process can be determined.

Vapor deposition modeling presents challenges in two aspects. First, except for some well-studied important reactions such as homogeneous and heterogeneous pyrolysis of silane, the kinetic data of gas-phase and surface reactions are limited. In this chapter, combined with experimental observation, theoretical methods are used to obtain insight into the chemical mechanisms and reaction kinetics of IVPE. Second, vapor deposition involves complex transport phenomena occurring on different length and time scales. The characteristic length and time for atomic surface diffusion and relaxation and chemical reactions are approximately 10^{-10} m and 10^{-12} s, respectively. The typical reactor size is about 10^{-1} m and the deposition time is 10^3–10^5 s. A simulation model accurately describing both the microscale molecular motion and the macroscale species transport remains elusive due to the tremendous memory and extremely fast central processing unit (CPU) required for such a model. Currently, most numerical models have focused only on particular scales, either macroscale or microscale. Different models and their corresponding scales are summarized in Table 37.2.

Among the various models, only continuum models have been extensively used for vapor-growth reactor design. More advanced models tend to predict

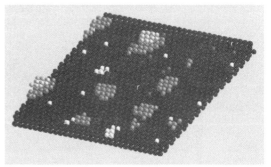

Fig. 37.2 Simulated Si(111) surface after deposition of 12% of C monolayer. *Black* and *white spheres* denote substrate atoms and vacancies created by thermal fluctuations, respectively. *Dark grey spheres* denote carbon atoms incorporated into the substrate. Atoms that are combined into the clusters are shown in *light grey* (*lighter tint* is used for upper cluster layers) (after CMS, with permission from Elsevier)

Table 37.3 Researchers and their models of GaN/AlN vapor growth

Authors	Models
Liu and *Edgar* [37.90, 91]	AlN sublimation growth: 2-D thermal convection; Stefan flow model; surface kinetics model
Karpov et al. [37.9–11, 14, 92, 93]	GaN sublimation growth: 2-D gas dynamics and gaseous species model
	AlN sublimation growth: Diffusive and transport kinetic model
	GaN MBE growth: Surface kinetic model
	GaN HVPE growth: 3-D gas flow and growth rate model
	GaN MOCVD growth: Surface kinetic model
Safvi et al. [37.94, 95]	GaN HVPE growth: Effects of flow rate, V/III ratio, and geometry on growth rate
	GaN MOCVD growth: Growth rate model
Theodoropoulos et al. [37.96]	GaN MOCVD growth: 2-D transport model; kinetic model
Aujol et al. [37.97]	GaN HVPE growth: Thermodynamic and kinetic model of growth rate
Dollet et al. [37.98]	AlN HVPE growth: Gas-phase and surface reaction analysis of $AlCl_3$ with NH_3; 2-D species transport modeling
Zhang et al. [37.29, 37, 72, 99–101]	AlN sublimation growth: 2-D induction heating and vapor transport; growth morphology and growth rate prediction
	AlN HVPE growth: Gas-phase and surface reactions modeling; growth optimization of geometric and operating conditions
	GaN IVPE growth: Thermodynamic and kinetic analysis and modeling of gas-phase and surface reactions; growth rate prediction

the microstructure and properties of materials. Figure 37.2 shows three-dimensional SiC clusters on a Si substrate modeled by the kinetic Monte Carlo (KMC) method [37.89]. Parameters used in the KMC model are estimated from molecular-dynamics simulation and by fitting data to experimental transmission electron microscopy (TEM) results. Formation of three-dimensional clusters and pits in the surface is successfully demonstrated.

These models, however, are still under development and their use is still limited to research purposes because of numerous approximations used in the models. A model combining two or more of the aforementioned models is called a multiscale model. A few reviews of multiscale modeling of vapor deposition thin-film growth are available in the open literature [37.102–106]. It is difficult for this method to consider the effect of feedback from the lower-scale model on the results of the upscale model [37.102].

Due to the lack of detailed reaction mechanisms, simulations of AlN/GaN vapor growth have been focused on the continuum modeling and solved gas flow, heat transfer, species diffusion, and chemical reactions in a vapor deposition reactor. A summary of the relevant research groups and their models is given in Table 37.3.

The continuum model will be used here to study the transport phenomena in the vapor-growth reactors for AlN/GaN growth. Gas-phase and surface reactions in the reactor will be studied and modeled to predict the species concentrations above the surface of the substrate accurately. The information obtained from the model will then be passed onto a surface deposition model to predict the AlN/GaN deposition rate distribution on the substrate.

37.2 Mathematical Models for AlN/GaN Vapor Deposition

Characteristics of chemical vapor deposition process are determined by gas hydrodynamics and chemical kinetics, which are affected by reactor geometry and process conditions. To reduce the complexity, the following assumptions are usually made for simulations:

1. The gas mixture is treated as a continuum
2. An ideal gas law is used
3. Gas flow is laminar
4. Gas mixture in the CVD reactor is radiative-transparent
5. Viscous dissipation of gas mixture is neglected.

It should be noted that these assumptions do not essentially limit the accuracy of the model for a wide range of vapor deposition process conditions.

37.2.1 Transport Equations

Based on these assumptions, the following equations can be used to describe the vapor deposition process:

Continuity
$$\frac{\partial \rho}{\partial t} + \nabla \cdot (\rho \mathbf{V}) = 0 . \quad (37.1)$$

Momentum
$$\frac{\partial}{\partial t}(\rho \mathbf{V}) + \nabla \cdot (\rho \mathbf{V}\mathbf{V})$$
$$= -\nabla p + \rho \mathbf{g} . \quad (37.2)$$

Energy
$$\frac{\partial}{\partial t}(\rho h) + \nabla \cdot (\rho \mathbf{V} h)$$
$$= \nabla \left(\frac{k}{C_p} \nabla h \right) + \tau_{ij} \frac{\partial u_i}{\partial x_j}$$
$$+ q'''_{\text{latent}} + q'''_{\text{radi}} + q'''_{\text{eddy}}$$
$$+ q'''_{\text{react}} . \quad (37.3)$$

Species transport (in mass fractions Y_i)
$$\frac{\partial}{\partial t}(\rho Y_i) + \nabla(\rho \mathbf{V} Y_i) = \nabla(\rho D \nabla Y_i) + \dot{w}_i . \quad (37.4)$$

Ideal gas law
$$p = \rho RT \sum_i \frac{Y_i}{W_i} . \quad (37.5)$$

Here ρ is the density, $\mathbf{g}$ is the gravitational acceleration vector, $\mathbf{V}$ is the gas velocity, p is the pressure, τ is the viscous stress tensor, T is the temperature, k is the thermal conductivity, D is the binary diffusion coefficient of reactant in carrier gas, Y_i is the mass fraction of species i, $\dot{w}_i$ is the gas-phase reaction rate of species i, W_i is the molecular weight of species i, and R is the gas constant. $q'''_{\text{latent}} = \Delta H_{\text{vs}} \dot{M}_{\text{vs}}$ is the latent heat of phase change with ΔH_{vs} as the specific latent heat and $\dot{M}_{\text{vs}}$ as the deposition rate; q'''_{radi} is the heat exchange rate due to radiation; $q'''_{\text{eddy}} = \frac{1}{2}\sigma_c \omega A_0 A_0^*$ is the power dissipation rate due to an induced eddy current in the susceptor, in which σ_c is the electrical conductivity of the susceptor, ω is the alternate current frequency in the induction coils, and A_0 is the induced magnetic potential; q'''_{react} is the heat release or absorption due to reactions. Calculation of q'''_{radi} will be detailed in the radiative heat transfer model.

Critical Condition for Radiation Dominance. Radiative heat transfer becomes important when temperature is high ($\approx 1000\,^\circ\text{C}$). Heat fluxes transferred between the hot and cold walls due to heat conduction, convection, and radiation are calculated analytically to show the importance of radiative heat transfer. Considering a typical reactor for GaN/AlN growth, the distance between two walls d is chosen as $0.1016\,\text{m}$ (5 inches), and the length of the walls is $0.508\,\text{m}$ (20 inches). The radiative emissivity of the walls is 0.25 (if the reactor is made of metal). Temperature for the cold wall is 300 K, and the hot-wall temperature varies from 300 to 2000 K. Nitrogen is running between the walls with a Reynolds number of 100. Such a system can be approximated as a one-dimensional (1-D) problem for a first analysis. Heat flux due to heat conduction, convection, and radiation can then be calculated. The conductive heat flux is

$$q'' = k \frac{\Delta T}{d} = k \frac{T_{\text{hot}} - T_{\text{cold}}}{d} . \quad (37.6)$$

The convective heat flux is [37.107]

$$q'' = h \Delta T = \frac{k \text{Nu}}{d} \Delta T$$
$$= 0.46 k \text{Re}_f^{0.5} \text{Pr}_f^{0.43} \left(\frac{\text{Pr}_f}{\text{Pr}_w} \right)^{0.25}$$
$$\times \left(\frac{d}{L} \right)^{0.4} (T_{\text{hot}} - T_{\text{cold}})/d . \quad (37.7)$$

The radiative heat flux is

$$q'' = \frac{\sigma \left(T_{\text{hot}}^4 - T_{\text{cold}}^4 \right)}{\frac{2(1-\varepsilon)}{\varepsilon} + \frac{1}{F_{\text{h-c}}}} , \quad (37.8)$$

where k is the thermal conductivity of nitrogen, h is the convective heat transfer of cold nitrogen on the hot wall surface, Nu is the Nusselt number, Re_f is the Reynolds number of the nitrogen flow, Pr_f is the Prandtl number with the reference temperature of the gas, Pr_w

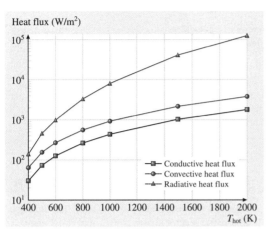

Fig. 37.3 Comparison of heat flux due to conduction, convection, and radiation

is the Prandtl number with the reference temperature of the hot wall, σ is the Stefan–Boltzmann constant (5.672×10^{-8} W/m² K⁴), and F_{h-c} is the view factor between the hot and cold walls, which is calculated to be 0.82. The heat fluxes calculated from (37.6–37.8) are summarized in Fig. 37.3.

It is found that the radiative heat flux is an order of magnitude higher than conduction or convection when the hot-wall temperature is 1000 K. Radiation will therefore be dominant when the temperature is higher than 1000 K.

The integro-differential radiative heat transfer equation for calculating the radiative heat flux can be written as follows

$$\frac{\mathrm{d}I(r,s)}{\mathrm{d}s} = -(\alpha_a + \sigma_s)I(r,s) + \alpha_a I_b(r)$$
$$+ \frac{\sigma_s}{4\pi} \int_0^{4\pi} I(r,s')\Phi(s \cdot s')\mathrm{d}\Omega', \quad (37.9)$$

where I is the radiation intensity, r is the position vector, s is the direction vector, α_a is the absorption coefficient, and σ_s is the scattering coefficient. I_b is the intensity of the black body, Φ is the phase function of energy transfer, and Ω' is the solid angle of the radiation beam.

Thermal radiation can be transported by the following four mechanisms:

1. From surface to surface
2. From surface to volume
3. From volume to surface
4. From volume to volume.

Equation (37.9) is applicable for an absorbing, emitting, and scattering medium. For some media, a simplified model can be used instead of (37.9). For inert gases, which can be considered as totally transparent to radiation, only surface-to-surface radiation should be considered instead of solving (37.9). It is therefore important to select an appropriate radiation model for a certain radiation problem to save calculation time. Optical thickness is usually used as a critical parameter to determine which radiation model should be used. The optical thickness L is defined as follows:

$$L = \alpha R, \quad (37.10)$$

where α is the absorption coefficient and R is the characteristic length of the medium. Some of the most commonly used radiation models and their applicable regimes in reference to optical thickness are summarized in Table 37.4, together with the radiation transportation considered in the model and the property of the radiative surface.

The first three models in Table 37.4 do not require (37.9) to be solved. The discrete ordinate model that needs to solve (37.9) will be used in most calculations presented in this chapter.

37.2.2 Growth Kinetics

In a typical chemical vapor deposition system, a thin film is formed by deposition from surface chemical reactions. Reactive molecules that contain atoms of the material to be deposited are introduced as a mixture of gases (usually diluted with inert carrier gases such as argon and nitrogen) into the temperature- and pressure-controlled environment of the reacting chamber in which the targets (wafers) on which deposition takes place are located. Heterogeneous reactions take place on the substrate. For example, in the case of epi-silicon production from silane pyrolysis, some initial portions of the overall reaction may occur in the gas phase, but the final stage of the reaction forming elemental silicon occurs on the wafer surface. Homogeneous gas-phase reactions may also occur, either preceding, parallel with, or in competition with the heterogeneous surface reactions. As the precursor gas approaches the wafer surface, it may react in the gas phase due to the high temperature ($> 400\,°$C at 1 atm). The reactions may form intermediate species, such as SiH_2, or a series of reactions may proceed to form el-

Table 37.4 Summary of radiation models

Radiation model	Optical thickness (m)	Radiation transport mechanisms	Radiation surface
P1 model [37.108, 109]	$L > 1$	(a) (c)	Gray
Rosseland model [37.108]	$L > 3$	(a) (c)	Gray
Surface-to-surface model [37.110]	$L \ll 1$	(c)	Gray
Discrete exchange factor model [37.111]	$0 < L < +\infty$	(a) (b) (c) (d)	Gray or non-gray
Discrete ordinate mode [37.112]	$0 < L < +\infty$	(a) (b) (c) (d)	Gray or non-gray
Discrete transfer model [37.113]	$0 < L < +\infty$	(a) (b) (c) (d)	Gray

emental silicon. In the latter case, silicon atoms may form small silicon grains in the gas phase. These grains then flow out of the system (sometimes causing an additional problem) or migrate to the wafer surface, where they form an irregular porous deposit, as well as loosely adherent particles. This gas-phase or homogeneous reaction does not form the dense uniform films needed for integrated-circuit applications and is usually suppressed in favor of heterogeneous reactions.

The sequence of events that may take place during vapor deposition process is shown in Fig. 37.4. It includes:

1. Reactant gases enter the reaction area inside the reactor by convection and diffusion.
2. Homogeneous gas-phase reactions take place, and intermediate species may be formed.
3. Species diffuse trough the boundary layer onto the surface.
4. Adsorption and diffusion of these species occur on the surface.
5. Heterogeneous reaction, nucleation, and lattice incorporation take place on the surface, leading to the formation of a solid film.
6. Desorption of adsorbed species away from the surface through the boundary layer.
7. Byproduct is transferred away from the reaction area in the reactor chamber.

Among these steps, the slowest will determine the deposition rate. A species concentration layer only exists when the reactor pressure is relatively high. If the gases pressure inside the reactor is low (i.e., in the range of mTorr), the species concentration boundary layer is no longer applicable, since the species transportation speed due to diffusion is much faster than the speed due to convection.

It is clear that many of the steps involved are transport related. The extent and role of the transport phenomena are, however, determined by process parameters such as substrate temperature, flow rate, and reactant partial pressure, as well as the chemistries involved. Depending on which of these is the dominant factor, the deposition process may be thermodynamical, diffusion or kinetics controlled.

In a thermodynamically controlled process, the mass transfer of species to and from the deposition zone is much slower than either mass transfer between the main flow and the substrate or mass transfer from the surface processes. Steps 1 and 7 are therefore rate-controlling steps. The process is assumed to proceed under thermodynamic equilibrium, and the deposition rate is generally determined by the equilibrium values of the partial pressures of species in the system.

In a diffusion- or mass-transport-controlled process, the rate-determining step is the diffusion or transport of the reactant gases to the substrate surface. The gas flow, heat transfer, and species diffusion play the dominant roles in determining the deposition characteristics.

Finally, in a kinetics-controlled system, the surface processes are not as fast as steps 1 and 7, nor 3 and 6. The rate-determining step is therefore the slowest of steps 4 and 5. Substrate temperature then

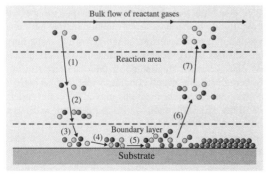

Fig. 37.4 Sequence of events during vapor deposition

plays an important role whereas the flow field has only a marginal influence. For some combinations of temperature and concentration, the homogeneous gas-phase reaction plays an important role and determines the deposition rate.

37.2.3 Numerical Solution

The simulation procedures are summarized as follows:

1. Grids are generated for calculating the electromagnetic field and heat and mass transfer, respectively.
2. The electromagnetic potentials are calculated. If the electrical/magnetic properties are dependent on temperature, this procedure will be repeated at a certain time interval until a steady-state temperature distribution is achieved. The calculated power dissipation distribution will be used as a source term for the energy transport calculation.
3. Pressure and velocity fields are calculated.
4. Energy transport and radiative heat transfer equations are solved iteratively at each time step until convergence is reached.
5. The species transport equations are solved for every species.
6. The thermoelastic stress distribution in the crystal may be solved based on the temperature field obtained in the crystal.

It should be noted that, in the program, steps 2–5 have to be repeated in an iterative way until the final variable fields stabilize.

37.3 Characteristics of AlN/GaN Vapor Deposition

The purpose of studying the characteristics of transport phenomena in a CVD system is to achieve fast and effective design of CVD reactors. In this section GaN growth using an IVPE system will be used as an example to identify the critical transport parameters. Heat and mass transfer will be analyzed theoretically for the following two reasons. First, this can provide useful information for the control of the transport process. Second, analytical results can be used to validate the accuracy of numerical simulation. Gas-phase and surface reactions will be analyzed thermodynamically and kinetically. Based on this analysis, a model for gas-phase and surface reactions can be provided.

37.3.1 Theoretical Analysis of Heat and Mass Transfer

Important Dimensionless Groups

An order-of-magnitude study provides a basic understanding of the complex heat and mass transport phenomena in a vapor-growth system. The important dimensionless groups and their definitions are listed in Table 37.5.

For a typical vapor deposition system, the Reynolds number (Re) is usually in the range of 1–100, which corresponds to laminar flow. The Grashof number is usually kept below 10^5. A large value of Gr indicates strong natural convection in the system, leading to a complicated flow pattern in the growth chamber. The ratio of the Grashof number Gr to the Reynolds number squared, Re^2, is used to quantify whether heat transfer is mainly controlled by natural or forced (inertia) convection. The regimes for natural convection, forced convection, and mixed convection are shown in Table 37.6 [37.114].

The value of the Prandtl number (Pr) is usually in the range of 0.1–0.7 for gas. The Schmidt number (Sc) is in the range of 0.01–1.0, depending strongly on pressure. At low pressure the species exhibits strong diffusion and the Schmidt number will lie towards the low end of this range. The thermal and mass Péclet numbers (Pe^T and Pe^M) are usually in the range of 1–100. The Damköhler numbers (Da) for the gas phase and surface are reaction dependent. A small value ($\ll 1$) means that the gas-phase or surface reactions are negligible. These dimensionless groups, along with the governing equations and appropriate boundary conditions, can provide the entire picture of a vapor deposition process.

Analytical Analysis of Gas Flow

For a conventional vapor-growth reactor, one of the key issues is maintaining a suitable growth environment near the substrate area, which is usually achieved by controlling the flow, temperature, and species concentration. To achieve an appropriate gas flow environment in the reactor, the relationship between the reactor diameter, total gas flux, and Reynolds number is studied. The Reynolds number can have the form

$$\mathrm{Re} = \frac{\rho Q d}{A \mu} , \qquad (37.11)$$

Table 37.5 Important dimensionless groups in a CVD reactor

Dimensionless group	Definition	Physical interpretation
Reynolds (Re)	$\dfrac{\rho_0 V_0 L_c}{\mu_0}$	Inertial forces/ viscous forces
Grashof (Gr)	$\dfrac{g\rho_0^2 L_c^3 (T_w - T_{in})}{(\mu_0^2 T_0)^2}$	Buoyancy forces/ viscous forces
Prandtl (Pr)	$\dfrac{\mu_0 C_{p0}}{k_0}$	Momentum diffusivity/ thermal diffusivity
Schmidt (Sc)	$\dfrac{\mu_c}{\rho_0 D}$	Momentum diffusivity/ species diffusivity
Thermal Péclet (Pe^T)	$Re \cdot Pr$	Convective heat transfer/ conductive heat transfer
Mass Péclet (Pe^M)	$Re \cdot Sc$	Convective mass transfer/ conductive mass transfer
Gas-phase Damköhler (Da^g)	$\dfrac{r_0^g L_c}{V_0}$	Typical time for flow/ typical gas-phase reaction time
Surface Damköhler (Da^s)	$\dfrac{r_0^s L_c}{V_0}$	Typical time for flow/ typical surface reaction time

$\rho_0, \mu_0, k_0, C_{p0}, D_0$, and β_0 are the values of ρ, μ, k, C_p, D, and β ($\beta = 1/T_0$ for ideal gas) at the average process gas temperature $T_0 = (T_w + T_{in})/2$, where T_w and T_{in} are the temperature on the reactor wall and the temperature at inlet of the reactor, respectively. L_c is the characteristic length, defined as the diameter of the reactor, V_0 is the characteristic velocity, defined as the inlet process gas velocity, r_0^g is the gas-phase reaction constant, and r_0^s is the surface reaction constant.

where ρ is the density of the mixing gases, Q is the total volume flow rate, d is the diameter of the reactor, A is the area, and μ is the dynamic viscosity. In the GaN growth system presented here, nitrogen and ammonia are the mass-dominant species in the reactor, and their molar ratio is assumed to be 1 : 1. The dynamic viscosities of nitrogen and ammonia are very similar. Sutherland's law is used to determine the dynamic viscosity $\mu = (AT^{3/2})/(B+T)$, where the coefficients are

Table 37.6 Criteria for natural and forced convection

Convection	Gr/Re^2 criteria
Natural	> 16
Mixed	$\in [0.3, 16]$
Forced	$\in [0, 0.3]$

$A = 1.461 \times 10^6$ and $B = 79.96$. The relationships of the volume flow rate and density with temperature and pressure can be expressed as $Q = Q_0(T/T_0)(P_0/P)$ and $\rho = \rho_0(T_0/T)(P/P_0)$, respectively.

Equation (37.11) can be rewritten as

$$Re_D = \frac{0.907 \rho_0 Q_0 (111.5 + T)}{dT^{3/2}}, \quad (37.12)$$

where ρ_0 (1.005 kg/m³) and Q_0 are the density and flow rate of the mixture under standard conditions (0 °C and 1 atm). For a typical vapor-growth system, the Reynolds number is usually kept in the range 1–100 to maintain a stable, laminar flow environment, which allows a sufficient residence time of reactants on the substrate for surface deposition. Figure 37.5 shows the relationship between the volume flow rate and reactor diameter at different Reynolds numbers. Experimental operating temperature of 1050 °C and pressure of 200 Torr are used in the calculation. It is shown in Fig. 37.5 that the total flow rate should be controlled at about 1 slm in order to achieve a Reynolds number of 10 for a reactor diameter of 5 cm.

Analytical Analysis of Heat and Mass Transfer

Besides forming a laminar-flow condition, a uniform species concentration distribution on the substrate is also preferred, as this is essential for deposition quality. Since gas flow is laminar in the reactor, there is no direct mixing. The transport of NH_3 from the near wall to the substrate surface will be mainly achieved through molecular diffusion in the stream. It is therefore impor-

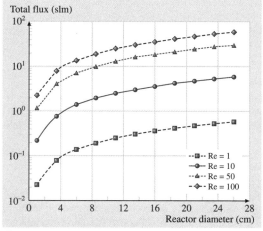

Fig. 37.5 Relationship between chamber diameter and gas flow rate with different Reynolds numbers

tant to allow ammonia sufficient time to diffuse to the substrate surface. This can be controlled by the total flow rate Q, the diameter of the reactor d, the diffusion coefficient of NH_3 in N_2 D, the distance between the substrate and the silica nozzle L, and the temperature and pressure in the reactor. To estimate the mixing time needed, or to select an appropriate substrate–nozzle distance (Fig. 37.1a) at different flow rates, it is assumed that the reactor tube is filled with N_2 with a flow rate of Q, and at $t = 0$ an NH_3 concentration of C_s is added on the wall of the reactor. According to [37.115], by solving the diffusion problem for a laminar flow in a rounded tube, the dimensionless average NH_3 concentration can be expressed as

$$\Delta' = \frac{C_s - C_{av}}{C_s - C_0} = 0.819 e^{-14.6272\beta''} + 0.0976 e^{-89.22\beta''} + 0.0135 e^{-212.2\beta''} + \ldots \quad (37.13)$$

where C_{av} is the averaged NH_3 concentration at a certain axial position X, $\beta'' = DX/4r^2\bar{U}_{av}$, D is the binary diffusion coefficient, r is the radius of the cylinder, $\bar{U}_{av}$ is the averaged gas velocity, and $C_0 = 0$ is the initial concentration of NH_3 in the tube. According to (37.13), when $\beta'' = 0.2$, $C_{av} = 0.95 C_s$ with an exposure length of X. It is therefore concluded that, when $\beta'' = 0.2$, different species have been mixed very well due to molecular diffusion. In the growth reactor, $\bar{U}_{av} = Q/A$, and the total exposure length, or the substrate–nozzle distance, is L. The following equation has to be satisfied for a uniform species concentration distribution on the substrate surface

$$\beta'' = 0.2 = DL \frac{A}{4r^2 Q} . \quad (37.14)$$

The diffusion coefficient of NH_3 in N_2 can be estimated from [37.90]

$$D = D_0 \left(\frac{T}{T_0}\right)^n \left(\frac{P_0}{P}\right) , \quad (37.15)$$

with $D_0 = 2.3 \times 10^{-5}$ m^2/s, $T_0 = 298$ K, $P_0 = 760$ Torr, and $n = 1.8$. The influence of species concentration on the diffusion coefficient can be neglected. Equation (37.14) can therefore be rewritten as

$$Q_0 = \frac{\pi D_0 \left(\frac{T}{T_0}\right)^{0.8}}{4\beta''} L , \quad (37.16)$$

with $\beta'' = 0.2$ and $T = 1373$ K. The relationship between the standard flow rate and the substrate–nozzle distance is given by

$$Q_0 = 1.62 L . \quad (37.17)$$

According to (37.17), when the total flow rate increases, a larger substrate–nozzle distance is required to achieve a uniform species concentration distribution on the substrate. For a typical flow rate of 3 slm, the corresponding substrate–nozzle distance for complete mixing is about 11.9 cm.

Temperature control in a vapor-growth system is critical. To obtain sufficiently high activation energy, reactants have to be preheated before they reach the substrate. The analytical method can be used to calculate the temperature variation of gases in the tubular reactor subject to different operational conditions such as flow rate, pressure and temperature of the furnace, and geometric conditions such as the diameter of the reactor. For the resistance heating system discussed here, we assume a constant temperature of $T_w = 1373$ K on the reactor wall and that the inlet gas mixture is composed of N_2 and NH_3 with a molar ratio of 1 : 1 and a temperature of $T_{in} = 300$ K.

The analytical solution of the problem can be shown to be [37.116]

$$T_{m(x)} = T_w - (T_w - T_{in}) \exp\left(-\frac{\alpha Nu}{r^2 \bar{U}_{av}} z\right) , \quad (37.18)$$

where $T_{m(x)}$ is the bulk longitudinal average temperature at position z, $\alpha_T = (k/(\rho C p))$ is the thermal diffusion coefficient of the mixture, and $Nu = hd/k$ is the Nusselt number. Solution of the above problem gives $Nu = 3.66$. Since the average velocity in the round tube is $\bar{U}_{av} = Q/A = Q_0(T_{av}/T_0)(P_0/P)/(\pi r^2)$,

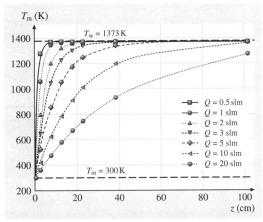

Fig. 37.6 Bulk mean temperature along the z-axis

(37.18) can be rewritten as

$$T_{m(z)} = T_w - (T_w - T_{in})$$
$$\times \exp\left(-\frac{3.66\alpha}{Q_0\left(\frac{T_{av}}{T_0}\right)\left(\frac{P_0}{P}\right)z}\right). \quad (37.19)$$

Equation (37.19) indicates that temperature is independent of the reactor diameter. With a reference pressure of 200 Torr and $T_{av} = 1000$ K, the thermal diffusion coefficients for NH_3 and N_2 are 8.49×10^{-4} and 6.22×10^{-4} m²/s, respectively. We used the averaged thermal diffusion of 7.36×10^{-4} m²/s in (37.19) to calculate the mean temperature distribution.

Figure 37.6 shows the bulk mean temperature variation along the z-axis, i.e., the distance from the gas inlet in the axial direction, under different total flow rates. The inlet gas temperature in this case is 300 K and the wall temperature is 1373 K (dashed line). For a reactor tube length of 102 cm, the gas can be heated to the reactor wall temperature at a flow rate of less than 10 slm. For a typical flow rate of 3.0 slm, the temperature of mixed gas can reach 1300 K within a heating length of 17.8 cm.

In experiments, a uniform temperature distribution on the substrate surface is essential to achieve uniform deposition and reduce the thermal stress in thick films. Numerical simulations can be conducted to optimize operating and geometric conditions.

37.3.2 Thermodynamic and Kinetic Analysis of Chemical Reactions

Thermodynamic and kinetic analysis of reactions allows us to predict the growth rate on the substrate in order to improve geometric and operating conditions for fast and uniform deposition. Assuming that reactions are in quasithermodynamic equilibrium, reaction constants can be calculated, and the quasi-equilibrium partial pressures of the reactants and products can be determined. The rate of the gas-phase reaction and surface deposition can be calculated. Thermodynamic calculations, however, only provide an upper limit. For the situation with high flow rate and limited reaction space, reactions in a vapor deposition system could be far from the equilibrium status. Kinetic analysis is therefore needed to understand how fast the reactions can proceed in the system. Kinetic data for gas-phase and surface reactions required for numerical simulations can be obtained from experiments. By solving the gas-phase reaction as a volumetric term and the surface reaction as a boundary condition on the substrate surface, the distributions of species concentration in the CVD reactor and the deposition rate on the substrate surface can be predicted under various operating conditions.

Thermodynamic Analysis of Gas-Phase and Surface Reactions

Prediction of Gas-Phase Reactions. Gas-phase reactions can be analyzed thermodynamically under different species concentrations, temperatures, and pressures. There are two methods to predict gas-phase reactions. In the first method, the species present in the system have to be specified, however no reaction step will be assigned. The concentration of difference species will be determined by minimizing the Gibbs energy of the system. In the second method, reaction steps will be assigned, and the reaction constants will be used to determine the reaction rates.

In the first method, chemical equilibrium is reached at constant temperature and pressure when the Gibbs energy is minimized. The Gibbs energy per unit mass of a system with N species can be written as follows:

$$G = \sum_1^N n_i \mu_i = \sum_{i=1}^N n_i \left(\mu_i^0 + RT \ln \frac{p_i}{p_0}\right), \quad (37.20)$$

where p_0 is the reference pressure of 1 atm and n_i is the molar amount of species i. Since the chemical potential μ_i is a function of temperature and pressure, the Gibbs energy is minimized at constant T and p for the correct combination of n_i. Elements must be conserved by the change in composition, which adds additional constraints to the system of the form

$$\sum_{i=1}^N a_{ji} n_i = b_j \quad \text{for} \quad j = 1, \ldots, M, \quad (37.21)$$

where a_{ji} is the number of atoms of element j in species i, M is the total number of elements in the system, and b_j is the total number of moles of element j per unit mass.

The composition that minimizes the Gibbs energy while satisfying the element balances is obtained by introducing the function

$$\Theta = G + \sum_{j=1}^M \lambda_j \left(\sum_{i=1}^N a_{ji} n_i - b_j\right). \quad (37.22)$$

The quantities λ_j are termed Lagrangian multipliers. Since (37.21) must be satisfied to conserve elements, the second term on the right-hand side vanishes and the composition that minimizes Θ also minimizes G.

Differentiating (37.22) with respect to n_i gives

$$\frac{\partial \Theta}{\partial n_i} = \mu_i - \sum_{j=1}^{M} a_{ji}\lambda_j \quad \text{for } i = 1, \ldots, N. \quad (37.23)$$

Differentiating (37.22) with respect to λ_i gives

$$\frac{\partial \Theta}{\partial \lambda_i} = b_i - \sum_{j=1}^{N} a_{ji} n_i \quad \text{for } i = 1, \ldots, N. \quad (37.24)$$

Setting (37.23) and (37.24) to zero generates $N + M$ equations which can be solved to give the composition at the chemical equilibrium.

Thermodynamic gas-phase reaction analysis in the second method makes calculations simpler than in the first method since the reaction steps are already known. The gas-phase reaction from I_2 and H_2 will be used as an example to demonstrate the analysis process. The gas-phase reaction from I_2 and H_2 is given by

$$\tfrac{1}{2}H_2(g) + \tfrac{1}{2}I_2(g) \rightleftharpoons HI(g). \quad (37.25)$$

The following equations can be obtained according to the above reaction

$$K = \exp\left(-\frac{\Delta G_r^0}{RT}\right) = \frac{P_{HI}^e}{\sqrt{P_{I_2}^e P_{H_2}^e}}, \quad (37.26)$$

$$P_{H_2}^0 - P_{H_2}^e = P_{I_2}^0 - P_{I_2}^e, \quad (37.27)$$

$$P_{HI}^e = 2\left(P_{I_2}^0 - P_{I_2}^e\right), \quad (37.28)$$

where $P_{H_2}^0$ and $P_{I_2}^0$ are the initial partial pressures of H_2 and I_2; $P_{H_2}^e$, $P_{I_2}^e$, and P_{HI}^e are the equilibrium partial pressures of H_2, I_2, and HI; K is the reaction equilibrium constant; and ΔG_r^0 is the Gibbs free energy change. The dependence of ΔG_r^0 on temperature is shown in Fig. 37.7 [37.117, 118].

With the given reaction temperature, the Gibbs free energy change can be obtained. The equilibrium constant K can, therefore, be calculated. The reaction equilibrium pressure of each species in reaction (37.25) can be solved from (37.26–37.28) when the initial partial pressures of H_2 and I_2 are given.

Thermodynamic Prediction of Surface Reactions

Based on the assumption that surface kinetic limitations occur at the stage of species adsorption/absorption on the substrate surface, a quasithermodynamic approach can be used to predict the surface deposition rate with different sticking coefficients, growth temperatures, and gas supersaturations [37.92]. Surface deposition of GaN from GaI and NH_3 is used as an example to predict the deposition rate

$$GaI(g) + NH_3(g) \rightleftharpoons GaN(s) + HI(g) + H_2(g). \quad (37.29)$$

Figure 37.8a and b show a schematic of GaN deposition on the substrate surface and a typical reactant species concentration boundary layer formed on the substrate due to surface reactions, respectively. The concentration of GaI before the mixing gas reaches the substrate area is defined as $C = C_0$. Due to the surface reaction of GaI with NH_3 on the substrate, the GaI concentration decreases to $C = C_w$ on the substrate surface. The concentration boundary layer χ is defined as the distance where the GaI concentration changes from C_0 in the bulk gas to C_w on the substrate surface.

Species are transported to the concentration boundary layer by the concentration gradient due to the surface deposition. Assuming that the decrease of species concentration from the bulk gas to the substrate surface is linear with the concentration boundary-layer thickness, a species mass flux in the gas phase can be expressed as

$$J_i = D_i M_i \frac{\partial C}{\partial n} = \frac{D_i M_i}{\chi}\left(C_i^0 - C_i^w\right) \quad [\text{kg/m}^2\text{s}], \quad (37.30)$$

where n is the coordinate normal to the substrate surface, C_i^0 and C_i^w are the species molar concentration in the bulk gas and on the deposition surface at the interface temperature, respectively, D_i is the

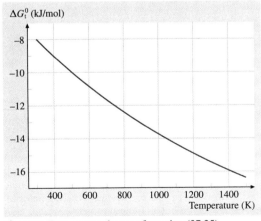

Fig. 37.7 Free energy change of reaction (37.25)

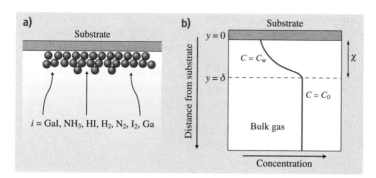

Fig. 37.8 (a) Schematic of GaN deposition on a substrate and **(b)** concentration boundary formed on the substrate due to surface reactions

diffusion coefficient of species i, M_i is the molecular weight of species i, and χ is the concentration boundary-layer thickness. According to the ideal gas law, $P_i = C_i RT$, (37.30) can be rewritten as

$$J_i = \frac{D_i M_i}{RT\chi}(P_i^0 - P_i^w) \quad [\text{kg/m}^2\text{s}], \quad (37.31)$$

where P_i^0 and P_i^w are the species partial pressure in the bulk gas and on the deposition surface. At the substrate surface, according to the Hertz–Knudsen law, the interface mass flux of species i is given by

$$J_i = \alpha_i \beta_i (P_i^w - P_i^e) \quad [\text{kg/m}^2\text{s}], \quad (37.32)$$

where α_i ($0 \leq \alpha_i \leq 1$) is the sticking coefficient of species i which accounts for all interfacial kinetics controlled by step and kink density, provided by two-dimensional nucleation and dislocations in the modern version of the Burton–Cabrera–Frank theory [37.119]; $\beta_i = \left(\frac{M_i}{2\pi RT}\right)^{1/2}$ is the thermodynamic factor of species i and P_i^e is the equilibrium partial pressure of species i on the substrate surface.

The following equations can be obtained according to reaction (37.29)

$$\frac{J_{\text{GaI}}}{M_{\text{GaI}}} = \frac{J_{\text{NH}_3}}{M_{\text{NH}_3}} = \frac{\rho_{\text{GaN}} V_g}{M_{\text{GaN}}}, \quad (37.33)$$

$$\frac{J_{\text{GaI}}}{M_{\text{GaI}}} + \frac{J_{\text{HI}}}{M_{\text{HI}}} = 0, \quad (37.34)$$

$$\frac{J_{\text{GaI}}}{M_{\text{GaI}}} + \frac{J_{\text{H}_2}}{M_{\text{H}_2}} = 0, \quad (37.35)$$

$$P_{\text{HI}}^e P_{\text{H}_2}^e = k_1 P_{\text{GaI}}^e P_{\text{NH}_3}^e. \quad (37.36)$$

According to (37.33–37.36), there are four unknowns, i.e., P_{HI}^e, P_{GaI}^e, $P_{\text{NH}_3}^e$, and $P_{\text{H}_2}^e$, and four equations. Given the wall partial pressures P_{HI}^w, P_{GaI}^w, $P_{\text{NH}_3}^w$, and $P_{\text{H}_2}^w$ and the sticking probabilities α_i of the species HI, GaI, NH$_3$, and H$_2$ on the substrate surface, the reaction rate can be obtained by solving (37.33–37.36).

Modeling of Gas–Phase and Surface Reactions Kinetically

Kinetic Modeling of Gas-Phase Reactions. A system that has a number N_r of gas-phase reactions involving a number N_{sp} of species can be expressed in general notation by

$$\sum_{i=1}^{N_{\text{sp}}} v'_{ij} \Lambda_i = \sum_{i=1}^{N_{\text{sp}}} v''_{ij} \Lambda_i, \quad j = 1, \ldots, N_r, \quad (37.37)$$

where v'_{ij} and v''_{ij} are the forward and backward stoichiometric coefficients for the ith species in the jth reaction. The above equation can be written more compactly as

$$\sum_{i=1}^{N_{\text{sp}}} v_{ij} \Lambda_i = 0, \quad j = 1, \ldots, N_r, \quad (37.38)$$

where $v_{ij} = v''_{ij} - v'_{ij}$. The stoichiometric coefficients are integers for elementary reactions and are normally 0, 1 or 2.

The molar production rate of species i due to gas-phase reaction is

$$\omega_i^g = \sum_{j=1}^{N_r} v_{ij} q_j, \quad (37.39)$$

where q_j is the rate-of-progress variable for the jth reaction, which can be expressed as

$$q_j = (k_f)_j \prod_{i=1}^{N_{\text{sp}}} c_i^{a'_{ij}} - (k_r)_j \prod_{i=1}^{N_{\text{sp}}} c_i^{a''_{ij}}, \quad (37.40)$$

where $(k_f)_j$ and $(k_r)_j$ are temperature-dependent forward and backward rate coefficients, c_i is the concentration of species i, and a'_{ij} and a''_{ij} are constants. For elementary reactions which obey the mass action law,

Table 37.7 An example of gas species, adsorbed species, and bulk species

Surface reactions	Gas species	Adsorbed species	Deposited species
$NH_3(g) \rightleftharpoons NH_{3\,ads}$	NH_3	$NH_{3\,ads}$	None
$GaI(g) + NH_{3\,ads} \rightleftharpoons GaN(s) + HI(g) + H_2$	GaI, HI, H_2	$NH_{3\,ads}$	GaN
$AlCl_3(g) + NH_3(g) \rightleftharpoons AlN(s) + 3HCl(g)$	$AlCl_3, NH_3, HCl$	None	AlN

$a'_{ij} = v'_{ij}$ and $a''_{ij} = v''_{ij}$, the rate coefficients are assumed to have the Arrhenius form of

$$k_f = AT^n \left(\frac{p}{p_{atm}}\right)^m e^{\frac{-E_{af}}{RT}}, \tag{37.41}$$

$$k_r = AT^n \left(\frac{p}{p_{atm}}\right)^m e^{\frac{-E_{ar}}{RT}}, \tag{37.42}$$

where A is pre-exponential constant, n indicates the temperature dependence, m is the exponent of the pressure dependency, E_{af} is the activation energy for the forward reaction, and E_{ar} is the activation energy for the backward reaction. All these constants are provided by experiment results of kinetic reaction analysis. For a typical reaction of the form

$$\alpha A + \beta B \rightarrow C + D, \tag{37.43}$$

the rate of the gas-phase reaction is expressed as

$$\dot{\omega}_g = A_p T^n \exp\left(\frac{-E_a}{RT}\right) \left(\frac{p}{p_{atm}}\right)^m \\ \times C_A^\alpha C_B^\beta \quad [\text{kmol/m}^3\text{s}], \tag{37.44}$$

where C_A and C_B are the molar concentrations of reactants A and B, and α and β are the concentration exponents of A and B, respectively.

Kinetic Modeling of Surface Reactions. The general form of the surface reaction is

$$\sum_{i=1}^{N_g} a'_{ij} A_i + \sum_{i=1}^{N_s} b'_{ij} B_i(s) + \sum_{i=1}^{N_b} c'_{ij} C_i(b)$$
$$= \sum_{i=1}^{N_g} a''_{ij} A_i + \sum_{i=1}^{N_s} b''_{ij} B_i(s) + \sum_{i=1}^{N_b} c''_{ij} C_i(b), \tag{37.45}$$

where a_{ij} is the gas species stoichiometric, b_{ij} is the adsorbed species stoichiometric, c_{ij} is the bulk species stoichiometric, N_g is the total number of gas-phase species, N_s is the total number of adsorbed species, and N_b is the total number of deposited species. For example, in Table 37.7 the gas species, adsorbed species, and bulk species are listed for each reaction.

The surface reaction rate corresponding to reaction (37.43) may be expressed as

$$\dot{\omega}_{sj} = k_{fj} \prod_{i=1}^{N_g} C_{aiw}^{a'_{ij}} \prod_{i=1}^{N_s} C_{bi(s)}^{b'_{ij}}$$
$$- k_{rj} \prod_{i=1}^{N_g} C_{aiw}^{a''_{ij}} \prod_{i=1}^{N_s} C_{bi(s)}^{b''_{ij}} \quad [\text{kmol/m}^2\text{s}], \tag{37.46}$$

where k_{fj} and k_{rj} are the forward and backward reaction rates, C_{aiw} is the gas-phase concentrations of species i at the surface and can be expressed as $C_{aiw} = \rho_w Y_i^w / M_i$, where ρ_w is the gas-phase mass density, Y_i^w is the gas-phase mass fraction of species i near the wall, and M_i is the molecular weight of gas species i. $C_{bi(s)}$ is the surface concentration of adsorbed species i and can be expressed as $C_{bi(s)} = \rho_s X_i$, where ρ_s is the surface site density and X_i is the surface site fraction of adsorbed species i. It is seen from (37.46) that the concentration dependence of bulk species is neglected.

Two different approaches, the sticking probability (or sticking coefficient) method and finite rate method, are usually used to estimate the surface reaction rate in numerical simulations. The sticking probability method calculates the reaction rate based on sticking probability and precursor thermal flux, while the finite rate chemistry uses the kinetic expression to evaluate the reaction rate. When using the sticking probability method, the surface rate becomes

$$\dot{\omega}_{sj} = \gamma_j J_A \prod_{i=1}^{N_s} (X_i)^{b'_{ij}}, \tag{37.47}$$

where the sticking probability γ_j is described in Arrhenius form as

$$\gamma_j = f(\theta^s)_j \exp\left(\frac{-E_{aj}(\theta^s)}{RT}\right), \tag{37.48}$$

where $f(\theta^s)_j$ is the function of the existing surface coverage of adsorbed species j, θ^s is the Langmuir definition of surface coverage of certain species or the fraction of sites that are occupied on the substrate surface, and E_{aj} is the activation energy for adsorption of species j.

The thermal flux of precursor A, or the incident rate of precursor onto the substrate (Hertz–Knudsen law) can be expressed as

$$J_A = \left(\frac{RT_w}{2\pi M_A}\right)^{1/2} C_{Aw}, \quad (37.49)$$

where C_{Aw} is the concentration of precursor A near the wall.

For the finite rate method, the surface reaction rate is the same as (37.47). For example, the reaction rate using a finite rate method can be expressed as

$$\dot{\omega}_s = A_s T^n \exp\left(\frac{-E_{as}}{RT}\right)\left(\frac{p}{p_{atm}}\right)^m C_{Aw}^\alpha C_{B(s)}^\beta, \quad (37.50)$$

for the reaction $\alpha A(g) + \beta B_{ads} \to C(s) + D(g)$.

37.4 Modeling of GaN IVPE Growth – A Case Study

In this section, a case study is presented to demonstrate optimization of GaN IVPE growth by computer modeling. In this example, the growth surface evolution mechanism will be presented and thermodynamic and kinetic methods will be used to predict gas-phase and surface reactions.

37.4.1 Scaling Analysis

The important dimensionless groups, their physical definitions, and calculated values in the vapor-growth reactor are listed in Table 37.8.

The values of ρ_0, μ_0, k_0, C_{p0}, and D_0 obtained from simulation results are 6.0×10^{-2} kg/m^3, 4.8×10^{-5} kg s/m, 1.3×10^{-1} J s/m K, 1740.0 J K/kg, and 1.0×10^{-3} m^2/s, respectively. L_c is selected as the diameter of the furnace reactor, which is 0.045 m, and V_0 is estimated as 0.67 m/s based on a total flow rate of 3.0 slm and a reactor pressure of 200 Torr.

According to Table 37.8, the flow is mainly laminar in the reactor; heat and mass transfer due to convection is much more important than conduction/diffusion. The ratio of the Grashof number Gr to the Reynolds number squared Re^2 is used to examine whether heat transfer is mainly controlled by natural convection, forced convection, or both. For the case presented here Gr/Re$^2 = 0.66$, indicating that both forced convection and natural convection are important.

37.4.2 Computational Issues

To better understand the heat and mass transfer inside the system, numerical simulations are conducted by solving governing equations together with gas-phase and surface reactions. A schematic of the system and numerical grids is shown in Fig. 37.9.

Since the growth cell is positioned in a uniform-temperature region, a constant temperature was as-

Table 37.8 Important dimensionless groups in the IVPE reactor

Dimensionless group	Value
Reynolds (Re)	37.7
Grashof (Gr)	940.0
Prandtl (Pr)	0.64
Schmidt (Sc)	0.73
Thermal Péclet (Pe^T)	24.2
Mass Péclet (Pe^M)	27.4
Gas-phase Damköhler (Da^g)	–
Surface Damköhler (Da^s)	–

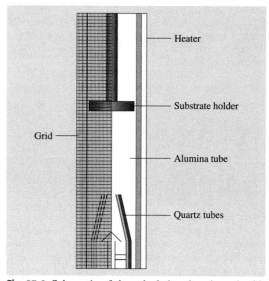

Fig. 37.9 Schematic of the calculation domain and grids for the system designed by *Mecouch* et al. [37.30, 72, 120]

Table 37.9

Wall on the graphite heater:	$V=0$, $T=1323\,\text{K}$, $\varepsilon=1.0$
Wall on the substrate holder:	$V=0$, $\varepsilon=0.65$
Top surface of the gallium source:	$\dot{w}_{\text{gallium}}=\dot{w}_1$, $T=700\,\text{K}$, $\varepsilon=0.2$;
Wall on the surface of the substrate:	$V=0$, $J_i=M_i\dot{w}_i^s$, $\varepsilon=0.65$
Wall on the alumina tube:	$V=0$, $\varepsilon=0.7$
Wall on the quartz surface:	$V=0$, $\varepsilon=0.14$
Inlet carrier gas (NH_3+N_2):	$V=V_1$, $T=700\,\text{K}$, $P_{\text{in}}=P_{\text{ref}}$, $Y_{NH_3}=Y_1$, $Y_{N_2}=Y_2$, $\varepsilon=0.15$
Inlet I_2, N_2, and H_2:	$V=V_2$, $T=700\,\text{K}$, $P_{\text{in}}=P_{\text{ref}}$, $Y_{I_2}=Y_3$, $Y_{N_2}=Y_4$, $Y_{H_2}=Y_5$, $\varepsilon=0.15$
Shield gas at the inlet (N_2):	$V=V_3$, $T=700\,\text{K}$, $P_{\text{in}}=P_{\text{ref}}$, $Y_{N_2}=Y_6$, $\varepsilon=0.15$
Outlet:	$T=300\,\text{K}$, $P_{\text{out}}=P_{\text{ref}}$, $\varepsilon=1.0$

signed to the surface of the furnace wall with emissivity of unity. The experimentally measured weight losses of iodine and Ga sources are converted to mass flow rates that are used as inlet gas flow rates. The surface reaction rate $\dot{w}_i^s$ after being converted to the mass flux J_i can be incorporated into the simulation as boundary conditions on the substrate surface.

General boundary conditions are prescribed as in Table 37.9.

37.4.3 Gas-Phase and Surface Reactions Analysis

Gas-Phase Reaction Analysis

In the GaN epitaxy growth system, the carrier gas mixture of N_2 and H_2 enters the I_2 source bubbler, where liquid I_2, heated by a jacket heater, vaporizes and will be transported to the reactor chamber by the carrier gas. H_2 and I_2 react to form HI. The gas mixture flowing over the gallium source surface includes N_2, H_2, I_2, and HI. Both I_2 and HI may react with Ga to form GaI_x, with $x=1$ or 3. From [37.26], it is deduced that GaI is the main product of the reactions between gallium and iodine gas in the reactor chamber being studied. It is found that GaI_3 is more stable thermodynamically at the operational conditions of $T=1323\,\text{K}$ and $P=200\,\text{Torr}$. On the substrate surface, NH_3 will be adsorbed and cracked to form activated nitrogen atoms that will react with GaI_x to grow GaN [37.121]. It is important to control the concentrations of NH_3 and GaI_x achieved on the substrate surface in experiments. Gas-phase species reaching the substrate area include NH_3, HI, H_2, N_2, GaI, GaI_3, and maybe I_2 and Ga. The gas pathways can be found in Fig. 37.1a.

Due to the high-temperature environment in the growth reactor ammonia may decompose to nitrogen and hydrogen gases before it reaches the substrate area. This will reduce the amount of ammonia reaching the substrate. The global equation for the ammonia gas phase decomposition can be expressed as

$$NH_3(g) \rightleftharpoons \tfrac{1}{2}N_2(g) + \tfrac{3}{2}H_2(g) \,. \tag{37.51}$$

This reaction is a mildly endothermic process ($\Delta H = 46\,\text{kJ/mol}$). Thermodynamically, almost all NH_3 is decomposed into N_2 and H_2 at temperature higher than $300\,°\text{C}$. In the actual experiments, the NH_3 is decomposed slowly and the decomposition rate strongly depends on the growth conditions and the equipment. This reaction is, in fact, far from the kinetic equilibrium state.

The Damköhler number of the gas phase is calculated for the ammonia gas-phase decomposition reaction to determine its importance under the aforementioned operational conditions. The physical interpretation of the gas-phase Damköhler number is the ratio of the gas-phase reaction time to the gas flow residence time. The gas flow residence time can be estimated from $t_F = L_F/\bar{U}_{\text{av}}$, where L_F is the distance between the gas inlet and the substrate, which is about $0.16\,\text{m}$. The averaged velocity is about $0.67\,\text{m/s}$ with a typical flow rate of $3.0\,\text{slm}$ under a pressure of $200\,\text{Torr}$ and a furnace wall temperature of $1323\,\text{K}$. The fluid flow residence time is about $0.24\,\text{s}$. The comprehensive mechanism for NH_3 pyrolysis can be found in [37.122]. The controlling step of the reactions is found to be

$$NH_3 + M \rightleftharpoons H_2 + H + M \,, \tag{37.52}$$

where M is the third-party molecule that serves as a catalyst. Reaction (37.52) is a second-order reaction with a reaction coefficient of $k_r = A\exp(-E_a/RT)$ and a reaction rate of $\dot{w}_g = k_r[NH_3][M]$, where the preexponential coefficient (molecular collision frequency) A is 2.2×10^{10}, and the reaction activation energy E_a is $3.93 \times 10^5\,\text{J/mol}$ [37.122]. $k_r = 2.55 \times 10^{-5}\,\text{m}^3/(\text{mol}\,\text{s})$ is calculated with the reference temperature of $1323\,\text{K}$. The third-party molecular concentration at temperature of $1323\,\text{K}$ and pressure of $200\,\text{Torr}$ is $1.17\,\text{mol/m}^3$, if the molar ratio of the mixing gas of NH_3 and N_2 is assumed to be $1:1$. The reaction rate is calculated as $r_0^g = k_r[M] = 2.98 \times 10^{-5}\,\text{s}^{-1}$. The gas-phase Damköhler number (Da) for ammonia pyrolysis is 7.12×10^{-6}, which is much smaller than 1. Based on this analysis, it can be concluded that ammonia gas-phase decomposition can be neglected under the discussed operational conditions. It should be noted that ammonia also tents to decompose on reactive solid surfaces. The heterogeneous reaction rate under the current operating conditions is, however, expected to be very low according to [37.9].

As well as the substrate surface, ammonia may also react with GaI_x in the gas phase. To reduce gas-phase reactions between NH_3 and GaI_x before their mixture reaches the substrate, shield gas N_2 runs between the inner and outer silica nozzle to prevent mixing in the area above the silica nozzle (Fig. 37.1a) in order to increase the NH_3 and GaI_x concentrations on the substrate surface. The gas-phase reactions of NH_3 and GaI_x are expected to be weak and can therefore be neglected due to the effect of the shield gas and short residence time of the reacting gases.

If the equilibrium vapor pressure of iodine has been reached in the iodine source bubbler, the molar concentration fraction of iodine $[I_2]$ can be written as

$$[I_2] = \frac{P_{I_2}^0}{P_{\text{total}}}. \tag{37.53}$$

The total mass reduction rate of I_2 in the source can be expressed as

$$\dot{m}_{I_2} = \rho_{I_2} Q_{\text{total}} \frac{P_{I_2}^0}{P_{\text{total}}}, \tag{37.54}$$

where Q_{total} is the total flow rate of the carrier gas.

The gas-phase reaction in the iodine source bubbler has been given in reaction (37.25). No reaction between N_2 and other gases are expected. Before reaction (37.25) starts, the initial partial pressures of H_2, I_2, HI, and N_2 are $P_{H_2}^0$, $P_{I_2}^0$, P_{HI}^0, and $P_{N_2}^0$, respectively, where the initial partial pressure of HI obviously equals zero. When the reaction equilibrium is reached in the source bubbler, the equilibrium partial pressure of the species can be expressed as $P_{H_2}^e$, $P_{I_2}^e$, P_{HI}^e, and $P_{N_2}^e$. The total pressure inside the bubbler is maintained at 1 atm and remains unchanged since the partial pressure reduction of H_2 and I_2 due to reaction equals the partial pressure production of HI in reaction (37.25). The partial pressure of N_2 will not change during reaction. Controlled by the mass flow controller, the molar ratio of H_2 and N_2 in the carrier gas is 9 : 91, and it is assumed to be the same in the bubbler. Therefore, at time

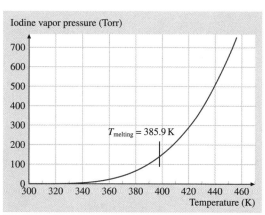

Fig. 37.10 Temperature dependence of iodine vapor pressure

zero, the follow equations can be obtained:

$$\sum P_i^0 = 1 = P_{H_2}^0 + P_{I_2}^0 + P_{N_2}^0, \tag{37.55}$$

$$\frac{P_{H_2}^0}{P_{N_2}^0} = \frac{9}{91}, \tag{37.56}$$

$$\frac{P_{I_2}^0}{\sum P_i^0} = [I_2]. \tag{37.57}$$

The initial vapor pressure of iodine over the source surface $P_{I_2}^0$ can be obtained using Fig. 37.10 [37.123].

The equilibrium expression of the reaction (37.25) and two additional equations obtained from reaction (37.25) are provided in (37.26–37.28).

Given the initial partial pressure of iodine vapor (or the source bubbler temperature), the reaction equilibrium pressure of each species in reaction (37.25) can be solved from (37.26–37.28) and (37.56–37.64).

Figure 37.11a shows the equilibrium partial pressures of different species at the inlet of the reactor under different iodine concentrations in the source. The highest iodine vapor pressure that can be achieved in the source bubbler is limited by the temperature of the valve connected to the bubbler. Gas temperature cannot exceed 150 °C for a regular needle valve, so the highest I_2 vapor pressure is about 300 Torr in the bubbler. When the equilibrium for reaction (37.25) is established before the mixture flows over the gallium source, the HI gas concentration will be around 0.1. When the I_2 concentration in the source is larger than 0.3, I_2 will be the main species to react with gallium, through GaI_x formation.

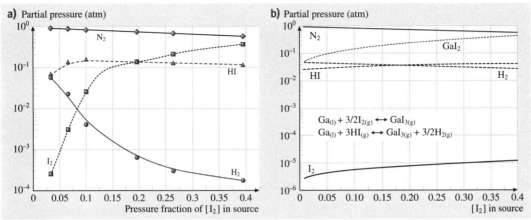

Fig. 37.11 (a) Equilibrium partial pressures with [I$_2$] in the source, and (b) equilibrium partial pressures above the liquid gallium at 1050 °C

The gas mixture coming out of the iodine source bubbler enters the reactor through the inner silica tube and reacts with gallium at the operating pressure of 200 Torr and growth temperature of 1050 °C. Instead of reacting with Ga vapor, both I$_2$ and HI will mainly react with gallium on its liquid surface, due to the very low gallium vapor pressure of 0.01 Torr under the given operating conditions. The reactions between I$_2$ and gallium can be written as

$$Ga_{(l)} + \tfrac{1}{2}I_{2(g)} \rightleftharpoons GaI_{(g)}, \quad (37.58)$$

$$Ga_{(l)} + \tfrac{3}{2}I_{2(g)} \rightleftharpoons GaI_{3(g)}, \quad (37.59)$$

$$GaI_{(l)} + I_{2(g)} \rightleftharpoons GaI_{3(g)}. \quad (37.60)$$

The reactions between gallium and HI can be written as

$$Ga_{(l)} + HI_{(g)} \rightleftharpoons GaI_{(g)} + \tfrac{1}{2}H_{2(g)}, \quad (37.61)$$

$$Ga_{(l)} + 3HI_{(g)} \rightleftharpoons GaI_{3(g)} + \tfrac{3}{2}H_{2(g)}, \quad (37.62)$$

$$GaI_{(g)} + 2HI_{(g)} \rightleftharpoons GaI_{3(g)} + H_{2(g)}. \quad (37.63)$$

At a typical growth temperature of 1050 °C, the calculated reaction equilibrium constants for reactions (37.58–37.60)) are 8900, 840 000, and 50, respectively. The equilibrium reaction constants are 3400, 13 000, and 4 for reactions (37.61–37.63), respectively. The formation of GaI$_3$ in both cases is strongly favored thermodynamically. It is, therefore, assumed that only GaI$_3$ will be formed at equilibrium. A group of equations similar to the formation of HI can be derived and the equilibrium partial pressure of species above the gallium source calculated.

Figure 37.11b shows the calculated partial pressures on the gallium source. Comparing Fig. 37.11a with Fig. 37.11b, it is revealed that the partial pressure of GaI$_3$ is seen to track the initial partial pressure of iodine.

The favorability of reactions (37.59) and (37.62) indicates that any GaI formed will further react either with I$_2$ or HI to form GaI$_3$ under equilibrium conditions. However, the thermodynamic calculation only provides an upper limit for what to expect. Given the small surface area of the gallium source (1.1×10^{-4} m^2) and the relatively high velocity of the carrier gas over the gallium source (on the order of 0.1 m/s), it is unlikely that there is sufficient time for equilibrium condition to be established above the gallium source. The thermodynamic model describing transport of iodine and gallium species has to be compared with the experimental data to evaluate its accuracy.

Reported by *Mecouch* et al. [37.72, 120], Fig. 37.12a shows the measured gallium loss rate compared with the measured iodine loss rate. The data points indicated by the diamond and square correspond to a carrier gas mixture of 9% H$_2$/91% N$_2$, whereas the triangle and circle correspond to carrier gas of pure hydrogen. The data were measured based on 20 h growth runs. Also two lines represent iodine transport as GaI (Ga : I$_2$ = 2 : 1) and GaI$_3$ (Ga : I$_2$ = 2 : 3), respectively. It is revealed that the measured iodine loss rate falls between the values calculated based on the two transport species. This indicates that there is enough iodine loss to account for GaI transport, but not enough iodine is lost to account for GaI$_3$ transport. It is concluded that the dominant trans-

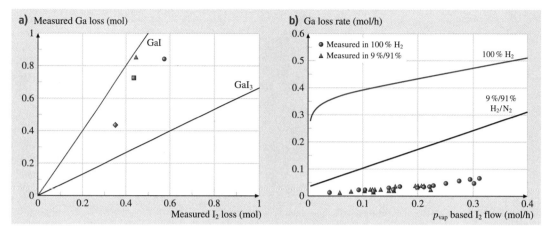

Fig. 37.12 (a) Measured gallium loss rate versus measured iodine loss rate and (b) gallium loss rate predicted from GaI$_3$-based thermodynamic equilibrium (*lines*) and measured during experiment growth runs (*bullets* and *triangles*) (after [37.72, 120])

port species is GaI, with some GaI$_3$ formed either on the Ga surface, or through subsequent reaction with the excess iodine species.

Figure 37.12b shows the predicted and measured rates of gallium loss versus the flow rates of I$_2$ [37.72, 120]. The solid lines represent the results obtained based on the thermodynamic equilibrium of the GaI$_3$ reaction, and diamonds and triangles represent the measured Ga loss with a pure H$_2$ carrier gas and a carrier gas mixture of 9% H$_2$/91% N$_2$, respectively. The experimental values fall well below the equilibrium prediction, and show no difference in measured loss rate of Ga between a carrier gas of pure H$_2$ and 9% H$_2$/91% N$_2$ mixture. This indicates that the source does not operate near the equilibrium as predicted by thermodynamics, and the reaction of HI formation does not significantly increase the transport of iodine. Thermodynamically calculated iodine loss rate is about 10–30 times greater than the measured iodine loss. Even if iodine is assumed to be transported from the source at the vapor equilibrium without any reaction to form HI, the calculated iodine flow rate is still 6–9 times higher than the observed loss rate from the iodine source. It is likely that the high flow rate of carrier gas and the crystal shape change in the iodine source as temperature fluctuates will both contribute to the difficulty in achieving equilibrium iodine vapor pressure in the source bubbler.

Based on the direct measurement of weight loss from both iodine and gallium sources after each experiment run, *Tassev* [37.26] reported that transport species in their reactor was GaI, with a slight excess of I$_2$. Furthermore, *Rolsten* [37.124] reported that GaI can be formed by heating either GaI$_3$ or the reaction product of gallium and iodine. With respect to these data and the relatively high flow rate of carrier gas and small volume for the reaction, it is assumed that the measured gallium loss was transported as GaI.

In the numerical model, only reaction (37.58) is therefore used to describe the gas-phase reaction for GaI$_x$ formation. Measured iodine and gallium mass reduction rates will be converted to the flow rate of I$_2$ and Ga, respectively. The gas-phase reaction rate of (37.58) can be described as

$$\dot{\omega}_g = A_p T^n \exp\left(\frac{-E_a}{RT}\right)\left(\frac{p}{p_{atm}}\right)^m \times [A]^\alpha [B]^\beta \quad [\text{kmol/m}^3\,\text{s}], \quad (37.64)$$

where [A] and [B] are the molar concentrations of reactants Ga and I$_2$, and α and β are the concentration exponents of [A] and [B], respectively. $E_a/R = 0$ K and $n = m = 0$ are assumed in the simulation. The value of A_p is used to determine the reaction rate and it is estimated as 4×10^8, corresponding to more than 95% conversion of experimentally weighed Ga loss to GaI. $\alpha = 1$ and $\beta = 0.5$ are assigned, corresponding to the stoichiometric coefficients of the reactants.

Surface Reaction Analysis

The overall surface reaction rate depends on the partial pressures of gas species such as GaI and NH$_3$, available free sites on the surface, surface concentrations of adsorbed species, surface diffusion coefficients, rate constants of individual reaction step, and surface characteristics. Unlike the gas-phase reactions, predicting reaction paths and rate constants is more difficult for the heterogeneous surface reactions since the interactions between gas-phase and surface entities are more complicated than those between gas-phase molecules. A semi-empirical approach is usually used to simulate the surface mechanisms and kinetics.

Tassev et al. [37.26] discussed surface reactions from GaI/NH$_3$. Based on the byproducts detected in the GaN reactor, the rate-limiting step for the growth of GaN is given as

$$\text{GaI}(g) + 2\text{NH}_3(g) \rightleftharpoons \text{GaN}(s) + \text{NH}_4\text{I}(g) + \text{H}_2(g) \,. \tag{37.65}$$

However, NH$_4$I detected in the cold reaction zone might be formed from NH$_3$ and HI at a temperature lower than 800 K since NH$_4$I decomposes above 800 K [37.118]. In the growth zone from silica nozzle to the substrate, the gas temperature will be close to 1323 K. It is therefore unlikely that NH$_4$I will be present. Considering different product gases which are stable at a temperature around 1323 K, the following analogs of element reactions to GaN growth from GaCl/NH$_3$ mixture in a halide vapor-phase epitaxy system [37.121] can be expressed:

$$V + \text{NH}_3 \rightleftharpoons \text{NH}_{3\,\text{ads}} \,, \tag{37.66}$$
$$\text{NH}_{3\,\text{ads}} \rightleftharpoons N + \tfrac{3}{2}\text{H}_{2(g)} \,, \tag{37.67}$$
$$N + \text{GaI} \rightleftharpoons \text{NGaI} \,, \tag{37.68}$$
$$2\text{NGaI} + \text{H}_{2(g)} \rightleftharpoons 2\text{NGa}-\text{IH} \,, \tag{37.69}$$
$$\text{NGa}-\text{IH} \rightleftharpoons \text{NGa} + \text{HI}_{(g)} \,, \tag{37.70}$$
$$2\text{NGaI} + \text{GaI}_{(g)} \rightleftharpoons 2\text{NGa}-\text{GaI}_3 \,, \tag{37.71}$$
$$2\text{NGa}-\text{GaI}_3 \rightleftharpoons 2\text{NGa} + \text{GaI}_3 \,, \tag{37.72}$$

where V is the vacant surface site of ammonia adsorption.

Since a very large NH$_3$ partial pressure is maintained over the substrate in the experiments, reaction (37.66) is close to thermodynamic equilibrium, which means that the NH$_{3\,\text{ads}}$ concentration on the substrate will be close to constant. The GaN deposition rate is mostly limited by reaction (37.68). This is supported by the experimental results that show the independence of GaN growth rate from the NH$_3$ partial pressure, but strong dependence on the GaI concentration. Since it is difficult to obtain the reaction constant for each elemental step, simplified overall surface reactions will be used in the simulations. Two overall reactions corresponding to the above element reactions are obtained as follows, corresponding to the most energetically favorable reactions for the iodine vapor-growth system:

$$\text{GaI}_{(g)} + \text{NH}_{3(g)} \rightleftharpoons \text{GaN}_{(s)} + \text{HI}_{(g)} + \text{H}_{2(g)} \,, \tag{37.73}$$
$$3\text{GaI}_{(g)} + 2\text{NH}_{3(g)} \rightleftharpoons 2\text{GaN}_{(s)} + \text{GaI}_{3(g)} + 3\text{H}_{2(g)} \,. \tag{37.74}$$

The following reaction might also be energetically favorable since GaI$_3$ is thermodynamically preferred

$$\text{GaI}_{3(g)} + \text{NH}_{3(g)} \rightleftharpoons \text{GaN}_{(s)} + 3\text{HI}_{(g)} \,. \tag{37.75}$$

Reaction (37.75) will also be considered in our numerical simulation model, though the thermodynamics analysis shows that the contribution of reaction (37.75) to the GaN deposition rate will be less than 2%.

Figure 37.13 shows the free energy of reaction for the above three surface reactions on the substrate surface. Since GaI is assumed to be the dominant species for gallium transport, reactions (37.73) and (37.74) are expected to be more important than reaction (37.75). It is revealed that the reaction free energies of the above three reactions are all positive at a typical growth temperature of 1323 K, which means that the equilibrium reaction constants are all smaller than 1, and none of the reactions are spontaneous. An appropriate effective supersaturation, however, can still be achieved by controlling the reactants' partial pressure on the substrate, which serves as the driving force for GaN deposition. Supersaturation of a typical vapor-growth system is

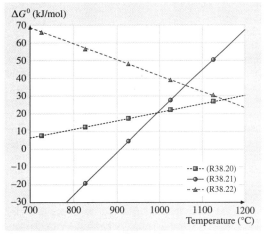

Fig. 37.13 Free energy of reaction for GaN surface deposition

about 5. For reaction (37.73), it can be expressed as

$$\sigma = \ln\left(\frac{P_{GaI} P_{NH_3}}{P_{HI} P_{H_2} K_{20}}\right), \quad (37.76)$$

where K_{20} is the equilibrium constant. At a growth temperature of 1323 K, this can be calculated as

$$K_{20} = \frac{P^e_{HI} P^e_{H_2}}{P^e_{GaI} P^e_{NH_3}} = \exp\left(\frac{-\Delta G^0_r}{RT}\right) = 0.12. \quad (37.77)$$

Since reactions (37.74) and (37.75) can be combined to form reaction (37.73), reaction (37.73) will therefore be used as the overall surface deposition step at first in the numerical model. The surface reaction rate for reaction (37.73) is determined by

$$\dot{w}_s = \delta A_p T^n \exp\left(\frac{-E_a}{RT}\right)\left(\frac{p}{p_{atm}}\right)^m \times [A]^\alpha [B]^\beta \quad [\text{kmol/m}^2\,\text{s}], \quad (37.78)$$

where δ is the deposition coefficient of GaN on the substrate surface. The rate constant r^s for reaction (37.73) has not been reported in the literature. An activation energy of 103 000 J/mol is reported in reference to the data from *Shintani* et al. [37.125] for the following surface reaction of GaN HVPE growth:

$$GaCl_{(g)} + NH_{3(g)} \rightleftharpoons GaN_{(s)} + HCl_{(g)} + H_{2(g)}. \quad (37.79)$$

The temperature range of their data is 860–1020 °C, which is similar to the IVPE growth temperature (1050–1100 °C). The activation energy of 1.03×10^5 J/mol is estimated for the reaction (37.73). $n = m = 0$ is assumed here. $\alpha = x$ and $\beta = 0$ are assigned for the concentration coefficients of GaI and NH$_3$. The value of α is set as undetermined since the concentration dependency of GaI should be 13 according to reactions (37.73–37.75). The ammonia concentration dependence is set to zero due to the fact that ammonia is always in excess on the substrate surface in the experiment. By matching the experimental data for the deposition rate with simulation results under different NH$_3$/GaI molar ratios, the value of δA_p and α can be determined using an optimization procedure, the detail of which will be covered in Sect. 37.4.8.

To determine the importance of reactions (37.73–37.75) in the GaN deposition rate, all three equations are included in the simulation. The contribution from individual reaction to the GaN growth rate will be determined under the assumption that all of the reactions are stoichiometric. No kinetic data for reactions (37.74–37.75) are available in the literature. *Shaw* et al. [37.126] reported that the activation energy of GaAs epitaxial growth is between 6.18×10^4 J/mol and 1.648×10^5 J/mol in the Ga–As–Cl$_3$ system due to surface adsorption, surface reaction, and surface diffusion. The activation energy for reaction (37.74) is therefore chosen as $E_a = 1.133 \times 10^5$ J/mol, which is the medium value of the above energy range. For reaction (37.74), $n = m = 0$, $\alpha = 3$, and $\beta = 0$, and the value of δA_p is again determined by matching the simulation results with the experiment ones. $E_a = 116\,396$ K is reported by *Dollet* et al. [37.98] for the following surface reaction of AlN epitaxial growth:

$$AlCl_{3(g)} + NH_{3(g)} \rightleftharpoons AlN_{(s)} + 3HCl_{(g)}, \quad (37.80)$$

which is used in the simulation as the activation energy for reaction (37.75). Also for reaction (37.75), we use $n = m = 0$, $\alpha = 1$, and $\beta = 0$, and $\delta A_p = 0.36$, as reported by *Cai* [37.29] for the reaction (37.80) of AlN epitaxial growth with a temperature of 1100 °C and pressure of 760 Torr.

Finally the GaN deposition rate can be calculated as

$$G = \frac{\dot{w}_s M_A}{\rho_A} \quad [\text{m/s or } 3.6 \times 10^9 \, \mu\text{m/h}], \quad (37.81)$$

where M_A and ρ_A are the molecular weight (84 kg/kmol) and density (6.15×10^3 kg/m^3) of GaN, respectively.

37.4.4 Geometrical and Operational Conditions Optimization

Many parameters pertinent to reactor geometry and mixed gas injection are important to optimal design of the vapor-growth reactor. In this section, the effects of geometrical configurations such as the diameter of the substrate holder and operating conditions such as the process and shield gas flow rate on mixing process, deposition rate distribution, and deposition uniformity on the substrate are studied.

37.4.5 Effect of Total Gas Flow Rate on Substrate Temperature

A uniform and sufficient high temperature on the substrate is required to achieve high growth rate, uniform film thickness, and good film quality. In the experiments conducted by *Mecouch* et al. [37.30, 72, 120], the temperature on the furnace wall is monitored using thermocouples. In situ observation of temperature achieved

Table 37.10 Inlet gas conditions for an experiment run by *Mecouch* et al. [37.30, 72, 120]

Inlet	Species	Initial volume flow rate (slm)	Adjusted volume flow rate (slm)
1	I_2	0.008	0.008
	H_2	0.026	0.026
	N_2	0.5	0.25
2	N_2	1.0	0.62
3	NH_3	2.0	1.0
	N_2	–	1.25

on the substrate surface, however, is difficult to realize. Temperature difference between the substrate surface and furnace wall, and the effect of total gas flow rate on the temperature achieved on the substrate surface will be investigated numerically. Gas species at each inlet and their flow rates used are listed in Table 37.10.

The initial inlet gas conditions are tested both experimentally and numerically. The measured gallium weight reduction rate of 0.0133 slm is used in the simulation. This flow rate corresponds to a molar ratio of Ga : I_2 = 1.7 : 1, which favors the formation of GaI in the gas phase. The temperature (1050 °C) of the heating unit and the reactor pressure (200 Torr) remain unchanged for all the simulations presented here. Gas-phase and surface reactions are not activated in the simulation unless stated, since they are expected to have insignificant influence on the gas flow and heat transfer in the reactor. Thermodynamic properties of different substances are listed in Table 37.11.

Figure 37.14 shows the streamline and temperature distributions in the simulated system. Figure 37.14a shows that a reverse flow is formed near the alumina tube wall, which means that the radial mixing of different species is enhanced. The gas-phase reactions of GaI/NH_3 in the center part and the GaN deposition on the alumina tube wall will therefore increase, and the GaN deposition rate will decrease on the substrate. In addition, the reverse flow might be unstable in the experiment, which could be larger or smaller

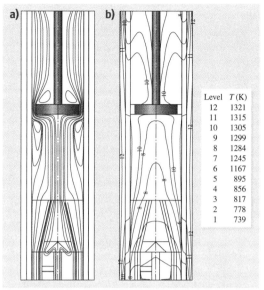

Fig. 37.14a,b Distributions of (**a**) streamlines and (**b**) temperature inside the VPE reactor designed by *Mecouch* et al. [37.30, 72, 120]

as time progresses. The GaN deposition quality on the substrate could therefore be worsened. For the above reasons, it is necessary to make sure that no big reverse flow is formed in the reactor. Figure 37.14b shows the temperature distribution for the simulated system. The temperature of the gas mixture will be heated to more than 1200 K above the silica nozzle. Attributed to such high temperature, chemical deposition may occur on the surfaces of the reactor wall, substrate, and substrate holder.

The velocity ratio of NH_3 from inlet 3, shield gas N_2 from inlet 2, and carrier gas from inlet 1 is 1 : 11 : 13 at the height of the silica nozzle outlet. It is concluded that the reverse flow is formed mainly due to a large gas velocity difference at the silica nozzle outlet. To prevent the reverse flow, adjusted inlet gas flow rates are

Table 37.11 Thermodynamic properties used in the simulation

Properties	All gases	Al_2O_3	GaN	Silica	BN
Specific heat (J/(kg K))	Mix JANAF method	900	490	710	1610
Density (kg/m³)	Ideal gas law	3900	6150	2198	1900
Thermal conductivity (W/(m K))	Mix kinetic theory	30	130	1.38	28
Dynamic viscosity (kg/(m s))	Mix kinetic theory	–	–	–	–
Diffusivity	Sc = 0.72	–	–	–	–
Adsorption coefficient	0	1	1	0.145	1

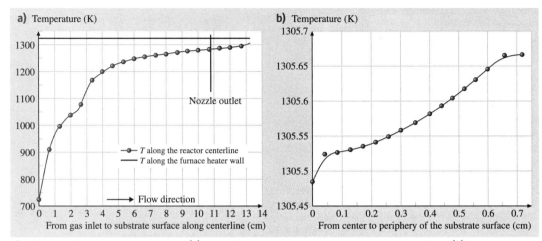

Fig. 37.15a,b Temperature distributions (**a**) along the reactor centerline and furnace heater wall and (**b**) from the periphery to the center on the substrate surface in the VPE reactor designed by *Mecouch* et al. [37.30, 72, 120]

used in the experiment (Table 37.10). The velocity ratio under the new inlet conditions is 1 : 6 : 6. Simulation results show that the reverse flow is eliminated under this condition.

Figure 37.15a shows the temperature distributions from inlet to the substrate surface along the reactor centerline and along the substrate surface, based on the adjusted inlet gas conditions. It is revealed that the mixing gases can be heated to about 1250 K when reaching the nozzle outlet, which further increases to 1300 K at the substrate area. Figure 37.15b shows the temperature distribution from the periphery to the center of the substrate surface. It is found that the temperature on the substrate surface will be high at the periphery but low in the middle. The temperature difference along the substrate surface is only about 0.18 °C, and the substrate temperature is about 17 °C lower than the temperature on the furnace heater.

By reducing or increasing the flow rate at each gas inlet with the same ratio, the effect of the total flow rate on the temperature achieved on the substrate was investigated. Figure 37.16 shows the averaged temperature and the largest temperature difference achieved on the substrate with different total flow rates. It is revealed that the substrate temperature drops from 1307 °C to 1283 °C as the flow rate changes from 1 to 10 slm. Temperature nonuniformity on the substrate also changes with total flow rate. The largest temperature difference is defined as the highest minus the lowest temperature on the substrate. It is revealed that a total flow rate of 3 slm gives the lowest temperature difference, or the best uniformity, on the substrate surface. In the experiments, the total flow rate is controlled to around 3 slm [37.72, 120].

Effect of the Shield Gas Flow Rate

The shield gas, N_2, is used to prevent Ga from mixing with NH_3 directly in the area above the silica nozzle in order to reduce the gas-phase reactions of GaI and NH_3. Four shield gas flow rates of 0.5, 0.8, 1.2, and 1.5 slm were studied. The only reaction activated is (37.58).

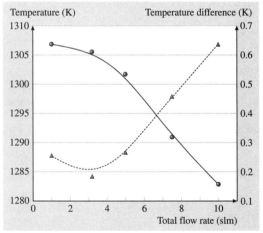

Fig. 37.16 Averaged temperature (*solid line*) and largest temperature difference (*dashed line*) on the substrate as functions of total flow rate

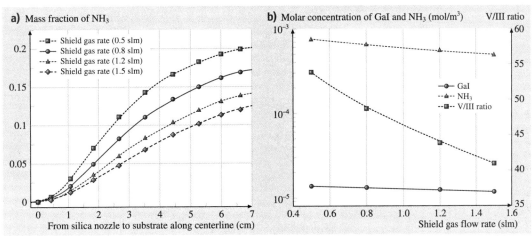

Fig. 37.17 (a) Mass fraction distribution of NH_3 along the centerline from the silica nozzle to the substrate and (b) averaged concentrations of GaI and H_3 on the substrate and V/III ratio achieved on the substrate as functions of the shield gas flow rate

Figure 37.17a shows the mass fraction change of NH_3 with the shield gas flow rate along the reactor centerline from the outlet of the silica nozzle to the substrate surface. It is seen that, at the silica nozzle outlet, the NH_3 concentration is zero, and then increases gradually with distance. The concentration of NH_3 with 1.5 slm shield gas flow rate is about half of that with 0.5 slm shield gas flow rate along the centerline. Figure 37.17b shows the averaged concentrations of GaI and NH_3 and the V/III ratio on the substrate surface. It is shown that the concentrations of *GaI* and NH_3 decrease as the shield gas flow rate increases. It is also observed that the effect of the shield gas flow rate on the NH_3 concentration is more significant than its effect on GaI concentration. When the shield gas flow rate increases from 0.5 to 1.5 slm, concentrations of GaI and NH_3 drop by 17% and 34%, respectively. This is attributed to the increase of the N_2 concentration on the substrate surface as the result of the increase of the shield gas flow rate. Furthermore, increasing the shield gas flow rate can reduce the residence time for NH_3 to diffuse into the center area, causing reduction of the NH_3 concentration and the V/III ratio.

Experiments were conducted to study the effect of the shield gas flow rate on the GaN deposition rate by *Mecouch* et al. [37.30]. Shield gas flow rates of 0.62 slm and 1.0 slm were used in their experiments. The flow rate of 0.62 slm corresponds to a velocity ratio of 1 : 1 between the shield gas and other gases at the outlet of the silica nozzle. It is revealed that the shield gas flow rate of 0.62 slm produces a higher GaN growth rate under different iodine concentrations in the source bubbler compared with the shield gas flow rate of 1.0 slm. A shield gas flow rate of 0.62 slm was therefore used in the subsequent experiments. Figure 37.18a,b shows the mass fraction distributions of GaI and NH_3 in the growth reactor. The gas-phase reaction between Ga and

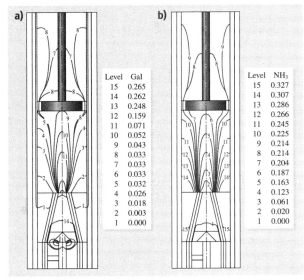

Fig. 37.18a,b Mass fraction distributions of (a) GaI and (b) NH_3 inside the VPE reactor designed by *Mecouch* et al. [37.30, 72, 120]

Table 37.12 Species concentrations on the substrate

	Ga	GaI	H$_2$	I$_2$	N$_2$	NH$_3$
Y_i	2.4×10^{-6}	0.042575	0.000855	0.005494	0.760977	0.190096
C_i (kmol/m^3)	2.16×10^{-9}	1.36×10^{-5}	2.69×10^{-5}	1.36×10^{-6}	1.71×10^{-3}	7.02×10^{-4}
P_i (Torr)	1.76×10^{-4}	$1.11 \times 10^{+0}$	$2.19 \times 10^{+0}$	1.11×10^{-1}	$1.39 \times 10^{+2}$	$5.73 \times 10^{+1}$

I$_2$ takes place mainly inside the inner silica nozzle. Since the gas flow in the reactor is laminar (Fig. 37.18b), the NH$_3$ has been transferred into the center area of the reactor by mass diffusion.

The mass fraction Y_i, the molar concentration C_i, and the partial pressure P_i of species i on the substrate are listed in Table 37.12 for the shield gas flow rate of 0.62 slm. It is seen that the partial pressure of Ga is close to zero, and the achieved partial pressure of GaI is around 1 Torr on the substrate at a reactor pressure of 200 Torr.

Effect of Silica Nozzle Angle

The GaN growth rate is directly proportional to the partial pressure of GaI achieved on the substrate. A silica nozzle is used in experiments to force more GaI to the substrate area [37.30]. Adding a silica nozzle will also help prevent ammonia from diffusing towards the gallium source area because ammonia reacts with Ga directly on the liquid gallium surface to form polycrystalline GaN through the following reaction:

$$2Ga(g) + 2NH_3(g) \rightleftharpoons 2GaN(s) + 3H_2(g). \quad (37.82)$$

This reaction (37.82) could terminate the transport of gallium and stop the GaN growth on the substrate.

Design of silica nozzle angle θ (Fig. 37.1a) is an important task. To test the effect of nozzle angle on concentration distributions of NH$_3$ and GaI between the outlet of silica nozzle and substrate surface, growth systems with four nozzle angles of 77°, 81°, 86°, and 90° were simulated. Figure 37.19a shows the mass fraction distribution of NH$_3$ along the reactor centerline from the silica nozzle outlet to the substrate surface. It is found that, at the silica nozzle outlet, the NH$_3$ concentration is close to 0 when using a nozzle with angle < 81°. The ammonia mass fraction is up to 0.03 at the outlet of the inner silica nozzle when its angle is > 81°, indicating that ammonia may diffuse into the inner silica tube. Using a nozzle with small angle, such as 77°, can significantly reduce the NH$_3$ concentration in the area above the silica nozzle. This influence will be weakened when the mixed gas approaches the substrate. Thus, the concentration of NH$_3$ on the substrate can be maintained at an appropriate level even with a small nozzle angle. Figure 37.19b shows the average molar concentrations of GaI and NH$_3$ on the substrate surface. The concentra-

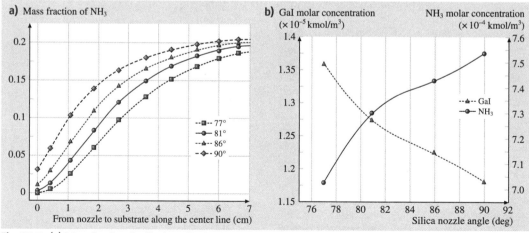

Fig. 37.19 (a) Mass fraction distribution of NH$_3$ along the centerline from the silica nozzle to the substrate and (b) averaged concentrations of GaI and NH$_3$ on the substrate under different silica nozzle angles

tion of NH$_3$ increases with the nozzle angle, while that of GaI decreases with the nozzle angle.

The silica nozzle used in the experiment has an angle of 77° [37.72]. This design can reduce the NH$_3$ concentration in the area above the silica nozzle outlet significantly, subsequently preventing NH$_3$ from diffusing into the inner silica nozzle and increasing the GaI concentration on the substrate. The simulation shows that using a nozzle angle of 77° can increase the GaI concentration on the substrate by about 10% compared with a silica nozzle angle of 90°.

Effect of Substrate–Nozzle Distance

The substrate–nozzle distance is one of the key parameters that affect the GaN growth rate and deposition quality. It controls the length and time allowed for different gases to mix with each other. An appropriate substrate–nozzle distance corresponds to a suitable V/III ratio above the substrate for fast and uniform GaN growth. The simulated substrate–nozzle distance ranges from 6.35 to 17.8 cm. Figure 37.20a shows the molar concentrations of GaI and NH$_3$ as a function of substrate–nozzle distance. The concentration of GaI decreases with substrate–nozzle distance until the mixing is completed at a substrate–nozzle distance of 13.0 cm. The error bars in Fig. 37.20a indicate the GaI concentration variation across the substrate surface. It is seen that, for a substrate–nozzle distance of 6.35 cm, the variation of GaI concentration on the substrate surface is small (around 0.6%). This indicates that species uniformity is not a problem. The concentration of NH$_3$ increases with the distance since more time is available for ammonia to diffuse into the center area of the reactor. The variation of the V/III ratio with substrate–nozzle distance is shown in Fig. 37.20b. When the substrate–nozzle distance is less than 13.0 cm, an increase of substrate–nozzle distance of about 2.5 cm corresponds to a V/III ratio increase of 10.

It is concluded that complete mixing of different species needs a substrate–nozzle distance larger than 13.0 cm. The species concentrations across the substrate surface is fairly uniform even with a small substrate–nozzle distance such as 6.35 cm. To achieve uniform GaN deposition with high growth rate, the substrate–nozzle distance of 6.35 cm is used. This distance can achieve uniform species concentrations across the substrate surface and a high GaI concentration of 1.4×10^{-5} kmol/m^3. The V/III ratio in this case will be about 50. For the growth of GaN from GaI$_{1-3}$/NH$_3$ or GaCl$_{1-3}$/NH$_3$, the concentration of NH$_3$ is maintained at a high value on the substrate, resulting in an appropriate supersaturation as the driving force for the surface deposition. For the current GaN vapor-growth system, the V/III ratio is kept above 50 for most experiments [37.30, 72, 120].

37.4.6 Effect of Substrate Rotation on Deposition Rate and Deposition Uniformity

In the experiment, the substrate is rotated by revolving the substrate holder to improve deposition uniformity,

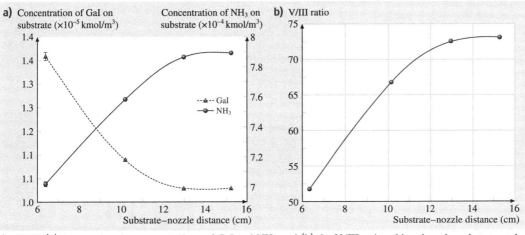

Fig. 37.20 (a) Averaged molar concentrations of GaI and NH$_3$ and (b) the V/III ratio achieved on the substrate under different substrate–nozzle distances

and consequently to improve deposition quality. It is shown by the simulation that temperature and species concentration distributions on the substrate are fairly uniform under the current geometrical and operational conditions. It is expected that the deposited GaN layer will be uniform. Experiments have been conducted using the gas inlet conditions listed in Table 37.10 for a growth time period of 1 h [37.30]. This experiment shows that the thickness of deposited GaN crystal layer varies from 35 to 75 μm/h on the substrate surface. To improve the deposition uniformity, the substrate was rotated in the experiments. The instantaneous GaN deposition rate distribution on the substrate will be averaged due to rotation and the deposition uniformity will be improved. Cross-sectional scanning electron microscopy (SEM) images of GaN layer with a substrate rotation speed of 1 rpm show that the deposition uniformity is significantly improved since the thickness difference in the GaN layer is reduced from 40 to 2 μm/h [37.30].

It should be noted that, in the simulation, since the entire system is assumed to be perfectly axisymmetric, the calculated temperature and species concentration distributions will be axisymmetric. In the experiment, the components in the growth reactor are mostly not axisymmetric. It is found experimentally that the deposition nonuniformity is mainly attributed to misalignment between the inner and outer silica nozzles and the seed holder. Deposition uniformity could therefore be improved by improving these alignments.

Another factor that could cause deposition nonuniformity across the substrate is reactant depletion in the species concentration boundary layer formed on the substrate due to surface reactions. Figure 37.8b shows that a typical boundary layer of the reactant species concentration is formed on the substrate due to surface reactions. The concentration of GaI before the mixing gas reaching the substrate area is defined as $C = C_0$. Due to the surface reactions (37.73–37.75) of GaI with NH$_3$ on the substrate, the GaI concentration drops to $C = C_w$ on the substrate surface. The concentration boundary layer thickness χ is defined as the distance where the GaI concentration changes from C_0 in the bulk gas to C_w on the substrate surface. The low-concentration layer will gradually accumulate as the gases flow from the center to the perimeter of the substrate. The growth rate is therefore expected to be low in the periphery area. To improve the deposition uniformity, one may reduce the concentration boundary-layer thickness. Substrate rotation can be used to reduce the thickness of the low-concentration layer. To estimate the rotation speed at which the boundary-layer thickness will be affected, the following equation is used to calculate the ratio of the Reynolds number due to forced convection and rotation:

$$\frac{\text{Forced convection}}{\text{Rotation}} = \frac{\text{Re}_d}{\text{Re}_\theta} = \frac{UD/v}{\omega d^2/v} = \frac{1500}{\varpi}, \quad (37.83)$$

where ω and ϖ are the rotation frequency in units of rps and rpm, respectively.

It is revealed that the rotation speed has to be on the order of 10^3 rpm for rotation to affect the boundary-layer thickness. Such a high rotation speed will cause the growth environment to become unstable near the substrate. In experiments, only a small rotation speed (1 rpm) is recommended in order to even out the instantaneous deposition rate on the periphery of the substrate.

37.4.7 Quasi-equilibrium Model for Deposition Rate Prediction

Predicting the GaN deposition rate is the most important issue for modeling. A thermodynamic surface reaction model (local model) combined with a numerical simulation model (global model) for mass transfer in the growth reactor are used to calculate the GaN deposition rates under different substrate temperatures. Since reactions (37.74–37.75) can be combined to form reaction (37.73), reaction (37.73) is assumed to prevail on the substrate when reaction equilibrium is reached.

Figure 37.8a shows a schematic of GaN deposition on the substrate surface. Species on the substrate include GaI, NH$_3$, HI, H$_2$, N$_2$, I$_2$, and Ga. From the global heat and mass transfer model, we can calculate the concentration of these seven species, which will be used as the initial species conditions. As shown in Fig. 37.8b, the transport of species into the concentration boundary layer is driven by the concentration gradient due to surface deposition.

According to (37.33–37.36), there are four unknowns, i.e., P_{HI}^{e}, $P_{\text{GaI}}^{\text{e}}$, $P_{\text{NH}_3}^{\text{e}}$, and $P_{\text{H}_2}^{\text{e}}$, and four equations. The partial pressures of each species, P_{HI}^{w}, $P_{\text{GaI}}^{\text{w}}$, $P_{\text{NH}_3}^{\text{w}}$, and $P_{\text{H}_2}^{\text{w}}$, can be obtained using the global simulation model, whereas the sticking probabilities of each species α_i are required. NH$_3$ will crack on the substrate surface to provide nitrogen atoms for GaN growth. The sticking coefficient of NH$_3$ on the substrate is defined as the ratio of the NH$_3$ flux incorporated as GaN to the total NH$_3$ flow rate incident on the substrate. This definition assumes that all NH$_3$ flux incident on the substrate is involved in the reaction process, i.e.,

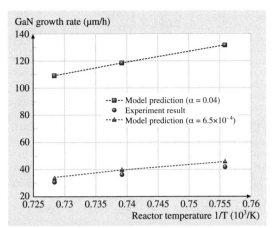

Fig. 37.21 Predicted and experimental results of GaN growth rate with ammonia sticking coefficient of 0.04 or 6.5×10^{-4} (after [37.72])

current system, the available surface sites for nitrogen atoms to deposit will be further reduced since it has to compete with other atoms on the substrate surface. The sticking coefficient will therefore be reduced further. Finally, since a much higher growth temperature is used, GaN decomposition on the substrate surface will be enhanced, causing the sticking coefficient of ammonia to drop further. By adjusting the ammonia sticking coefficient, it is found that, when the value is lowered to the order of 10^{-4}, the GaN growth rate from prediction and experiment matches very well, as shown in Fig. 37.21 for an ammonia sticking coefficient of 6.5×10^{-4}.

37.4.8 Kinetic Deposition Model

no saturation of surface sites occurs on the substrate. *Mesrine* et al. studied the efficiency of NH_3 as the nitrogen source for GaN molecular-beam epitaxy [37.127]. It was found that, at temperature of 830 °C and pressure of 10^{-5} Torr, the sticking coefficient of NH_3, α_{NH_3} is about 0.04. The sticking coefficient at temperature of 1050 °C is expected to be slightly higher, therefore 0.04 is used as the sticking coefficient of NH_3 in the simulation. The sticking coefficient of GaI is assumed to be unity. The sticking probabilities of H_2 and HI are also assumed to be unity, considering the high reactivity of these species under high growth temperature. In Fig. 37.21 the square line and the black dots indicate the experimental and predicted GaN deposition rates, respectively. The predicted results are 3–4 times the experiment ones. The difference between the experiments and simulations is attributed to overprediction of the sticking coefficient of ammonia in the modeling, which is due to several reasons. First, the assumption that all the incident ammonia flux on the substrate is involved in the reaction process is not correct for the current GaN growth system. To study the ammonia sticking coefficient in the MBE system, *Mesrine* et al. [37.127] controlled the ammonia flow rate low enough to ensure that the above assumption is valid. However, in the GaN growth system, the ammonia flow rate is high and it is found experimentally that the GaN deposition rate is independent of the ammonia flow rate, as discussed in the next section. The efficiency of the ammonia deposition coefficient will be lowered greatly in this case. Second, since a much higher pressure, 200 Torr, is used in the

Combined with the experimental data, the GaN deposition rate at a constant growth temperature of 1050 °C was tested numerically while the flow rates at the inlet were changed to obtain different reactant concentrations and V/III ratios on the substrate. More importantly, simulations were performed to test the surface reaction pathway of GaN growth and the contribution of different reactions to the final GaN growth rate.

The energetically preferred surface reactions on the substrate for GaN deposition have been obtained previously. Reaction (37.73) was first used as the overall surface reaction step on the substrate in the simulation. Figure 37.22 shows the experimental data for the GaN deposition rate at different ammonia flow rates and iodine vapor fractions [37.30, 72]. The highest growth rate

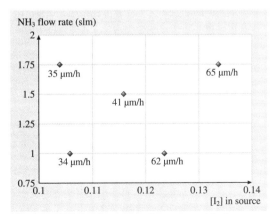

Fig. 37.22 Effect of ammonia flow rate and iodine vapor fraction in the source on the GaN deposition rate (after [37.30, 72])

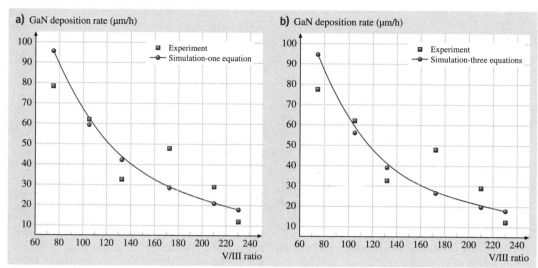

Fig. 37.23a,b GaN deposition rate as a function of the V/III ratio for (**a**) one surface reaction and (**b**) three surface reactions (after [37.72])

achieved is 65 μm/h. It is found that the GaN growth rate is independent of the NH$_3$ flow rate but dependent on the GaI concentration.

Experiments were conducted to examine the effect of V/III ratio on the GaN deposition rate [37.72, 120]. The V/III ratio was varied by maintaining the NH$_3$ flow rate at 1 slm while changing the I$_2$ concentration in the source. Using the same operating conditions as in the experiment, simulations were performed with reaction (37.73) as the boundary condition. The surface reaction rate coefficient δA_p and the concentration dependency of GaI α were determined by fitting the simulation results to the experimental data. Figure 37.23a shows the variation of the GaN deposition rate with the V/III ratio.

The surface reaction rate of reaction (37.73) used to obtain the numerical results in Fig. 37.23a is determined by the following equation

$$\dot{w} = 6.5 \times 10^9 \exp\left(\frac{-12\,390}{T}\right)[\text{GaI}]^{2.2}\ . \tag{37.84}$$

It is concluded that, when one overall surface reaction step is used to predict the GaN deposition rate, the reaction rate coefficient is 6.5×10^9 and the GaI concentration dependency is 2.2.

When reactions (37.73–37.75) are used, the reaction rate coefficient of each reaction step is determined by matching the simulation results to the experimental data. Figure 37.23b shows the GaN deposition rate calculated using all three surface reactions and the deposition rates from the experiment. The surface reactions and their reaction rates used here are summarized in Table 37.13.

The calculated GaN deposition rate with reactions (37.73) and (37.74) is presented in Table 37.14. Since GaI is considered as the only medium for Ga source transport, the contribution of reaction (37.75) to the deposition rate is found to be less than 0.1% for all cases and is therefore not listed in Table 37.14. In the experiment, the contribution of reaction (37.75) to the GaN deposition rate might be larger. It is, however, not expected to be as significant as reactions (37.73)

Table 37.13 Surface reaction rates obtained by matching the simulation data with the experimental data

Reaction		Rate expression (kmol/m^2 s)
(37.73)	GaI(g) + NH$_3$(g) $\rightleftharpoons$ GaN(s) + HI(g) + H$_2$(g)	$0.07 \exp(-12\,390/T)[\text{GaI}]$
(37.74)	3GaI(g) + 2NH$_3$(g) $\rightleftharpoons$ 2GaN(s) + GaI$_3$(g) + 3H$_2$(g)	$5.88 \times 10^9 \exp(-13\,630/T)[\text{GaI}]^3$
(37.75)	GaI$_3$(g) + NH$_3$(g) $\rightleftharpoons$ GaN(s) + 3HI(g)	$0.36 \exp(-14\,000/T)[\text{GaI}_3]$

Table 37.14 GaN deposition rate and rate contribution from reactions (37.73) and (37.74)

V/III ratio	GaN deposition rate (μm/h)		Growth rate contribution (%)	
	Reaction (37.73)	Reaction (37.74)	Reaction (37.73)	Reaction (37.74)
75	21.80	73.10	22.97	38.03
105	17.62	38.61	31.33	68.67
132	15.04	24.01	38.52	61.48
172	12.54	13.89	47.41	52.59
210	10.86	9.03	54.60	45.40
230	10.16	7.40	57.86	42.14

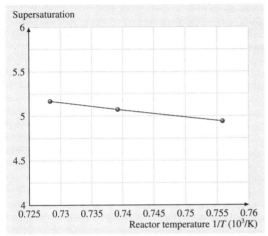

Fig. 37.24 Supersaturation of GaN growth under different reactor temperatures

and (37.74). The contribution of reaction (37.73) to the total GaN deposition rate increases with the V/III ratio, while for reaction (37.74) it decreases with the V/III ratio.

The supersaturation of GaN growth on the substrate can be calculated once the partial pressure of each species is obtained. Figure 37.24 shows the calculation results for the supersaturation with different reactor

Table 37.15 Sticking probability of reactant i with different V/III ratios

V/III ratio	Sticking probability		
	GaI ($\times 10^{-3}$)	NH_3 ($\times 10^{-6}$)	GaI_3 ($\times 10^{-3}$)
75	2.66	6.50	4.15
105	2.07	4.10	3.12
132	1.75	3.02	2.69
172	1.43	2.08	2.07
210	1.22	1.55	1.86
230	1.13	1.38	1.77

temperatures. The supersaturation is around 5. This is typical for an epitaxy growth system that needs a relatively low supersaturation to prevent the reactant partial pressures from deviating excessively from their equilibrium values. The supersaturation decreases only slightly with temperature, which means that the GaN deposition is diffusion controlled under the current operating conditions.

Table 37.15 summarizes the sticking probability of the reacting species deduced from various simulations. The calculated sticking probabilities of the order of 10^{-3}–10^{-6} are reasonable for the saturated compound. The sticking probability of species i is defined by

$$S_i = f(\theta^s)_i \exp\left[\frac{-E_{ai}(\theta^s)}{RT}\right] = \frac{R_i}{F_i}, \qquad (37.85)$$

where $f(\theta^s)_i$ is the function of the existing surface coverage of adsorbed species i, θ^s is the Langmuir definition of surface coverage of certain species, or the fraction of sites which are occupied on the substrate surface, E_{ai} is the activation energy for adsorption of species i,

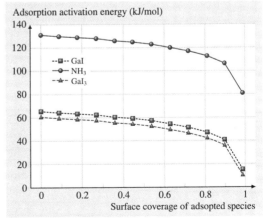

Fig. 37.25 Activation energy of adsorption with surface coverage of 0–0.99

R_i (kmol/m²s) is the rate of adsorption of species i chemically, and $F_i = P_i^w(2\pi RTM_i)^{1/2}$ [kmol/m²s] is the incident flux rate of species i onto the substrate.

It is observed that the sticking probability decreases with increasing V/III ratio. When the I_2 concentration decreases in the source, the species reaction rate decreases more significantly compared with its incident flux onto the substrate. The sticking probability of ammonia is three orders of magnitude smaller than that of GaI or GaI$_3$, because the ammonia concentration is kept much higher than that of GaI on the substrate.

To determine the activation energy for adsorption of species i, it is assumed that the sticking probability is directly proportional to the concentration of vacant surface sites $(1-\theta^s)$. This assumption is a reasonable first approximation for nondissociative adsorption. The activation energy of adsorption is assumed to be independent of the surface coverage. The sticking probability is, thus, revised as

$$S_i = (1-\theta^s)\exp\left(\frac{-E_{ai}}{RT}\right). \tag{37.86}$$

Figure 37.25 shows the calculated activation energy of adsorption for the three species for surface coverage of 0–0.99. The activation energy drops as the surface coverage increases. The averaged activation energies of adsorption for GaI, NH$_3$, and GaI$_3$ by calculation are 5.72×10^4, 1.24×10^5, and 5.28×10^4 J/mol, respectively.

37.5 Surface Evolution of GaN/AlN Film Growth from Vapor

In this section, sublimation and HVPE growth of AlN will be used as an example to investigate the surface evolution of thin film during growth. Since large-diameter AlN single crystal is not available in nature, the goal of research is to find a way to produce large, thick AlN films/ingots, which can then be sliced for use as substrates. The growth conditions of two systems are compared in Table 37.16. Figure 37.26 shows the AlN crystals grown by sublimation and HVPE techniques.

Srolovitz et al. [37.128] studied the surface evolution of film growth in which surface diffusion determines the crystal grain size and morphology. Figure 37.27 shows the deposition geometry in which a one-dimensional partial differential equation describes the evolution of an arbitrary initial surface profile $h(x,0)$ under the joint influence of constant uniform deposition flux rate J of finite-size atoms of radius r and surface diffusion. C_k is the surface curvature and V_s is the velocity of surface diffusion.

By assuming a sinusoidal perturbation, i.e., $h(x,t) = \sin(x,t)$, the surface profile can be described as

$$h(x,t) = e^{(rJk^2 - D_e k^4)t}\sin(kx), \tag{37.87}$$

where $D_e = (D_s\sigma_s\Omega^2\varepsilon)/(k_B T)$, D_s is the surface diffusivity of the atoms, σ_s is the isotropic surface energy density, Ω is the atomic volume, ε is the number of atoms per unit area, $k_B T$ is the thermal energy, and k is the wavenumber, relating to the wavelength λ by $\lambda = 2\pi/k$. It is revealed by (37.88) that, if $\delta Jk^2 - D_e k^4 > 0$, $e^{(rJk^2 - D_e k^4)t} > 1$ is obtained, and the initial sinusoidal perturbation will be enlarged, which means that the growth will be unstable and growth quality will be deteriorated. For stable growth, the wavelength of the initial perturbation λ has to be smaller than the effective diffusion length λ_e, which is defined as

$$\lambda_e = \sqrt{\frac{4\pi D_e}{rJ}}. \tag{37.88}$$

Table 37.16 Growth comparison for sublimation and HVPE systems

Growth system	Sublimation	HVPE
Surface reaction	$Al(g) + \frac{1}{2}N_2(g) \rightleftharpoons AlN(s)$	$AlCl_3(g) + NH_3(g) \rightleftharpoons AlN(s) + 3HCl(g)$
Substrate	AlN seed	Sapphire/SiC
Growth temperature	2240 °C	900–1200 °C
Growth rate	0.2–0.5 mm/h along c axis	5–75 μm/h
Crystal property	Transparent, DCRC 20 arcsec	Opaque, DCRC 1000 arcsec

Fig. 37.26 (a) An AlN crystal grown by sublimation method (after [37.70]). (b) An AlN crystal grown by HVPE method (after [37.130]) (with permission from Elsevier) ▶

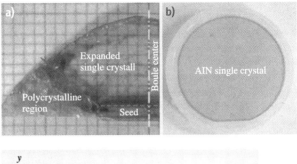

To obtain stable growth, one can either reduce the initial perturbation wavelength or increase the effective diffusion wavelength. The initial perturbation wavelength is related to the substrate and supersaturation. The effective diffusion length is proportional to the effective diffusion coefficient D_e and inversely proportional to the atom radius r and the growth flux rate. *Collazo* et al. [37.129] studied the effect of process conditions such as surface temperature T and the growth flux rate J on the effective diffusion length λ_e and the effect of process conditions such as supersaturation on the initial perturbation wavelength λ. They found that changing the process conditions cannot significantly affect the values of λ_e and λ.

It is concluded here that the substrate plays a critical role in determining the quality of AlN crystal when the crystal is grown large and thick. For HVPE growth using a foreign substrate such as sapphire or silicon carbide, the wavelength of the surface perturbation will be larger than the effective surface diffusion length due to the lattice mismatch between the AlN crystal and substrate when the thickness of the crystal is large.

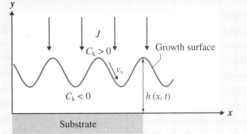

Fig. 37.27 Deposition geometry (after [37.128])

Therefore the film growth will be unstable, and consequently, low-quality crystals will be obtained.

37.6 Concluding Remarks

Transport phenomena in chemical vapor deposition of III nitrides are complicated due to gas flow, coupled convection and radiation, conjugate heat and mass transfer, homogeneous gas-phase and heterogeneous surface reactions, and the coexistence of multiple length and time scales. The multiplicity of governing parameters, complex geometric configurations, complicated boundary conditions, and the lack of information on gas-phase and surface reaction mechanisms make this process difficult to predict and control.

It is the authors' intention to provide a systematical procedure for those who are new to the field of vapor-growth process modeling. In this chapter an attempt has been made to present a comprehensive description of the fundamental theory as well as an extensive overview of the state of the art in vapor-growth process modeling. The main focus has been on continuum modeling of macroscopic gas flow, homogeneous gas-phase reactions inside the reactor chamber, heterogeneous surface reactions on the substrate surface, heat transfer, and species transport in reactors for GaN/AlN vapor crystal growth.

Gas velocity, temperature and pressure, and species concentration distributions in a CVD reactor are correlated. The main difficulty in accurately predicting species concentration lies in the lack of detailed information for gas-phase and surface reactions. By using GaN IVPE growth as an example, mathematic models for thermodynamic and kinetic gas-phase and surface reactions have been described. Thermodynamic models can be used to predict gas-phase and surface reactions with or without detailed reactions steps. The results, however, only give the upper limit of what one can expect. By comparing the results with available experiment observations, one may determine the extent to which the gas-phase and surface reactions deviate from the equilibrium assumption. Empirical models were used in this chapter to simulate gas-phase and surface reaction kinetically. The detailed multistep gas-phase and surface reaction mechanisms can only be obtained from the experiments. The current models can be used to predict species transport in the reactor in order to identify the critical controlling parameters for fast and uniform deposition.

Film instability during GaN/AlN growth has been analyzed, in which surface diffusion determines the crystal grain size and morphology. The key parameters, such as the lattice mismatch between the substrate and deposited film, the effective diffusion length, and the supersaturation above the substrate, have been identified and their effects on film instability during growth quantified. It is found that, for growth of GaN/AlN bulk crystals, use of native substrate is the key for high deposition quality when the deposited film thickness becomes very large.

References

37.1 E.J. Tarsa, B. Heying, X.H. Wu, P. Fini, S.P. DenBaars, J.S. Speck: Homoepitaxial growth of GaN under Ga-stable and N-stable conditions by plasma-assisted molecular beam epitaxy, J. Appl. Phys. **82**, 5472–5479 (1997)

37.2 C.R. Elsass, I.P. Smorchkova, H.Y. Ben, E. Haus, C. Poblenz, P. Fini, K. Maranowski, P.M. Petroff, S.P. DenBaars, U.K. Mishra, J.S. Speck, A. Saxler, S. Elhamri, W.C. Mitchel: Electron transport in AlGaN/GaN heterostructures grown by plasma-assisted molecular beam epitaxy, Jpn. J. Appl. Phys. Part 2-Letters **39**, L1023–L1025 (2000)

37.3 A. Hierro, A.R. Arehart, B. Heying, M. Hansen, J.S. Speck, U.K. Mishra, S.P. DenBaars, S.A. Ringel: Capture kinetics of electron traps in MBE-grown n-GaN, Phys. Status Solidi b **228**, 309–313 (2001)

37.4 R.F. Davis, A.M. Roskowski, E.A. Preble, J.S. Speck, B. Heying, J.A. Freitas, E.R. Glaser, W.E. Carlos: Gallium nitride materials – progress, status, and potential roadblocks, Proc. IEEE **90**, 993–1005 (2002)

37.5 P. Waltereit, H. Sato, C. Poblenz, D.S. Green, J.S. Brown, M. McLaurin, T. Katona, S.P. DenBaars, J.S. Speck, J.H. Liang, M. Kato, H. Tamura, S. Omori, C. Funaoka: Blue GaN-based light-emitting diodes grown by molecular-beam epitaxy with external quantum efficiency greater than 1.5%, Appl. Phys. Lett. **84**, 2748–2750 (2004)

37.6 C. Poblenz, P. Waltereit, J.S. Speck: Uniformity and control of surface morphology during growth of GaN by molecular beam epitaxy, J. Vac. Sci. Technol. B **23**, 1379–1385 (2005)

37.7 A. Corrion, F. Wu, T. Mates, C.S. Gallinat, C. Poblenz, J.S. Speck: Growth of Fe-doped GaN by RF plasma-assisted molecular beam epitaxy, J. Cryst. Growth **289**, 587–595 (2006)

37.8 A. Corrion, C. Poblenz, P. Waltereit, T. Palacios, S. Rajan, U.K. Mishra, J.S. Speck: Review of recent developments in growth of AlGaN/GaN high-electron mobility transistors on 4H-SiC by plasma-assisted molecular beam epitaxy, IEICE Trans. Electron. **E89C**, 906–912 (2006)

37.9 A.S. Segal, A.V. Kondratyev, S.Y. Karpov, D. Martin, V. Wagner, M. Ilegems: Surface chemistry and transport effects in GaN hydride vapor phase epitaxy, J. Cryst. Growth **270**, 384–395 (2004)

37.10 S.Y. Karpov, D.V. Zimina, Y.N. Makarov, B. Beaumont, G. Nataf, P. Gibart, M. Heuken, H. Jurgensen, A. Krishnan: Modeling study of hydride vapor phase epitaxy of GaN, Phys. Status Solidi a **176**, 439–442 (1999)

37.11 A.S. Segal, S.Y. Karpov, Y.N. Makarov, E.N. Mokhov, A.D. Roenkov, M.G. Ramm, Y.A. Vodakov: On mechanisms of sublimation growth of AlN bulk crystals, J. Cryst. Growth **211**, 68–72 (2000)

37.12 M.V. Bogdanov, S.E. Demina, S.Y. Karpov, A.V. Kulik, M.S. Ramm, Y.N. Makarov: Advances in modeling of wide-bandgap bulk crystal growth, Cryst. Res. Technol. **38**, 237–249 (2003)

37.13 E.N. Mokhov, A.D. Roenkov, Y.A. Vodakov, S.Y. Karpov, M.S. Ramm, A.S. Segal, Y.A. Makarov, H. Helava: Growth of AlN bulk crystals by sublimation sandwich method. In: *Silicon Carbide and Related Materials*, ed. by P. Bergman, E. Janzén (Trans Tech, Zurich 2002) pp. 979–982

37.14 S.Y. Karpov, D.V. Zimina, Y.N. Makarov, E.N. Mokhov, A.D. Roenkov, M.G. Ramm, Y.A. Vodakov: Sublimation growth of AlN in vacuum and in a gas atmosphere, Phys. Status Solidi a **176**, 435–438 (1999)

37.15 P.G. Baranov, E.N. Mokhov, A.O. Ostroumov, M.G. Ramm, M.S. Ramm, V.V. Ratnikov, A.D. Roenkov, Y.A. Vodakov, A.A. Wolfson, G.V. Saparin, S.Y. Karpov, D.V. Zimina, Y.N. Makarov, H. Juergensen: Current status of GaN crystal growth by sublimation sandwich technique, MRS Internet J. Nitride Semicond. Res. **3**, 50 (1998)

37.16 C.H. Wei, Z.Y. Xie, L.Y. Li, Q.M. Yu, J.H. Edgar: MOCVD growth of cubic GaN on 3C-SiC deposited on Si(100) substrates, J. Electron. Mater. **29**, 317–321 (2000)

37.17 Z.Y. Xie, C.H. Wei, L.Y. Li, J.H. Edgar, J. Chaudhuri, C. Ignatiev: Effects of surface preparation on epitaxial GaN on 6H-SiC deposited via MOCVD, MRS Internet J. Nitride Semicond. Res. **4S1**, G3.39 (1999)

37.18 B.S. Sywe, J.R. Schlup, J.H. Edgar: Fourier-transform infrared spectroscopic study of pre-deposition reactions in metalloorganic chemical

37.19 J.H. Edgar, L.H. Robins, S.E. Coatney, L. Liu, J. Chaudhuri, K. Ignatiev, Z. Rek: A comparison of aluminum nitride freely nucleated and seeded on 6H-silicon carbide, Silicon Carbide Relat. Mater. **338-3**, 1599–1602 (2000)

37.20 Y. Shi, B. Liu, L.H. Liu, J.H. Edgar, E.A. Payzant, J.M. Hayes, M. Kuball: New technique for sublimation growth of AlN single crystals, MRS Internet J. Nitride Semicond. Res. **6**, 1–10 (2001)

37.21 J.H. Edgar, L. Liu, B. Liu, D. Zhuang, J. Chaudhuri, M. Kuball, S. Rajasingam: Bulk AlN crystal growth: self-seeding and seeding on 6H-SiC substrates, J. Cryst. Growth **246**, 187–193 (2002)

37.22 B. Liu, J.H. Edgar, Z. Gu, D. Zhuang, B. Raghothamachar, M. Dudley, A. Sarua, M. Kuball, H.M. Meyer: The durability of various crucible materials for aluminum nitride crystal growth by sublimation, MRS Internet J. Nitride Semicond. Res. **9**, 6 (2004)

37.23 Z. Gu, L. Du, J.H. Edgar, E.A. Payzant, L. Walker, R. Liu, M.H. Engelhard: Aluminum nitride-silicon carbide alloy crystals grown on SiC substrates by sublimation, MRS Internet J. Nitride Semicond. Res. **10**, 5 (2005)

37.24 Z. Gu, J.H. Edgar, D.W. Coffey, J. Chaudhuri, L. Nyakiti, R.G. Lee, J.G. Wen: Defect-selective etching of scandium nitride crystals, J. Cryst. Growth **293**, 242–246 (2006)

37.25 M. Callahan, M. Harris, M. Suscavage, D.F. Bliss, J. Bailey: Synthesis and growth of gallium nitride by the chemical vapor reaction process (CVRP), MRS Internet J. Nitride Semicond. Res. **4**, 10 (1999)

37.26 V. Tassev, D.F. Bliss, M. Suscavage, Q.S. Paduano, S.Q. Wang, L. Bouthillette: Iodine vapor phase growth of GaN: dependence of epitaxial growth rate on process parameters, J. Cryst. Growth **235**, 140–148 (2002)

37.27 M. Suscavage, L. Bouthillette, D.F. Bliss, S.Q. Wang, C. Sung: New iodide method for growth of GaN, Phys. Status Solidi a **188**, 477–480 (2001)

37.28 D.F. Bliss, V.L. Tassev, D. Weyburne, J.S. Bailey: Aluminum nitride substrate growth by halide vapor transport epitaxy, J. Cryst. Growth **250**, 1–6 (2003)

37.29 D. Cai, L.L. Zheng, H. Zhang, V.L. Tassev, D.F. Bliss: Modeling of aluminum nitride growth by halide vapor transport epitaxy method, J. Cryst. Growth **276**, 182–193 (2005)

37.30 W.J. Mecouch, B.J. Rodriguez, Z.J. Reitmeier, J.S. Park, R.F. Davis, Z. Sitar: Initial stages of growth of gallium nitride via iodine vapor phase epitaxy, Mater. Res. Soc. Symp. Proc. **831**, E3.23.1. (2005)

37.31 R. Schlesser, Z. Sitar: Growth of bulk AlN crystals by vaporization of aluminum in a nitrogen atmosphere, J. Cryst. Growth **234**, 349–353 (2002)

37.32 D. Zhuang, Z.G. Herro, R. Schlesser, B. Raghothamachar, M. Dudley, Z. Sitar: Seeded growth of AlN crystals on nonpolar seeds via physical vapor transport, J. Electron. Mater. **35**, 1513–1517 (2006)

37.33 D. Zhuang, Z.G. Herro, R. Schlesser, Z. Sitar: Seeded growth of AlN single crystals by physical vapor transport, J. Cryst. Growth **287**, 372–375 (2006)

37.34 R. Dalmau, R. Schlesser, B.J. Rodriguez, R.J. Nemanich, Z. Sitar: AlN bulk crystals grown on SiC seeds, J. Cryst. Growth **281**, 68–74 (2005)

37.35 V. Noveski, R. Schlesser, B. Raghothamachar, M. Dudley, S. Mahajan, S. Beaudoin, Z. Sitar: Seeded growth of bulk AlN crystals and grain evolution in polycrystalline AlN boules, J. Cryst. Growth **279**, 13–19 (2005)

37.36 V. Noveski, R. Schlesser, S. Mahajan, S. Beaudoin, Z. Sitar: Growth of AlN crystals on AlN/SiC seeds by AlN powder sublimation in nitrogen atmosphere, MRS Internet J. Nitride Semicond. Res. **9**, 2 (2004)

37.37 B. Wu, R.H. Ma, H. Zhang, M. Dudley, R. Schlesser, Z. Sitar: Growth kinetics and thermal stress in AlN bulk crystal growth, J. Cryst. Growth **253**, 326–339 (2003)

37.38 E. Silveira, J.A. Freitas, G.A. Slack, L.J. Schowalter, M. Kneissl, D.W. Treat, N.M. Johnson: Depth-resolved cathodoluminescence of a homoepitaxial AlN thin film, J. Cryst. Growth **281**, 188–193 (2005)

37.39 J.C. Rojo, G.A. Slack, K. Morgan, B. Raghothamachar, M. Dudley, L.J. Schowalter: Report on the growth of bulk aluminum nitride and subsequent substrate preparation, J. Cryst. Growth **231**, 317–321 (2001)

37.40 L.J. Schowalter, J.C. Rojo, N. Yakolev, Y. Shusterman, K. Dovidenko, R.J. Wang, I. Bhat, G.A. Slack: Preparation and characterization of single-crystal aluminum nitride substrates, MRS Internet J. Nitride Semicond. Res. **5**, W6.7 (2000)

37.41 G. Koley, M.G. Spencer: Scanning Kelvin probe microscopy characterization of dislocations in III-nitrides grown by metalorganic chemical vapor deposition, Appl. Phys. Lett. **78**, 2873–2875 (2001)

37.42 I. Jenkins, K.G. Irvine, M.G. Spencer, V. Dmitriev, N. Chen: Growth of solid-solutions of aluminum nitride and silicon-carbide by metalorganic chemical vapor-deposition, J. Cryst. Growth **128**, 375–378 (1993)

37.43 K. Wongchotigul, N. Chen, D.P. Zhang, X. Tang, M.G. Spencer: Low resistivity aluminum nitride: Carbon (AlN:C) films grown by metal organic chemical vapor deposition, Mater. Lett. **26**, 223–226 (1996)

37.44 C.M. Zetterling, M. Ostling, K. Wongchotigul, M.G. Spencer, X. Tang, C.I. Harris, N. Nordell, S.S. Wong: Investigation of aluminum nitride grown by metal-organic chemical-vapor deposition on silicon, J. Appl. Phys. **82**, 2990–2995 (1997)

37.45 K. Wongchotigul, S. Wilson, C. Dickens, J. Griffin, X. Tang, M.G. Spencer: Growth of aluminum nitride with superior optical and morphological properties. In: *Silicon Carbide, III-Nitrides and Related*

37.46 H. Hirayama: Quaternary InAlGaN-based high-efficiency ultraviolet light-emitting diodes, J. Appl. Phys. **97**, 091101 (2005)

37.47 S. Krukowski, C. Skierbiszewski, P. Perlin, M. Leszczynski, M. Bockowski, S. Porowski: Blue and UV semiconductor lasers, Acta Phys. Polonica B **37**, 1265–1312 (2006)

37.48 T.P. Chow, R. Tyagi: Wide bandgap compound semiconductors for superior high-voltage unipolar power devices, IEEE Trans. Electron. Dev. **41**, 1481–1483 (1994)

37.49 S.J. Pearton, C.R. Abernathy, B.P. Gila, A.H. Onstine, M.E. Overberg, G.T. Thaler, J. Kim, B. Luo, R. Mehandru, F. Ren, Y.D. Park: Recent advances in gate dielectrics and polarised light emission from GaN, Optoelectron. Rev. **10**, 231–236 (2002)

37.50 S.H. Kim, J.H. Ko, S.H. Ji, Y.S. Yoon: Crystallinity effect of AlN thin films on the frequency response of an AlN/IDT/Si surface acoustic wave device, J. Korean Phys. Soc. **49**, 199–202 (2006)

37.51 K. Katahira, H. Ohmori, Y. Uehara, M. Azuma: ELID grinding characteristics and surface modifying effects of aluminum nitride (AlN) ceramics, Int. J. Mach. Tool. Manuf. **45**, 891–896 (2005)

37.52 Y.G. Gao, D.A. Gulino, R. Higgins: Effects of susceptor geometry on AlN growth on Si(111) with a new MOCVD reactor, MRS Internet J. Nitride Semicond. Res. **4S1**, G3.53 (1999)

37.53 M. Morita, S. Isogai, N. Shimizu, K. Tsubouchi, N. Mikoshiba: Aluminum nitride epitaxially grown on silicon – orientation relationships, Jpn. J. Appl. Phys. **20**, L173–L175 (1981)

37.54 J.D. Brown, R. Borges, E. Piner, A. Vescan, S. Singhal, R. Therrien: AlGaN/GaN HFETs fabricated on 100 mm GaN on silicon (111) substrates, Solid-State Electron. **46**, 1535–1539 (2002)

37.55 K. Kim, S.K. Noh: Reactor design rules for GaN epitaxial layer growths on sapphire in metal-organic chemical vapour deposition, Semicond. Sci. Technol. **15**, 868–874 (2000)

37.56 R.P. Parikh, R.A. Adomaitis: An overview of gallium nitride growth chemistry and its effect on reactor design: application to a planetary radial-flow CVD system, J. Cryst. Growth **286**, 259–278 (2006)

37.57 W.E. Hoke, A. Torabi, R.B. Hallock, J.J. Mosca, T.D. Kennedy: Reaction of molecular beam epitaxial grown AlN nucleation layers with SiC substrates, J. Vac. Sci. Technol. B **24**, 1500–1504 (2006)

37.58 R.M. Feenstra, Y. Dong, C.D. Lee, J.E. Northrup: Recent developments in surface studies of GaN and AlN, J. Vac. Sci. Technol. B **23**, 1174–1180 (2005)

37.59 S.H. Cho, K. Hata, T. Maruyama, K. Akimoto: Optical and structural properties of GaN films grown on c-plane sapphire by ECR-MBE, J. Cryst. Growth **173**, 260–265 (1997)

37.60 A. Usui, H. Sunakawa, A. Sakai, A.A. Yamaguchi: Thick GaN epitaxial growth with low dislocation density by hydride vapor phase epitaxy, Jpn. J. Appl. Phys. Part 2 – Letters **36**, L899–L902 (1997)

37.61 A. Dollet, Y. Casaux, M. Matecki, R. Rodriguez-Clemente: Chemical vapour deposition of polycrystalline AlN films from $AlCl_3 - NH_3$ mixtures: II – surface morphology and mechanisms of preferential orientation at low-pressure, Thin Solid Films **406**, 118–131 (2002)

37.62 M. Callahan, B.G. Wang, K. Rakes, D.F. Bliss, L. Bouthillette, M. Suscavage, S.Q. Wang: GaN single crystals grown on HVPE seeds in alkaline supercritical ammonia, J. Mater. Sci. **41**, 1399–1407 (2006)

37.63 Y. Kumagai, T. Yamane, A. Koukitu: Growth of thick AlN layers by hydride vapor-phase epitaxy, J. Cryst. Growth **281**, 62–67 (2005)

37.64 M. Tanaka, S. Nakahata, K. Sogabe, H. Nakata, M. Tobioka: Morphology and x-ray diffraction peak widths of aluminum nitride single crystals prepared by the sublimation method, Jpn. J. Appl. Phys. Part 2 – Letters **36**, L1062–L1064 (1997)

37.65 C.M. Balkas, Z. Sitar, T. Zheleva, L. Bergman, R. Nemanich, R.F. Davis: Sublimation growth and characterization of bulk aluminum nitride single crystals, J. Cryst. Growth **179**, 363–370 (1997)

37.66 S. Kurai, K. Nishino, S. Sakai: Nucleation control in the growth of bulk GaN by sublimation method, Jpn. J. Appl. Phys. Part 2-Letters **36**, L184–L186 (1997)

37.67 Y. Naoi, K. Kobatake, S. Kurai, K. Nishino, H. Sato, M. Nozaki, S. Sakai, Y. Shintani: Characterization of bulk GaN grown by sublimation technique, J. Cryst. Growth **190**, 163–166 (1998)

37.68 I.K. Shmagin, J.F. Muth, J.H. Lee, R.M. Kolbas, C.M. Balkas, Z. Sitar, R.F. Davis: Optical metastability in bulk GaN single crystals, Appl. Phys. Lett. **71**, 455–457 (1997)

37.69 L.J. Schowalter, J.C. Rojo, G.A. Slack, Y. Shusterman, R. Wang, I. Bhat, G. Arunmozhi: Epitaxial growth of AlN and $Al_{0.5}Ga_{0.5}N$ layers on aluminum nitride substrates, J. Cryst. Growth **211**, 78–81 (2000)

37.70 D. Zhuang, Z.G. Herro, R. Schlesser, Z. Sitar: Seeded growth of AlN single crystals by physical vapor transport, J. Cryst. Growth **287**, 372–375 (2006)

37.71 W.J. Mecouch: Preparation and characterization of thin, tomically clean GaN(0001) and AlN(0001) films and the depostion of thick GaN films via iodine vapor phase growth, Mater. Sci. Eng. Dep. (North Carolina State Univ. 2005)

37.72 D. Cai, W.J. Mecouch, L.L. Zheng, H. Zhang, Z. Sitar: Thermodynamic and kinetic study of transport and reaction phenomena in gallium nitride epitaxy growth, Int. J. Heat Mass Transf. **51**, 1264–1280 (2008)

37.73 R.J. Molnar, W. Gotz, L.T. Romano, N.M. Johnson: Growth of gallium nitride by hydride vapor-phase epitaxy, J. Cryst. Growth **178**, 147–156 (1997)

37.74 T. Paskova, E.M. Goldys, B. Monemar: Hydride vapour-phase epitaxy growth and cathodoluminescence characterisation of thick GaN films, J. Cryst. Growth **203**, 1–11 (1999)

37.75 K. Ohno, K. Esfarjani, Y. Kawazoe: *Computational Materials Science: From ab Initio to Monte Carlo Methods* (Springer, Berlin 1999)

37.76 W. Hergert, A. Ernst, M. Däne: *Computational Materials Science: From Basic Principles to Material Properties* (Springer, Berlin 2004)

37.77 D. Raabe: *Computational Materials Science: The Simulation of Materials Microstructures and Properties* (Wiley-VCH, Weinheim 1998)

37.78 C.R.A. Catlow: *Computational Materials Science* (NATO Science, Amsterdam 2003)

37.79 M.A. Gosálvez, R.M. Nieminen: Target-rate kinetic Monte Carlo method for simulation of nanoscale structural evolution, J. Comput. Theor. Nanosci. **1**, 303–308 (2004)

37.80 C.C. Battaile, D.J. Srolovitz, J.E. Butler: A kinetic Monte Carlo method for the atomic-scale simulation of chemical vapor deposition: application to diamond, J. Appl. Phys. **82**, 6293–6300 (1997)

37.81 C.C. Battaile, D.J. Srolovitz: Kinetic Monte Carlo simulation of chemical vapor deposition, Annu. Rev. Mater. Res. **32**, 297–319 (2002)

37.82 C. Cavallotti, A. Barbato, A. Veneroni: A combined three-dimensional kinetic Monte Carlo and quantum chemistry study of the CVD of Si on Si(100) surfaces, J. Cryst. Growth **266**, 371–380 (2004)

37.83 C. Cavallotti, M. DiStanislao, D. Moscatelli, A. Veneroni: Materials computation towards technological impact: The multiscale approach to thin films deposition, Electrochem. Acta **50**, 4566–4575 (2005)

37.84 Y. Akiyama: Modeling thermal CVD, J. Chem. Eng. Jpn. **35**, 701–713 (2002)

37.85 L.C. Musson, P. Ho, S.J. Plimpton, R.C. Schmidt: Feature length-scale modeling of LPCVD and PECVD MEMS fabrication processes, Microsystem Technol. Micro. Nanosyst. Info. Storage Proc. Syst. **12**, 137–142 (2005)

37.86 H.A. Al-Mohssen, N.G. Hadjiconstantinou: Arbitrary-pressure chemical vapor deposition modeling using direct simulation Monte Carlo with nonlinear surface chemistry, J. Comput. Phys. **198**, 617–627 (2004)

37.87 K.F. Jensen: Transport phenomena in vapor phase epitaxy reactors. In: *Handbook of Crystal Growth*, ed. by F.J.T. Hurle (Elsevier, Amsterdam 1994) pp. 541–

37.88 C.R. Kleijn: Chemical vapor deposition processes. In: *Computational Modeling in Semiconductor Processing*, ed. by M. Meyyappan (Artech House, Northwood 1995) pp. 97–216

37.89 A.A. Schmidt, V.S. Kharlamov, K.L. Safonov, Y.V. Trushin, E.E. Zhurkin, V. Cimalla, O. Ambacher, J. Pezoldt: Growth of three-dimensional SiC clusters on Si modelled by KMC, Comput. Mater. Sci. **33**, 375–381 (2005)

37.90 L.H. Liu, J.H. Edgar: Transport effects in the sublimation growth of aluminum nitride, J. Cryst. Growth **220**, 243–253 (2000)

37.91 L.H. Liu, J.H. Edgar: A global growth rate model for aluminum nitride sublimation, J. Electrochem. Soc. **149**, G12–G15 (2002)

37.92 S.Y. Karpov, V.G. Prokofyev, E.V. Yakovlev, R.A. Talalaev, Y.N. Makarov: Novel approach to simulation of group-III nitrides growth by MOVPE, MRS Internet J. Nitride Semicond. Res. **4**, 4 (1999)

37.93 S.Y. Karpov, R.A. Talalaev, Y.N. Makarov, N. Grandjean, J. Massies, B. Damilano: Surface kinetics of GaN evaporation and growth by molecular-beam epitaxy, Surf. Sci. **450**, 191–203 (2000)

37.94 S.A. Safvi, N.R. Perkins, M.N. Horton, R. Matyi, T.F. Kuech: Effect of reactor geometry and growth parameters on the uniformity and material properties of GaN/sapphire grown by hydride vapor-phase epitaxy, J. Cryst. Growth **182**, 233–240 (1997)

37.95 S.A. Safvi, J.M. Redwing, M.A. Tischler, T.F. Kuech: GaN growth by metalorganic vapor phase epitaxy – a comparison of modeling and experimental measurements, J. Electrochem. Soc. **144**, 1789–1796 (1997)

37.96 C. Theodoropoulos, T.J. Mountziaris, H.K. Moffat, J. Han: Design of gas inlets for the growth of gallium nitride by metalorganic vapor phase epitaxy, J. Cryst. Growth **217**, 65–81 (2000)

37.97 E. Aujol, J. Napierala, A. Trassoudaine, E. Gil-Lafon, R. Cadoret: Thermodynamical and kinetic study of the GaN growth by HVPE under nitrogen, J. Cryst. Growth **222**, 538–548 (2001)

37.98 A. Dollet, Y. Casaux, G. Chaix, C. Dupuy: Chemical vapour deposition of polycrystalline A1N films from $AlCl_3 - NH_3$ mixtures: analysis and modelling of transport phenomena, Thin Solid Films **406**, 1–16 (2002)

37.99 B. Wu, R.H. Ma, H. Zhang, V. Prasad: Modeling and simulation of AlN bulk sublimation growth systems, J. Cryst. Growth **266**, 303–312 (2004)

37.100 B. Wu, H. Zhang: Isotropic and anisotropic growth models for the sublimation vapour transport process, Model. Simul. Mater. Sci. Eng. **13**, 861–873 (2005)

37.101 D. Cai, L.L. Zheng, H. Zhang, V.L. Tassev, D.F. Bliss: Modeling of gas phase and surface reactions in an aluminum nitride growth system, J. Cryst. Growth **293**, 136–145 (2006)

37.102 A. Dollet: Multiscale modeling of CVD film growth – a review of recent works, Surf. Coat. Technol. **177**, 245–251 (2004)

37.103 A.V. Vasenkov, A.I. Fedoseyev, V.I. Kolobov, H.S. Choi, K.H. Hong, K. Kim, J. Kim, H.S. Lee,

37.103 J.K. Shin: Computational framework for modeling of multi-scale processes, J. Comput. Theor. Nanosci. **3**, 453–458 (2006)
37.104 S.T. Rodgers, K.F. Jensen: Multiscale modeling of chemical vapor deposition, J. Appl. Phys. **83**, 524–530 (1998)
37.105 K.F. Jensen, S.T. Rodgers, R. Venkataramani: Multiscale modeling of thin film growth, Curr. Opin. Solid State Mater. Sci. **3**, 562–569 (1998)
37.106 H.N.G. Wadley, A.X. Zhou, R.A. Johnson, M. Neurock: Mechanisms, models and methods of vapor deposition, Prog. Mater. Sci. **46**, 329–377 (2001)
37.107 W.M. Roshsenow, J.P. Hartnett: *Handbook of Heat Tranfer* (McGraw-Hill, New York 1975)
37.108 P. Cheng: 2-dimensional radiating gas flow by a moment method, AIAA Journal **2**, 1662–1664 (1964)
37.109 R. Siegel, J.R. Howell: *Thermal Radiation Heat Transfer* (Hemisphere, Washington 1992)
37.110 M.F. Modest: *Radiative Heat Transfer* (McGraw-Hill, New York 1993)
37.111 M.H.N. Naraghi, J.C. Huan: An N-bounce method for analysis of radiative-transfer in enclosures with anisotropically scattering media, J. Heat Transf. Trans. ASME **113**, 774–777 (1991)
37.112 G.D. Raithby, E.H. Chui: A finite-volume method for predicting a radiant-heat transfer in enclosures with participating media, J. Heat Transf. Trans. ASME **112**, 415–423 (1990)
37.113 M.G. Carvalho, T. Farias, P. Fotes: Predicting radiative heat transfer in absorbing, emmiting, scattering media using the discrete transfer method. In: *Fundamentals of Radiation Heat Transfer*, ed. by W.A. Fiveland (ASME HTD, New York 1991) pp. 17–26
37.114 L.S. Yao: Free and forced-convection in the entry region of a heated vertical channel, Int. J. Heat Mass Transf. **26**, 65–72 (1983)
37.115 T.K. Sherwood, R.J. Pigford, C.R. Wilke: *Mass Transfer* (McGraw-Hill, New York 1975)
37.116 A. Bejan: *Convective Heat Transfer* (Wiley-Interscience, Hoboken 2004)
37.117 O. Knacke, O. Kubaschewski, K. Hesselmann: *Thermochemical Properties of Inorganic Substances I* (Springer, Berlin 1991)
37.118 I. Barin: *Thermochemical Data of Pure Substances* (VCH, Weinheim 1989)
37.119 A.A. Chernov: *Modern Crystallography III. Crystal Growth* (Springer, Berlin 1984)
37.120 W.J. Mecouch: Preparation and characterization of thin, atomically clean GaN(0001) and AlN(0001) films and the depostion of thick GaN films via iodine vapor phase growth, Ph. D. Thesis (North Carolina State Univ. 2005)
37.121 R. Cadoret, E. Gil-Lafon: GaAs growth mechanisms of exact and isoriented {001} faces by the chloride method in H_2: Surface diffusion, spiral growth, HCl and $GaCl_3$ desorption mechanisms, J. Phys. I France **7**, 889–907 (1997)
37.122 D.F. Davidson, K. Kohse-Hoinghaus, A.Y. Chang, R.K. Hanson: A pyrolysis mechanism for ammonia, Int. J. Chem. Kinetics **22**, 513–535 (1990)
37.123 D.R. Stull: Vapor pressure of pure substances – organic compounds, Ind. Eng. Chem. **39**, 517–540 (1947)
37.124 R.F. Rolsten: *Iodide Metals and Metal Iodides* (Wiley, New York 1961)
37.125 A. Shintani, S. Minagawa: Kinetics of epitaxial-growth of GaN using Ga, HCl and NH_3, J. Cryst. Growth **22**, 1–5 (1974)
37.126 D.W. Shaw: Influence of substrate temperature on gaas epitaxial deposition rates, J. Electrochem. Soc. **115**, 405 (1968)
37.127 M. Mesrine, N. Grandjean, J. Massies: Efficiency of NH_3 as nitrogen source for GaN molecular beam epitaxy, Appl. Phys. Lett. **72**, 350–352 (1998)
37.128 D.J. Srolovitz, A. Mazor, B.G. Bukiet: Analytical and numerical modeling of columnar evolution in thin-films, J. Vac. Sci. Technol. A **6**, 2371–2380 (1988)
37.129 R. Collazo, R. Dalmau, Z. Herro, D. Zhuang, Z. Sitar: Is HVPE fundamentally inferior to PVT of AlN, ONR MURI: III–nitride crystal growth and wafering meeting (Arizona 2006)
37.130 O. Kovalenkov, V. Soukhoveev, V. Ivantsov, A. Usikov, V. Dmitriev: Thick AlN layers grown by HVPE, J. Cryst. Growth **281**, 87–92 (2005)

38. Continuum-Scale Quantitative Defect Dynamics in Growing Czochralski Silicon Crystals

Milind S. Kulkarni

The vast majority of modern microelectronic devices are built on monocrystalline silicon substrates produced from crystals grown by the Czochralski (CZ) and float-zone (FZ) processes. Silicon crystals inherently contain various crystallographic imperfections known as microdefects that often affect the yield and performance of many devices. Hence, quantitative understanding and control of the formation and distribution of microdefects in silicon crystals play a central role in determining the quality of silicon substrates. These microdefects are primarily aggregates of intrinsic point defects of silicon (vacancies and self-interstitials) and oxygen (silicon dioxide). The distribution of microdefects in a CZ crystal is determined by the complex dynamics, influenced by various reactions involving the intrinsic point defects and oxygen, and their transport. The distribution of these microdefects can also be strongly influenced and controlled by the addition of impurities such as nitrogen to the crystal. In this chapter, significant developments in the field of defect dynamics in growing CZ and FZ crystals are reviewed. The breakthrough discovery of the *initial point defect incorporation* in the vicinity of the melt–crystal interface, made in the early 1980s, allows a simplified quantification of CZ and FZ defect dynamics. Deeper insight into the formation and growth of microdefects was provided in the last decade by various treatments of the aggregation of oxygen and the intrinsic point defects of silicon. In particular, rigorous quantification of the aggregation of intrinsic point defects using the classical nucleation theory, a recently developed lumped model that captures the microdefect distribution by representing the actual population of microdefects by an equivalent population of identical microdefects, and another rigorous treatment involving the Fokker–Planck equations are discussed in detail.

The industrially significant dynamics of growing CZ crystals free of large microdefects is also reviewed. Under the conditions of large microdefect-free growth, a moderate vacancy supersaturation develops in the vicinity of the lateral surface of a growing crystal, leading to the formation of oxygen clusters and small voids, at lower temperatures. The vacancy incorporation near the lateral surface of a crystal, or the *lateral incorporation of vacancies*, is driven by the interplay among the Frenkel reaction, the diffusion of the intrinsic point defects, and their convection.

A review of CZ defect dynamics with a particular focus on the growth of large microdefect-free crystals is presented and discussed.

38.1 The Discovery of Microdefects 1283

38.2 Defect Dynamics in the Absence
ของ Impurities 1284
    38.2.1 The Theory
        of the Initial Incorporation
        of Intrinsic Point Defects 1284
    38.2.2 The Quantification
        of the Microdefect Formation 1290

38.3 Czochralski Defect Dynamics
in the Presence of Oxygen 1304
    38.3.1 Reactions in Growing CZ Crystals... 1304
    38.3.2 The Model 1305
    38.3.3 Defect Dynamics in
        One-Dimensional Crystal Growth.. 1308
    38.3.4 Defect Dynamics in
        Two-Dimensional Crystal Growth . 1310

38.4 Czochralski Defect Dynamics
in the Presence of Nitrogen.................... 1313
    38.4.1 The Model 1313
    38.4.2 CZ Defect Dynamics in
        One-Dimensional Crystal Growth.. 1316

38.4.3 CZ Defect Dynamics in
Two-Dimensional Crystal Growth . 1318
38.5 **The Lateral Incorporation of Vacancies
in Czochralski Silicon Crystals** 1321
    38.5.1 General Defect Dynamics:
A Brief Revisit 1322
    38.5.2 Defect Dynamics Under Highly
Vacancy-Rich Conditions 1323
    38.5.3 Defect Dynamics
Near the Critical Condition........... 1324

38.6 **Conclusions**.. 1328
    38.6.1 CZ Defect Dynamics
in the Absence of Impurities........ 1329
    38.6.2 CZ Defect Dynamics
in the Presence of Oxygen 1330
    38.6.3 CZ Defect Dynamics
in the Presence of Nitrogen 1330
    38.6.4 The Lateral Incorporation
of Vacancies 1331

References 1332

In modern microelectronics industry, the majority of devices are fabricated on silicon substrates produced from silicon crystals grown by the Czochralski (CZ) process. The float-zone (FZ) process is also used to produce a small fraction of the silicon substrates used in the modern industry. In the CZ process, a silicon crystal is continuously pulled from a silicon melt placed in a quartz crucible, as shown in Fig. 38.1. A CZ crystal has a conical top called the *crown*, a conical bottom called the *endcone*, and a cylindrical section called the *body*, which provides substrates for the fabrication of devices. In the FZ process, a molten silicon zone is allowed to form and solidify along the entire length of a polycrystalline silicon body to form a silicon single crystal. Silicon crystals grown prior to the late 1950s contained thermomechanically induced dislocations. A breakthrough discovery by *Dash* in the late 1950s allowed crystal growth free of these dislocations [38.2, 3]. In modern crystals, free of thermomechanically induced dislocations, various crystallographic imperfections known as microdefects, aggregated defects, and grown-in bulk defects can form and grow [38.4–15]. Microdefects vary in size from a few nanometers to > 200 nm. Microdefects of this size significantly affect the performance of the continuously shrinking modern microelectronic devices. This chapter addresses key developments in the crystal growth industry in the understanding and quantification of the physics of the formation of microdefects, popularly termed *defect dynamics*.

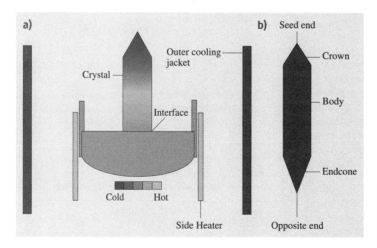

Fig. 38.1 (a) A schematic representation of the Czochralski (CZ) crystal growth process, (b) a grown CZ crystal (after [38.1])

38.1 The Discovery of Microdefects

Various types of microdefects are formed in CZ and FZ crystals. However, two types of microdefects commonly occur and hence are of critical significance. These are the aggregates of the intrinsic point defects of silicon: self-interstitials and vacancies. Vacancies are formed by silicon atoms missing from the silicon crystal lattice, whereas self-interstitials are silicon atoms that are not part of the lattice. Self-interstitial aggregates in FZ crystals were first observed in the 1960s and were termed *A defect swirls* (*A defects*) and *B defect swirls* (*B defects*), although the origin of these defects was not clearly known [38.4, 6–9]. Later, A defects were identified as dislocation loops and B defects were presumed to be globular self-interstitial aggregates [38.10–13]. In the 1980s, vacancy aggregates in FZ crystals were first reported and later termed *D defects* [38.14, 15]. More recent studies have identified D defects as octahedral voids [38.17, 18].

The discovery of the described microdefects was followed by various attempts to explain their formation. The dependence of the type of microdefects formed on the crystal growth rate was reported in various papers [38.6, 8, 14, 15]. A and B defects were observed at lower growth rates while D defects were observed at higher growth rates (Fig. 38.2). A unifying and acceptable analysis of the microdefect distribution in silicon crystals, however, was not available until 1982, when *Voronkov* provided the first groundbreaking explanation of defect dynamics in both FZ and CZ growth [38.19].

Various impurities present in a silicon crystal can influence the defect dynamics. CZ crystals are grown

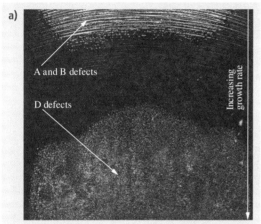

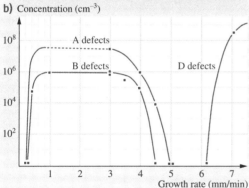

Fig. 38.2 (a) X-ray transmission topograph of a copper-decorated longitudinal section of a pedestal-pulled crystal, showing the dependence of the microdefect type on the growth rate. The growth rate varies from 4 to 7 mm/min from top to bottom. Note that the morphological details of microdefects are not revealed by the decoration of microdefects by the precipitation of copper. (b) The dependence of the type and density of microdefects on the crystal growth rate (after [38.14], © Elsevier 1981)

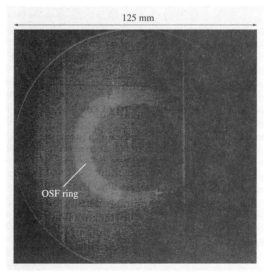

Fig. 38.3 An x-ray topograph (positive image) of a wafer showing the annular region containing OSFs. The wafer was sliced from a CZ crystal and treated at 1100 °C for 1 h in steam (after [38.16] © Jpn. J. Appl. Phys.)

from a melt contained in a quartz crucible, which is a source of oxygen. Hence, CZ crystals contain oxygen, unlike FZ crystals. Oxygen in CZ crystals typically precipitates as particles of silicon dioxide, which are popularly termed *oxygen precipitates* or *oxygen clusters*. Oxygen clusters in a growing CZ crystal are typically very small. These clusters facilitate the formation of stacking faults in crystals subjected to selective heat treatments that generate self-interstitials. Hence, oxygen clusters are typically identified by these stacking faults, known as oxidation-induced stacking faults (OSFs) (Fig. 38.3) [38.16]. As microdefects in silicon substrates produced from CZ crystals can adversely affect the performance of devices built on them, the development of crystal growth processes that reduce the size of microdefects in growing crystals is of industrial significance. Various studies have shown that the microdefect distribution in CZ crystals can be influenced in the presence of nitrogen, in general, and that the size of D defects (voids) can be reduced, in particular [38.20]. Dopants such as boron also affect CZ defect dynamics [38.21–24].

38.2 Defect Dynamics in the Absence of Impurities

Considering the complexity of CZ defect dynamics, first CZ and FZ defect dynamics in the absence of impurities are discussed in this chapter, followed by CZ defect dynamics in the presence of nitrogen and oxygen.

38.2.1 The Theory of the Initial Incorporation of Intrinsic Point Defects

Voronkov was the first to provide a satisfactory and well-accepted explanation for the quality of the microdefect distributions observed in silicon crystals [38.19]. According to Voronkov's theory, an interplay between the transport of the intrinsic point defects in a growing crystal in the vicinity of the melt–crystal interface and the Frenkel reaction involving the mutual annihilation or the recombination of vacancies and self-interstitials plays a key role in determining the final microdefect distribution. Voronkov analyzed this interplay by quantifying the Frenkel reaction dynamics and the intrinsic point defect balances in a growing crystal. His groundbreaking work is briefly discussed below.

The Frenkel Reaction Dynamics in a Growing Crystal

The Frenkel reaction involves the reversible annihilation of vacancies and self-interstitials by their recombination and the spontaneous generation of vacancies and self-interstitials from silicon lattice atoms

$$i + v \rightleftharpoons Si, \tag{38.1}$$

where i is a self-interstitial, v is a vacancy, and Si is a silicon lattice atom. The net rate of recombination of vacancy and self-interstitial pairs is equal and opposite to the net rate of formation of self-interstitials or vacancies, which is given by

$$-r_i = -r_v = k_{i \leftrightarrow v}(C_i C_v - C_{i,e} C_{v,e}), \tag{38.2}$$

where r is the net rate of formation of any species per unit volume, C is the concentration of any species, and $k_{i \leftrightarrow v}$ is the reaction rate constant known as the recombination constant. The subscript "i" denotes self-interstitials, "v" vacancies, "e" equilibrium conditions, and "i ↔ v" the interaction between self-interstitials and vacancies. The intrinsic point defects are assumed to exist at equilibrium at the melt–crystal interface. The concentration of the intrinsic point defects, however, drops significantly in a crystal near the interface, as the decreasing temperature in the crystal facilitates rapid recombination of self-interstitials and vacancies. Hence, the intrinsic point defects diffuse from the interface into the crystal. In addition, the continuous growth of the crystal, or the crystal pulling, facilitates the transport of the intrinsic point defects by their convection. Compared with the timescale of the transport, the Frenkel reaction can be assumed to be instantaneous and, hence, in equilibrium. Thus, the following relationship remains valid

$$C_i C_v = C_{i,e} C_{v,e}. \tag{38.3}$$

The equilibrium concentrations of the intrinsic point defects are functions of the crystal temperature and are expressed as

$$C_{i,e} = C_{i,0} \exp\left(-\frac{E_{\text{form},i}}{k_B T}\right), \tag{38.4}$$

$$C_{v,e} = C_{v,0} \exp\left(-\frac{E_{\text{form},v}}{k_B T}\right), \tag{38.5}$$

where E_{form} is the formation energy (the subscript "form" denotes the formation energy) and k_B is the

Boltzmann constant. The temperature profile in a crystal, in the vicinity of the melt–crystal interface and at a fixed radial location, is reasonably represented by a linear approximation of the inverse of the temperature

$$\frac{1}{T} = \frac{1}{T_{s/l}} + \frac{1}{T_{s/l}^2} G z, \quad (38.6)$$

where T is the temperature, G is the magnitude of the axial temperature gradient at the interface at any radial location, and z is the distance from the interface. The subscript "s/l" represents the conditions at the interface. Using (38.6), (38.4) and (38.5) can be written as

$$C_{i,e} = C_{i,e}(T_{s/l}) \exp\left(-\frac{E_{form,i} G z}{k_B T_{s/l}^2}\right), \quad (38.7)$$

$$C_{v,e} = C_{v,e}(T_{s/l}) \exp\left(-\frac{E_{form,v} G z}{k_B T_{s/l}^2}\right). \quad (38.8)$$

Thus, (38.3) describing the Frenkel reaction equilibrium takes the form

$$C_i C_v = C_{i,e}(T_{s/l}) C_{v,e}(T_{s/l}) \exp\left(\frac{-2z}{l}\right), \quad (38.9)$$

where the *characteristic recombination length l* is expressed as

$$l = \frac{2 k_B T_{s/l}^2}{(E_{form,i} + E_{form,v}) G}. \quad (38.10)$$

The discussed Frenkel reaction dynamics can be coupled with the overall intrinsic point defect balance in a growing crystal to describe the microdefect distributions observed in CZ and FZ crystals.

The Intrinsic Point Defect Balance and the Initial Incorporation

Driven by the Frenkel reaction, the concentrations of both the intrinsic point defect species dramatically drop in the vicinity of the melt–crystal interface. This intrinsic point defect concentration drop, in turn, drives the diffusion of both species from the interface, where they exist at equilibrium, into the crystal. In addition, the intrinsic point defects are also transported by the physical growth of the crystal, with respect to a fixed coordinate system. Crystal growth at a fixed rate and through a fixed temperature field with respect to a fixed coordinate system can be assumed to take place under a steady state. The excess intrinsic point defect flux, defined as the difference between the flux of vacancies and the flux of self-interstitials, is not explicitly affected by the Frenkel reaction. This excess flux is fixed for a given crystal growth condition

$$j_{iv} = \left(-D_v \frac{dC_v}{dz} + V C_v\right) - \left(-D_i \frac{dC_i}{dz} + V C_i\right), \quad (38.11)$$

where j_{iv} is the excess intrinsic point defect flux, D is the diffusivity, and V is the magnitude of the axial crystal pull rate. The effects of radial diffusion are ignored in (38.11).

When the vacancy flux is greater than the self-interstitial flux or when the excess intrinsic point defect flux is positive, vacancies are the surviving dominant species in the growing crystal and self-interstitials are annihilated to very low concentrations, within a short distance from the melt–crystal interface. Vacancy supersaturation increases at lower temperatures to drive the formation of D defects. Vacancies are termed the *incorporated dominant intrinsic point defects* under this condition. When the self-interstitial flux is greater than the vacancy flux or when the excess intrinsic point defect flux is negative, self-interstitials are the dominant species in the crystal and vacancies are annihilated to very low concentrations, within a short distance from the interface. The crystal becomes supersaturated with self-interstitials, leading to the formation of B and A defects by the aggregation of self-interstitials at lower temperatures. Self-interstitials are termed the *incorporated dominant intrinsic point defects* under this condition.

When the vacancy flux is equal to the self-interstitial flux or when the excess intrinsic point defect flux is equal to zero, both the intrinsic point defect species remain in comparable concentrations in the growing crystal and annihilate each other to very low concentrations. Under this condition, defined as the *critical condition*, no detectable microdefects are formed in the crystal at any temperature. At the critical condition, (38.11) is satisfied when both the intrinsic point defect concentrations show the same dependence on z, if the diffusivities of both the species are fixed and temperature independent. This assumption is approximate but acceptable within a narrow range of temperatures close to the interface. Then, using (38.9), the intrinsic point defect concentrations in a crystal growing under the critical condition are defined as

$$C_i = C_{i,0} \exp\left(\frac{E_{form,i} + E_{form,v}}{2 k_B T_{s/l}} - \frac{E_{form,i}}{k_B T_{s/l}} - \frac{E_{form,i} + E_{form,v}}{2 k_B T}\right), \quad (38.12)$$

$$C_{\text{v}} = C_{\text{v},0} \exp\left(\frac{E_{\text{form,i}} + E_{\text{form,v}}}{2k_{\text{B}} T_{\text{s/l}}}\right.$$
$$\left.- \frac{E_{\text{form,v}}}{k_{\text{B}} T_{\text{s/l}}} - \frac{E_{\text{form,i}} + E_{\text{form,v}}}{2k_{\text{B}} T}\right). \quad (38.13)$$

Using (38.11–38.13), the critical condition is analytically derived as

$$(V/G)_{\text{c}} = \frac{E_{\text{form,i}} + E_{\text{form,v}}}{2k_{\text{B}} T_{\text{s/l}}^2}$$
$$\times \frac{D_{\text{i}}(T_{\text{s/l}}) C_{\text{i,e}}(T_{\text{s/l}}) - D_{\text{v}}(T_{\text{s/l}}) C_{\text{v,e}}(T_{\text{s/l}})}{C_{\text{v,e}}(T_{\text{s/l}}) - C_{\text{i,e}}(T_{\text{s/l}})}, \quad (38.14)$$

where the subscript "c" denotes the critical value at zero excess intrinsic point defect flux.

Using this analysis, Voronkov explained the dependence of the microdefect distribution in a crystal on its growth rate (Fig. 38.2). At very high V, the convection dominates the diffusion. He hypothesized that the flux of vacancies into the crystal is higher than the flux of self-interstitials when the species convection is relatively appreciable, because the concentration of vacancies is higher than the concentration of self-interstitials at the interface. The crystal remains vacancy rich as the temperature drops. At very low V, the diffusion dominates the convection; Voronkov hypothesized that the flux of self-interstitials is greater than the flux of vacancies when the species diffusion is relatively appreciable, because self-interstitials diffuse faster than vacancies at higher temperatures. Thus, self-interstitials become the dominant incorporated species within a short distance from the interface, while vacancies are effectively annihilated. The dominant incorporated species eventually nucleates to form the appropriate microdefects at lower temperatures.

The competition between the intrinsic point defect convection and the intrinsic point defect diffusion is quantified not just by the crystal pull rate, but by the ratio of the crystal pull rate to the magnitude of the axial temperature gradient V/G. The convection of the intrinsic point defects increases with the crystal pull rate. The diffusion flux of an intrinsic point defect species increases with the increasing magnitude of its concentration gradient, which is driven by the temperature gradient near the interface. Thus, vacancies become the dominant incorporated intrinsic point defect species at higher V/G and self-interstitials become the dominant incorporated species at lower V/G. At the *critical* V/G, the flux of vacancies is equal to the flux of self-interstitials and there is no dominance of either intrinsic point defect species as both species mutually annihilate each other to very low concentrations.

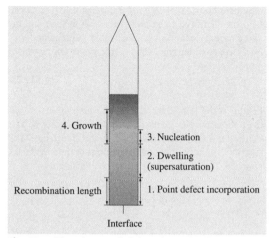

Fig. 38.4 The phases of defect dynamics in a growing crystal according to Voronkov's theory (not to scale) (after [38.25])

Table 38.1 Key properties of various species participating in reactions in growing CZ crystals and the recombination rate constant (after [38.1])

Property set I
D_{i} (cm^2/s) $= 0.19497 \exp\left(\dfrac{-0.9 \text{ (eV)}}{k_{\text{B}} T}\right)$
D_{v} (cm^2/s) $= 6.2617 \times 10^{-4} \exp\left(\dfrac{-0.4 \text{ (eV)}}{k_{\text{B}} T}\right)$
$C_{\text{i,e}}$ (cm^{-3}) $= 6.1859 \times 10^{26} \exp\left(\dfrac{-4.0 \text{ (eV)}}{k_{\text{B}} T}\right)$
$C_{\text{v,e}}$ (cm^{-3}) $= 7.59982 \times 10^{26} \exp\left(\dfrac{-4.0 \text{ (eV)}}{k_{\text{B}} T}\right)$
$k_{\text{i} \leftrightarrow \text{v}}$ (cm^3/s) $= 1.2 \times 10^{-6} [D_{\text{i}}(T) + D_{\text{v}}(T)]$ $\times \exp\left[-\dfrac{0.61 + \left(-2.30 + 7.38 \times 10^{-3} T\right) k_{\text{B}} T}{k_{\text{B}} T}\right]^{*}$
λ_{i} (eV) $= 2.95^{**}$
λ_{v} (eV) $= 1.85$

* Reported by *Sinno* et al. [38.26]; the enthalpic barrier (0.61) is ignored in the simulations discussed in this chapter [38.1, 27]. An accurate estimation of this parameter is not necessary, because the Frenkel reaction dynamics is fast
** For two-dimensional (2-D) simulations, a value of 2.85 eV is used

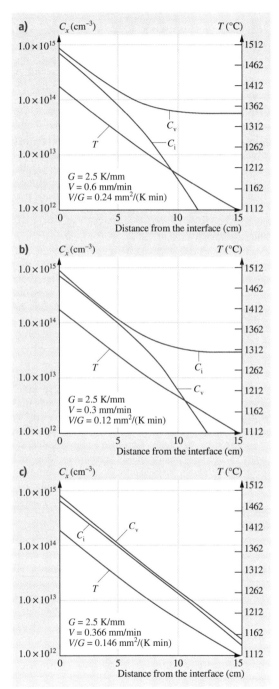

Fig. 38.5a–c The profiles of the vacancy concentration, the self-interstitial concentration, and the temperature in a Czochralski crystal growing at (**a**) a higher V/G, (**b**) a lower V/G, (**c**) close to the critical V/G (after [38.25])

The process of establishing the key intrinsic point defect concentration field in a growing CZ crystal in the vicinity of the melt–crystal interface is termed the *initial incorporation*. The initial incorporation takes place within a short distance from the interface, which scales with the characteristic recombination length l. When the operating V/G is either higher or lower than the critical V/G, the dominance of one intrinsic point defect species is established within the recombination length; when the operating V/G is closer to the critical V/G, however, both the intrinsic point defect species are incorporated in comparable concentrations and continue to recombine and annihilate each other beyond the recombination length, without forming large microdefects. A schematic representation of Voronkov's theory is shown in Fig. 38.4. Voronkov's theory can be verified by quantifying the intrinsic point defect concentration profiles ($C_x(z)$, x denoting either i or v) in a CZ crystal growing at various V/G, as shown in Fig. 38.5a–c. These computations were accomplished by assuming that both the diffusivities and the equilibrium concentrations of the intrinsic point defects vary with the temperature and by accounting for the Frenkel reaction kinetics [38.25]. The properties of the intrinsic point defects used in these simulations are listed in Table 38.1. There is significant uncertainty in the estimation of the properties of the intrinsic point defects. The parameters describing the Frenkel reaction kinetics, in particular, are not well known. An accurate estimation of the kinetic parameters is not necessary because the Frenkel reaction dynamics is very fast and, hence, reaction equilibrium prevails. It must be noted that the intrinsic point defect concentration profiles computed assuming the constant (fixed) intrinsic point defect diffusivities give slightly different results.

The evolution of the concentrations of the intrinsic point defects in a growing crystal strongly depends on the properties of the intrinsic point defects, especially the formation and migration energies. As shown by (38.12) and (38.13), at the critical V/G, the concentration of each intrinsic point defect species is equal to its equilibrium concentration at any axial location z, when the formation energy of a vacancy is equal to the formation energy of a self-interstitial, the migration energies of both species are zero, and the temperature profile in the crystal is given by (38.6). Deviations from these conditions establish the nonequilibrium intrinsic point defect concentrations in a crystal growing at the critical V/G. The dynamics of crystal growth near the critical condition and the effects of the properties of the intrinsic point defects on this dynamics are discussed in detail in Sect. 38.5.

The significance of the discussed discovery of the initial intrinsic point defect incorporation in growing silicon crystals extends beyond the field of CZ and FZ defect dynamics; it remains valuable in the prediction of the properties of the intrinsic point defects of silicon by requiring that the difference $C_{v,e}(T_{s/l}) - C_{i,e}(T_{s/l})$ and the difference $D_i(T_{s/l})C_{i,e}(T_{s/l}) - D_v(T_{s/l})C_{v,e}(T_{s/l})$ be positive.

Validation and Limits of the Theory of One-Dimensional Initial Incorporation

Voronkov's theory of the intrinsic point defect incorporation has been validated by many experimental observations reported before and after its publication, in spite of a few initial questions raised by *Tan* and *Gösele*, among others [38.28]. The dependence of the microdefect type in FZ crystals on the crystal growth rate was reported in [38.6, 8, 14, 15]. By varying the pull rate of various CZ crystals, *Sadamitsu* et al. showed that the microdefect quality in the crystals shifted from vacancy type (vacancy aggregates) to self-interstitial type (self-interstitial aggregates) [38.29]. This study also revealed a radial variation in the microdefect distribution. This was explained by the radial variation in the temperature field, or the radial variation of G. In a typical CZ crystal, G monotonically increases and V/G monotonically decreases along the radial position. Thus, the central regions in many CZ crystals exhibit D defects and the peripheral regions exhibit A defects, as determined by V/G. The narrow microdefect-free region between the region of vacancy aggregates and the region of self-interstitial aggregates is known as the v/i boundary. The microdefect distributions in CZ crystals showing the v/i boundary were reported by many in the last decade [38.30–33]

The surface of a growing CZ crystal acts as a source or a sink of the intrinsic point defects and, hence, induces their radial diffusion; in addition, the radial variation in the temperature field also causes radial diffusion of the intrinsic point defects. The microdefect distributions away from the surfaces in rapidly pulled crystals in which the radial diffusion effects can be ignored, however, are very well explained by the one-dimensional initial incorporation theory. In a silicon crystal, the region containing abundant vacancies or vacancy aggregates is termed the *v-rich region* and the region containing abundant self-interstitials or the self-interstitial aggregates is termed the *i-rich region*. The position of the boundary between the v-rich region and the i-rich region established after the initial incorporation remains essentially the same even after the aggregation of the intrinsic point defects in a rapidly pulled CZ crystal. Hence, by following the v/i boundary separating the vacancy aggregates and the self-interstitial aggregates, in such crystals, the critical V/G can be experimentally determined. Typically, a narrow annular region inside the v-rich region of a CZ crystal near the v/i boundary exhibits another type of microdefects known as oxidation-induced stacking faults (OSFs), after a treatment with selective heat cycles. These OSFs are formed by the oxides of silicon, formed during the crystal growth, in the regions grown at V/G slightly above the critical V/G (Figs. 38.3 and 38.6). The incorporated vacancies in this region are too low in concentration to nucleate at higher temperatures ($\approx 1100\,°C$) to form D defects during crystal growth; they survive at lower temperatures ($\approx 1000–850\,°C$) to facilitate the formation of silicon oxide (primarily silicon dioxide) particles. Oxygen required for this oxidation in CZ crystals comes from the quartz crucible used in the CZ process. The specific volume of an oxide particle is greater than the specific volume of silicon. Hence, the formation and growth of the oxide particles generates compressive stresses. Relief of this stress can take place by the consumption of vacancies and the ejection of self-interstitials from the silicon lattice. During crystal growth, the oxide particles in the OSF region are formed essentially by facilitation by vacancies. The growth of the oxide particles by the ejection of self-interstitials can take place after the depletion of free vacancies at lower temperatures. These compressed oxide particles in a silicon wafer facilitate the growth of stacking faults or OSFs after selective heat treatments that inject self-interstitials into the silicon wafer. *Hasebe* et al. were among the first to report the presence of the annular region of OSFs termed the *OSF ring* in CZ crystals (Fig. 38.3) [38.16]. Often the location of the *OSF ring*, because of its vicinity to the v/i boundary, is used to mark the critical V/G. The CZ defect dynamics in the presence of oxygen describing the formation and growth of oxygen clusters is discussed later in this chapter.

The radial diffusion of the intrinsic point defects in a CZ crystal, both driven by the variation in the crys-

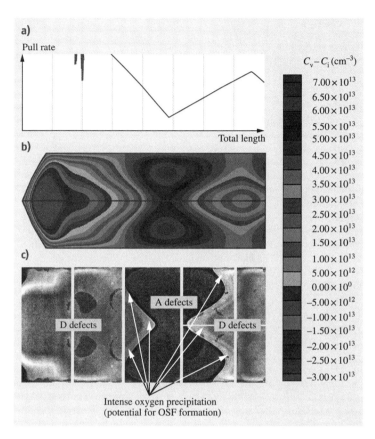

Fig. 38.6 (a) The pull-rate profile used to pull the experimental crystal. (b) The predicted excess intrinsic point defect concentration field in the crystal. (c) The experimentally observed microdefect distribution. The radial variation of G at the interface was more than 140%. Note: The simulation results are not completely mesh insensitive (after [38.25])

tal temperature field and induced by the lateral surface of the crystal, cannot be ignored in modern CZ growth. Thus, a more complete picture of the intrinsic point defect dynamics is described by considering the transport of the intrinsic point defects in an axisymmetric crystal growing through a temperature field at pseudo-steady state

$$\frac{\partial C_i}{\partial t} + V \frac{\partial C_i}{\partial z} = \nabla \cdot (D_i \nabla C_i)$$
$$- k_{i \leftrightarrow v}(C_i C_v - C_{i,e} C_{v,e}), \quad (38.15)$$

$$\frac{\partial C_v}{\partial t} + V \frac{\partial C_v}{\partial z} = \nabla \cdot (D_v \nabla C_v)$$
$$- k_{i \leftrightarrow v}(C_i C_v - C_{i,e} C_{v,e}). \quad (38.16)$$

Nakamura et al. and *Sinno* et al. were among the first to solve the two-dimensional intrinsic point defect dynamics in growing CZ crystals to clearly define the final microdefect distributions [38.26, 34].

The equations describing the intrinsic point defect distributions in a CZ crystal can be further simplified by assuming the Frenkel reaction equilibrium. The evolution of the excess intrinsic point defect concentration $C_v - C_i$ is described without an explicit use of the Frenkel reaction kinetics as

$$\frac{\partial (C_v - C_i)}{\partial t} + V \frac{\partial C_v}{\partial z} - V \frac{\partial C_i}{\partial z}$$
$$= \nabla \cdot (D_v \nabla C_v) - \nabla \cdot (D_i \nabla C_i). \quad (38.17)$$

The intrinsic point defect concentration field is then predicted by the solution of (38.17) and (38.3). The concentrations of the intrinsic point defects at the boundary of the crystal including the melt–crystal interface can be assumed to be at equilibrium. The excess intrinsic point defect concentration field can be mapped to the final microdefect distribution. A comparison between the predicted excess intrinsic point defect concentration field and the experimentally observed microdefect distribution in a crystal pulled by a varying rate in MEMC Electronic Materials is shown in Fig. 38.6 [38.25]. The crystal temperature field used in the simulation was first predicted by the commercial software MARC and then corrected using the experimentally measured interface [38.35–37]. The heat transport dynamics was assumed to be very fast, and the melt was assumed to be an effective solid [38.35–37]. The shape of the interface changes with the crystal pull rate. It is reasonable, however, to assume that the interface remains fixed when the pull rate variation is moderate or that there exists a locally fixed effective interface when the pull rate variation is significant. Typically, a representative interface measured in the center of the region of interest suffices. The simulation was performed using one such experimentally measured interface. The properties of the intrinsic point defects used in the simulation are listed in Table 38.1. Figure 38.6a shows the actual crystal pull rate profile, and Fig. 38.6b shows the predicted excess intrinsic point defect concentration field in the crystal. The positive excess intrinsic point defect concentration represents vacancy aggregates or D defects, and the negative excess intrinsic point defect concentration represents self-interstitial aggregates (A and B defects). The experimental microdefect distribution is determined by the method of copper decoration and Secco etching and is shown in Fig. 38.6c. The reader is referred to *Kulkarni* et al. for the details of the microdefect decoration by the employed characterization technique [38.38–40]. The excess intrinsic point defect concentration field shown in Fig. 38.6 is influenced by the varying pull rate, the temperature field in the crystal, and the radial intrinsic point defect diffusion, to some extent. The v/i boundary in the crystal is clearly identified. The intensity of the oxygen precipitation is very high very close to the v/i boundary, indicating the potential for OSF formation. In rapidly pulled crystals, the observed microdefect distribution, the location of the OSF region, and the location of the v/i boundary can be quantified by the radial variation of G and the axial variation of the pull rate. Several studies [38.30–33] have directly or indirectly quantified the location of the v/i boundary in CZ crystals. It must be noted that the simulation results shown in Fig. 38.6 are not completely mesh insensitive at the level of discretization used. This inaccuracy is corrected in the computations discussed in the later sections of this chapter.

38.2.2 The Quantification of the Microdefect Formation

The quantification of the intrinsic point defect concentration field in a CZ crystal provides only a qualitative information of the final microdefect distribution [38.19, 26, 34, 41–45]. As microelectronic devices are very sensitive to the size of microdefects, an accurate quantification of the microdefect distribution in CZ crystals is essential. Capturing the distribution of microdefects in a CZ crystal is quite rigorous and involves the solution of a set of complex equations. The formation and growth of microdefects in an element of silicon takes place over a finite period of time; thus, a population of the microdefects formed at various moments of the elapsed time period exists in the element. Hence, various approximations are made in capturing the entire microdefect distribution in a growing CZ crystal [38.1, 27, 46–54].

Based on the research reported in the literature, three broad approaches for the quantification of the microdefect distribution in a CZ crystal can be identified. The first approach involves the application of the classical nucleation theory for the formation of stable nuclei of microdefects followed by their diffusion-limited growth, the second approach involves representing the population of microdefects present at any location by an equivalent population of identical microdefects, and the third approach involves the application of the Fokker–Planck equation to create a continuum of the microdefect size. *Voronkov* and *Falster* were the first to publish the first approach [38.46]. For the sake of simplicity, however, they ignored axial and radial diffusion of the intrinsic point defects during the formation and growth of microdefects. Thus, in effect, the formation of microdefects was treated in an isolated element of silicon following a predetermined decrease in the temperature. *Kulkarni* et al. [38.1] augmented this model and quantified the CZ defect dynamics including the axial diffusion effects [38.1]. A detailed two-dimensional treatment of the CZ defect dynamics to capture the microdefect distribution by the approach of *Kulkarni* et al. [38.1] remains computationally unattractive. To address this shortcoming, *Kulkarni* and *Voronkov* [38.53] developed the *lumped model*, which represents a population of microdefects of varying sizes at any location in a CZ crystal by an equivalent population of identical microdefects and captures the two-dimensional distribution of microdefects.

The quantification of the CZ defect dynamics by a more rigorous approach can be accomplished by a direct treatment of the reactions between the nucleating monomers and microdefects of various sizes. This approach is computationally impractical, considering the size of mature microdefects. The Fokker–Planck equations (FPE) are popularly used to address the quantification of such systems. By the Fokker–Planck approach, a microdefect population at any location in a CZ crystal is quantified by a *size* coordinate. *Sinno* and *Brown* [38.47], *Mori* [38.27], *Wang* and *Brown* [38.48], and *Brown* et al. [38.49] accomplished the quantification of the defect dynamics in CZ crystals using the Fokker–Planck formulation.

For the sake of continuity, the quantifications of the defect dynamics in CZ crystals by the first and the second approach are discussed first in this chapter. The third approach, employing the Fokker–Planck equations, is discussed last.

First Approach: CZ Defect Dynamics Using the Classical Nucleation Theory

All microdefects are treated as clusters of the intrinsic point defects. The formation of microdefects takes place by a series of reactions of the following type [38.55–57]:

$$P_{m_x} + x \rightleftharpoons P_{(m+1)_x}, \quad \{x = i, v\}. \quad (38.18)$$

In reaction (38.18), P is a cluster of monomers or an intrinsic point defect species x. A cluster of m intrinsic point defects of type x is represented by P_{m_x}. At a given temperature and intrinsic point defect concentration, the total free energy change associated with the formation of a cluster containing m intrinsic point defects from a supersaturated solution is given by

$$\Delta F_x(m_x) = -m_x k_B T \ln \frac{C_x}{C_{x,e}} + \lambda_x m_x^{2/3}, \quad (38.19)$$

where F is the free energy and λ is the surface energy coefficient for the cluster. Note that the subscript x (for i and v) denotes the intrinsic point defect x (as in C_x) as well as the clusters containing the intrinsic point defect x (as in F_x and λ_x), depending on the variable. The first term on the right-hand side of (38.19) is the bulk (volume) free energy change (per m intrinsic point defects) associated with the intrinsic point defect supersaturation, and the second term is the cluster surface energy associated with the formation of the new cluster surface. The number of intrinsic point defects in the so-called *critical cluster* is obtained by maximizing the free energy change with respect to m

$$m_x^* = \left(\frac{2\lambda_x}{3k_B T \ln \frac{C_x}{C_{x,e}}}\right)^3. \quad (38.20)$$

The maximum free-energy change associated with the formation of the critical clusters containing m^* intrinsic point defects is interpreted as the nucleation barrier and is given by the substitution of m^* into (38.19)

$$\Delta F_x(m_x^*) = \Delta F_x^* = \frac{4}{27} \frac{\lambda_x^3}{\left(k_B T \ln \frac{C_x}{C_{x,e}}\right)^2}. \quad (38.21)$$

The classical nucleation theory gives the rate of formation of stable supercritical nuclei or clusters per unit volume, defined as the *nucleation rate*, as a function of various properties of the critical clusters and the intrinsic point defects as [38.57]

$$J_x = \eta_{x \leftrightarrow x}^* \times Z_x \times \phi_{x,e}^*, \quad (38.22)$$

where J_x is the nucleation rate of clusters of type x, or x-clusters, $\eta_{x \leftrightarrow x}^*$ is the attachment frequency of monomer x to the critical clusters of x, $\phi_{x,e}^*$ is the equilibrium density of the critical clusters, and Z_x is the Zheldovich factor. Clusters are assumed to be spherical in shape. The relevance and accuracy of this assumption are discussed in the following section. Using the classical expressions for the Zheldovich factor and the equilibrium concentration of the critical clusters, the nucleation rate is given as

$$J_x = \left[4\pi D_x C_x \times R_x\left(m_x^*\right)\right]$$
$$\times \left[\{12\pi k_B T \times \Delta F_x\left(m_x^*\right)\}^{-\frac{1}{2}} \left(k_B T \ln \frac{C_x}{C_{x,e}}\right)\right]$$
$$\times \left[\rho_x \exp\left(-\frac{\Delta F_x\left(m_x^*\right)}{k_B T}\right)\right]. \quad (38.23)$$

Here R is the radius (size) of a cluster, D is the diffusivity of any species, and ρ is the site density for nucleation. The first term in the square brackets on the right-hand side of (38.23) is the diffusion-limited intrinsic point defect attachment frequency to a critical cluster, the second term is the Zheldovich factor, and the third term is the equilibrium concentration of the critical clusters.

The Model. In this study, all microdefects are treated as spherical aggregates of either vacancies or self-interstitials and are termed *clusters*. Vacancy aggregates are termed v-clusters and represent D defects; self-interstitial aggregates are termed i-clusters and represent A and B defects. Since D defects are known to be octahedral voids, their approximation as spherical clusters is reasonably accurate [38.17, 18]. A defects are dislocation loops [38.10–13]. They presumably form, however, from globular self-interstitial aggregates such as B defects [38.12, 19]. Therefore, the approximation of A defects as spherical clusters provides their density with a representative accuracy; globular B defects are quite accurately approximated as spherical clusters. All clusters are assumed to be immobile. The effect of impurities is ignored in this treatment.

The microdefect distribution in a CZ crystal is symmetric about the axis of growth. Thus, it suffices to treat the CZ defect dynamics by an axisymmetric model using the cylindrical coordinates r and z, where r is the radial coordinate and z is the axial coordinate. At any location (r, z) in the crystal, at any given time t, a population of clusters formed at various locations (r, ξ) at various moments of the elapsed time τ exists. The size (radius) R of these clusters is then a function of r, z, τ, and t. The clusters, once formed, are assumed to grow by a diffusion-limited kinetics. The growth equations for i-clusters and v-clusters are thus given by

$$\frac{\partial R_i^2(r,z,\tau,t)}{\partial t} + V\frac{\partial R_i^2(r,z,\tau,t)}{\partial z}$$
$$= \frac{2D_i}{\psi_i^i}(C_i - C_{i,e}), \quad (38.24)$$

$$\frac{\partial R_v^2(r,z,\tau,t)}{\partial t} + V\frac{\partial R_v^2(r,z,\tau,t)}{\partial z}$$
$$= \frac{2D_v}{\psi_v^v}(C_v - C_{v,e}), \quad (38.25)$$

where ψ_x^x is the concentration of any intrinsic point defect species, denoted by the superscript, in a cluster of any type, denoted by the subscript. The relationship among z, ξ, and τ is determined by the pull rate profile

$$\xi = z - \int_\tau^t V \, d\tau', \quad (38.26)$$

where τ' is the time between τ and t. The *cross-interaction* between i-clusters and vacancies and between v-clusters and self-interstitials is ignored, because the cluster growth is affected primarily by the dominant intrinsic point defect species. The dynamics at moderately low temperatures is strongly influenced by oxygen, which is beyond the scope of this model. The rate of the consumption of the intrinsic point defects x by x-clusters at any location at any time, q_x^x, is obtained by integrating the contributions from all the clusters present at the location

$$q_i^i = 4\pi D_i (C_i - C_{i,e}) \int_0^t R_i(r,z,\tau,t) J_i(r,\xi,\tau) \, d\tau, \quad (38.27)$$

$$q_v^v = 4\pi D_v (C_v - C_{v,e}) \int_0^t R_v(r,z,\tau,t) J_v(r,\xi,\tau) \, d\tau. \quad (38.28)$$

The balance of the intrinsic point defects must account for the change in the intrinsic point defect concentration by convection, diffusion, the Frenkel reaction, their consumption by the diffusion-limited growth of the formed clusters, and their consumption by their nucleation.

$$\frac{\partial C_i}{\partial t} + V\frac{\partial C_i}{\partial z} = \nabla(D_i \nabla C_i)$$
$$- k_{i\leftrightarrow v}(C_i C_v - C_{i,e} C_{v,e})$$
$$- 4\pi D_i (C_i - C_{i,e})$$
$$\times \int_0^t R_i(r,z,\tau,t) J_i(r,\xi,\tau) \, d\tau$$
$$- J_i(r,z,t) m_i^*, \quad (38.29)$$

$$\frac{\partial C_v}{\partial t} + V\frac{\partial C_v}{\partial z} = \nabla(D_v \nabla C_v)$$
$$- k_{i\leftrightarrow v}(C_i C_v - C_{i,e} C_{v,e})$$
$$- 4\pi D_v (C_v - C_{v,e})$$
$$\times \int_0^t R_v(r,z,\tau,t) J_v(r,\xi,\tau) \, d\tau$$
$$- J_v(r,z,t) m_v^*. \quad (38.30)$$

The consumption of intrinsic point defects by their nucleation events only is negligible, and is ignored. The nucleation rates of both vacancies and self-interstitials are given by the classical nucleation theory

$$J_i = \left[4\pi R(m_i^*) D_i C_i\right]$$
$$\times \left[\{12\pi k_B T \Delta F_i(m_i^*)\}^{-\frac{1}{2}} \left(k_B T \ln \frac{C_i}{C_{i,e}}\right)\right]$$
$$\times \left[\rho_i \exp\left\{-\frac{\Delta F_i(m_i^*)}{k_B T}\right\}\right], \quad (38.31)$$

$$J_v = [4\pi R_x (m_x^*) D_v C_v]$$
$$\times \left[\{12\pi k_B T \Delta F_v(m_x^*)\}^{-\frac{1}{2}} \left(k_B T \ln \frac{C_v}{C_{v,e}} \right) \right]$$
$$\times \left[\rho_v \exp \left\{ -\frac{\Delta F_v(m_x^*)}{k_B T} \right\} \right]. \quad (38.32)$$

The discussed set of equations must be solved for a moving crystal. The transient domain of the computation is described by the shape of the crystal as a function of time

$$\Omega(r, z, t) = 0. \quad (38.33)$$

The initial height of the crystal is assumed to be either zero or negligible. Equilibrium conditions are assumed on all crystal surfaces, on the basis of a fast surface kinetics. This assumption is valid in most CZ growth conditions [38.26, 34]. The initial size of the clusters formed at any location (r, ξ) is approximately described by the size of the critical clusters.

The cumulative density of all x-clusters in a population present at any location, n_x, can be explicitly defined using the solution of the described equations

$$n_i = \int_0^t J_i(r, \xi, \tau) \mathrm{d}\tau, \quad (38.34)$$

$$n_v = \int_0^t J_v(r, \xi, \tau) \mathrm{d}\tau. \quad (38.35)$$

The volumetric average radius of the cluster population at any location is given by

$$R_{i,\text{avg,vol}} = \left(\frac{\int_0^t R_i^3 (r, z, \tau, t) J_i(r, \xi, \tau) \mathrm{d}\tau}{\int_0^t J_i(r, \xi, \tau) \mathrm{d}\tau} \right)^{1/3}, \quad (38.36)$$

$$R_{v,\text{avg,vol}} = \left(\frac{\int_0^t R_v^3 (r, z, \tau, t) J_v(r, \xi, \tau) \mathrm{d}\tau}{\int_0^t J_v(r, \xi, \tau) \mathrm{d}\tau} \right)^{1/3}. \quad (38.37)$$

The subscript "avg,vol" denotes the volumetric average value.

Typically, the described set of equations can be solved in a quasistationary temperature field at any given height of the crystal. The CZ heat transport dynamics is much faster than the CZ defect dynam-

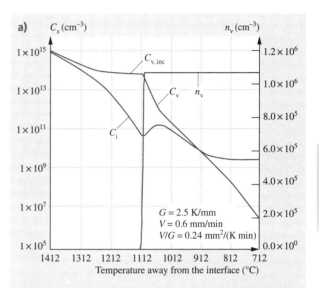

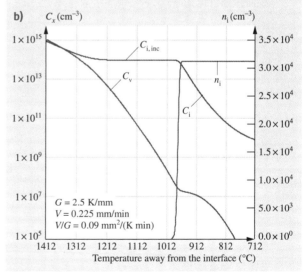

Fig. 38.7a,b The profiles of the concentrations of the intrinsic point defect species and the cluster densities as functions of the crystal temperature in a simulated (a) vacancy-rich crystal growing under steady state, (b) self-interstitial-rich crystal growing under steady state (after [38.1]). Note that the Frenkel reaction equilibrium remains valid until the cluster growth is complete and that the cross-interactions (i with v-clusters and v with i-clusters) are not included in the model

ics [38.26, 27, 34, 48, 49, 53, 54]. The temperature field in a growing crystal can be quantified accurately by the quasi-steady-state approximation [38.35–37]. In most cases, the assumption of one representative temperature field independent of the crystal height suffices [38.53, 54].

The accuracy of the model predictions strongly depends on the accuracy of the parameters describing the key properties of the intrinsic point defects. There is significant uncertainty in the reported values of the formation and migration energies of the intrinsic point defects. The parameters describing the Frenkel reaction kinetics, in particular, are highly approximate. Various studies have reported the acceptable parameters describing the CZ defect dynamics [38.1, 26, 34, 58–64]. In this section, the results obtained by *Kulkarni* et al. by solving the described equations using the properties of the intrinsic point defects listed in Table 38.1 are discussed [38.1]. The Frenkel reaction rate constant reported by *Sinno* et al. [38.26, 27] was used by *Kulkarni* and coworkers in their study [38.1]; the enthalpic barrier, however, was set to zero [38.1, 27]. An accurate estimation of this rate constant is not necessary, because the Frenkel reaction dynamics is very fast, leading to reaction equilibrium in the relevant temperature range.

It must be noted that the discussed model quantifies continuum-scale CZ defect dynamics. The effects of oxygen on the defect dynamics are also not included in the model.

Results and Discussion. *Voronkov* and *Falster* solved the described model by assuming that the effects of the axial and radial diffusion of the intrinsic point defects after the initial incorporation are negligible [38.46]. *Kulkarni* et al. solved the model describing both the steady-state and unsteady-state defect dynamics including axial diffusion effects [38.1]. The basic aspects of their study are discussed in this section.

Kulkarni et al. used a representative temperature profile described by (38.6) for their steady-state simulations [38.1]. A representative value for G of 2.5 K/mm was used. For the sake of simplicity, the units popularly applied in the crystal growth industry are used hereafter to describe key variables. Vacancies are incorporated as the dominant species at high V/G, as shown in Fig. 38.7a, and self-interstitials are the dominant incorporated species at low V/G, as shown in Fig. 38.7b. Figure 38.7 also shows that v-clusters and i-clusters are formed within a narrow range of temperature known as the *nucleation temperature range*. The nucleation temperature is defined as the temperature at which the nucleation rate is at its maximum. The predicted nucleation temperature of vacancies for the conditions studied is around 1100 °C and that for self-interstitials is around 950 °C. As reported by *Kulkarni*, the Frenkel reaction equilibrium prevails in the crystal until the cluster growth is complete [38.1]. At lower temperatures, when the intrinsic point defect concentrations are too low to affect the cluster distribution, the model predicts a deviation from the Frenkel reaction equilibrium. The model does not focus on an accurate quantification of the very low residual intrinsic point defect concentrations. Hence, it is not necessary to account for the Frenkel reaction kinetics in CZ growth. A set of simulations like this at a fixed G and varying V can capture the effect of V/G on the initial incorporation (Fig. 38.8). As shown in Fig. 38.8, the critical V/G is around 0.15 mm²/(K min). It must be noted that various other groups report slightly different values of the critical V/G [38.26, 30–33]. Figure 38.8 also shows that the nucleation temperature increases with

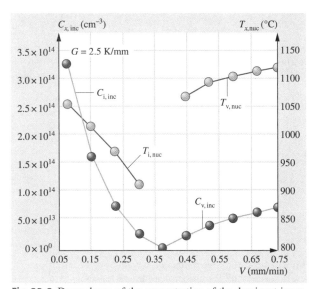

Fig. 38.8 Dependence of the concentration of the dominant incorporated point defect species and the nucleation temperature on the pull rate at a fixed G (subscript "inc" denotes the incorporated concentration and "nuc" denotes nucleation) (after [38.1])

increasing concentration of the incorporated dominant intrinsic point defect species, because the temperature at which the intrinsic point defect supersaturation is high enough to drive the nucleation events increases with increasing intrinsic point defect concentration. The cluster size distribution is influenced by an interplay between the formation of new clusters, which is driven by the dominant intrinsic point defect supersaturation, and the consumption of the intrinsic point defects by the existing clusters, which decreases the intrinsic point defect supersaturation. The conditions that allow the nucleation of the intrinsic point defects at a higher rate, before the intrinsic point defect concentration decreases by the cluster growth, lead to the formation of a large number of clusters, which remain very small in size; the conditions allowing the rapid growth of the formed clusters quickly reduce the intrinsic point defect concentration and the nucleation rate, leading to the formation of a small number of clusters that grow very large in size. More specifically, a higher cooling rate through the nucleation range leads to the evolution of smaller clusters at higher densities, whereas a higher incorporated intrinsic point defect concentration leads to the evolution of larger clusters at lower densities. This interplay is captured in the size distributions of the mature cluster populations in the simulated CZ crystals grown under varying conditions, as shown in Fig. 38.9. Using their simplified model, *Voronkov* and *Falster* quantified this interplay in terms of the cooling rate through the nucleation range and the concentration of the incorporated dominant intrinsic point defect species [38.46]

$$n_{x,\text{approx}} \propto \left(\frac{1}{D_x T_{\text{nuc},x}^2}\right)^{\frac{3}{2}} Q_{\text{nuc},x}^{\frac{3}{2}} C_{x,\text{nuc}}^{-\frac{1}{2}} \quad (38.38)$$

$$R_{x,\text{avg,approx}} \propto \left(D_x T_{\text{nuc},x}^2\right)^{\frac{1}{2}} \left(\frac{C_{x,\text{nuc}}}{Q_{\text{nuc},x}}\right)^{\frac{1}{2}}, \quad (38.39)$$

where Q is the cooling rate, given by the product of the local pull rate and the magnitude of the axial temperature gradient. The subscript "nuc" denotes the conditions at nucleation, and the subscript "approx" denotes an approximate value. *Kulkarni* et al. showed that the predictions of (38.38) and (38.39) agree quite well with the predictions of their rigorous model [38.1].

Many modern CZ processes enforce unsteady-state conditions in crystal growth by varying the crystal pull rate. *Kulkarni* et al. [38.1] captured the salient fea-

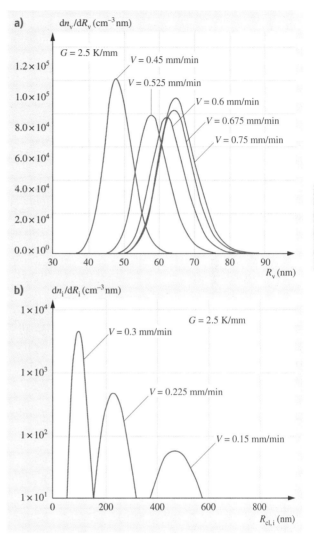

Fig. 38.9a,b The mature size distributions of (**a**) v-clusters and (**b**) i-clusters in various simulated crystals grown under steady states at various pull rates (after [38.1])

tures of unsteady-state CZ growth by simulating the growth of a crystal pulled at the varying rate shown in Fig. 38.10a. The body of the simulated crystal was grown first by continuously decreasing the pull rate and then continuously increasing the pull rate. The predicted cluster type and the cluster density variation are shown in Fig. 38.10b. Each element of the crystal undergoes

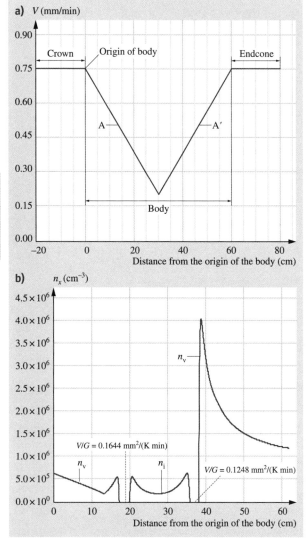

Fig. 38.10 (a) The crystal pull-rate profile used to understand the CZ defect dynamics under unsteady-state conditions. (b) The cluster density profile in the crystal grown using the pull-rate profile shown in (a) (after [38.1])

defect flux associated with the variation of the crystal pull rate. When the pull rate continuously decreases, an element of silicon moves away from the melt–crystal interface at a decreasing rate, allowing a more efficient diffusion of self-interstitials from the interface. In addition, at a given time and for a given pull rate, the driving force for self-interstitial diffusion at the interface is slightly higher than that for steady-state growth, and the driving force for the vacancy diffusion is slightly lower than that for steady-state growth; this dynamics is caused by the prior higher pull rate. Hence, the critical V/G increases when the pull rate continuously decreases. Conversely, the exact opposite effects explain the decrease of the critical V/G when the pull rate continuously increases. These shifts of the critical V/G were quantified by *Kulkarni* et al. for engineering applications as functions of the rate of the change of the pull rate with respect to crystal length [38.1]

$$\left(\frac{V}{G}\right)_{\mathrm{c,-slope}} = \left(\frac{V}{G}\right)_{\mathrm{c}} - 7.85\,(\mathrm{mm^2/K})$$
$$\times \left(\frac{\mathrm{d}V}{\mathrm{d}L}\right) (\mathrm{min}^{-1})\,, \qquad (38.40)$$

$$\left(\frac{V}{G}\right)_{\mathrm{c,+slope}} = \left(\frac{V}{G}\right)_{\mathrm{c}} - 13.745\,(\mathrm{mm^2/K})$$
$$\times \left(\frac{\mathrm{d}V}{\mathrm{d}L}\right) (\mathrm{min}^{-1})\,. \qquad (38.41)$$

The subscripts "−slope" and "+slope" indicate the decreasing pull rate and the increasing pull rate, respectively.

Finally, *Kulkarni* et al. [38.1] also showed that the predictions of the applied model agree reasonably well with experimental observations, as shown in Fig. 38.11. The D defect density was experimentally determined in a crystal grown under unsteady-state conditions by the method of copper decoration and Secco etching [38.38–40]. These studies clearly establish the validity of the applied model.

The model applied for the quantification of CZ defect dynamics discussed so far predicts the size distributions of all populations of clusters at all locations in a CZ crystal. The model requires the solution of the integro-differential equations for the intrinsic point defect concentration fields and of the cluster growth equations describing the evolution of the cluster populations at all locations in a CZ crystal. As this model provides rigorous quantification of the CZ defect dynamics, it is termed the *rigorous model*.

initial incorporation and nucleation under varying conditions, resulting in the interesting cluster distribution shown in Fig. 38.10. The most striking features of this study are the predicted shifts of the critical V/G, induced by the unsteady-state crystal growth. These shifts are caused by the variation of the excess intrinsic point

Fig. 38.11 (a) Comparison between the model predicted and experimentally determined approximate v-cluster density profiles in an experimental crystal. *Circles* indicate experimental data points. The profile of the volume-averaged v-cluster size in the crystal is also shown. (b) The v-cluster size distribution at two chosen axial locations in the crystal (after [38.1]) ▶

Second Approach: The Quantification of the CZ Defect Dynamics by the Lumped Model

The rigorous model can be simplified by representing a population of clusters of varying sizes at any given location in a CZ crystal by an equivalent population of identical clusters, as first shown by *Kulkarni* and *Voronkov* [38.53]. Thus, the complex rigorous model is reformulated by explicitly introducing the density and the average size of clusters in a population

$$n_i(r, z, t) = \int_0^t J_i(r, \xi, \tau) d\tau, \quad (38.42)$$

$$n_v(r, z, t) = \int_0^t J_v(r, \xi, \tau) d\tau, \quad (38.43)$$

$$\langle R_i \rangle = \frac{\int_0^t R_i(r, z, \tau, t) J_i(r, \xi, \tau) d\tau}{n_i}, \quad (38.44)$$

$$\langle R_v \rangle = \frac{\int_0^t R_v(r, z, \tau, t) J_v(r, \xi, \tau) d\tau}{n_v}, \quad (38.45)$$

where the brackets $\langle \cdots \rangle$ indicate the average value. Note that these average radii are different from the volumetric average radii defined in (38.36) and (38.37). The total consumption rate (per unit volume) of the intrinsic point defects by this population is now given as

$$q_i^i = 4\pi D_i (C_i - C_{i,e}) \langle R_i \rangle n_i, \quad (38.46)$$

$$q_v^v = 4\pi D_v (C_v - C_{v,e}) \langle R_v \rangle n_v. \quad (38.47)$$

The essential aspect of the simplified model is to replace the average radius of the cluster population by the square root of the average of the squares of the radii of

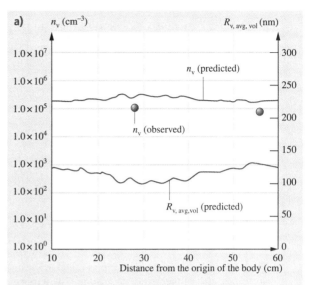

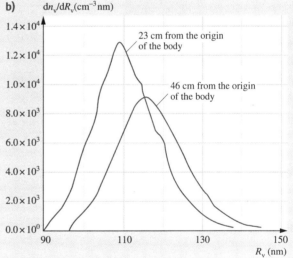

all clusters in the population

$$\langle R_i \rangle \approx \langle R_i^2 \rangle^{\frac{1}{2}} = \left(\frac{\int_0^t R_i^2(r, z, \tau, t) J_i(r, \xi, \tau) d\tau}{n_i} \right)^{1/2}$$

$$= \left(\frac{U_i}{n_i} \right)^{1/2}, \quad (38.48)$$

$$\langle R_{\mathrm{v}}\rangle \approx \langle R_{\mathrm{v}}^{2}\rangle^{\frac{1}{2}} = \left(\frac{\int_{0}^{t} R_{\mathrm{v}}^{2}(r,z,\tau,t) J_{\mathrm{v}}(r,\xi,\tau)\,\mathrm{d}\tau}{n_{\mathrm{v}}} \right)^{1/2}$$

$$= \left(\frac{U_{\mathrm{v}}}{n_{\mathrm{v}}} \right)^{1/2}, \quad (38.49)$$

where

$$U_{\mathrm{i}}(r,z,t) = \int_{0}^{t} R_{\mathrm{i}}^{2}(r,z,\tau,t) J_{\mathrm{i}}(r,\xi,\tau)\,\mathrm{d}\tau, \quad (38.50)$$

$$U_{\mathrm{v}}(r,z,t) = \int_{0}^{t} R_{\mathrm{v}}^{2}(r,z,\tau,t) J_{\mathrm{v}}(r,\xi,\tau)\,\mathrm{d}\tau. \quad (38.51)$$

The new auxiliary variable U is proportional to the total surface area of the cluster population. The intrinsic point defect consumption rate per unit volume is now rewritten as

$$q_{\mathrm{i}}^{\mathrm{i}} = 4\pi D_{\mathrm{i}}(C_{\mathrm{i}} - C_{\mathrm{i,e}}) \langle R_{\mathrm{i}}^{2}\rangle^{1/2} n_{\mathrm{i}}$$
$$= 4\pi D_{\mathrm{i}}(C_{\mathrm{i}} - C_{\mathrm{i,e}})(U_{\mathrm{i}} n_{\mathrm{i}})^{1/2}, \quad (38.52)$$
$$q_{\mathrm{v}}^{\mathrm{v}} = 4\pi D_{\mathrm{v}}(C_{\mathrm{v}} - C_{\mathrm{v,e}}) \langle R_{\mathrm{v}}^{2}\rangle^{1/2} n_{\mathrm{v}}$$
$$= 4\pi D_{\mathrm{v}}(C_{\mathrm{v}} - C_{\mathrm{v,e}})(U_{\mathrm{v}} n_{\mathrm{v}})^{1/2}. \quad (38.53)$$

The intrinsic point defect balances (38.29) and (38.30) are written using (38.52) and (38.53).

$$\frac{\partial C_{\mathrm{i}}}{\partial t} + V \frac{\partial C_{\mathrm{i}}}{\partial z} = \nabla(D_{\mathrm{i}}\nabla C_{\mathrm{i}})$$
$$- k_{\mathrm{i}\leftrightarrow\mathrm{v}}(C_{\mathrm{i}} C_{\mathrm{v}} - C_{\mathrm{i,e}} C_{\mathrm{v,e}})$$
$$- 4\pi D_{\mathrm{i}}(C_{\mathrm{i}} - C_{\mathrm{i,e}})(U_{\mathrm{i}} n_{\mathrm{i}})^{1/2}, \quad (38.54)$$

$$\frac{\partial C_{\mathrm{v}}}{\partial t} + V \frac{\partial C_{\mathrm{v}}}{\partial z} = \nabla(D_{\mathrm{v}}\nabla C_{\mathrm{v}})$$
$$- k_{\mathrm{i}\leftrightarrow\mathrm{v}}(C_{\mathrm{i}} C_{\mathrm{v}} - C_{\mathrm{i,e}} C_{\mathrm{v,e}})$$
$$- 4\pi D_{\mathrm{v}}(C_{\mathrm{v}} - C_{\mathrm{v,e}})(U_{\mathrm{v}} n_{\mathrm{v}})^{1/2}. \quad (38.55)$$

Thus the intrinsic point defect balances are described by (38.54) and (38.55) without the knowledge of the formation and growth histories of the cluster populations at any location. If the Frenkel reaction equilibrium is assumed, (38.54) and (38.55) are replaced by (38.3) and (38.56), obtained by subtracting (38.54) from (38.55)

$$\frac{\partial (C_{\mathrm{v}} - C_{\mathrm{i}})}{\partial t} + V \frac{\partial (C_{\mathrm{v}} - C_{\mathrm{i}})}{\partial z}$$
$$= \nabla(D_{\mathrm{v}}\nabla C_{\mathrm{v}}) - \nabla(D_{\mathrm{i}}\nabla C_{\mathrm{i}})$$
$$- 4\pi D_{\mathrm{v}}(C_{\mathrm{v}} - C_{\mathrm{v,e}})(U_{\mathrm{v}} n_{\mathrm{v}})^{1/2}$$
$$+ 4\pi D_{\mathrm{i}}(C_{\mathrm{i}} - C_{\mathrm{i,e}})(U_{\mathrm{i}} n_{\mathrm{i}})^{1/2}. \quad (38.56)$$

The evolution of the auxiliary variable U is derived using the cluster growth (38.24) and (38.25) with the definitions (38.50) and (38.51)

$$\frac{\partial U_{\mathrm{i}}}{\partial t} + V \frac{\partial U_{\mathrm{i}}}{\partial z} = \frac{2 D_{\mathrm{i}} n_{\mathrm{i}}}{\psi_{\mathrm{i}}^{\mathrm{i}}}(C_{\mathrm{i}} - C_{\mathrm{i,e}}), \quad (38.57)$$

$$\frac{\partial U_{\mathrm{v}}}{\partial t} + V \frac{\partial U_{\mathrm{v}}}{\partial z} = \frac{2 D_{\mathrm{v}} n_{\mathrm{v}}}{\psi_{\mathrm{v}}^{\mathrm{v}}}(C_{\mathrm{v}} - C_{\mathrm{v,e}}). \quad (38.58)$$

The initial size of the formed clusters is assumed to be zero in the derivation of (38.57) and (38.58). This assumption is accurate and does not the affect the predictions of the model. The evolution of the density of clusters is directly obtained using the classical nucleation theory

$$\frac{\partial n_{\mathrm{i}}}{\partial t} + V \frac{\partial n_{\mathrm{i}}}{\partial z} = J_{\mathrm{i}}, \quad (38.59)$$

$$\frac{\partial n_{\mathrm{v}}}{\partial t} + V \frac{\partial n_{\mathrm{v}}}{\partial z} = J_{\mathrm{v}}. \quad (38.60)$$

The representative size of clusters at any location $\mathcal{R}$ is given as

$$\mathcal{R}_{\mathrm{i}} = \langle R_{\mathrm{i}}^{2}\rangle^{1/2} = \left(\frac{U_{\mathrm{i}}}{n_{\mathrm{i}}}\right)^{1/2}, \quad (38.61)$$

$$\mathcal{R}_{\mathrm{v}} = \langle R_{\mathrm{v}}^{2}\rangle^{1/2} = \left(\frac{U_{\mathrm{v}}}{n_{\mathrm{v}}}\right)^{1/2}. \quad (38.62)$$

The defect dynamics in a CZ crystal is now quantified by the intrinsic point defect concentration C, the auxiliary variable U, and the cluster density n. These variables are described by a set of partial differential equations without the necessity to quantify the formation and growth histories of clusters. The simplified model eliminates the elapsed time, introduced in the rigorous model to describe the size distribution of a cluster population at any given location in a CZ crystal, as an independent variable. Hence, this simplified model is computationally attractive. As the simplified model represents the population of clusters of varying sizes at a given location in the CZ crystal by an equivalent population of identical clusters, it is termed the *lumped model* by *Kulkarni* and *Voronkov* [38.53].

The accuracy of the lumped model is verified by a comparison of its predictions with the predictions of the rigorous model, for many different crystal growth conditions, as shown in Fig. 38.12a. In addition, the lumped model is validated by experimental observations (Fig. 38.12b). Finally, as shown in Fig. 38.13, the two-dimensional microdefect distribution in a crystal pulled at a varying rate is predicted reasonably well by the lumped model. It must be noted that the mesh discretization used for the computation is relatively coarse. The inaccuracies associated with the mesh discretization are reduced in the simulations discussed in Sect. 38.3. In the discussed simulations, the lumped model was solved assuming the Frenkel reaction equilibrium. The method of copper decoration and Secco etching was used for the experimental determination of the microdefect distribution [38.38–40]. These studies establish the lumped model as a valuable engineering tool for the development of new CZ crystal growth processes.

Third Approach: The Quantification of the CZ Defect Dynamics by the Discrete Rate Equations and the Fokker–Planck Equation

The models described in the previous sections apply the classical nucleation theory to predict the formation of stable clusters and a diffusion-limited growth kinetics to quantify the growth of these clusters. This approach works very well for Czochralski crystal growth. A more rigorous treatment of CZ defect dynamics, however, must account for all reactions involved in the intrinsic point defect aggregation, as described by reaction (38.18). Considering the large size of microdefects in CZ crystals, this approach requires the solution of an impractically large number of equations. *Sinno* and *Brown* [38.47], *Mori* [38.27], and *Brown* et al. [38.49] quantified the defect dynamics in CZ crystals by applying a mixed approach involving the solution of a set of equations derived treating smaller clusters as discrete particles and the solution of a set of Fokker–Planck equations derived from the discrete equations for larger clusters. In this section, the contributions of this work to the field of CZ defect dynamics are discussed.

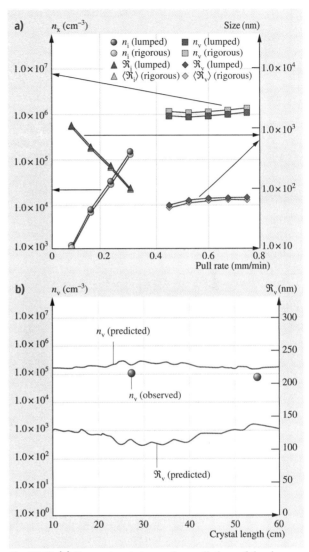

Fig. 38.12 (a) Comparison between the predictions of the rigorous model and the predictions of the lumped model. (b) Comparison between the predictions of the lumped model with the experimental observations (after [38.53])

The Discrete Rate Equations. Reaction set (38.18) defines the series of reactions driving the aggregation of vacancies and self-interstitials. In this reaction set, clusters of the same size and type are treated as a separate species. The cross-interactions between v-clusters and

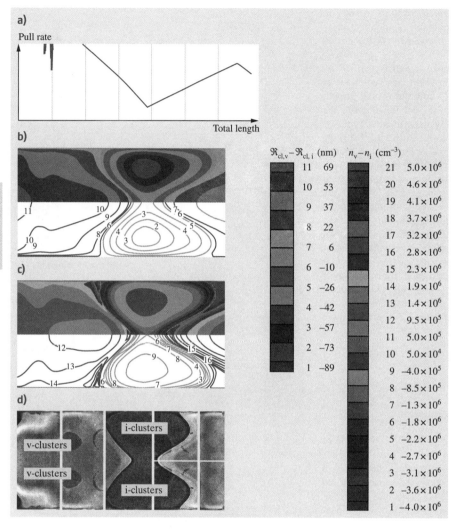

Fig. 38.13 (a) The pull-rate profile used to pull the experimental CZ crystal. (b) The predicted $\mathfrak{R}_v - \mathfrak{R}_i$. (c) The predicted $n_v - n_i$. (d) The observed defect distribution in the crystal. Positive values indicate the v-cluster size and density and the magnitudes of the negative values indicate the i-cluster size and density. v-clusters and i-clusters do not coexist (after [38.53])

self-interstitials and i-clusters and vacancies are ignored in this chapter.

In the previous sections, the size of a cluster was defined as its radius. In this section, the size of a cluster is defined by the number of intrinsic point defects in it rather than by its radius, following the work of *Sinno* and *Brown* [38.47], *Mori* [38.27], *Wang* and *Brown* [38.48], and *Brown* et al. [38.49]. In an element of silicon, the overall rate of formation of clusters of size m, Φ_m, is given by the difference between the net volumetric flux coming from the clusters of size $m-1$ to the clusters of size m, I_m, and the net volumetric flux going from clusters of size m to clusters of size $m+1$, $I_{(m+1)}$

$$\Phi_{m_x} = I_{m_x} - I_{(m+1)_x} \, . \tag{38.63}$$

The subscript "m_x" indicates an x-cluster containing m intrinsic point defects of type x. Note that J_x defines the nucleation rate of stable x-clusters, as defined in the previous sections, and Φ_{m_x} defines the net formation rate of x-clusters of size m. The net volumetric flux coming from the clusters containing $m-1$ intrinsic point defects to the clusters containing m intrinsic point defects is defined by the growth rate of the former and the dissolution rate of the latter

$$I_{m_x} = g_{(m-1)_x} \phi_{(m-1)_x} - d_{m_x} \phi_{m_x} \, , \tag{38.64}$$

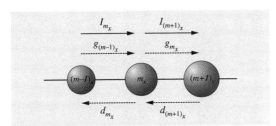

Fig. 38.14 The relationship among the nucleation flux I, the growth rate g, and the dissolution rate d (after [38.27])

where ϕ is the density of clusters, g is the growth rate of a cluster, and d is the dissolution rate. The relationship among I, g, and d is shown schematically in Fig. 38.14. *Sinno* and *Brown* [38.47], *Mori* [38.27], *Wang* and *Brown* [38.48], and *Brown* et al. [38.49] define the cluster growth rate and the dissolution rate by incorporating the kinetic interactions between monomers and clusters. In this section, these rates are defined assuming a diffusion-limited interaction between clusters and monomers, for the sake of consistency with the discussion in other sections of this chapter. The growth rate of a cluster containing m monomers of type x is given by the diffusion-limited attachment frequency of monomers to the cluster

$$g_{m_x} = 4\pi R_{m_x} D_x C_x , \qquad (38.65)$$

where R_m is the radius of a cluster of size m. The cluster dissolution rate is given by its thermodynamic relationship with the cluster growth rate

$$d_{m_x} = \frac{g_{(m-1)_x} \phi_{(m-1)_x,e}}{\phi_{m_x,e}} . \qquad (38.66)$$

The equilibrium concentration of clusters of size m, in a solution of a given composition, at a given temperature, is given by

$$\phi_{m_x,e} = \rho_x e^{-\left(\frac{\Delta F_{m_x}}{k_B T}\right)} . \qquad (38.67)$$

The total free energy change associated with the formation of a cluster of size m, from a solution of a given composition, at a given temperature, is

$$\Delta F_{m_x} = -m_x k_B T \ln \frac{C_x}{C_{x,e}} + \Gamma_{m_x} , \qquad (38.68)$$

where Γ_m is the formation energy of a cluster of size m. The formation energy of a large unstrained cluster of size m is simply approximated as $\lambda m^{2/3}$. Using (38.67) and (38.68), (38.66) is written as

$$d_{m_x} = \frac{g_{(m-1)_x}}{\left(\frac{C_x}{C_{x,e}}\right) e^{\frac{(\Gamma_{(m-1)_x} - \Gamma_{m_x})}{k_B T}}} . \qquad (38.69)$$

These equations define the diffusion limited growth rates and dissolution rates of all clusters in a Czochralski crystal.

The Cluster Balance Using Discrete Rate Equations. With the discrete formulation, the cluster conservation equations in an axisymmetric CZ crystal are written as follows

$$\frac{\partial \phi_{m_x}(r,z,t)}{\partial t} + V \frac{\partial \phi_{m_x}(r,z,t)}{\partial z}$$
$$= I_{m_x}(r,z,t) - I_{(m+1)_x}(r,z,t) , \qquad (38.70)$$

for $m_x \geq 2$.

In (38.70), the density of clusters of the same size and type is treated as a dependent variable. This equation must be solved with the intrinsic point defect balances. The quantification of the defect dynamics, in CZ crystals containing microdefects as large as 100–200 nm, using the discrete rate equations is computationally impractical. Therefore, reasonable approximations based on the discrete rate equations are necessary for a practical solution of the problem.

The Fokker–Planck Equation. The Fokker–Planck equation (FPE) is derived from the discrete rate equations by a Kramers–Moyal expansion treating m as a continuous independent variable [38.27, 47]. The FPE reduces the number of equations defining the CZ defect dynamics. By this formulation, the size distribution of clusters is written as a continuous function of m

$$\frac{\partial f_x(r,z,t,m_x)}{\partial t} + V \frac{\partial f_x(r,z,t,m_x)}{\partial z}$$
$$= -\frac{\partial}{\partial m_x}\left[A_x(r,z,t,m_x) f_x(r,z,t,m_x) \right.$$
$$\left. - B_x(r,z,t,m_x) \frac{\partial f_x(r,z,t,m_x)}{\partial m_x} \right] . \qquad (38.71)$$

The cluster density determined by the FPE is written as f and the subscript x defines the type of the cluster. A is termed the *drift coefficient* and B is termed the *diffusion coefficient* following the generalized transport equation written in m-space, and they are related to the discrete rate equations as

$$A_x(r,z,t,m_x) = g_x(r,z,t,m_x) - d_x(r,z,t,m_x)$$
$$- \frac{\partial B_x(r,z,t,m_x)}{\partial m_x} , \qquad (38.72)$$

$$B_x(r,z,t,m_x) = \frac{g_x(r,z,t,m_x) + d_x(r,z,t,m_x)}{2} . \qquad (38.73)$$

The Fokker-Plank equation thus reduces the number of equations describing the cluster growth. Using this formulation, a simplified model can now be developed.

The Model. The Fokker–Planck equation is accurate when clusters are large. Hence, an accurate model involves describing the CZ defect dynamics by the discrete rate equations for smaller clusters and by the Fokker–Planck equation for larger clusters. Thus, the cluster balances are written as follows

$$\frac{\partial \phi_{m_i}(r,z,t)}{\partial t} + V\frac{\partial \phi_{m_i}(r,z,t)}{\partial z}$$
$$= I_{m_i}(r,z,t) - I_{(m+1)_i}(r,z,t),$$
$$\text{for } (m_{\text{dis}} \geq m_i \geq 2), \quad (38.74)$$

$$\frac{\partial \phi_{m_v}(r,z,t)}{\partial t} + V\frac{\partial \phi_{m_v}(r,z,t)}{\partial z}$$
$$= I_{m_v}(r,z,t) - I_{(m+1)_v}(r,z,t),$$
$$\text{for } (m_{\text{dis}} \geq m_v \geq 2), \quad (38.75)$$

$$\frac{\partial f_i(r,z,t,m_i)}{\partial t} + V\frac{\partial f_i(r,z,t,m_i)}{\partial z}$$
$$= -\frac{\partial}{\partial m_i}\left[A_i(r,z,t,m_i) f_i(r,z,t,m_i) \right.$$
$$\left. - B_i(r,z,t,m_i)\frac{\partial f_i(r,z,t,m_i)}{\partial m_i} \right],$$
$$\text{for } (m_{\text{dis}} < m_i \leq m_{\text{max}}), \quad (38.76)$$

$$\frac{\partial f_v(r,z,t,m_v)}{\partial t} + V\frac{\partial f_v(r,z,t,m_v)}{\partial z}$$
$$= -\frac{\partial}{\partial m_v}\left[A_v(r,z,t,m_v) f_v(r,z,t,m_v) \right.$$
$$\left. - B_v(r,z,t,m_v)\frac{\partial f_v(r,z,t,m_v)}{\partial m_v} \right],$$
$$\text{for } (m_{\text{dis}} < m_v \leq m_{\text{max}}). \quad (38.77)$$

The subscript "dis" denotes the maximum cluster size treated by the discrete rate equations, and the subscript "max" denotes the maximum cluster size quantified by the Fokker–Planck equations. The intrinsic point defect balances following this approach are written as

$$\frac{\partial C_i}{\partial t} + V\frac{\partial C_i}{\partial z} = \nabla\cdot(D_i \nabla C_i)$$
$$- k_{i\leftrightarrow v}(C_i C_v - C_{i,e}C_{v,e}) - q_i^i, \quad (38.78)$$

$$\frac{\partial C_v}{\partial t} + V\frac{\partial C_v}{\partial z} = \nabla\cdot(D_v \nabla C_v)$$
$$- k_{i\leftrightarrow v}(C_i C_v - C_{i,e}C_{v,e}) - q_v^v, \quad (38.79)$$

where the intrinsic point defect consumption rates by clusters are given by

$$q_i^i = \left(\frac{\partial}{\partial t} + V\frac{\partial}{\partial z}\right)$$
$$\times \left[\sum_{m_i=2}^{m_i=m_{\text{dis}}} m_i \phi_{m_i} + \int_{m_i=m_{\text{dis}}+1}^{m_i=m_{\text{max}}} m_i f_i \, dm_i \right], \quad (38.80)$$

$$q_v^v = \left(\frac{\partial}{\partial t} + V\frac{\partial}{\partial z}\right)$$
$$\times \left[\sum_{m_v=2}^{m_v=m_{\text{dis}}} m_v \phi_{m_v} + \int_{m_v=m_{\text{dis}}+1}^{m_v=m_{\text{max}}} m_v f_v \, dm_v \right]. \quad (38.81)$$

The first term in the square brackets on the right-hand side of (38.80) accounts for the consumption of self-interstitials by the discrete i-clusters, and the second term accounts for the consumption of self-interstitials by the self-interstitial *FP-cluster*; (38.81) describes the vacancy consumption. The coupled model using the discrete rate equations for smaller clusters and the Fokker–Planck equation for larger clusters is still computationally expensive. As the model uses the described coupled approach, it is termed the *discrete–continuous* model in this chapter. For further details, the reader is referred to [38.27, 47–49]. As noted earlier, these researchers account for the kinetic interactions between monomers and clusters.

Results. Although the unsteady-state discrete–continuous model is described in this chapter, only the quantification of the steady-state CZ defect dynamics, at a fixed pull rate, involving the cluster growth has been reported in the literature thus far. The results obtained by the solution of the discrete–continuous model agree well with the results described by the rigorous model and the lumped model. The initial incorporation and the effects of the cooling rate and the incorporated dominant intrinsic point defect concentration on the cluster size distribution are captured very well. The evolution of the intrinsic point defect and the microdefect concentration profiles in a CZ crystal, as predicted by *Wang* and *Brown*, are shown in Fig. 38.15 [38.48]. The two-dimensional intrinsic point defect concentration fields and the microdefect distributions captured by the discrete–continuous model reported by *Brown* et al. are shown in Fig. 38.16 [38.49]. The physics of the CZ defect dynamics is quantified accurately by the applied model.

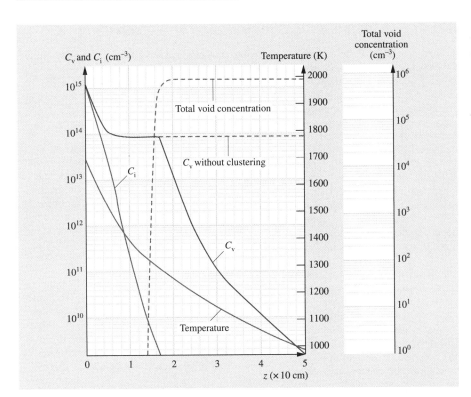

Fig. 38.15 The evolution of the intrinsic point defects and the microdefects as predicted by the discrete–continuous model (after [38.48], © Elsevier 2001)

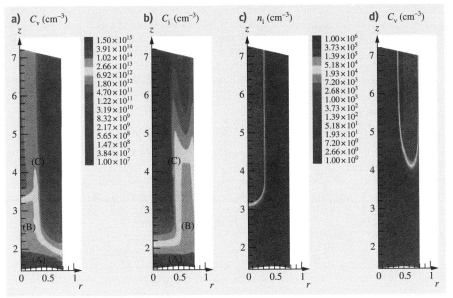

Fig. 38.16a–d Simulation results for steady-state crystal growth at $V = 0.6$ mm/min, showing (**a**) vacancy concentration, (**b**) self-interstitial concentration, (**c**) the total v-cluster density (> 50 nm diameter), and (**d**) the total i-cluster density (> 50 nm diameter). The *upper-case letters* in the figures are used by *Brown* et al. to describe the physics of the CZ defect dynamics and should not be mistaken for the figure labels, which are in lower case (after [38.49], © Elsevier 2001)

38.3 Czochralski Defect Dynamics in the Presence of Oxygen

As discussed so far, the aggregates of the intrinsic point defects commonly exist in silicon crystals grown by both the Czochralski process and the float zone (FZ) process. In addition, CZ crystals contain oxide particles, primarily silicon dioxide, termed oxygen clusters in this chapter. Oxygen clusters form only in CZ crystals, because CZ crystals, during their growth, incorporate oxygen in appreciable concentration from the crucible containing the silicon melt. Oxygen clusters in a growing CZ crystal are typically very small. These clusters facilitate the formation of stacking faults in the crystal subjected to selective heat treatments that generate self-interstitials (Fig. 38.3). Hence, oxygen clusters are popularly identified by these stacking faults known as oxidation-induced stacking faults (OSFs) [38.16].

The dynamics of the formation of various microdefects in CZ crystals is affected by many reactions involving the intrinsic point defects and oxygen, and their transport. This defect dynamics in the absence of oxygen has been discussed in detail in Sect. 38.2. The reported research on the direct quantification of the CZ defect dynamics in the presence of oxygen, in particular, and oxygen cluster formation in monocrystalline silicon, in general, are discussed in this section [38.48, 59, 65–70].

38.3.1 Reactions in Growing CZ Crystals

The essential aspect of understanding CZ defect dynamics is the quantification of the kinetics of all relevant reactions in a growing crystal. The Frenkel reaction involving the intrinsic point defects and silicon, the reactions involving vacancies and oxygen, and the aggregation reactions forming all microdefects influence the CZ defect dynamics.

Reactions Involving No Aggregation
The Frenkel reaction discussed in Sect. 38.1 and the reactions involving vacancies and interstitial oxygen (simply, oxygen) do not directly produce microdefects. The Frenkel reaction involves the mutual annihilation of a vacancy v and a self-interstitial i by their recombination to produce a silicon lattice atom Si and the backward production of a pair of a vacancy and a self-interstitial from a silicon lattice atom (reaction 38.1). Oxygen participates in a series of reversible reactions with vacancies and complexes of vacancies and oxygen in a growing CZ crystal. The following reactions involving vacancies and oxygen are of primary importance [38.59]:

$$v + O \rightleftharpoons vO, \quad (38.82)$$
$$vO + O \rightleftharpoons vO_2, \quad (38.83)$$

where vO and vO_2 are vacancy–oxygen complexes. Each forward or reverse reaction listed above is considered to be an elementary reaction. The net rate of formation of a reacting species is given by the summation of the rates of formation of the species by each elementary reaction. At equilibrium, the net rate of production of any species is zero.

The Nucleation of the Intrinsic Point Defects and Oxygen
Octahedral voids or D defects are formed by the aggregation of vacancies. Globular B defects are formed by the aggregation of self-interstitials. A defects presumably form by the transformation of B defects. These microdefects are modeled as clusters of intrinsic point defects. The thermodynamics and kinetics of the formation of these microdefects is discussed in Sect. 38.2.

Oxygen clusters (O-clusters) are modeled as spherical aggregates of oxygen (silicon dioxide) [38.70]. The specific volume of an oxygen cluster is greater than that of silicon. Thus, the formation of an oxygen cluster is associated with the generation of stress. In the presence of vacancies, however, the clusters relieve stress by the consumption of vacancies. Oxygen cluster formation proceeds through a series of reactions involving oxygen and vacancies. Hence, this series of reactions is written as

$$O + P_{m_O} + \gamma v + \tfrac{1}{2}\text{Si} \rightleftharpoons P_{(m+1)_O}, \quad (38.84)$$

where γ is the number of vacancies absorbed per oxygen atom participating in the reaction. It must be noted that an oxygen cluster containing m oxygen atoms also contains $m/2$ silicon atoms.

The volume (bulk) free-energy change associated with the formation of an oxygen cluster containing m oxygen atoms in an isolated element of silicon at a fixed temperature and composition is given by the contributions from the oxygen supersaturation and the vacancy supersaturation. Thus, the total free-energy change associated with the formation of an oxygen cluster containing m oxygen atoms is

$$\Delta F_O(m_O) = \left[-m_O k_B T \ln \frac{C_O}{C_{O,e}} - \gamma m_O k_B T \ln \frac{C_v}{C_{v,e}} \right] + \left[\lambda_O m_O^{2/3} \right].$$
(38.85)

The first term in the square brackets on the right-hand side of (38.85) is the volume (bulk) free energy change, and the second term is the energy required to form the surface of an oxygen cluster containing m oxygen atoms. The subscript "O" denotes both oxygen and oxygen clusters depending on the variable.

The formation kinetics of oxygen clusters is quite complex. An oxygen cluster undergoes morphological changes as it grows. This chapter does not address the details of these morphological changes. A broad macroscopic understanding of the oxygen cluster distribution is obtained by assuming these clusters to be spherical. The number of oxygen atoms in the critical cluster is obtained by maximizing the free-energy change ΔF with respect to m. The net rate of formation of stable oxygen clusters is obtained using the classical nucleation theory. Typically, it is accurate to assume that the formation rate of stable oxygen clusters is described by the diffusion-limited attachment of oxygen atoms to the critical oxygen clusters. For the sake of completeness, however, the attachment frequency is described both by the oxygen diffusion-limited mechanism and the vacancy diffusion-limited mechanism, depending on the ratio of $D_v C_v$ to $D_O C_O$. Hence, the formation rate of stable oxygen clusters per unit volume of silicon, or the oxygen nucleation rate, is given as

$$J_O = \left[\eta^*_{O \leftrightarrow O} \right] \left[\{ 12\pi k_B T \times \Delta F_O \left(m_O^* \right) \}^{-1/2} \right.$$

$$\left. \times \left(k_B T \ln \frac{C_O}{C_{O,e}} + \gamma k_B T \ln \frac{C_v}{C_{v,e}} \right) \right]$$

$$\times \left[\rho_O \exp \left(-\frac{\Delta F_O \left(m_O^* \right)}{k_B T} \right) \right],$$
(38.86)

$$\eta^*_{O \leftrightarrow O} = \begin{cases} 4\pi R \left(m_O^* \right) D_O C_O & \left| D_v C_v \geq \gamma D_O C_O \right. \\ \\ \dfrac{4\pi R \left(m_O^* \right) D_v C_v}{\gamma} & \left| D_v C_v < \gamma D_O C_O \right. \end{cases},$$
(38.87)

where $\Delta F(m^*)$ is the free-energy change associated with the formation of a critical cluster containing m^* monomers; the subscript "O" denotes oxygen or oxygen clusters and the superscript asterisk denotes the critical clusters. $\eta^*_{O \leftrightarrow O}$ is the attachment frequency of oxygen atoms to a critical oxygen cluster. The second term in the square brackets on the right-hand side of (38.86) is the Zheldovich factor and the third term is the equilibrium concentration of the critical oxygen clusters. The discussed kinetics can now be applied in the development of the equations governing the CZ defect dynamics.

38.3.2 The Model

The model quantifying the CZ defect dynamics must account for the balances of all species, cluster formation, and cluster growth. All microdefects are approximated as spherical clusters. As discussed before, D defects are termed v-clusters, A and B defects are termed i-clusters, and the aggregates of oxygen (silicon dioxide) are termed O-clusters. At any given location of a growing CZ crystal, at a given time, one or more than one population of clusters formed at various other locations during the elapsed time period can exist. The clusters are assumed to be immobile; thus, they are only carried convectively from one location to the next by the physical movement of the growing crystal. In addition, there is a spatial distribution of these populations. A rigorous treatment of the spatial distribution of these cluster populations is computationally expensive. *Kulkarni* and *Voronkov* developed a lumped model that represents a population of clusters at any given location by an equivalent population of identical clusters [38.53]. Later, *Kulkarni* applied this model to quantify the CZ defect dynamics in the presence of oxygen [38.70]. In this chapter, this research reported by Kulkarni is discussed in detail.

The Governing Equations
The balance of self-interstitials includes their transport and their consumption by the Frenkel reaction and i-clusters

$$\frac{\partial C_i}{\partial t} + V \frac{\partial C_i}{\partial z} = \nabla \left(D_i \nabla C_i \right)$$

$$+ [k_{i \leftrightarrow v}(C_{i,e} C_{v,e} - C_i C_v)] - q_i^i.$$
(38.88)

The term in the square brackets in (38.88) is the net rate of formation (negative consumption rate) of self-interstitials per unit volume by the Frenkel reaction.

Vacancies are consumed by both v-clusters and O-clusters. In addition, vacancies participate in reactions with self-interstitials, oxygen, and vO. Hence, the vacancy balance is written as

$$\frac{\partial C_v}{\partial t} + V\frac{\partial C_v}{\partial z} = \nabla(D_v \nabla C_v)$$
$$+ [k_{i\leftrightarrow v}(C_{i,e}C_{v,e} - C_i C_v)$$
$$- k_{v\leftrightarrow O}C_v C_O + k_{vO}C_{vO}]$$
$$- q_v^v - q_O^v. \qquad (38.89)$$

The rate constant for an elementary forward or an elementary reverse reaction is denoted by k. The subscripts of k indicate the reactants involved in a forward or a reverse reaction: "k_x" denotes the rate constant for the elementary reaction involving only x and "$k_{x\leftrightarrow y}$" indicates the elementary reaction involving x and y, where x and y represent the reacting species. q_x^y is the volumetric consumption rate of species y, denoted by the superscript, by the clusters containing species x, denoted by the subscript. The term in the square brackets in (38.89) is the net rate of vacancy production by reactions (38.1) and (38.82). The species vO is considered to be immobile. It is not directly consumed by clusters. Thus, the vO species balance must account only for the convection and reactions (38.82) and (38.83)

$$\frac{\partial C_{vO}}{\partial t} + V\frac{\partial C_{vO}}{\partial z} = (k_{v\leftrightarrow O}C_v C_O - k_{vO}C_{vO}$$
$$- k_{vO\leftrightarrow O}C_{vO}C_O + k_{vO_2}C_{vO_2}). \qquad (38.90)$$

The species vO_2 is also considered to be immobile and it is also not directly consumed by clusters. It participates only in reaction (38.83)

$$\frac{\partial C_{vO_2}}{\partial t} + V\frac{\partial C_{vO_2}}{\partial z} = (k_{vO\leftrightarrow O}C_{vO}C_O - k_{vO_2}C_{vO_2}). \qquad (38.91)$$

Oxygen is in abundance

$$\frac{\partial C_O}{\partial t} + V\frac{\partial C_O}{\partial z} = 0. \qquad (38.92)$$

It is evident from (38.88–38.91) that the balance of the *excess total vacancy concentration*, defined as the difference between the sum of the concentrations of all species containing vacancies (v, vO, and vO_2) and the concentration of self-interstitials, $C_v + C_{vO} + C_{vO_2} - C_i$, is not explicitly affected by nonaggregation reactions (38.1), (38.82), and (38.83); this balance is written

as

$$\frac{\partial (C_v + C_{vO} + C_{vO_2} - C_i)}{\partial t}$$
$$+ V\frac{\partial (C_v + C_{vO} + C_{vO_2} - C_i)}{\partial z}$$
$$= \nabla(D_v \nabla C_v) - \nabla(D_i \nabla C_i) - q_v^v - q_O^v + q_i^i. \qquad (38.93)$$

Assuming the reaction equilibrium for reactions (38.1), (38.82), and (38.83), the species balances (38.88–38.91) are defined by (38.93), (38.3), and the reaction equilibria [38.59]

$$\frac{C_{vO}}{C_v} = \sqrt{\frac{C_{vO_2,e}}{C_{v,e}}}, \qquad (38.94)$$

$$\frac{C_{vO_2}}{C_v} = \frac{C_{vO_2,e}}{C_{v,e}}. \qquad (38.95)$$

In CZ crystals, only oxygen nucleation facilitated by vacancies is of primary interest. The formation of O-clusters by ejection of self-interstitials is negligible. Once formed in the presence of vacancies, O-clusters initially grow by consuming vacancies without ejecting self-interstitials; later, when the vacancy concentration decreases, O-clusters can grow by ejection of self-interstitials. For the sake of simplicity, the growth of O-clusters by ejection of self-interstitials is ignored. When the vacancy concentration is sufficiently high, the O-cluster growth is assumed to be limited by the consumption of oxygen by the clusters; when the vacancy concentration is relatively low, the O-cluster growth is assumed to be limited by the consumption of vacancies by the clusters. Thus, O-clusters do not grow when vacancies are at equilibrium concentration. These approximations accurately quantify the density of O-clusters but underpredict their size by ignoring their growth by ejection of self-interstitials under vacancy-lean conditions. The assumptions used in the model are self-consistent, however, and provide meaningful insights into the CZ defect dynamics. If desired, the upper limit of O-cluster size can be quantified by simply assuming oxygen diffusion-limited cluster growth under all conditions; this assumption is not used in the formulation of the discussed model, although it can be implemented without much effort.

The diffusion-limited volumetric consumption rates of vacancies, self-interstitials, and oxygen by various clusters are defined following the methodology developed by *Kulkarni* and *Voronkov* [38.53]

$$q_i^i = 4\pi D_i (C_i - C_{i,e}) (U_i n_i)^{1/2}, \tag{38.96}$$

$$q_v^v = 4\pi D_v (C_v - C_{v,e}) (U_v n_v)^{1/2}, \tag{38.97}$$

$$q_O^v = \begin{cases} \gamma 4\pi D_O (C_O - C_{O,e}) (U_O n_O)^{1/2} \\ \quad |D_v(C_v - C_{v,e})| \geq \gamma D_O (C_O - C_{O,e}) \\ 4\pi D_v (C_v - C_{v,e}) (U_O n_O)^{1/2} \\ \quad |D_v(C_v - C_{v,e})| < \gamma D_O (C_O - C_{O,e}) \end{cases}. \tag{38.98}$$

The evolution of the auxiliary variable U, which is proportional to the surface area of the cluster population, is described by the cluster growth equation

$$\frac{\partial U_i}{\partial t} + V \frac{\partial U_i}{\partial z} = \frac{2 D_i n_i}{\psi_i^i} (C_i - C_{i,e}), \tag{38.99}$$

$$\frac{\partial U_v}{\partial t} + V \frac{\partial U_v}{\partial z} = \frac{2 D_v n_v}{\psi_v^v} (C_v - C_{v,e}), \tag{38.100}$$

$$\frac{\partial U_O}{\partial t} + V \frac{\partial U_O}{\partial z}$$
$$= \begin{cases} \dfrac{2 D_O n_O}{\psi_O^O} (C_O - C_{O,e}) \\ \quad |D_v(C_v - C_{v,e})| \geq \gamma D_O (C_O - C_{O,e}) \\ \dfrac{2 D_v n_O}{\gamma \psi_O^O} (C_v - C_{v,e}) \\ \quad |D_v (C_v - C_{v,e})| < \gamma D_O (C_O - C_{O,e}) \end{cases}. \tag{38.101}$$

The total cluster density is directly obtained by the classical nucleation theory

$$\frac{\partial n_i}{\partial t} + V \frac{\partial n_i}{\partial z} = J_i, \tag{38.102}$$

$$\frac{\partial n_v}{\partial t} + V \frac{\partial n_v}{\partial z} = J_v, \tag{38.103}$$

$$\frac{\partial n_O}{\partial t} + V \frac{\partial n_O}{\partial z} = J_O. \tag{38.104}$$

The representative radius of a cluster population at any location is given as

$$\mathcal{R}_i = \left(\frac{U_i}{n_i} \right)^{1/2}, \tag{38.105}$$

$$\mathcal{R}_v = \left(\frac{U_v}{n_v} \right)^{1/2}, \tag{38.106}$$

$$\mathcal{R}_O = \left(\frac{U_O}{n_O} \right)^{1/2}. \tag{38.107}$$

The domain of computation is transient, because a CZ crystal is continuously pulled. The equation describing this domain transience must be solved with the discussed equations. Vacancies and self-interstitials are assumed to exist at equilibrium on all crystal surfaces including the melt–crystal interface; the concentrations of the vO and vO$_2$. species are determined by the re-

Table 38.2 Key properties of various species, including oxygen and vacancy–oxygen complexes, participating in reactions in growing CZ crystals

Property set II	Property set III
D_i (cm^2/s) $= 0.19497 \exp\left(\dfrac{-0.9\,(\text{eV})}{k_B T} \right)$	D_i (cm^2/s) $= 4 \times 10^{-3} \exp\left(\dfrac{-0.3\,(\text{eV})}{k_B T} \right)$
D_v (cm^2/s) $= 6.2617 \times 10^{-4} \exp\left(\dfrac{-0.4\,(\text{eV})}{k_B T} \right)$	D_v (cm^2/s) $= 2 \times 10^{-3} \exp\left(\dfrac{-0.38\,(\text{eV})}{k_B T} \right)$
D_O (cm^2/s) $= 1.3 \times 10^{-1} \exp\left(\dfrac{-2.53\,(\text{eV})}{k_B T} \right)$	D_O (cm^2/s) $= 1.3 \times 10^{-1} \exp\left(\dfrac{-2.53\,(\text{eV})}{k_B T} \right)$
$C_{i,e}$ (cm^{-1}) $= 6.1759 \times 10^{26} \exp\left(\dfrac{-4.0\,(\text{eV})}{k_B T} \right)$	$C_{i,e}$ (cm^{-1}) $= 4.725 \times 10^{27} \exp\left(\dfrac{-4.3492\,(\text{eV})}{k_B T} \right)$
$C_{v,e}$ (cm^{-1}) $= 7.52 \times 10^{26} \exp\left(\dfrac{-4.0\,(\text{eV})}{k_B T} \right)$	$C_{v,e}$ (cm^{-1}) $= 1.2 \times 10^{27} \exp\left(\dfrac{-4.12\,(\text{eV})}{k_B T} \right)$
C_O (cm^{-1}) $= 9 \times 10^{22} \exp\left(\dfrac{-1.52\,(\text{eV})}{k_B T} \right)$	C_O (cm^{-1}) $= 9 \times 10^{22} \exp\left(\dfrac{-1.52\,(\text{eV})}{k_B T} \right)$
$C_{vO_2,e}$ (cm^{-1}) $= \dfrac{C_O^2}{5 \times 10^{22}} \exp\left(\dfrac{-0.5\,(\text{eV})}{k_B T} \right)$	$C_{vO_2,e}$ (cm^{-1}) $= \dfrac{C_O^2}{5 \times 10^{22}} \exp\left(\dfrac{-0.5\,(\text{eV})}{k_B T} \right)$
λ_i (eV) $= 2.75 – 2.85$*	λ_i (eV) $= 2.75 – 2.85$*
λ_v (eV) $= 1.75$	λ_v (eV) $= 1.75$
λ_O (eV) $= 1.7$	λ_O (eV) $= 1.7$
$\gamma = 0.42$	$\gamma = 0.42$

* Values between 2.75–2.85 eV give acceptable results. The simulations presented in this chapter were performed using $\lambda_i = 2.75$ eV

action equilibria. As the final size of a cluster is far greater than its critical size, the initial size of clusters upon their formation is assumed to be zero. The initial length of a growing crystal is assumed to be finite but negligible. The discussed model describes the defect dynamics in CZ crystals growing under both steady as well as unsteady states.

There is a considerable uncertainty in the parameters describing the properties of many species participating in the CZ defect dynamics. Particularly, properties of self-interstitials are not very well known. Reported self-interstitial migration energies vary from 0.95 to 0.3 eV [38.1, 26, 34, 58–64]. As there are many parameters describing the properties of various species, a reasonably accurate prediction of the general characteristics of the observed microdefect distributions is possible for many different sets of values of these parameters. Two sets of properties, listed in Table 38.2, were used for simulations discussed in this section. The intrinsic point defect properties listed under property set II were derived from the property set proposed by *Kulkarni* et al., on the basis of further fine-tuning to fit experimental data [38.1, 70]. The intrinsic point defect properties proposed by *Voronkov* and *Falster* were fine-tuned to derive property set III [38.64, 70]. The formation energy of vO_2 species and the surface energies of all clusters were tuned to predict experimental data and well-accepted nucleation ranges of self-interstitials, vacancies, and oxygen. Both property sets yield qualitatively similar results.

38.3.3 Defect Dynamics in One-Dimensional Crystal Growth

A CZ crystal growing at a fixed rate through a fixed temperature field remains in a steady state, with respect to a fixed coordinate system, far from the regions formed in the beginning of the growth. A solution of the one-dimensional version of the developed model assuming only axial variation of the microdefect distribution provides insights into the basics of the CZ defect dynamics in the presence of oxygen. For these simulations, the crystal is assumed to grow through a temperature profile described by the linear dependence of $1/T$ on z, according to (38.6).

As discussed in Sect. 38.2, *Voronkov* described the conditions leading to the formation of various microdefects in growing FZ and CZ crystals in the early 1980s, in the absence of oxygen [38.19]. In the interest of continuity and in the context of understanding the influence of oxygen on the CZ defect dynamics, salient features of this theory are discussed again in this section. According to Voronkov's theory, an interplay between the Frenkel reaction and the transport of the intrinsic point defects of silicon determines the concentration fields of the intrinsic point defects in the vicinity of the melt–crystal interface. Vacancies and self-interstitials are assumed to exist at equilibrium at the interface. The temperature drop in the crystal in the vicinity of the interface drives the recombination of vacancies and self-interstitials, decreasing their concentrations. The developed concentration gradients drive the diffusion of vacancies and self-interstitials from the interface into the crystal. The vacancy concentration at the interface is higher than the concentration of self-interstitials, whereas self-interstitials diffuse faster. Thus, when the convective transport is relatively appreciable, vacancies remain the dominant species in the crystal; when the diffusion is relatively appreciable, self-interstitials are replenished at a higher rate from the interface and become the dominant species. Voronkov approximately quantified the relative effect of the convection over the diffusion by the ratio of V to G (V/G). At higher V/G, convection is appreciable; at lower V/G, diffusion is appreciable; at the criti-

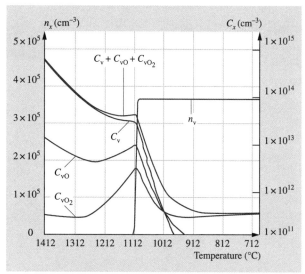

Fig. 38.17 Evolution of the concentrations of various reacting species and the density of v-clusters in a CZ crystal growing at a high rate ($G = 3.5$ K/mm, $V = 0.7$ mm/min, $C_O = 6.25 \times 10^{17}$ cm^{-3} or 12.5 ppma) (after [38.70])

cal V/G, the flux of vacancies is equal to the flux of self-interstitials. This analysis does not take into account the effect of oxygen. Oxygen introduces reactions involving vacancies and oxygen into this dynamics. In the presence of oxygen, free vacancies for the recombination with self-interstitials are supplied from the dissociation of vO and vO$_2$ species as well as from the interface, which is an infinite source. Hence, the presence of oxygen shifts the balance of this dynamics in favor of vacancies.

The evolution of the concentrations of v, vO, and vO$_2$ species as functions of the temperature in a CZ crystal growing at very high V/G, or under highly vacancy-rich conditions, is shown in Fig. 38.17. Near the interface, where the recombination rate is significant, concentrations of all three species decrease, as both free vacancies (v) and bound vacancies (bound as vO and vO$_2$) participate in the recombination reaction; the participation of free vacancies in the recombination is direct, whereas the participation of bound vacancies results through the coupling of reactions (38.1), (38.82), and (38.83). Once the recombination rate decreases, the *total vacancy concentration*, $C_v + C_{vO} + C_{vO_2}$, remains essentially constant. The *bound vacancy concentration*, $C_{vO} + C_{vO_2}$, increases with decreasing temperature because of a shift in the reaction equilibrium. Free vacancies, however, remain dominant and nucleate at around 1100 °C. The growth of voids predominantly consumes all vacancy species, as shown in Fig. 38.17. It must be noted, however, that the residual total vacancy concentration left at lower temperatures remains appreciable because of the binding between vacancies and oxygen.

At close to the critical yet moderately vacancy-rich condition, the free vacancy concentration does not remain high enough to form voids at higher temperatures, in a growing CZ crystal. As the temperature drops further, the concentration of bound vacancies increases. Under these conditions, free vacancies facilitate O-cluster formation at lower temperatures. The formation and growth of O-clusters predominantly consumes both free and bound vacancies, as shown in Fig. 38.18. The predicted total vacancy concentration at lower temperatures is approximate, because of the assumptions discussed in the previous section.

Conditions leading to the growth of crystals free of large v-clusters and i-clusters are desired in many microelectronic applications. Hence, the range of the pull rate within which a CZ crystal free of large clusters can be grown at different oxygen concentrations

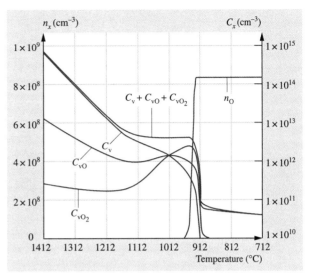

Fig. 38.18 Evolution of the concentrations of various reacting species and the density of O-clusters in a CZ crystal growing close to but moderately above the critical condition ($G = 3.5$ K/mm, $V = 0.48$ mm/min, $C_O = 6.25 \times 10^{17}$ cm^{-3} or 12.5 ppma) (after [38.70])

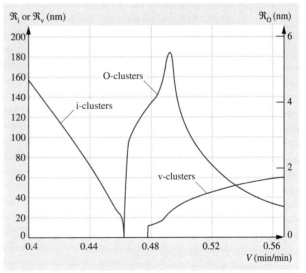

Fig. 38.19 The simulated sizes of various clusters as functions of the CZ crystal pull rate (after [38.70])

is of primary interest in industrial crystal growth. This range can be determined by a series of simulations at different oxygen concentrations at different fixed pull rates. These simulations, however, are computationally expensive. An approximate pull rate range that allows the growth of a crystal without large v-clusters and i-clusters, at a given oxygen concentration, can be determined by simulating the growth at a continuously decreasing rate such that the microdefect distribution in the crystal continuously shifts as a function of the pull rate. A direct but approximate correlation between the microdefect distribution and the pull rate can thus be obtained. Figure 38.19 shows one such simulation, defining the microdefect distribution as a function of the pull rate for a given oxygen concentration and temperature profile. It must be noted that the region free of large v-clusters and i-clusters contains O-clusters. O-clusters are quite large in the vicinity of the boundary between v-clusters and O-clusters, as they are formed at higher temperatures in the presence of a relatively higher vacancy concentration. The size of O-clusters decreases as the vacancy concentration during their formation decreases. A series of such simulations at different oxygen concentrations shows how oxygen affects the range of the pull rate within which the growth of a crystal free of large v-clusters and i-clusters is possible. Oxygen clearly expands this range because of the binding between vacancies and oxygen, as shown in Fig. 38.20. In this figure, v-clusters greater than 20 nm in radius are defined as large. It must be noted that this definition is arbitrary. Figure 38.20 also shows how V/G defining the boundary between O-clusters and i-clusters, known as the v/i boundary, shifts with the oxygen concentration in the discussed one-dimensional crystals growing through the temperature profile defined by (38.6). As discussed before, the presence of vO and vO_2 species near the interface increases the total vacancy concentration available for the recombination with self-interstitials, thus decreasing the V/G marking the v/i boundary; in effect, the crystal becomes marginally more vacancy rich in the presence of oxygen. The surface energy of voids is assumed to be a constant and independent of the oxygen concentration in all these simulations.

The series of one-dimensional simulations discussed in this section establishes the salient effects of oxygen on the CZ defect dynamics. All simulations discussed in this section are performed using property set III.

38.3.4 Defect Dynamics in Two-Dimensional Crystal Growth

The radial variation of the temperature field in a growing CZ crystal and the radial diffusion of the intrinsic point defects, induced by the lateral surface of the crystal and the radial variation of the intrinsic point defect concentration, introduce a two-dimensional variation of the microdefect distribution in the crystal. In addition, variation of the crystal pull rate, commonly observed in modern CZ processes, introduces an axial variation of the microdefect distribution. Hence, it is necessary to validate the discussed model by a comparison of its predictions with the microdefect distribution observed in a crystal grown under an unsteady state representing a variety of possible conditions in modern CZ growth. An experimental crystal was pulled by the varying rate shown in Fig. 38.21. The crystal was cut longitudinally and the microdefect distribution was characterized by the method of copper decoration followed by etching [38.38–40]. The crystal was assumed to grow through a fixed temperature field predicted by the commercial software MARC, using the algorithm developed by *Virzi* [38.37]. As shown in Fig. 38.21, v-clusters are observed in the regions of crystal grown at higher rates, and i-clusters are observed in the regions grown at lower rates. The dense bands at the

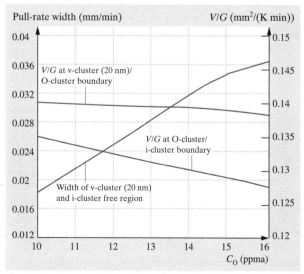

Fig. 38.20 The sensitivity of the microdefect distribution near the critical condition to the oxygen concentration. Note: The i-cluster and large v-cluster free region includes v-clusters smaller than 20 nm (radius) and O-clusters (after [38.70])

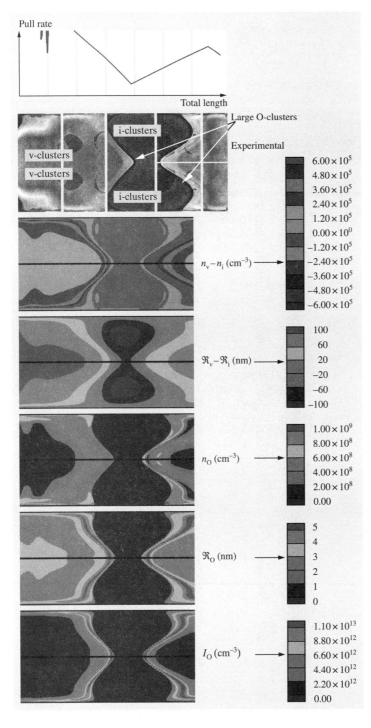

Fig. 38.21 Comparison of the predictions of the model, using property set III, with the experimental observations. The dense bands along the v/i boundary, near the edge of the region containing v-clusters, indicate intense oxygen precipitation (note: scales truncated for clarity) (after [38.70])

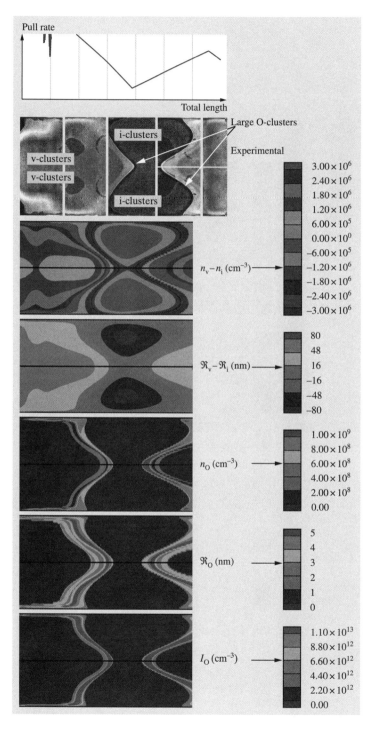

Fig. 38.22 Comparison of the predictions of the developed model, using property set II, with the experimental observations. The dense bands along the v/i boundary, near the edge of the region containing v-clusters, indicate intense oxygen precipitation (note: scales truncated for clarity) (after [38.70])

edge of the region containing large v-clusters, close to the featureless region in the vicinity of i-clusters, indicate the presence of large O-clusters at higher densities. This observation is only qualitative, as all O-clusters are not revealed by the applied characterization technique. The featureless band or region very close to the region of i-clusters can contain unobservable, small microdefects.

The microdefect distributions in the crystal, predicted using property set III, are also shown in Fig. 38.21. As i-clusters and v-clusters do not coexist, a distribution of $\mathcal{R}_v - \mathcal{R}_i$ simultaneously shows the distribution of both types of clusters; a positive value of $\mathcal{R}_v - \mathcal{R}_i$ indicates the size of v-clusters and a negative value of $\mathcal{R}_v - \mathcal{R}_i$ indicates the size of i-clusters. Similarly, the distribution of the densities of i-clusters and v-clusters are quantified by the distribution of $n_v - n_i$. The size and density distribution of O-clusters are shown separately in Fig. 38.21. It must be noted that the lumped model predicts the representative size of an entire population of microdefects present at any given location. In crystal regions formed under the appreciable influence of the convection of the intrinsic point defects, the dominant clusters formed are v-clusters; in regions formed under the appreciable influence of the diffusion of the intrinsic point defects, the dominant clusters formed are i-clusters; near the critical but moderately vacancy-rich condition, O-clusters are dominant. Since the lumped model applies the continuum-scale kinetics and thermodynamics in the prediction of the representative size and density of any given cluster population, it typically predicts negligible but finite oxygen precipitation even under highly vacancy-rich conditions. *Kulkarni* implemented various algorithms to ignore this negligible oxygen precipitation in the developed computer program that solves the model equations; but depending on the operating conditions and the species property set used for the simulation, a negligible but finite oxygen precipitation is predicted by the computer program even under highly vacancy-rich conditions [38.70]. This is evident in Fig. 38.21, which shows a negligible but definite presence of O-clusters in the regions grown under appreciable influence of convection of the intrinsic point defects. Both the size (radius) and density of O-clusters must be studied simultaneously to quantify the intensity of oxygen precipitation. The intensity of oxygen precipitation (I_O) can be measured by the amount of oxygen precipitated per unit volume of the crystal, which is given as

$$I_O = \tfrac{4}{3}\pi \mathcal{R}_O^3 \psi_O^O n_O . \tag{38.108}$$

As shown in Fig. 38.21, the intensity of oxygen precipitation is very high in the regions formed marginally above the critical condition and is negligible in the regions dominated by v-clusters and i-clusters.

The microdefect distributions, in the same crystal shown in Fig. 38.21, predicted using the property set II are shown in Fig. 38.22. The model captures the spatial distributions of various microdefects quite well. Although the predicted densities and sizes of various microdefects are not directly compared with the experimental data, the predictions are consistent with the reported experimental data [38.38, 39, 48].

38.4 Czochralski Defect Dynamics in the Presence of Nitrogen

As microdefects in silicon substrates produced from CZ crystals can adversely affect the performance of devices built on them, the development of crystal growth processes to reduce the size of the microdefects in growing crystals is of industrial significance. Various studies have shown that the microdefect distribution in CZ crystals can be influenced by the presence of nitrogen in general, and that the size of voids can be reduced, in particular [38.20, 71–75]. Various mechanisms to describe the effect of nitrogen on the CZ defect dynamics have been proposed [38.20, 72, 74, 75]. *Voronkov* and *Falster* quantified the effect of nitrogen on the void distribution in an isolated element of a growing CZ crystal [38.74]. An accurate model for the prediction of the effect of nitrogen on CZ defect dynamics, however, must account for the diffusion of mobile reacting species in a growing crystal. It must also incorporate reactions involving vacancies, self-interstitials, oxygen, and complexes of vacancies with oxygen and nitrogen, along with the formation of various microdefects in CZ crystals growing under both steady as well as unsteady states. This chapter discusses such a model, proposed and solved by *Kulkarni*, which quantifies both steady-state and unsteady-state CZ defect dynamics in the presence of nitrogen and oxygen [38.75].

38.4.1 The Model

An accurate quantification of CZ defect dynamics requires a treatment of a series of elementary reactions describing the evolution of the populations of various clusters, which is computationally impractical. Therefore, reasonable approximations are required for the quantification of the microdefect distribution in a CZ crystal. The lumped model proposed by *Kulkarni* and *Voronkov* that approximates a population of microdefects of various sizes present at a given location in the crystal as an equivalent population of identical microdefects is adopted for the treatment of the growth of microdefects [38.53]. All microdefects are approximated as spherical clusters. The stable clusters are assumed to form according to the classical nucleation theory and to grow by diffusion-limited kinetics. The cluster formation and growth dynamics is treated simultaneously with the reactions involving vacancies, self-interstitials, oxygen, nitrogen, and complexes of oxygen and nitrogen.

Reactions in Growing CZ Crystals

The primary reaction influencing the CZ defect dynamics is the Frenkel reaction, which involves the mutual annihilation of a vacancy and a self-interstitial to produce a silicon lattice atom, and the generation of a pair of a vacancy and a self-interstitial from a lattice silicon atom, as shown in reaction (38.1). Oxygen participates in reactions with vacancies and the complexes of vacancies and oxygen, according to (38.82) and (38.83). Nitrogen exists as both a monomer and a dimer, and interacts with vacancies [38.74]

$$N + N \rightleftharpoons N_2, \quad (38.109)$$
$$v + N \rightleftharpoons vN. \quad (38.110)$$

The microdefect distributions in CZ crystals are affected by defect dynamics at fairly higher temperatures, at which nitrogen predominantly exists as N, N_2, and vN; hence, the presence of any other species containing nitrogen can be ignored [38.74]. The species N and N_2 are assumed to exist in the interstitial form in silicon, and vN can be viewed to exist in the substitutional form. The equilibria for reactions (38.109) and (38.110) are given by

$$\frac{C_N^2}{C_{N_2}} = K_{N_2}^{N \leftrightarrow N}, \quad (38.111)$$

$$\frac{C_v C_N}{C_{vN}} = K_{vN}^{v \leftrightarrow N}, \quad (38.112)$$

where K is the equilibrium constant; the subscripts specify the reaction type.

The aggregation of the intrinsic point defects, vacancies and self-interstitials, and the aggregation of oxygen proceed as discussed in Sects. 38.2 and 38.3, respectively. The thermodynamics and kinetics of formation of all clusters are also discussed in Sects. 38.2 and 38.3.

The Governing Equations

The governing equations describe the balances of all species participating in the CZ defect dynamics. The total vacancy concentration, defined as the sum of the concentration of free vacancies (v), and of the cumulative concentration of bound vacancies (vO, vO_2, and vN), is not affected by reactions (38.82), (38.83), (38.109), and (38.110); both free vacancies and self-interstitials are consumed by the Frenkel reaction at an equal rate and are generated at an equal rate. Hence, the rate of change of the *total excess vacancy concentration*, defined as the difference between the total vacancy concentration and the concentration of self-interstitials, is affected only by the transport of individual species, and their consumption by the clusters

$$\frac{\partial \left(C_v + C_{vO} + C_{vO_2} + C_{vN} - C_i \right)}{\partial t}$$
$$+ V \frac{\partial \left(C_v + C_{vO} + C_{vO_2} + C_{vN} - C_i \right)}{\partial z}$$
$$= \nabla \left(D_v \nabla C_v \right) - \nabla \left(D_i \nabla C_i \right) - q_v^v - q_O^v + q_i^i . \quad (38.113)$$

All species except for vacancies and self-interstitials are assumed to be immobile. It must be noted that the formation of N_2 from N (reaction 38.109) requires N to be a mobile species. The diffusivity of N is implicitly assumed to be high enough to validate the assumption of equilibrium of reaction (38.109), but the diffusion flux of N is assumed to be negligible compared with the convective flux of N, in the overall species balance. These are reasonable simplifying assumptions. Equation (38.113) can also be derived from individual balances for all species. The overall nitrogen balance is given by

$$\frac{\partial \left(C_N + 2C_{N_2} + C_{vN} \right)}{\partial t} + V \frac{\partial \left(C_N + 2C_{N_2} + C_{vN} \right)}{\partial z} = 0 . \quad (38.114)$$

The assumption of equilibrium of reactions (38.1), (38.82), (38.83), (38.109), and (38.110) does not necessitate the solution of the individual rate equations. Equations (38.3), (38.94), (38.95), (38.111), and (38.112), which describe the reaction equilibria, and equations (38.113), (38.114), and (38.92) define the balances of i, v, O, vO, vO_2, N, N_2, and vN.

As discussed before, O-clusters are formed in the presence of vacancies without the ejection of silicon atoms as self-interstitials. Once formed, the clusters grow by the consumption of vacancies. At higher vacancy concentrations, the O-cluster growth is assumed to be determined by the diffusion-limited consumption of oxygen by the clusters. At lower vacancy concentrations, the O-cluster growth is assumed to be determined by the diffusion-limited consumption of vacancies by the clusters. The growth of O-clusters by the ejection of silicon atoms as self-interstitials, at lower vacancy concentrations, is ignored in the model. These approximations accurately predict the density of O-clusters but underpredict their size. If desired, the upper limit on the size of O-clusters can be computed by assuming the oxygen diffusion-limited growth of the clusters under all conditions. This assumption is not made in the discussed model, although it can be incorporated without much effort. The volumetric consumption rates of different species by different clusters (i, v, and O), and the average sizes and densities of the cluster populations are given by (38.96–38.107) in Sect. 38.3, as per the methodology developed by *Kulkarni* and *Voronkov* [38.53].

The domain of the calculation is transient, because a CZ crystal is continuously pulled. The described equations must be solved with the equations defining the domain transience, the temperature field in the crystal, and the segregation of nitrogen between the melt and the crystal at the melt–crystal interface.

The heat transport dynamics is assumed to be faster than the defect dynamics. Hence, quasi-steady-state assumptions for the calculation of the temperature field suffice. The temperature field is first calculated by the commercial software MARC, using the algorithms developed by *Virzi*, and then corrected using the experimentally measured interface shape [38.37, 53, 54].

The concentration field of oxygen in a crystal is determined by the process conditions. Hence, the concentration of oxygen in a crystal at the melt–crystal interface is directly assigned on the basis of the process conditions. The concentrations of all vacancy–oxygen complexes are defined by the reaction equilibria, and the intrinsic point defects (vacancies and self-interstitials) are assumed to be at their respective equilibrium concentrations at the interface.

The concentration of nitrogen in a crystal at the melt–crystal interface can be computed by assuming the thermodynamic segregation of nitrogen between the crystal and a well-mixed melt. If it is assumed that nitrogen exists as a monomer in the melt, the thermodynamic segregation of monomeric nitrogen between the crystal and the melt defines its concentration in the crystal, and the concentrations of N_2 and vN species in the crystal are defined by the reaction equilibria (38.111) and (38.112), respectively. The evolution of the nitrogen concentration in the well-mixed melt as a function of time is calculated by following the loss of silicon by solidification and by recognizing that the melt is the only source of nitrogen in all forms (N, N_2, and vN) in the crystal. It is a popular industrial practice, however, to define the relationship between the total nitrogen content in a crystal at the melt–crystal interface and the total nitrogen content in the melt (from which the crystal grows) using an apparent or effective segregation coefficient, which is just an engineering parameter that describes experimental observations. It is reiterated that this apparent segregation coefficient is not a thermodynamic parameter. This approach does not require any explicit assumptions on the state of nitrogen in a melt or a crystal and provides a simple yet reasonably accurate relationship between the total nitrogen content in the crystal and that in the melt. For a given total nitrogen content in the melt, the concentrations of all nitrogen-containing species in the crystal at the interface are defined by the apparent segregation of total nitrogen between the melt and the crystal, the stoichiometric relationship between the total nitrogen concentration and the concentrations of individual nitrogen containing species, and the reaction equilibria (38.111) and (38.112). The evolution of the nitrogen concentration in the well-mixed melt is quantified by following the silicon and nitrogen losses to the growing crystal across the interface. The results discussed in this section are generated using the apparent segregation coefficient, in accordance with popular industrial practice.

The accuracy of the predictions of the model depends on the accuracy of the parameters describing the properties of all reacting species and the parameters defining the thermodynamics of the discussed reactions. The parameters proposed by *Kulkarni* et al. for the simulation of the CZ defect dynamics in the presence of oxygen are used in this study [38.70]. The equilibrium constants defining the equilibria of reactions (38.109) and (38.110), $K_{N_2}^{N \leftrightarrow N}$ and $K_{vN}^{v \leftrightarrow N}$, are adopted from *Voronkov* and *Falster* [38.74]. The apparent segregation coefficient of nitrogen determining its segregation between a crystal and a melt is adopted from *Yatsurugi* et al. and others [38.20, 76, 77]. There is significant uncertainty in the parameters defining the thermodynamics of the reactions and the properties of all species involved in the CZ defect dynamics.

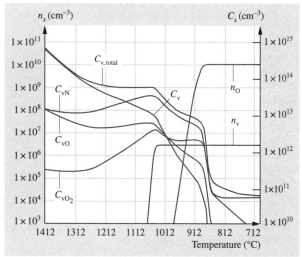

Fig. 38.23 Evolution of the concentrations of various reacting species in a nitrogen-doped CZ crystal growing at a high rate ($G = 3.9$ K/mm, $V = 0.8$ mm/min, $C_{N,total} = 2 \times 10^{14}$ cm^{-3}, $C_O = 7 \times 10^{17}$ or 14 ppma)

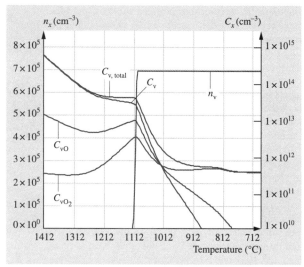

Fig. 38.24 Evolution of the concentrations of various reacting species in a CZ crystal growing at a high rate ($G = 3.9$ K/mm, $V = 0.8$ mm/min, $C_{N,total} = 0$ cm^{-3}, $C_O = 7 \times 10^{17}$ or 14 ppma). Note the difference between the *scale* of the density axis in this figure and in Fig. 38.23

Hence, the parameter set used must be interpreted as just a suggestion.

The developed model can capture the microdefect distributions in the presence of oxygen and nitrogen in CZ crystals growing under steady as well as unsteady states.

38.4.2 CZ Defect Dynamics in One-Dimensional Crystal Growth

The important effects of nitrogen on the CZ defect dynamics are better demonstrated by the simulation of the growth of one-dimensional crystals showing no radial variation in the microdefect distributions. Further simplification can be achieved by simulating steady-state crystal growth. A crystal growing at a fixed rate through a fixed temperature field reaches a steady state far from the regions formed in the beginning of the growth. For simulation of crystal growth at steady state, the crystal is assumed to grow through a fixed representative temperature profile.

Nitrogen affects the reaction dynamics near the melt–crystal interface by interacting with vacancies according to reaction (38.110). The vN species produced by this interaction provides free vacancies (v) by dissociation, for the annihilation of self-interstitials. As free vacancies are also available from the interface, which is an infinite source, the interaction of nitrogen with vacancies shifts the balance of this dynamics in favor of vacancies. The crystal becomes richer in vacancies in the presence of nitrogen. Oxygen also shifts this balance in favor of vacancies by producing vO and vO$_2$ species, which provide free vacancies by their dissociation; but the effect of oxygen is weaker than that of nitrogen [38.59, 70, 74, 75].

The evolution of the concentrations of key reacting species, when the convective transport is appreciable, as functions of the temperature along the length of a nitrogen-doped CZ crystal growing from a very large melt (effectively infinite), is shown in Fig. 38.23. Since the melt is very large, the total nitrogen concentration, $C_N + 2C_{N_2} + C_{vN}$ or $C_{N,total}$ in the crystal is effectively constant. Near the interface, all species containing vacancies participate in the recombination with self-interstitials; the participation of free vacancies is direct, and the participation of vN, vO, and vO$_2$ species is indirect through their dissociation. After the self-interstitial concentration drops significantly, the total vacancy concentration, $C_v + C_{vO} + C_{vO_2} + C_{vN}$ or $C_{v,total}$ remains essentially constant until v-clusters are formed in appreciable numbers at lower temperatures.

Free vacancies directly participate in the formation and growth of v-clusters; the bound vacancies, vO, vO$_2$, and vN species, participate indirectly by their dissociation. The strong binding of vacancies with nitrogen reduces the free vacancy concentration before appreciable nucleation, compared with that in the absence of nitrogen, decreasing the vacancy nucleation temperature. The free vacancy consumption rate by v-clusters is low in the presence of lower vacancy concentration, facilitating their formation in appreciably higher densities. As the supply of total vacancies is fixed at the onset of the vacancy nucleation, the formation of v-clusters at higher densities reduces their size. The effect of nitrogen on the CZ defect dynamics is studied effectively by a comparison of the evolution of various reacting species in a growing CZ crystal in the presence of nitrogen (Fig. 38.23) with that in the absence of nitrogen (Fig. 38.24), under otherwise identical conditions.

A distinct effect of nitrogen is the facilitation of the formation of O-clusters in the presence of voids at appreciable densities. The strong binding between vacancies and nitrogen maintains appreciably high concentrations of bound and free vacancy species at lower temperatures at which vacancy-facilitated O-cluster formation and growth mark the CZ defect dynamics. After their formation, O-clusters become the dominant vacancy sinks compared with v-clusters, as shown in Fig. 38.23. In the absence of nitrogen, however, vN species is also absent, and both weakly bound (vO and vO$_2$) and free vacancies are consumed at higher temperatures primarily by v-clusters, suppressing the formation of O-clusters (Fig. 38.24). In a nitrogen-doped crystal exhibiting moderate dominance of vacancies near the interface, the free vacancy concentration does not remain high enough to form v-clusters at higher temperatures; at lower temperatures, however, O-clusters are formed in the absence of v-clusters (Fig. 38.25). It must be noted that the total vacancy concentration, even in the presence of clusters, remains negligible but finite in a fully grown CZ crystal because of the binding of vacancies with oxygen at lower temperatures.

The pull rate marking the boundary between O-clusters and i-clusters, known as the v/i boundary, represents the critical operating condition. The quantification of the variation of this critical pull rate with the nitrogen concentration, for otherwise fixed crystal growth conditions, is important. For a given temperature field in a crystal-puller, this pull rate can be quantified

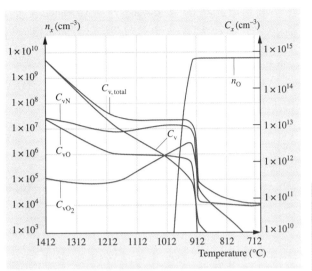

Fig. 38.25 Evolution of the concentrations of various reacting species in a nitrogen-doped CZ crystal growing close to the critical condition ($G = 3.9$ K/mm, $V = 0.50$ mm/min, $C_{N,total} = 2 \times 10^{14}$ cm^{-3}, $C_O = 7 \times 10^{17}$ or 14 ppma)

by a series of crystal growth simulations at various fixed or constant pull rates. These computations are numerous and can be expensive. An approximate variation of the cluster distribution in a crystal as a function of its pull rate can be captured by predicting this distribution in the crystal pulled by either continuously increasing or decreasing rate as shown in Fig. 38.26. One such simulation provides the approximate correlation between the cluster distribution in a crystal and its pull rate. The crystal growth conditions marking the v/i boundary can also be quantified in terms of the popular but approximate growth parameter V/G. The V/G marking this boundary, known as the critical V/G, continuously decreases with increasing nitrogen concentration (Fig. 38.27). The total vacancy concentration near the interface increases with increasing nitrogen concentration because of the strong binding between vacancies and nitrogen, resulting in a decrease in the critical V/G. This shift is particularly significant at higher nitrogen concentrations.

The conditions that allow the growth of CZ crystals in the absence of large clusters are also of interest in the crystal growth industry. Therefore, the quantification of the range of the crystal pull rate or V/G within which CZ crystals can grow free of large v-clusters and i-clusters is important. The large cluster-free range is bracketed by the v/i boundary and the boundary between O-clusters and large v-clusters, arbitrarily defined

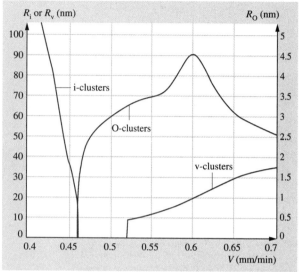

Fig. 38.26 The simulated sizes of various clusters as functions of the CZ crystal pull rate ($G = 3.9$ K/mm, $C_{N,total} = 2 \times 10^{14}$ cm^{-3}, $C_O = 7 \times 10^{17}$ or 14 ppma)

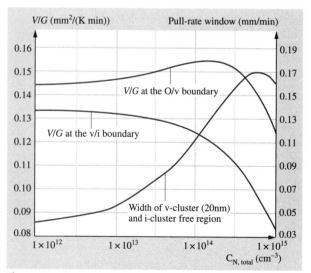

Fig. 38.27 The sensitivity of the cluster distributions near the critical condition to the nitrogen concentration. Note: The v-cluster and i-cluster free region includes v-clusters smaller than 20 nm ($G = 3.9$ K/mm, $C_O = 7 \times 10^{17}$ or 14 ppma)

to be 20 nm large in radius; the latter boundary is termed the O/v boundary. It must be noted that the crystals free of such large clusters contain O-clusters, as discussed before. The size of O-clusters increases with increasing concentration of the free vacancies available during their formation, as shown in Fig. 38.26. The V/G marking the O/v boundary increases with the nitrogen concentration first, reaches a maximum, and then decreases; the increase in this V/G at lower nitrogen concentration, in spite of the decrease in the V/G marking the v/i boundary, is caused by the strong binding between v and N that reduces the free vacancy concentration near the vacancy nucleation temperature range. At moderately higher nitrogen concentrations, however, the shift in the V/G marking the v/i boundary is too strong because of a significant increase in the total vacancy concentration near the interface, forcing a decrease in the V/G defining the O/v boundary. The pull rate range and the V/G range allowing the large cluster-free crystal growth still increase until the nitrogen concentration is significantly high. In the presence of nitrogen at very high concentrations, the total vacancy concentration in the crystal near the interface dramatically increases with the nitrogen concentration such that the free vacancy concentration around the vacancy nucleation temperature increases. Hence, the pull rate range (as well as V/G range) within which crystals free of large clusters can be grown also marginally decreases with increasing nitrogen concentration, at very high nitrogen concentration, as shown in Fig. 38.27. This predicted increase in the size of v-clusters with increasing nitrogen concentration does not agree with reported data [38.75, 78]. Therefore, the model predictions are far less accurate at very high nitrogen doping levels.

The salient features of the CZ defect dynamics and the strengths and weaknesses of the discussed model are thus adequately described by the discussed simulations of one-dimensional crystal growth.

38.4.3 CZ Defect Dynamics in Two-Dimensional Crystal Growth

The observed microdefect distributions in a CZ crystal vary along its axis and radius. This radial variation is caused by the radial variation of the temperature field and the radial diffusion of the intrinsic point defects induced by the temperature field and the lateral surface of a growing crystal. Axial variations are primarily induced by the change in the pull rate and, in the case of nitrogen-doped crystals, by the axial variation in the nitrogen concentration. Quantification of these two-dimensional microdefect distributions requires a solution of the discussed model for two-dimensional crystal growth.

Kulkarni predicted the microdefect distribution in a crystal, at the end of its growth, pulled at a fixed rate through a fixed temperature field, shown in Fig. 38.28 [38.75]. The oxygen concentration in the crystal is assumed to be 16 ppma. The nitrogen concentration varies from 8.25×10^{13} to 6.7×10^{14} cm^{-3}, from one end (left-hand side) of the crystal to the other (right-hand side), and from 8.4×10^{13} to 3.9×10^{14} cm^{-3}, along the central axis within the cylindrical body of the crystal. The crystal growth at a fixed rate well above the pull rate at the critical condition is assumed. The crystal grows under an appreciable influence of convection. The central region of the crystal contains larger v-clusters and relatively smaller O-clusters; the peripheral region is dominated by smaller v-clusters, relative to those in the center, and larger O-clusters, relative to those in the center. This variation, away from the lateral surface of the crystal, is primarily caused by the radial variation of the temperature field; the relative effect of the axial convection with respect to the axial diffusion becomes weaker along the radial position of the crystal, away from the lateral surface, because of the radial variation of the temperature field. The defect distribution in the crystal near the lateral surface is strongly influenced by the surface-induced diffusion of the intrinsic point defects, details of which will be discussed in Sect. 38.5. The vacancy-rich region, defined as the region marked by the presence of v-clusters, O-clusters or any vacancy containing species, expands in the crystal as the nitrogen concentration increases. It must be noted that the microdefect distributions in the crystal when the crystal is still in contact with the melt at the end of its growth are shown in Fig. 38.28. Hence, the vacancy nucleation

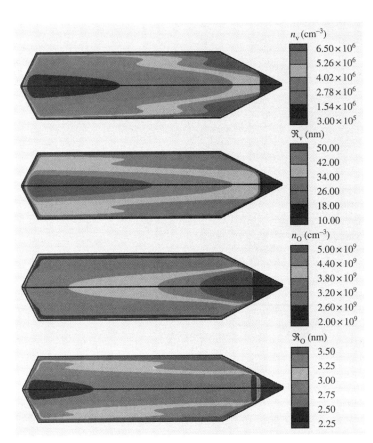

Fig. 38.28 The model predicted cluster distributions in a CZ crystal, pulled at a fixed rate well above the pull rate at the critical condition, at the end of its growth. Note: The melt–crystal interface is not flat. $C_{N,total}$ varies from 8.25×10^{13} to 6.7×10^{14} cm^{-3}, from one end (*left-hand side*) of the crystal to the other (*right-hand side*), and from 8.4×10^{13} to 3.9×10^{14} cm^{-3}, along the central axis within the cylindrical body of the crystal; $C_O = 16$ ppma. The crystal cooling rates are moderate. The *legend* for each plot is shown to the right

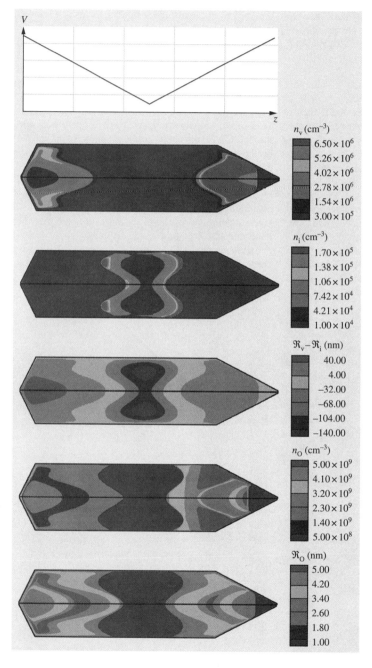

Fig. 38.29 The predicted cluster distributions in a simulated crystal at the end of its growth under transient conditions. Note: The melt–crystal interface is not flat. $C_{N,total}$ varies from 8.25×10^{13} to 6.7×10^{14} cm^{-3}, from one end (*left-hand side*) of the crystal to the other (*right-hand side*), and from 8.4×10^{13} to 3.9×10^{14} cm^{-3}, along the central axis within the cylindrical body of the crystal; $C_O = 16$ ppma. The *legend* for each plot is shown on the *right*

front and the O-cluster formation front are also visible near the right end of the crystal.

The capability of the discussed model can be demonstrated by predicting the microdefect distribution in the crystal pulled at the varying rate shown in Fig. 38.29. The oxygen and nitrogen concentrations in the crystal were set at the same level as those set for the simulation of the crystal discussed in Fig. 38.28 [38.75]. Hence, the effect of varying pull rate can be observed by a comparison between the cluster distributions shown in Fig. 38.28 and those shown in Fig. 38.29. The varying pull rate generates a distribution of all discussed clusters in this chapter. As i-clusters and v-clusters do not coexist, the simultaneous distributions of these clusters can be quantified by the distributions of $\mathcal{R}_v - \mathcal{R}_i$ and $n_v - n_i$. Positive values indicate the distribution of v-clusters and negative values indicate the distribution of i-clusters. Considering their differences, however, the density distributions of i-clusters and v-clusters are shown separately in Fig. 38.29. At a higher pull rate, v-clusters and O-clusters are formed; at a lower pull rate, i-clusters are formed; at a pull rate marking the moderate dominance of vacancies closer to the v/i boundary, only large O-clusters are formed. It must be noted that the microdefect distributions in the crystal at the end of its growth but still in contact with the melt are shown in Fig. 38.29. Hence, the vacancy nucleation front and the oxygen nucleation front are visible near the opposite end of the crystal.

The discussed model captures the salient features of the typical two-dimensional microdefect distributions observed in CZ crystals doped with nitrogen. In the presence of nitrogen at very high concentrations, however, the model predictions are less accurate.

38.5 The Lateral Incorporation of Vacancies in Czochralski Silicon Crystals

A crystal pulled at a very high rate, well above the critical condition, to meet industrial throughput goals, contains voids everywhere except very close to the lateral surface, where the surface-induced diffusion of both the intrinsic point defect species reduces the vacancy concentration at higher temperatures and allows the formation of oxygen clusters, during crystal growth (Fig. 38.30a). It must be noted that the Frenkel reaction couples the vacancy concentration with the self-interstitial concentration; hence, both vacancies that diffuse toward the lateral surface and self-interstitials that diffuse from the surface reduce the vacancy concentration near the surface. The temperature field of a typical growing CZ crystal shows significant radial variation; G increases significantly along the crys-

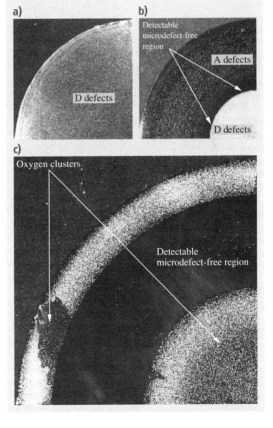

Fig. 38.30 (a) Typical microdefect distribution observed by copper precipitation followed by etching in a wafer produced from a CZ crystal grown at a very high rate well above the critical condition. (b) Typical microdefect distribution observed by copper precipitation followed by etching in a wafer produced from a CZ crystal grown through a radially highly nonuniform temperature field; note that a fraction of the crystal is grown above and the rest below the critical condition. (c) Typical microdefect distribution observed by copper precipitation followed by etching in a wafer produced from a CZ crystal grown close to the critical condition through a relatively radially uniform temperature field (Note: only $\frac{1}{4}$ of the wafer is shown) (after [38.79]) ▶

tal radius. Hence, many Czochralski crystals pulled at a moderate rate contain a central core of D defects and a peripheral region of A defects, as shown in Fig. 38.30b. The narrow detectable microdefect-free region close to the lateral surface of the crystal is also a result of the surface-induced diffusion of self-interstitials and vacancies, during its growth; while fast-diffusing self-interstitials dominantly contribute to their depletion near the surface, vacancies that diffuse from the surface also contribute to the depletion of self-interstitials through Frenkel reaction. Figure 38.30b also shows a very narrow detectable microdefect-free region termed the v/i boundary between the region of D defects and the region of A defects.

Recent advances in device industry demand either the reduction in the size of microdefects or their elimination from the CZ crystal. To address this requirement, many modern crystals are grown under a more radially uniform temperature field close to the critical condition. Hence, typical microdefects formed in these crystals are oxygen clusters and B defects. In practice, a perfect radial uniformity of the temperature field is not achieved; G marginally increases along the radius of a typical crystal. A crystal grown close to the critical condition, however, always exhibits a peripheral region, close to the surface, containing either oxygen clusters or both oxygen clusters and small voids, irrespective of the central microdefect distribution, as shown in Fig. 38.30c. *Kulkarni* described the physics of the evolution of microdefects in this region, known as the *edge ring* or the *peripheral ring* [38.79]. The peripheral ring loses its distinct appearance at higher crystal pull rates. This section addresses the evolution of the microdefect distribution in CZ crystals pulled at various rates, from the critical condition to highly vacancy-rich conditions.

38.5.1 General Defect Dynamics: A Brief Revisit

The theory of the initial point defect incorporation describes the defect dynamics in a growing crystal in the absence of oxygen [38.19]. According to this theory, the formation of microdefects in an oxygen-less crystal is driven by the supersaturation of crystal with either vacancies or self-interstitials. The concentrations of the intrinsic point defects in a growing crystal, away from its lateral surface, evolve primarily by an interplay between the species transport from the melt–crystal interface, both by convection and diffusion, and the Frenkel reaction. The intrinsic point defects of silicon exist at their respective equilibrium concentrations at the interface. The recombination of vacancies and self-interstitials in a growing crystal, driven by the decreasing temperature with increasing distance from the interface, generates their concentration gradients, inducing their diffusion from the interface into the crystal. The species convection is induced by the physical growth of the crystal. Self-interstitials diffuse faster than vacancies near the interface at higher temperatures but the vacancy concentration is higher than the self-interstitial concentration at the interface, where the equilibrium conditions prevail. Thus, when the species diffusion is relatively appreciable, or *below the critical condition*, the self-interstitial flux into the crystal from the interface is greater than the vacancy flux and the concentration of self-interstitials increases, away from the interface, with increasing effect of diffusion relative to convection. When the species convection is relatively appreciable, or *above the critical condition*, the vacancy flux is greater than the self-interstitial flux and the concentration of vacancies, away from the interface, increases with increasing effect of convection relative to diffusion. The dominant intrinsic point defect species annihilates the other species to very low concentrations and aggregates to form microdefects, at lower temperatures. At the critical condition, the vacancy flux is equal to the self-interstitial flux and the two intrinsic point defect species annihilate each other to very low concentrations, without forming large microdefects. This dynamics, defined ignoring the radial diffusion of the reacting species and the effects of oxygen, essentially establishes the incorporated concentration fields of the intrinsic point defects in a growing crystal, prior to the formation of microdefects.

Vacancies also interact with oxygen present in CZ silicon, producing various vacancy–oxygen complexes. The sum of the concentrations of all vacancy-containing species ($C_{v,\text{total}}$), defined as the total vacancy concentration, is relevant in considering the interactions between vacancies and self-interstitials. Free vacancies (v) directly participate in the annihilation of self-interstitials through the Frenkel reaction; bound vacancies (vO and vO_2) are also available for the annihilation of self-interstitials by dissociation. Because the total vacancy concentration available for the annihilation of self-interstitials increases in the presence of oxygen, the crystal becomes richer in vacancies with increasing oxygen concentration. Bound vacancies are assumed to be immobile; hence they affect the CZ defect dynamics only through reactions and physical movement of the crystal.

The model discussed in Sect. 38.3 describes the CZ defect dynamics in the presence of oxygen, and, hence, was applied by *Kulkarni* in his study on the peripheral ring formation. The results of this study are discussed below [38.79].

38.5.2 Defect Dynamics Under Highly Vacancy-Rich Conditions

The simulation of the defect dynamics in a CZ crystal growing close to the critical condition through a radially uniform temperature field can provide insights into the formation of the peripheral ring. To understand the difference between the evolution of microdefects in a CZ crystal growing at a very high rate well above the critical condition and that in a crystal growing close to the critical condition, the simulations of defect dynamics under both conditions must be performed.

Figure 38.31a shows the radial concentration profiles of key reacting species (C_x) in a crystal growing at a very high rate through a radially uniform temperature field. In the interior or bulk of the crystal away from the lateral surface, vacancy supersaturation and self-interstitial undersaturation develop. Since the equilibrium conditions prevail on the lateral surface of the crystal, the self-interstitial concentration gradually increases and the vacancy concentration gradually decreases along the radius of the crystal in the vicinity of the surface, at a given axial location. Both the diffusion of vacancies toward the surface and the diffusion of self-interstitials from the surface into the crystal contribute to the depletion of vacancies near the surface; self-interstitials, which diffuse faster than vacancies at higher temperatures, reduce the vacancy concentration by recombining with vacancies. Although the coupled transport of all species and reactions involving all species determine the species concentration fields in the crystal, the Frenkel reaction and the transport of self-interstitials and vacancies primarily influence their concentration fields, at higher temperatures.

As the vacancy supersaturation increases in the crystal interior (bulk) with decreasing temperature, voids are formed at around 1100 °C. Very close to the lateral surface, however, the radial diffusion of the intrinsic point defects maintains a lower concentration of vacancies. Vacancies in lower concentrations escape consumption by void formation at higher temperatures; they facilitate the formation of oxygen clusters at lower temperatures, around 950 °C. Figure 38.31b shows the radial profile of the amount of vacancies absorbed by voids, per unit volume of the crystal ($C_{v-v\text{-clusters}}$), and

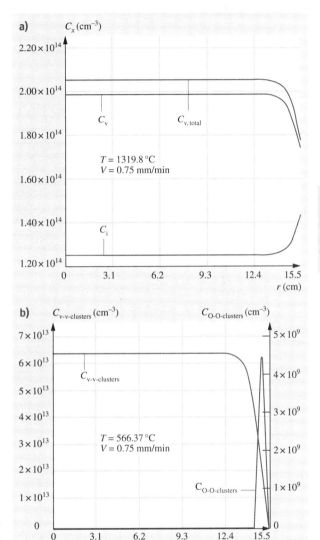

Fig. 38.31 (a) The radial profiles, close to the interface, of the concentrations of key reacting species in a CZ crystal growing well above the critical condition through a radially uniform temperature field. (b) The radial profiles of the concentration of vacancies in voids, per unit volume of crystal, and of oxygen in oxygen clusters, per unit volume of crystal, in a CZ crystal growing well above the critical condition through a radially uniform temperature field

of the amount of oxygen atoms absorbed by oxygen clusters, per unit volume of the crystal ($C_{O-O\text{-clusters}}$). There is negligible formation of oxygen clusters close to the lateral crystal surface when the pull rate is very high, as shown in Fig. 38.31b.

38.5.3 Defect Dynamics Near the Critical Condition

A large microdefect-free crystal can be typically grown if the intrinsic point defect supersaturation in the crystal prior to the onset of nucleation is very low and radially uniform. This is achieved by growing the crystal through a relatively radially uniform temperature field, at close to the critical condition, at which an acceptable balance between the total vacancy flux (convection and diffusion of total vacancies) and the self-interstitial flux (convection and diffusion of self-interstitials) is achieved, in the vicinity of the interface. This discussion

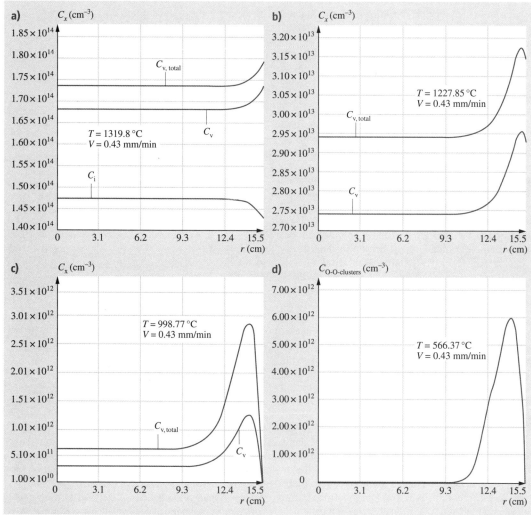

Fig. 38.32a–d The radial profiles of the concentrations of key reacting species in a CZ crystal growing close to the critical condition, through a radially uniform temperature field: (**a**) close to the interface, (**b**) at 1227.85 °C, (**c**) and at 998.77 °C. (**d**) The radial profiles of the concentration of oxygen in oxygen clusters, per unit volume of the crystal, in a CZ crystal growing close to the critical condition through a radially uniform temperature field

focuses on the growth of such marginally vacancy-rich or self-interstitial-rich crystals growing close to the critical condition. As total vacancies and self-interstitials remain in comparable concentrations under these growth conditions, the continuous interplay between the Frenkel reaction and the species transport remains relevant over the entire length of the crystal prior to the microdefect formation. Hence, the concentration fields of the intrinsic point defects and other species are sensitive to the relative shifts in their equilibrium concentrations and diffusivities with temperature.

Figure 38.32a shows the radial profiles of the concentrations of vacancies, self-interstitials, and total vacancies close to the interface of a crystal growing close to the critical condition through a radially uniform temperature field. The vacancy concentration in the bulk or the interior region of the crystal, away from the lateral surface, is higher than the self-interstitial concentration, but the crystal is marginally undersaturated with vacancies and marginally supersaturated with self-interstitials; the equilibrium conditions, however, prevail on the crystal surface. Therefore, self-interstitials diffuse towards the lateral surface from the crystal interior and vacancies diffuse from the lateral surface into the crystal. At higher temperatures, self-interstitials diffuse faster than vacancies and, hence, significantly affect the vacancy concentration field through the Frenkel reaction.

As the temperature decreases in the crystal, the diffusivities of both vacancies and self-interstitials decrease, along with their equilibrium concentrations, which prevail on the lateral surface. The Frenkel reaction remains at equilibrium, as shown by *Kulkarni* et al. [38.1]. All other homogeneous reactions are also assumed to be at equilibrium [38.70]. As the temperature decreases, the influence of the lateral surface-induced diffusion decreases and vacancies that were incorporated from the lateral surface at higher temperatures become the dominant species in the vicinity of the surface, annihilating self-interstitials to lower concentrations. Moreover, both the vacancy concentration at the lateral surface and the vacancy concentration in the crystal interior are lower than the vacancy concentration established in the vicinity of the surface. Hence, a relatively distinct vacancy supersaturation and a self-interstitial undersaturation in the vicinity of the lateral surface are established, as shown in Fig. 38.32b. In short, the interplay among decreasing equilibrium concentrations of all species, the decreasing effect of the species diffusion, and the fast reaction dynamics establishes a self-interstitial-lean and vacancy-rich ring near

the lateral surface. At the stage shown in Fig. 38.32b, self-interstitials diffuse to this region from the lateral

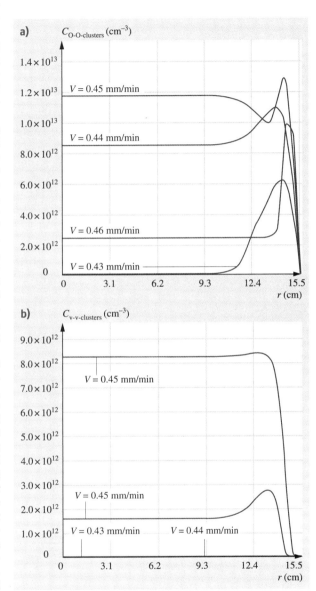

Fig. 38.33a,b The predicted radial profiles of the concentrations of (**a**) oxygen in oxygen clusters, per unit volume of crystal and (**b**) vacancies in voids, per unit volume of crystal, in CZ crystals growing close to the critical condition through a radially uniform temperature field, as functions of the pull rate. The predictions shown in this figure were generated using property set II

surface as well as from the crystal center and vacancies diffuse from this region, both radially inward and outward. Figure 38.32b shows that the interior (bulk) of the crystal is still undersaturated with vacancies when the region near the lateral surface is supersaturated with vacancies at around 1228 °C.

As the temperature drops further, the vacancy undersaturation in the crystal interior (bulk) decreases, because of a fast reduction in the equilibrium vacancy concentration. Eventually, the crystal can become marginally supersaturated with vacancies in the interior. The region near the lateral surface, however, becomes distinctly more supersaturated with vacancies, as shown in Fig. 38.32c.

In essence, the equilibrium conditions prevail on the surface of the growing crystal; near the critical condition, in the first stage of cooling at higher temperatures, the difference between the vacancy concentration and the self-interstitial concentration on the lateral crystal surface is much higher than that in the crystal interior (bulk), driving the vacancy enrichment of the near-surface region relative to the crystal interior; the vacancy supersaturation builds in the near-surface region with the decreasing temperature, because of the decreasing intrinsic point defect diffusivities and equilibrium concentrations.

It must be noted that, as the temperature decreases, the ratio of the bound vacancy concentration to the free vacancy concentration increases. In spite of the vacancy supersaturation established by the lateral surface, the vacancy concentration is not high enough to form large voids. Free and bound vacancies survive high temperatures to facilitate the formation of oxygen clusters (and small voids, in a few cases) at lower temperatures. The radial profile of the concentration of oxygen contained in oxygen clusters, per unit volume of the crystal, is shown in Fig. 38.32d. As can be observed, there is a distinct ring of oxygen clusters near the lateral surface of the crystal, which is the peripheral ring.

The peripheral ring becomes richer in vacancies but loses its distinct structure with increasing crystal pull rate. The microdefect distribution near the critical condition in a crystal as a function of the crystal pull rate is shown in Fig. 38.33. Figure 38.33a shows the radial profiles of $C_{O-O-clusters}$, and Fig. 38.33b shows the $C_{v-v-clusters}$ profiles. The competition between the void formation and the oxygen cluster formation for vacancies is evident as the pull rate increases. The peripheral ring cannot be formed when the relative effect of species convection is significant, because the vacancy supersaturation is maintained everywhere in the crystal under such condition, resulting in the diffusion of vacancies toward the lateral surface close to the interface, as already shown in Fig. 38.31. These results were generated

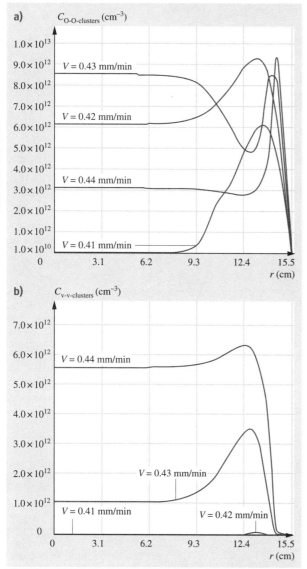

Fig. 38.34a,b The predicted radial profiles of the concentrations of (a) oxygen in oxygen clusters, per unit volume of crystal and (b) vacancies in voids, per unit volume of crystal, in CZ crystals growing close to the critical condition through a radially uniform temperature field, as functions of the pull rate. The predictions shown in this figure were generated using property set III

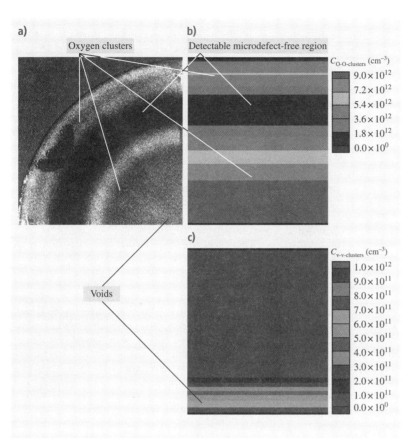

Fig. 38.35 (a) Observed microdefect distribution in a wafer showing a peripheral ring (Note: only 1/4 of the wafer is shown). (b) Predicted $C_{O-O\text{-clusters}}$ distribution in a longitudinal cross-section of the crystal showing a peripheral ring, (c) Predicted $C_{v-v\text{-clusters}}$ distribution in a longitudinal cross-section of the crystal showing a peripheral ring. (Note: the central core of oxygen clusters and voids surrounded by the featureless region is a result of a marginal radial variation in the temperature field)

using property set II; qualitatively similar results were generated using property set III (Fig. 38.34).

A comparison between the predicted microdefect distribution and the observed microdefect distribution in a crystal containing a peripheral ring is shown in Fig. 38.35. The simulation was performed using property set III. The central region of the crystal contains oxygen clusters and small voids. There is a detectable microdefect-free region between the central region containing oxygen clusters and the peripheral ring. This distribution shows that the crystal growth occurred through a radially nonuniform temperature field. However, the presence of the peripheral ring implies that the temperature field was sufficiently uniform.

Influence of the Intrinsic Point Defect Properties on the Edge Ring

There is uncertainty in the properties of many species influencing the CZ defect dynamics. Although the defect dynamics describing the evolution of the peripheral ring is driven by many parameters determining the properties of all participating species, the formation and migration energies of the intrinsic point defects have a particularly stronger influence.

The effect of the properties of the intrinsic point defects on the formation of the peripheral ring in a CZ crystal can be explained by considering the defect dynamics in the absence of oxygen, for simplicity. In the interior region of a CZ crystal growing through a uniform temperature field, the effects of radial species diffusion are negligible; thus, the evolution of the intrinsic point defect concentrations in this region near the melt–crystal interface is explained by the (axial) one-dimensional theory of initial incorporation [38.19]. According to this theory, the concentrations of the intrinsic point defects in an oxygen-less crystal growing at the critical condition through the temperature field approximated by a linear dependence of $1/T$ on z (38.6) evolve according to (38.12) and (38.13), provided that the diffusivities of both species are temperature independent. In a narrow range of temperature near the interface, the assumption of constant species diffusivities, and a linear $1/T$ profile are acceptable, to some extent. Thus, (38.12) and (38.13) can be used as analytical guidelines for evaluating the evolution of the intrinsic point defect concentrations near the interface in a typical Czochralski crystal growing at the critical condition.

According to (38.4), (38.5), (38.12), and (38.13), both of the intrinsic point defect species are at their respective equilibrium concentrations in the interior of the crystal growing at the critical condition, if their formation energies are identical. The equilibrium conditions prevail on the crystal surface as well, because of the fast surface kinetics. Since there is no radial concentration gradient in the crystal under this condition, there is no surface-induced vacancy enrichment at the higher temperatures, which is required for the formation of the peripheral ring.

It is necessary to establish the vacancy undersaturation and self-interstitial supersaturation in the crystal interior (bulk) at higher temperatures to drive the near-surface vacancy enrichment. Moreover, the intensity of the peripheral ring increases with increasing difference between the equilibrium vacancy concentration and the vacancy concentration in the interior of the crystal, at higher temperatures, at the critical condition. This difference increases with decreasing vacancy formation energy, as given by (38.5) and (38.13). This difference also depends on the variation of the intrinsic point defect diffusivities with the temperature. Therefore, the assumption of constant diffusivities of the intrinsic point defects made by *Voronkov* must be relaxed [38.19]. At the critical condition, the self-interstitial flux is equal to the vacancy flux, i.e., the excess intrinsic point defect flux is zero.

When the self-interstitial migration energy is higher than the vacancy migration energy, the relative decrease in the self-interstitial diffusivity is higher than the relative decrease in the vacancy diffusivity, per unit length, in the vicinity of the interface, relatively reducing the replenishment of self-interstitials by diffusion from the interface. Hence, the relative self-interstitial diffusion flux decreases, at any given location in the vicinity of the interface, with increasing self-interstitial migration energy. As the relative self-interstitial diffusion flux decreases, proper changes in the other components of the excess intrinsic point defect flux (38.11) are required to satisfy the necessary equality of the vacancy flux and the self-interstitial flux, at the critical condition. These changes involve a reduction in the vacancy flux by reductions in the critical pull rate (for a given temperature field) and the vacancy concentration. As a result, the established vacancy concentration decreases in the crystal interior with increasing self-interstitial migration energy.

The effect of the formation and migration energies of the intrinsic point defects on the peripheral ring dynamics is multifold. The difference between the vacancy concentration and the self-interstitial concentration on the lateral surface of the crystal at higher temperatures increases with decreasing vacancy formation energy. If the migration energy of a self-interstitial is higher than that of a vacancy, the lateral diffusion of self-interstitials from the surface into the near-surface crystal region, at a later stage of cooling, becomes less significant, which maintains a higher vacancy concentration within the peripheral ring. The near-surface vacancy concentration field is influenced by the diffusion of both self-interstitials and vacancies, coupled with the Frenkel reaction, and not by the vacancy diffusion alone. It must be noted that the migration energy of a self-interstitial is higher than the migration energy of a vacancy, according to property set II and that the formation energy of a vacancy is lower than that of a self-interstitial, according to property set III. It is also important to note that the presence of oxygen and a deviation from the assumed crystal temperature field described by the linear variation of $1/T$ along the crystal axis also influence the dynamics of the formation of the peripheral ring, in addition to the formation and migration energies of the intrinsic point defects.

38.6 Conclusions

The quality of modern microelectronic devices built on silicon substrates manufactured by the Czochralski (CZ) and float-zone (FZ) crystal growth processes is determined by the distribution of the crystallographic imperfections known as microdefects present in the substrates. Microdefects are formed by the aggregation of the intrinsic point defects of silicon, vacancies (v) and self-interstitials (i), and by the vacancy-assisted aggregation of oxygen (O) with silicon. Understanding the dynamics of the formation of these aggregates requires the quantification of this dynamics.

38.6.1 CZ Defect Dynamics in the Absence of Impurities

Early attempts at the quantification of defect dynamics were restricted to understanding the distribution of the intrinsic point defects in growing crystals, without accounting for the formation of microdefects, or the effects of oxygen. The theory of initial incorporation, proposed in 1982, explains and quantifies the conditions leading to the formation of microdefects [38.19]. According to this theory, in a growing CZ crystal, an interplay among the diffusion and convection of the intrinsic point defects and the Frenkel reaction driven by decreasing temperature near the vicinity of the melt–crystal interface establishes a dominant intrinsic point defect species, which nucleates at lower temperatures to form its aggregates. Vacancies, which exist at a higher concentration at the melt–crystal interface, remain dominant when the species convection is relatively appreciable; self-interstitials, which are fast diffusers in the higher temperature range, become the dominant species when the species diffusion is relatively appreciable. Vacancies form vacancy aggregates known as D defects; self-interstitials form self-interstitial aggregates known as A and B defects. This interplay is quantified in terms of the ratio of the crystal pull rate (V) to the magnitude of the axial temperature gradient at the interface (G), V/G, at any radial location of the crystal. Thus, D defects are formed at a radial location with higher V/G, A and B defects are formed at lower V/G, and no large microdefects are formed at V/G closer to its critical value. The radial distribution of microdefects in many crystals is reasonably explained by the radial variation of G.

The one-dimensional theory of initial incorporation does not effectively capture the effects of the radial diffusion of the intrinsic point defects induced by the radial variation in the temperature field and by the lateral surface of a crystal. Therefore, in the 1990s, many successful efforts were made to quantify the two-dimensional intrinsic point defect concentration fields in growing axisymmetric CZ crystals, to more accurately predict the microdefect distributions. These efforts, however, did not capture the effects of oxygen or the microdefect size distribution, as the actual events of the intrinsic point defect aggregation were not quantified.

Successful quantification of the microdefect distributions in CZ crystals, ignoring the effects of oxygen, was made only in and after the late 1990s. Capturing the CZ defect dynamics including the aggregation of the intrinsic point defects involves the quantification of the microdefect populations everywhere in a crystal. A microdefect population, at any given time and location in a crystal, consists of the microdefects formed at various locations during the elapsed time period. Therefore, a direct numerical treatment of the defect dynamics is impractical. Three popular approaches emerged to address the computational complexity of the CZ defect dynamics, in and after the late 1990s. All three approaches approximate the microdefects as spherical clusters; vacancy clusters or v-clusters represent D defects and self-interstitial clusters or i-clusters represent A and B defects.

The first approach, in its most rigorous form, requires the application of the classical nucleation theory to quantify the formation of the stable cluster nuclei and the prediction of the final microdefect distribution in a CZ crystal by quantifying the formation and growth histories of all clusters in the cluster populations everywhere in the crystal. Thus, the microdefect distribution is a function of location, time, and elapsed time. As the quantification of the CZ defect dynamics by the most rigorous form of this approach is computationally expensive, it is applied in its various reduced forms. An elegant treatment of the cluster formation and growth in an isolated element of a CZ crystal, cooled at a rate determined by its time–temperature history during the crystal growth, was first published in 1998. This treatment also quantifies the effect of the interesting interplay between the conditions facilitating the nucleation of the intrinsic point defects and the conditions facilitating the growth of the formed clusters on the final average size and the density of the clusters. This approximation, however, does not take into account the effect of the axial diffusion of the intrinsic point defects. The most elaborate model based on the first approach, reported in 2004, treats only the one-dimensional (axial) CZ defect dynamics. This model, however, is quite rigorous and addresses both the steady-state and unsteady-state defect dynamics. The shifts in the critical V/G associated with unsteady-state crystal growth were quantified for the first time by the unsteady-state model in 2004. The basic model used in the quantification of the defect dynamics by the first approach is termed the *rigorous model*.

The second approach followed in the quantification of the CZ defect dynamics approximates a population of clusters present at any location in a crystal by another population of identical clusters. This is accomplished by approximating the average radius of the clusters in a population by the square root of the average of the squared radii of the clusters and by quantifying an auxiliary variable proportional to the total surface area of the clusters. The model based on this approach is termed the *lumped model*. The lumped model is computationally more attractive and effectively quantifies the two-dimensional microdefect distributions in CZ crystals.

The third approach followed in capturing the CZ defect dynamics couples the direct treatment of the reactions taking part in the aggregation events, for smaller clusters, and the Fokker–Plank formulation, for larger clusters. The formation of clusters and their growth is initially quantified by treating clusters of the same size and type as a species participating in the reactions. The discrete reaction rate equations derived by applying the classical nucleation theory are used for capturing the evolution of smaller clusters. The concentration (density) of all clusters of the same type exceeding a threshold size is treated as a continuous variable, which is a function of location, time, and the cluster size itself, by using the Fokker–Plank formulation. The Fokker–Plank equation (FPE) for larger clusters is derived from the discrete rate equations by a Kramers–Moyal expansion, treating the cluster size as a continuous independent variable. This approach is termed the *discrete–continuous approach* and is computationally very expensive.

38.6.2 CZ Defect Dynamics in the Presence of Oxygen

CZ crystals contain oxygen, which influences the defect dynamics. One-dimensional steady-state CZ defect dynamics in the presence of oxygen in a growing CZ crystal showing no radial variations has been modeled by both the Fokker–Planck formulation and the lumped model. The unsteady-state oxygen-influenced CZ defect dynamics accounting for the two-dimensional variations in microdefect distributions in growing CZ crystals, however, has been quantified only by the lumped model. The presence of oxygen primarily generates two bound vacancy species, vO and vO$_2$, and the aggregates of oxygen. The aggregates of vacancies are modeled as spherical v-clusters; the aggregates of self-interstitials are modeled as spherical i-clusters; and the aggregates of oxygen, primarily silicon dioxide, are modeled as spherical O-clusters.

The key element of the model describing the oxygen-influenced CZ defect dynamics is the vacancy-assisted formation of O-clusters. Effectively, all large O-clusters in the CZ growth are formed by absorbing vacancies, as the specific volume of O-clusters is greater than that of silicon. The growing O-clusters directly consume only free vacancies (v); as the free vacancy concentration decreases, however, more free vacancies are generated by the disassociation of vO and vO$_2$ species. Thus, both free vacancies and vacancies bound in vO and vO$_2$ species are consumed.

The type of microdefect formed in a given region in a crystal depends on the concentration of the intrinsic point defects and of vO and vO$_2$ species established a short distance away from the melt–crystal interface. In the regions marked by a high free vacancy concentration, voids or v-clusters are formed at higher temperatures by the nucleation of vacancies. The v-cluster growth consumes both free and bound vacancies. In the regions marked by a moderate free vacancy concentration, v-cluster formation is suppressed at higher temperatures; free and bound vacancies are consumed by the formation and growth of O-clusters. The binding between vacancies and oxygen allows survival of vacancies in the bound form at very low concentrations at lower temperatures even in the presence of voids and O-clusters. In the regions marked by the dominance of self-interstitials, i-clusters are formed. The concentration fields of the intrinsic point defects in the vicinity of the interface are established primarily by the interplay between the Frenkel reaction and the intrinsic point defect transport. Oxygen increases the effective vacancy concentration available for the recombination with self-interstitials by increasing the concentration of vO and vO$_2$ species and marginally aids the conditions leading to the survival of vacancies as the dominant intrinsic point defect species, for fixed crystal growth conditions. The increase in the pull rate range within which crystals free of large v-clusters and i-clusters can be grown, with increasing oxygen concentration, is also predicted and explained by the reported lumped model. This behavior is caused by a strong binding between vacancies and oxygen.

38.6.3 CZ Defect Dynamics in the Presence of Nitrogen

The microdefect distribution in a CZ crystal is strongly influenced by impurities such as nitrogen. A reasonably good approximation of the dynamics of microdefect formation in growing CZ crystals in the presence of nitrogen and oxygen requires the quantification of the Frenkel reaction, the interactions between oxygen and vacancies and between nitrogen and vacancies, along with the aggregation events. Vacancies and self-interstitials annihilated by the recombination near the interface are partly supplied by their transport from the interface and by the dissociation of different reacting species. Nitrogen binding with vacancies is much stronger than that of oxygen with vacancies. The complexes of vacancies with nitrogen in particular, and oxygen to some extent, provide free vacancies by dissociation for the annihilation of self-interstitials near the melt–crystal interface. These free vacancies are available in addition to the vacancies from the interface, which is an infinite vacancy source. This dynamics increases the total concentration of vacancies in all forms, defined as the total vacancy concentration, and establishes relatively vacancy-rich conditions compared with those in the absence of nitrogen. In addition, nitrogen reduces the free vacancy concentration by strong binding at lower temperatures, reducing the vacancy nucleation temperature. The formation of vacancy aggregates in the presence of lower vacancy concentration results in an increase in their density at the expense of their size. The strong binding between vacancies and nitrogen provides an appreciable supply of vacancies for the facilitation of the formation and growth of the aggregates of oxygen with silicon at lower temperatures. Hence, vacancy aggregates and oxygen aggregates coexist in a wide range of crystal growth conditions in the presence of nitrogen.

The defect dynamics in nitrogen-doped CZ crystals, showing two-dimensional variations in the microdefect distribution and growing under both steady and unsteady states, has been quantified by the lumped model. The model approximates all aggregates as spherical clusters. Classical nucleation theory captures the formation of all clusters with a reasonable accuracy. The formed clusters grow by diffusion-limited kinetics. Vacancy clusters and self-interstitial clusters are formed by homogeneous nucleation of vacancies and self-interstitials, respectively. Oxygen clusters, because of their higher specific volume, are formed by the facilitation by vacancies. The growth of oxygen clusters is limited by the diffusion of oxygen when vacancies are in abundance and by the diffusion of vacancies when vacancies are scarce.

The conditions leading to the formation of different microdefects in CZ crystal growth have been quantified. The effects of varying pull rate and the nitrogen concentration are also captured. The model predictions agree well with the reported microdefect distributions in the presence of moderate and high nitrogen concentrations. In the presence of very high nitrogen concentrations, however, the model predictions are less accurate.

38.6.4 The Lateral Incorporation of Vacancies

A Czochralski crystal grown through a relatively radially uniform temperature field close to the critical condition is effectively free of large microdefects, but exhibits a prominent region of oxygen clusters (and small voids, in a few cases) near the lateral surface. Because of its ring-like appearance, the region is termed the peripheral ring. The peripheral ring is formed because of the vacancy supersaturation induced by the lateral surface of the crystal. The vacancies in moderate abundance facilitate the formation of oxygen clusters (and, possibly, small voids) at lower temperatures, giving the peripheral ring its characteristic appearance.

The equilibrium conditions prevail on the surface of a growing crystal; near the critical condition, in the first stage of cooling at higher temperatures, the difference between the vacancy concentration and the self-interstitial concentration on the lateral crystal surface is much higher than that in the crystal interior (bulk), driving the vacancy enrichment of the near-surface region relative to the crystal interior (bulk); the vacancy supersaturation builds in the near-surface region with the decreasing temperature, because of the decreasing intrinsic point defect equilibrium concentrations and diffusivities.

The moderate supersaturation of vacancies prior to the formation of oxygen clusters and small voids in the peripheral ring, when the bulk of the crystal is relatively vacancy lean, is driven by the Frenkel reaction, lateral surface-induced diffusion, and the crystal movement itself. This dynamics is termed the *lateral incorporation*. The lateral incorporation generates the *peripheral ring* only in crystals grown close to the critical condition.

References

38.1 M.S. Kulkarni, V.V. Voronkov, R. Falster: Quantification of unsteady-state and steady-state defect dynamics in the Czochralski growth of monocrystalline silicon, J. Electrochem. Soc. **151**, G663–G678 (2004)

38.2 W.C. Dash: Silicon crystals free of dislocations, J. Appl. Phys. **29**, 736–737 (1958)

38.3 W.C. Dash: Growth of silicon crystals free from dislocations, J. Appl. Phys. **30**, 459–474 (1959)

38.4 T. Abe, T. Samizo, S. Maruyama: Etch pits observed in dislocation free silicon crystals, Jpn. J. Appl. Phys. **5**, 458–459 (1966)

38.5 A.J.R. de Kock: Vacancy clusters in dislocation-free silicon, Appl. Phys. Lett. **16**, 100–102 (1970)

38.6 A.J.R. de Kock: The elimination of vacancy-cluster formation in dislocation-free silicon crystals, J. Electrochem. Soc. **118**, 1851–1856 (1971)

38.7 A.J.R. de Kock: Microdefects in dislocation-free silicon and germanium crystals, Acta Electron. **16**, 303 (1973)

38.8 A.J.R. de Kock, P.J. Roksnoer, P.G.T. Boonen: Effect of growth parameters on formation and elimination of vacancy clusters in dislocation-free silicon crystals, J. Cryst. Growth **22**, 311–320 (1974)

38.9 A.J.R. de Kock, P.J. Roksnoer, P.G.T. Boonen: Formation and elimination of growth striations In dislocation-free silicon crystals, J. Cryst. Growth **28**, 125–137 (1975)

38.10 P.M. Petroff, A.J.R. de Kock: Characterization of swirl defects in floating-zone silicon crystals, J. Cryst. Growth **30**, 117–124 (1975)

38.11 P.M. Petroff, A.J.R. de Kock: The formation of interstitial swirl defects in dislocation-free floating-zone silicon crystals, J. Cryst. Growth **36**, 4–12 (1976)

38.12 H. Föll, U. Gösele, B.O. Kolbesen: The formation of swirl defects in silicon by agglomeration of self-interstitials, J. Cryst. Growth **40**, 90–108 (1977)

38.13 A.J.R. de Kock, W.M. van de Wijert: The effect of doping on the formation of swirl defects in dislocation-free Czochralski-grown silicon crystals, J. Cryst. Growth **49**, 718–734 (1980)

38.14 P.J. Roksnoer, M.M.B. van den Boom: Microdefects in a non-striated distribution in floating-zone silicon crystals, J. Cryst. Growth **53**, 563–573 (1981)

38.15 P.J. Roksnoer: The mechanism of formation of microdefects in silicon, J. Cryst. Growth **68**, 596–612 (1984)

38.16 M. Hasebe, Y. Takeoka, S. Shinoyama, S. Naito: Formation process of stacking faults with ringlike distribution in CZ-Si wafers, Jpn. J. Appl. Phys. **28**, L1999–L2002 (1989)

38.17 M. Kato, T. Yoshida, Y. Ikeda, Y. Kitagawara: Transmission electron microscope observation of "IR scattering defects" in As-grown Czochralski Si crystals, Jpn. J. Appl. Phys. **35**, 5597–5601 (1996)

38.18 T. Ueki, M. Itsumi, T. Takeda: Octahedral void defects observed in the bulk of Czochralski silicon, Appl. Phys. Lett. **70**, 1248–1250 (1997)

38.19 V.V. Voronkov: The mechanism of swirl defects formation in silicon, J. Cryst. Growth **59**, 625–643 (1982)

38.20 M. Iida, W. Kusaki, M. Tamatsuka, E. Iino, M. Kimura, S. Muraoka: Effects of light element impurities on the formation of grown-in defects free region of Czochralski silicon single crystal, Proc. Electrochem. Soc. **99-1**, 499–510 (2000)

38.21 E. Dornberger, D. Gräf, M. Suhren, U. Lambert, P. Wagner, F. Dupret, W. von Ammon: Influence of boron concentration on the oxidation-induced stacking fault ring in Czochralski silicon crystals, J. Cryst. Growth **180**, 343–352 (1997)

38.22 T. Sinno, H. Susanto, R.A. Brown, W. von Ammon, E. Dornberger: Boron-retarded self-interstitial diffusion in Czochralski growth of silicon crystals and its role in oxidation-induced stacking-fault ring dynamics, Appl. Phys. Lett. **75**, 1544–1546 (1999)

38.23 V.V. Voronkov, R. Falster: Dopant effect on point defect incorporation into growing silicon crystal, J. Appl. Phys. **87**, 4126–4129 (2000)

38.24 K. Terashima, H. Noguchi: The effects of boron impurity on the extended defects in CZ silicon crystals grown under interstitial rich conditions, J. Cryst. Growth **237-239**, 1663–1666 (2002)

38.25 M.S. Kulkarni: A selective review of the quantification of defect dynamics in growing Czochralski silicon crystals, Ind. Eng. Chem. Res. **44**, 6246–6263 (2003)

38.26 T. Sinno, R.A. Brown, W. von Ammon, E. Dornberger: Point defect dynamics and the oxidation-induced stacking-fault ring in Czochralski-grown silicon crystals, J. Electrochem. Soc. **145**, 302–318 (1998)

38.27 T. Mori: Modeling the Linkages Between Heat Transfer and Microdefect Formation in Crystal Growth: Examples of Czochralski Growth of Silicon and Vertical Bridgman Growth of Bismuth Germanate. Ph.D. Thesis (Massachusetts Institute of Technology, Massachusetts 2000)

38.28 T.Y. Tan, U. Gösele: Point defects, diffusion processes, and swirl defect formation in silicon, Appl. Phys. A **37**, 1–17 (1985)

38.29 S. Sadamitsu, S. Umeno, Y. Koike, M. Hourai, S. Sumita, T. Shigematsu: Dependence of the grown-in defect distribution on growth rates in Czochralski silicon, Jpn. J. Appl. Phys. **32**, 3675–3681 (1993)

38.30 W. von Ammon, E. Dornberger, H. Oelkrug, H. Weidner: The dependence of bulk defects on the

38.31 axial temperature gradient of silicon crystals during Czochralski growth, J. Cryst. Growth **151**, 273–277 (1995)

38.31 E. Dornberger, W. von Ammon: The dependence of ring-like distributed stacking faults on the axial temperature gradient of growing Czochralski silicon crystals, J. Electrochem. Soc. **143**, 1648–1653 (1996)

38.32 E. Dornberger, W. von Ammon, N. Van den Bogaert, F. Dupret: Transient computer simulation of a CZ crystal growth process, J. Cryst. Growth **166**, 452–457 (1996)

38.33 R. Falster, V.V. Voronkov, J.C. Holzer, S. Markgraf, S. McQuaid, L. Mulèstagno: Intrinsic point defects and reactions in the growth of large silicon crystals, Proc. Electrochem. Soc. **98**(1), 468–489 (1998)

38.34 K. Nakamura, T. Saishoji, T. Kubota, T. Iida, Y. Shimanuki, T. Kotooka, J. Tomioka: Formation process of grown-in defects in Czochralski grown silicon crystals, J. Cryst. Growth **180**, 61–72 (1997)

38.35 P.A. Ramachandran, M.P. Dudukovic: Simulation of temperature distribution in crystals grown by Czochralski method, J. Cryst. Growth **71**, 399–408 (1985)

38.36 R.K. Srivastava, P.A. Ramachandran, M.P. Dudukovic: Interface shape in Czochralski grown crystals: Effect of conduction and radiation, J. Cryst. Growth **73**, 487–504 (1985)

38.37 A. Virzi: Computer modelling of heat transfer in Czochralski silicon crystal growth, J. Cryst. Growth. **112**, 699–722 (1991)

38.38 M.S. Kulkarni, J. Libbert, S. Keltner, L. Mulèstagno: A theoretical and experimental analysis of macrodecoration of defects in monocrystalline silicon, J. Electrochem. Soc. **149**(2), G153–G165 (2002)

38.39 M.S. Kulkarni: A review and unifying analysis of defect decoration and surface polishing by chemical etching in silicon processing, Ind. Eng. Chem. Res. **42**, 2558–2588 (2003)

38.40 M.S. Kulkarni, H.F. Erk: Acid-based etching of silicon wafers: mass-transfer and kinetic effects, J. Electrochem. Soc. **147**, 176–188 (2000)

38.41 R.A. Brown, D. Maroudas, T. Sinno: Modelling point defect dynamics in the crystal growth of silicon, J. Cryst. Growth **137**, 12–25 (1994)

38.42 N.I. Puzanov, A.M. Eidenzon, D.N. Puzanov: Modelling microdefect distribution in dislocation-free Si crystals grown from the melt, J. Cryst. Growth **178**, 468–478 (1997)

38.43 E. Dornberger, W. von Ammon, J. Virbulis, B. Hanna, T. Sinno: Modelling of transient point defect dynamics in Czochralski silicon crystals, J. Cryst. Growth **230**, 291–299 (2001)

38.44 M. Okui, M. Nishimoto: Effect of the axial temperature gradient on the formation of grown-in defect regions in Czochralski silicon crystals; reversion of the defect regions between the inside and outside of the Ring-OSF, J. Cryst. Growth **237**, 1651–1656 (2002)

38.45 V.V. Kalaev, D.P. Lukanin, V.A. Zabelin, Y.N. Makarov, J. Virbulis, E. Dornberger, W. von Ammon: Calculation of bulk defects in CZ Si growth: impact of melt turbulent fluctuations, J. Cryst. Growth **250**, 203–208 (2003)

38.46 V.V. Voronkov, R. Falster: Vacancy-type microdefect formation in Czochralski silicon, J. Cryst. Growth **194**, 76–88 (1998)

38.47 T. Sinno, R.A. Brown: Modeling microdefect formation in Czochralski silicon, J. Electrochem. Soc. **146**, 2300–2312 (1999)

38.48 Z. Wang, R.A. Brown: Simulation of almost defect-free silicon crystal growth, J. Cryst. Growth **231**, 442–447 (2001)

38.49 R.A. Brown, Z. Wang, T. Mori: Engineering analysis of microdefect formation during silicon crystal growth, J. Cryst. Growth **225**, 97–109 (2001)

38.50 E. Dornberger, J. Virbulis, B. Hanna, R. Hölzl, E. Daub, W. von Ammon: Silicon crystals for future requirements of 300 mm wafers, J. Cryst. Growth **229**, 11–16 (2001)

38.51 E. Dornberger, D. Temmler, W. von Ammon: Defects in silicon crystals and their impact on DRAM device characteristics, J. Electrochem. Soc. **149**, G226–G231 (2002)

38.52 K. Kitamura, J. Furukawa, Y. Nakada, N. Ono, Y. Shimanuki, A.M. Eidenzon, N.I. Puzanov, D.N. Puzanov: Radial distribution of temperature gradients in growing CZ-Si crystals and its application to the prediction of microdefect distribution, J. Cryst. Growth **242**, 293–301 (2002)

38.53 M.S. Kulkarni, V.V. Voronkov: Simplified two-dimensional quantification of the microdefect distributions in silicon crystals grown by the Czochralski process, J. Electrochem. Soc. **152**, G781–G786 (2005)

38.54 M.S. Kulkarni, J.C. Holzer, L.W. Ferry: The agglomeration dynamics of self-interstitials in growing Czochralski silicon crystals, J. Cryst. Growth **284**, 353–368 (2005)

38.55 D. Turnbull, J.C. Fisher: Rate of nucleation in condensed systems, J. Chem. Phys. **17**, 71–73 (1949)

38.56 A.S. Michaels: *Nucleation Phenomena* (American Chemical Society, Washington 1966)

38.57 D. Kashchiev: *Nucleation, Basic Theory with Applications* (Butterworth-Heinemann, Oxford 2000)

38.58 V.V. Voronkov, R. Falster: Vacancy and self-interstitial concentration incorporated into growing silicon crystals, J. Appl. Phys. **86**, 5975–5982 (1999)

38.59 V.V. Voronkov, R. Falster: Intrinsic point defects and impurities in silicon crystal growth, J. Electrochem. Soc. **194**, G167–G174 (2002)

38.60 S.M. Hu: Nonequilibrium point defects and diffusion in silicon, Mater. Sci. Eng. **R13**(3/4), 105–192 (1994)

38.61 P.M. Fahey, P.B. Griffin, J.D. Plummer: Point defects and dopant diffusion in silicon, Rev. Mod. Phys. **61**, 289–384 (1989)

38.62 U. Gösele, D. Conrad, P. Werner, Q.Y. Tong, R. Gafiteanu, T.Y. Tan: Point defects, diffusion and gettering in silicon, Mater. Res. Soc. Symp. Proc. **469**, 13 (1997)

38.63 R. Falster, V.V. Voronkov, F. Quast: On the properties of the intrinsic point defects in silicon: a perspective from crystal growth and wafer processing, Phys. Status Solidi (b) **222**, 219–244 (2000)

38.64 V.V. Voronkov, R. Falster: Parameters of intrinsic point defects in silicon based on crystal growth, wafer processing, self- and metal-diffusion, ECS Trans. **2**(2), 61–75 (2006)

38.65 J. Esfandyari, C. Schmeiser, S. Senkader, G. Hobler, B. Murphy: Computer simulation of oxygen precipitation in Czochralski-grown silicon during HI-LO-HI anneals, J. Electrochem. Soc. **143**, 995–1001 (1996)

38.66 V.V. Voronkov, R. Falster: Grown-in microdefects, residual vacancies and oxygen precipitation bands in Czochralski silicon, J. Cryst. Growth **204**, 462–474 (1999)

38.67 V.V. Voronkov, R. Falster: Strain-induced transformation of amorphous spherical precipitates into platelets: application to oxide particles in silicon, J. Appl. Phys. **89**, 5965–5971 (2001)

38.68 V.V. Voronkov, R. Falster: Nucleation of oxide precipitates in vacancy-containing silicon, J. Appl. Phys. **91**, 5802–5810 (2002)

38.69 K. Sueoka, M. Akatsuka, M. Okui, H. Katahama: Computer simulation for morphology, size, and density of oxide precipitates in CZ silicon, J. Electrochem. Soc. **150**, G469–G475 (2003)

38.70 M.S. Kulkarni: Defect dynamics in the presence of oxygen in growing Czochralski silicon crystals, J. Cryst. Growth **303**, 438–448 (2007)

38.71 K. Nakai, Y. Inoue, H. Yokota, A. Ikari, J. Takahashi, K. Kitahara, Y. Ohta, W. Ohashi: Oxygen precipitation in nitrogen-doped Czochralski-grown silicon crystals, J. Appl. Phys. **89**, 4301–4309 (2001)

38.72 W. von Ammon, R. Hölzl, J. Virbulis, E. Dornberger, R. Schmolke, D. Gräf: The impact of nitrogen on the defect aggregation in silicon, J. Cryst. Growth **226**, 19–30 (2001)

38.73 A. Karoui, F.S. Karoui, G.A. Rozgonyi, M. Hourai, K. Sueoka: Structure, energetics, and thermal stability of nitrogen-vacancy-related defects in nitrogen doped silicon, J. Electrochem. Soc. **150**, G771–G777 (2003)

38.74 V.V. Voronkov, R. Falster: The effect of nitrogen on void formation in Czochralski crystals, J. Cryst. Growth **273**, 412–423 (2005)

38.75 M.S. Kulkarni: Defect dynamics in the presence of nitrogen in growing czochralski silicon crystals, J. Cryst. Growth **310**, 324–335 (2008)

38.76 Y. Yatsurugi, N. Akijama, Y. Endo, T. Nozaki: Concentration, solubility, and equilibrium distribution coefficient of nitrogen and oxygen in semiconductor silicon, J. Electrochem. Soc. **120**, 975–979 (1973)

38.77 V.V. Voronkov, M. Porrini, P. Collareta, M.G. Pretto, R. Scala, R. Falster, G.I. Voronkova, A.V. Batunina, V.N. Golovina, L.V. Arapkina, A.S. Guliaeva, M.G. Milvidski: Shallow thermal donors in nitrogen-doped silicon, J. Appl. Phys. **89**, 4289–4293 (2001)

38.78 J. Takahashi, K. Nakai, K. Kawakami, Y. Inoue, H. Yokota, A. Toshikawa, A. Ikari, W. Ohashi: Microvoid defects in nitrogen- and/or carbon doped Czochralski-grown silicon crystals, Jpn. J. Appl. Phys. **42**, 363–370 (2003)

38.79 M.S. Kulkarni: Lateral incorporation of vacancies in Czochralski silicon crystals, J. Cryst. Growth **310**, 3183–3191 (2008)

39. Models for Stress and Dislocation Generation in Melt Based Compound Crystal Growth

Vishwanath (Vish) Prasad, Srinivas Pendurti

A major issue in the growth of semiconductor crystals is the presence of line defects or dislocations. Dislocations are a major impediment to the usage of III–V and other compound semiconductor crystals in electronic, optical, and other applications. This chapter reviews the origins of dislocations in melt-based growth processes and models for stress-driven dislocation multiplication. These models are presented from the point of view of dislocations as the agents of plastic deformation required to relieve the thermal stresses generated in the crystal during melt-based growth processes. Consequently they take the form of viscoplastic constitutive equations for the deformation of the crystal taking into account the microdynamical details of dislocations such as dislocation velocities and interactions. The various aspects of these models are dealt in detail, and finally some representative numerical results are presented for the liquid encapsulated Czochralski (LEC) growth of InP crystals.

39.1 Overview... 1335
39.2 **Crystal Growth Processes** 1336
    39.2.1 Czochralski Technique................ 1336
39.3 **Dislocations in Semiconductors Materials** 1337
    39.3.1 Deleterious Effects of Dislocations 1337
    39.3.2 Origin of Dislocations................... 1338
39.4 **Models for Dislocation Generation** 1339
    39.4.1 CRSS-Based Elastic Models........... 1340
    39.4.2 Viscoplastic Models..................... 1342

39.5 **Diamond Structure of the Crystal**............ 1343
39.6 **Deformation Behavior of Semiconductors** 1346
    39.6.1 Stage of Upper and Lower Yield Points 1347
39.7 **Application of the Haasen Model to Crystal Growth**................................ 1350
39.8 **An Alternative Model** 1351
    39.8.1 Different Types of Dislocations 1351
    39.8.2 Dislocation Glide Velocity 1352
    39.8.3 Dislocation Multiplication........... 1355
    39.8.4 Work Hardening......................... 1357
    39.8.5 The Initial Dislocation Density...... 1358
39.9 **Model Summary and Numerical Implementation** 1360
    39.9.1 Summary of the Model................ 1360
    39.9.2 Numerical Implementation.......... 1361
39.10 **Numerical Results**................................. 1362
    39.10.1 Strength of Convection in the Melt and Gas..................... 1362
    39.10.2 Temperature Boundary Condition. 1362
    39.10.3 A Sample Case............................ 1363
    39.10.4 Effect of Gas Convection and Radiation 1368
    39.10.5 Melt Convection and Rotation Reynolds Numbers 1369
    39.10.6 Control of Encapsulation Height... 1371
    39.10.7 The Cool-Down Period 1371
    39.10.8 The [$1\bar{1}1$] Growth Axis................... 1372
    39.10.9 Summary of the Calculations and Some Comparisons............... 1373
39.11 **Summary** 1374
References 1375

39.1 Overview

Semiconductors are the most important materials of our age. Currently, silicon-based technology accounts for more than 95% of the total semiconductor market. In silicon technologies, high operating speed is achieved primarily by scaling down device dimensions. Smaller dimensions enable greater packing on a single chip, which enhances its capability and efficiency. However, reducing the size of the device has many limitations. Firstly, the parasitic capacitance of the device does not scale down linearly with the device dimension. Secondly, interconnects may pose a problem, and lithographic techniques can place a limit on the size reduction as well. Lastly the minimum size of a device is ultimately limited by the induced electric fields, which cause the breakdown of the p–n junction. An alternative is the use of III–V materials in high-speed electronics. These materials have higher electron and hole mobilities than Si; for example, at room temperature, the electron mobility in GaAs is five times larger than that in Si, and hence it has a better high-frequency response. Therefore devices using III–V materials are expected to be faster than those using Si. Apart from high-frequency response, III–V-based devices have better thermal characteristics and lower threshold voltages than other devices. Better thermal characteristics would enable the manufacture of identical components in smaller form factors. For example, a power amplifier based on InP would be smaller, more compact, and more efficient than power amplifiers based on other technologies, and its lower threshold voltage would allow the amplifier to work at a lower supply voltage.

In the area of optoelectronics and photonics, III–V materials, particularly InP, are the industry leaders. The development of low-transmission-loss quartz optical fibers, with optimum transmission characteristics in the $1.1-1.6\,\mu$m wavelength region, has led to growing demand for InP bulk material, as InP is increasingly being used as the substrate material for GaInAsP lasers, which produce light of this wavelength. The high-speed and optoelectronic characteristics of InP facilitate integrating high-speed transistors, with photoreceivers and diodes, in close proximity on a single chip. This technology would eliminate interconnects and enable the creation of communications networks operating at $40\,\text{Gb/s}$ and beyond. At such high speeds the monolithic integration of transistors and photoreceivers is crucial as the presence of even small interconnects would seriously degrade integrated circuit (IC) performance. Thus clearly, III–Vs are the materials of the future. The present limited use of InP is not due to limitations in device technology, but rather to the lack of availability of high-quality defect-free InP substrates of large diameter. Commercially, dislocation-free Si substrates are available up to a diameter of 300 mm. GaAs substrates are available up to 150 mm, and InP up to 75 mm; however, the GaAs and InP substrates suffer from a variety of defects, the most serious of which are line defects or dislocations.

39.2 Crystal Growth Processes

The bulk of modern-day requirements of Si and III–Vs are met by techniques involving pulling single crystals from their melt – Czochralski and its variants, Bridgman and its variations, float-zone melting, etc. A brief description of the Czochralski technique is provided below.

39.2.1 Czochralski Technique

This technique was developed by *Czochralski* [39.1] and later perfected by *Teal* and *Little* [39.2]. The required equipment consists of a rotating crucible (Fig. 39.1), which contains the charge material to be crystallized surrounded by a heater capable of melting the charge and maintaining it in a molten condition. A pull rod, with a seed crystal attached to its bottom and rotating, usually in an opposite direction to that of the crucible, is mounted coaxially, and lowered until the end of the seed touches the melt. The melt temperature is carefully adjusted until a meniscus is supported by the end of the seed. Once a thermal steady state is achieved the pull rod is carefully rotated and lifted, while the melt crystallizes on the seed. The diameter of the crystal is increased from that of the seed to a desired value by careful variation of the heater power and pulling rate. The crystal and the crucible are rotated during the whole process to maintain radial homogeneity. The whole crystal growth assembly is placed in a closed water-cooled chamber, which is evacuated and filled with an inert gas (generally argon) at low pressure; this is to ensure that the system is shielded from the effect of harmful atmospheric gases and temperature fluctuations of the ambient. III–V growth presents additional problems in the form of the decomposition of the melt, since the V element is volatile and escapes from the hot melt. The problem of decomposition has been overcome through a novel modification – the LEC technique (Fig. 39.2). If a high pressure is not required to maintain

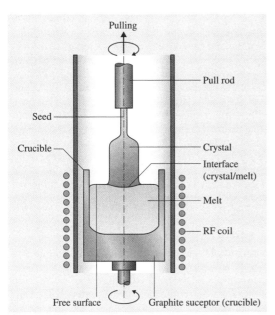

Fig. 39.1 Schematic of a low-pressure Czochralski system

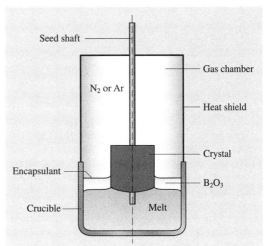

Fig. 39.2 Schematic of a high-pressure liquid-encapsulated Czochralski system

the chemical equilibrium (i. e., the vapor pressure of the volatile component is not high), only minor modifications of the standard Czochralski system are required. In the case of GaAs, for example, it involves charging the crucible with a quantity of boric oxide, a low-melting-point glass, in addition to the polycrystalline gallium arsenide. On heating the crucible, the boric oxide softens and flows over the charge. Since it wets the crucible and is immiscible with and lighter than the melt, it encapsulates the charge. An inert gas is introduced into the growth chamber, at a pressure higher than the dissociation pressure of the GaAs melt, preventing loss of As from the melt surface. This suppresses arsenic bubble formation. Being essentially insoluble in the boric oxide, arsenic cannot diffuse through it. The crystal growing is performed in the usual manner by lowering the seed crystal through the encapsulant layer, until it contacts the melt interface; pulling is then initiated in the usual way.

The pressure of the inert gas in the growth chamber depends on the dissociation pressure of the melt and may range from about 1 atm for GaAs, through 40 atm for InP, to 70 atm for compounds such as GaP, and much higher pressures for ZnO, etc. The complexity and technological challenges associated with the LEC technique increase greatly with the pressure, which is high – 40 atm – for InP.

39.3 Dislocations in Semiconductors Materials

Dislocations are line defects in semiconductors, which adversely affect the characteristic semiconductor properties such as carrier density, mean lifetime of minority carriers, and carrier mean free path. Large single crystals of traditional elemental semiconductor materials such as Si and Ge can be grown virtually dislocation free using the necking technique pioneered by *Dash* [39.3]. However, growth of large-diameter single crystals of III–V and II–VI compound crystals is prone to the problem of dislocations regardless of the technique used. This problem is particularly severe in the growth of large crystals, and is one of the limiting factors in the growth of large-diameter high-quality GaAs and InP single crystals.

39.3.1 Deleterious Effects of Dislocations

The presence of dislocations introduces insurmountable problems in the use of such defective crystals for device applications. A high dislocation density in the crystal introduces a large number of etch pits on the surfaces of wafers made from it. The use of wafers for device manufacture involves the etching and deposition of transistors, the dimensions of which are of the same order as that of the etch pits. If a transistor is engraved near an etch pit, its characteristic behavior is very different from that of a transistor etched in nondefective material. Hence it is imperative for device applications that wafers are uniform and largely free of defects.

Dislocations are detrimental for optoelectronic applications. Dislocations have been directly implicated in the case of degradation of GaAs-based light-emitting devices such as the GaAs–AlGaAs light-emitting diode (LED) [39.4]. The imperfect lattice near the dislocation core enhances nonradiative recombination between majority and minority carriers. Such recombinations near dislocation cores, instead of producing photons, only produce local lattice excitations and thermal energy, reducing luminescence. For example Roedel et al. [39.5] plotted luminescence efficiency versus dislocation density for 45 individual $Al_xGa_{1-x}As$:Si LEDs and obtained a curve directly documenting the decrease in efficiency with etch pit density on the original wafer from which the device was fabricated. The degradation is of three different broad categories, and the following are directly quoted from [39.4]

(1) A rapid degradation that occurs after a short duration of operation, (2) a slow degradation occurring during long term operation; and (3) a catastrophic degradation process that occurs at high optical power densities because of mirror facet damage.

The first two modes of degradation are closely linked to nonradiative recombination near the dislocation cores. Moreover the thermal energy generated during such recombinations leads to faster dislocation rearrangements and movements involving dislocation climb and glide, and leads to the development of more dislocation structure, more defective material, and degradation of luminescence with time. Reference [39.4] provides a more detailed discussion on this topic.

39.3.2 Origin of Dislocations

In most works thermal stresses are cited as the chief reason behind dislocations. This is because the stresses drive the multiplicative mechanisms of dislocations, which begin to operate as the small number of dislocations initially present begin to glide under the influence of thermal stresses. While this is certainly true, this is a matter of dislocation multiplication from *existing dislocations* and does not explain the nucleation of the first *existing dislocations* from which others multiply. Hence one has to differentiate between *dislocation multiplication* and *dislocation nucleation*. Dislocation nucleation is not possible solely due to thermal stresses; it has been estimated that stresses of the order of $\mu/2\pi$, where μ is the theoretical shear modulus of the material, are required for the nucleation of dislocations solely under their influence. The crystal growth process does not involve thermal stresses of this magnitude. The nucleation of dislocations is generally attributed to the following mechanisms, many or all of which might be operating in tandem.

Agglomeration of Point Defects

During crystal growth, there exists a high equilibrium concentration of point defects in a given material element, when it is near the crystal–melt interface, because of the high temperature close to the melting point. During the growth process, the material element moves away from the interface, and cools down, giving rise to supersaturation of point defects. These point defects agglomerate into clusters, which in order to minimize their elastic strain energy collapse into various structures. Chapter 38 in this volume details the dynamics of the agglomeration and the defects formed therein, under various conditions, for Czochralski growth of Si. One of the minimum-energy structures into which the agglomerations of point defects can collapse is a planar disc in a (111) plane. Such a disc is equivalent to a stacking fault bound by a Frank partial dislocation loop. The disc grows by the climb of the bounding partial dislocation, and once it reaches a critical size, a Shockley partial may be spontaneously nucleated, removing the stacking fault and transforming the Frank partial into a loop of a perfect dislocation [39.6, Chap. 10]. For example, a Frank dislocation loop in the (111) plane dissociates as

$$\tfrac{a}{3}[111] + \tfrac{a}{6}[\bar{2}11] \to \tfrac{a}{2}[011] \,.$$

The steps of point defect agglomeration – collapse into discs and Shockley nucleation – are governed by kinetics determined by the thermal history and the material. The perfect dislocation loop formed by the above process is not a glide dislocation, since its Burgers vector does not lie in the plane of the dislocation loop. However, a fraction of these loops become mobile glide dislocation loops by the escape of a segment onto a glide plane, which contains the Burgers vector. Refer to [39.7] for a more detailed explanation. In compound semiconductors, the situation is complicated by the sublattices of the diamond structure being occupied by atoms of different elements. In these compounds, the formation of a vacancy (interstitial) disc involves equal numbers of vacancies (interstitials) of both sublattices, which is difficult, since the concentration of point defects on the two sublattices is generally different, depending on factors such as stoichiometry, etc. *Lee* et al. [39.8] have identified mechanisms by which the above problem may be overcome, to nucleate glide dislocations. *Lagowski* et al. [39.9] have experimentally revealed the effect of doping on the nucleation of glide dislocations, through aggregation of point defects.

Surface Damage

Damage on the surface of the growing crystal due to indentation, scratching, grinding, and impingement of hard particles leads to the nucleation of dislocations. All of these processes involve the action of large local stresses causing the relative displacement of two layers of the crystal: the nucleation of a dislocation. This mode of nucleation might not be important in crystal growth, but it has been the experience of crystal growers that small particles from the melt on the surface of the growing crystal produce such dislocations [39.10].

Foreign Particles or Precipitates

The presence of a second phase in the form of foreign particles, precipitates, etc., leads to misfit due to differential thermal expansion of the particle and the surrounding matrix. This gives rise to very high local stresses and consequently misfit dislocations. Usually, these misfit dislocations are in the form of glide dislocations in the glide plane of maximum shear stress, as discussed by *Ashby* and *Johnson* [39.11] and *Alexander* [39.7]. Ashby and Johnson show that the radius of the original nucleated loop should be greater than $R_c = \mu b/\tau$, where τ is the effective shear stress and b is the Burgers vector, so that the loop can continue to grow and expand under the influence of the shear stress. In III–V compounds, inclusions from the B_2O_3 encapsulant or precipitates of one phase due to lack of stoichiometry might act as foreign particles, thus nucleating glide dislocations.

Chemical Inhomogeneity

Chemical inhomogeneity may arise due to doping or impurity atoms and nonstoichiometry. These might lead to lattice mismatch, giving rise to misfit dislocations. Sometimes, the phenomenon of constitutional supercooling might arise in the melt near the solidification interface, giving rise to instabilities in the interface. In such cases, the planar interface splits into cells. Impingement of different parts of the interface (cells) might give rise to misfit dislocations. However, constitutional supercooling is expected to be rare in semiconductor single-crystal growth, due to the low concentration of impurity/dopant.

Use of a Defective Seed

If the seed crystal contains dislocations, these are likely to glide into the newly grown crystal. These seed dislocations would then multiply under the effect of thermal stresses. The use of a defective seed can be overcome by the use of a narrow necking (*Dash* process) [39.3].

Thermal Stresses

The most important cause for the presence of dislocations in as-grown semiconductor crystals is thermal stresses. Since the presence of temperature gradients in the crystal is necessary, crystal growth is an inherently nonequilibrium process, involving heat transfer through the crystal. The crystal acts as a medium through which heat, both released heat of fusion and that from the melt, is transferred to the ambient. The heat enters the crystal at the crystal–melt interface, and leaves through the external surfaces by radiation and convection. Thus the core of the crystal is hotter than the periphery, and the bottom of the growing crystal is hotter than the top. These temperature gradients give rise to differential expansion of different parts of the crystal, and consequently to thermal strains and stresses. These thermal stresses are the cause of plastic deformation, through the movement and multiplication of existing dislocations, resulting in unacceptable final dislocation densities in as-grown crystals.

39.4 Models for Dislocation Generation

Before we dwell on the mathematical models, a word on notation: vectors are represented as bold-face italic and second-order tensors as bold-face upright; indicial notation for vectors and tensors and the Einstein summation convention are not used, except in Sect. 39.2 or unless specifically mentioned.

Since thermal stress in the growing crystal is considered as the primary reason for the presence of dislocations in as-grown crystals, models for dislocation generation take a solid mechanics approach. Towards this, efforts are made to characterize the mechanistic response of the crystal to loading through appropriate constitutive equations. Depending on the type of constitutive behavior, the modeling approach falls into two broad categories: the crystal is considered as (1) an elastic crystalline medium and (2) a viscoplastic crystalline medium.

39.4.1 CRSS-Based Elastic Models

In this approach the crystal is assumed to be a linear elastic continuum. The crystal usually being a cubic material, three elastic coefficients are prescribed to characterize its elastic behavior. Isotropic behavior is also assumed sometimes. The thermal loading during the growth is specified through a coefficient of thermal expansion and the consequent thermal strain. With the help of minimal boundary conditions designed to prevent rigid-body motion, the elastic problem can be solved at any given thermal loading (which can be obtained from the temperature field) during the growth of the crystal. Solution of the elastic problem yields information on the displacements, the total, thermal, and elastic strains, and the elastic stresses within the crystal. Once the elastic stresses are obtained, these are used to estimate the plastic deformation and whether a particular area of the crystal will have a high or low dislocation density.

Dislocations in III–V materials with zincblende structure glide in the family of four $\{111\}$ planes along the three $\langle 110 \rangle$ directions. Hence plastic deformation takes place along the 12 $\{111\} - \langle 110 \rangle$ slip systems. *Jordan* et al. [39.12] postulated that plastic deformation starts if the resolved elastic shear stress on any of these 12 systems exceeds a critical value, which they termed the critical resolved shear stress (CRSS). Thus this model is similar in spirit to the rate-independent non-hardening plasticity models, which have a fixed yield surface in stress space. The equations and the methodology applied by *Jordan* et al. [39.12] are summarized next.

The total strain (the symmetrical part of the gradient of the displacements) at any given instant in the growth history is split into an elastic component, which is a function of the stress, and a thermal component, which is function of the temperature as follows

$$\varepsilon = \varepsilon(\sigma, T) = \tfrac{1}{2}\left(\nabla \boldsymbol{u} + (\nabla \boldsymbol{u})^{\mathrm{T}}\right), \tag{39.1}$$

and

$$\varepsilon(\sigma, T) = \varepsilon^{\mathrm{el}}(\sigma) + \varepsilon^{\mathrm{th}}(T). \tag{39.2}$$

The elastic strain using the regular equations of linear elasticity is related to the stresses as

$$\sigma = \mathbf{C} : \varepsilon^{\mathrm{el}}, \tag{39.3}$$

where $\mathbf{C}$ is the fourth-order elastic modulus tensor. Similarly the thermal strain is expressed as

$$\varepsilon^{\mathrm{th}} = \alpha(T - T_0)\boldsymbol{\delta}, \tag{39.4}$$

where α is the coefficient of thermal expansion, T_0 is some reference temperature, and $\boldsymbol{\delta}$ represents the Kronecker delta tensor. The principle that completes the equations is the mechanical equilibrium in the crystal, which assuming negligible effects of body forces such as gravity is expressed as

$$\nabla \cdot \sigma = 0. \tag{39.5}$$

For a given temperature field and assuming boundary conditions just sufficient to prevent rigid-body motion, (39.1–39.5) represent the familiar linear elastic problem and can be solved using any numerical technique to obtain the total strains, elastic strains, and stresses in the crystal.

The thermal–elastic stresses thus calculated are then resolved along the $\{111\}\langle 110 \rangle$ systems. These slip systems are summarized in Table 39.1, and are defined in

Table 39.1 Summary of the slip systems in elemental and III–V semiconductors

Slip system number (k)	$\sqrt{2}s_k$	$\sqrt{3}n_k$
1	[101]	[11$\bar{1}$]
2	[011]	[11$\bar{1}$]
3	[1$\bar{1}$0]	[11$\bar{1}$]
4	[110]	[1$\bar{1}$1]
5	[011]	[1$\bar{1}$1]
6	[10$\bar{1}$]	[1$\bar{1}$1]
7	[101]	[1$\bar{1}$$\bar{1}$]
8	[110]	[1$\bar{1}$$\bar{1}$]
9	[01$\bar{1}$]	[1$\bar{1}$$\bar{1}$]
10	[10$\bar{1}$]	[111]
11	[0$\bar{1}$1]	[111]
12	[1$\bar{1}$0]	[111]

terms of the slip plane normal vector $\boldsymbol{n}$ and the slip direction vector $\boldsymbol{m}$, thus

$$\tau^i = \boldsymbol{m}_i \cdot \boldsymbol{\sigma} \cdot \boldsymbol{n}_i \,,$$

where τ^i is the resolved shear stress on slip system i.

If the resolved shear stress along a slip system exceeds the critical resolved shear stress (CRSS), then that system is deemed active, contributing to the plastic deformation and multiplication of dislocation density. Towards this the *excess shear stress* on a given slip system can be calculated from

$$\tau^i_{\text{excess}} = \left|\tau^i\right| - \tau_{\text{CRSS}} \,. \tag{39.6}$$

Jordan et al. [39.12] postulated that the total dislocation density in a crystal is proportional to the sum of the excess shear stress on the first five slip systems, when the excess shear stresses are arranged in descending order. This follows from the fact that, although the plastic strain tensor has nine components, its symmetry and the incompressibility of the plastic strain mean that only five of the components are independent. Hence slip on five slip systems is sufficient to describe the plastic strain. In regions where this excess shear stress is zero, the dislocation density will be zero. Thus the total dislocation density N at a given location in the crystal is described as

$$N \propto \sum_{i=1}^{5} \tau^i_{\text{excess}} \,. \tag{39.7}$$

Jordan et al. [39.12] presented the first detailed stress calculations after calculating the thermal field from a quasisteady heat transfer model. This analysis provided the first fundamental description of dislocation generation in III–V compounds. Jordan et al. in their heat transfer analysis made a few fundamental assumptions for the calculation of the temperature field, the most important of which are as follows:

1. The crystal is modeled as a cylindrical boule, which is pulled at a constant rate from the melt held at a constant temperature T_f.
2. The solid–liquid interface is planar and is at the temperature T_f.
3. The temperature of the ambient around the boule is a constant $T_a < T_f$.
4. Heat loss from the lateral and top surfaces by radiation and natural convection follows Newton's law of cooling.

An analytical solution for the temperature field in the crystal was obtained by solving the conduction equation in the crystal, along with the above assumptions. The plane-strain solution of classical thermoelasticity was then adopted to determine the stresses and calculate the dislocation densities. *Jordan* et al. [39.12] compared their calculated dislocation densities with a computerized scan of the etch pit density of a KOH-etched (100) GaAs wafer; the main feature of the dislocation distribution such as fourfold symmetry in the etch pit density, observed experimentally, was replicated by their calculations. Similar calculations were also performed for a $\langle 111 \rangle$ LEC-pulled InP wafer. In this case the sixfold defect distributions determined experimentally somewhat resembled the dislocation contours predicted by the model. These calculations firmly established the role of thermal stresses in dislocation generation and multiplication in crystals grown from the melt.

Following the pioneering work of *Jordan* et al. [39.12], *Kobayashi* and *Iwaki* [39.13] performed similar calculations without neglecting the axial displacements; their calculations are slightly closer to experimental data than the calculations of Jordan and coworkers. *Duseaux* [39.14] repeated the calculations with a more realistic geometry by including seed and shoulder regions of the growing crystal. These early calculations along with those by *Lambropoulos* [39.15], *Schvezov* et al. [39.16], and *Meduoye* et al. [39.17, 18] assumed a known geometry of the crystal, usually with a planar melt–crystal interface, and the thermal calculations were performed with an assumed convection coefficient for heat transfer from the melt to the crystal and from the crystal to the gas. Calculations by *Motakef* et al. [39.19], *Dupret* et al. [39.20], and *Bornside* [39.21], while still neglecting melt convection, used global radiation and conduction models which included both the melt and the crystal. The interface shape in these calculations was not assumed, but was obtained from the calculations. *Zou* et al. [39.22] coupled a finite element technique, performing an isotropic linear thermal–elastic analysis on the growing crystal, with a finite volume technique for the thermal and flow analysis of the growth process. This work incorporated a number of features, including convection in the melt and the gas, radiation, etc. It should be noted that most of these works treat the crystal as linear elastic and, after calculating the stresses, performed some sort of excess shear stress calculations, following Jordan et al. Thus, there is a large amount of literature devoted to calculating dislocation densities through a thermal–elastic model, using the approach of Jordan and coworkers.

This approach, though original and successful, has several shortcomings. For example, it treats the crystal as an elastic/perfectly plastic continuum, which is untenable for elemental and compound semiconductors. The deformation of semiconductors is a highly rate-dependent process; the response of such a material to stress is not instantaneous, but depends on the kinetics of the dislocation movement and multiplication, which are strong, time-dependent functions of temperature. The so-called CRSS proposed by Jordan is not a material property and is a function of the local dislocation configuration, temperature, doping, and other factors. The dislocation configuration and temperature are functions of growth history, and change within a given material element during its growth history. All the above analyses take one temperature field as an input; it is not very clear which temperature at which instant of the growth history is the appropriate one. It is however true that a CRSS such as locking stress does exist in certain materials such as GaAs [39.23] and Si due to the presence of impurities, which lock the dislocations. A certain minimum shear stress is then required to *unlock* these dislocations, but once the unlocking takes place the material again behaves in a rate-dependent manner.

39.4.2 Viscoplastic Models

The alternative approach to modeling dislocation density evolution is the use of viscoplastic constitutive equations. Once the thermal history of the crystal from seed to as-grown state is obtained, the procedure is to integrate these viscoplastic constitutive equations in time, while maintaining equilibration of stresses at all times to obtain the final dislocation density. The advantage of this approach is that the final dislocation densities and state of the as-grown crystal is a functional of the initial state of the seed, and the entire temperature history during the growth and the cool-down period of the crystal after growth. This approach also has the potential to give an estimate of the residual stresses in the cooled crystal.

While specific constitutive equations to be used will be discussed in the following sections, a brief description of the basic viscoplasticity formulation, described in detail in *Lubliner* [39.24], with some adaptations to present requirements is given here. As is customary in standard plasticity formulations, the internal structure, or the state of the crystal, at any point, is hypothesized as being represented by a set of internal variables denoted by ξ_α or the vector $\boldsymbol{\xi}$. The mechanical strains at a given position are completely determined by the troika of internal variables, stress components, and temperature at a given position.

$$\boldsymbol{\varepsilon} = \boldsymbol{\varepsilon}(\boldsymbol{\sigma}, \boldsymbol{\xi}, T). \tag{39.8}$$

Additionally, for small-strain viscoplastic constitutive formulations, the following tenets are usually adapted.

Additive Decomposition of Strain Tensor
One assumes that the strain tensor can be decomposed into an elastic, plastic, and thermal part, according to the relation

$$\boldsymbol{\varepsilon}(\boldsymbol{\sigma}, T, \boldsymbol{\xi}) = \boldsymbol{\varepsilon}^{\text{el}}(\boldsymbol{\sigma}) + \boldsymbol{\varepsilon}^{\text{pl}}(\boldsymbol{\xi}) + \boldsymbol{\varepsilon}^{\text{th}}(T), \tag{39.9}$$

where the elastic strain components are assumed to be a function of the stress tensor only, the plastic strain components are assumed to be a function of the internal variables only, and finally the thermal strain component is a function of temperature only.

(Elastic) Stress Response
The stress tensor σ_{ij} is related to the elastic strain $\varepsilon_{ij}^{\text{el}}$ through the existence of a stored energy functional W, according to the hyperelastic relationship

$$\boldsymbol{\sigma} = \frac{\partial W(\boldsymbol{\varepsilon}^{\text{el}})}{\partial \boldsymbol{\varepsilon}^{\text{el}}}.$$

For linear elasticity, the stored energy functional W assumes a quadratic form in the elastic strain, i.e., $W = \frac{1}{2} \boldsymbol{\varepsilon}^{\text{el}} : \mathbf{C} : \boldsymbol{\varepsilon}^{\text{el}}$, where $\mathbf{C}$ is the fourth-order tensor of elastic moduli, which is assumed constant. Then the relations between the stress components and the elastic strain tensor reduce to

$$\boldsymbol{\sigma} = \mathbf{C} : [\boldsymbol{\varepsilon} - \boldsymbol{\varepsilon}^{\text{pl}} - \boldsymbol{\varepsilon}^{\text{th}}]. \tag{39.10}$$

Evolution of Internal Variables and Plastic Strain
Evolution equations are proposed for the internal variables and plastic strain tensor over time. These evolution equations for $\{\boldsymbol{\varepsilon}^{\text{pl}}, \boldsymbol{\xi}\}$ are termed the flow and hardening rules, respectively. The relations are of the form

$$\dot{\boldsymbol{\varepsilon}}^{\text{pl}} = \mathbf{r}(\boldsymbol{\sigma}, \boldsymbol{\xi}, T), \tag{39.11a}$$

$$\dot{\boldsymbol{\xi}} = \mathbf{h}(\boldsymbol{\sigma}, \boldsymbol{\xi}, T). \tag{39.11b}$$

The relation (39.11b) is more basic and (39.11a) can be derived from it, because of the functional dependence of plastic strain solely on the internal variables according to (39.9).

Isotropic Thermal Strain Response

The thermal strain response is assumed to be isotropic and of the form

$$\varepsilon^{\text{th}} = \alpha(T - T_0)\delta, \quad (39.12)$$

where α is the coefficient of thermal expansion, T_0 is some reference temperature, and δ is the Kronecker delta tensor.

Equations (39.9–39.12) form the outline of an internal-variable-based viscoplastic constitutive model, set in a small deformation framework, and assuming linear elastic response coupled with isotropic thermal expansion. If the internal variables adopted are some measures of dislocation densities, and the evolution (39.11) is based on dislocation dynamics, the outline presented above has the potential to predict the final defect densities as integrated functions of the growth history. The procedure is essentially one of integrating (39.11) through the growth history, while ensuring the equilibration of stresses and fulfillment of boundary conditions on the surface of the crystal. Specific instances of constitutive relations to model the deformation behavior of semiconductor materials are presented in subsequent sections of this chapter. Prior to these, a brief description of the structure of the diamond crystal will be provided, as this has a great bearing on the deformation characteristics of semiconductor materials, at least, those that crystallize in this structure.

39.5 Diamond Structure of the Crystal

The peculiarities in the plasticity of elemental and III–V semiconductor materials arise from their structures. These materials crystallize in the diamond cubic structure or sphalerite structure for compound III–V semiconductors. These structures are illustrated in Figs. 39.3 and 39.4 (adapted from [39.25]), which show three-dimensional views of the diamond and sphalerite structure, respectively. It can be seen that each atom in these structures is tetrahedrally coordinated, i.e., each atom has exactly four neighbors, and the bonds between these neighbors are in the $\langle 111 \rangle$ directions. A projection of the three-dimensional structure on a $(1\bar{1}0)$ plane is shown in Fig. 39.5. In this view the (111) atomic planes are stacked in the sequence $AaBbCcAaBbCc\ldots$ It can be seen from a straightforward comparison of Figs. 39.3 and 39.5 that the double and single lines in Fig. 39.5 mean one or two interatomic bonds. There are two different interatomic plane distances between the (111) planes – from the tetrahedral geometry, it can be calculated that the perpendicular distance between the a and the A planes is thrice the distance between the A and the b planes. The diamond lattice can also be described as just two interpenetrating face-centered cubic (fcc) lattices, or alternatively as an fcc structure with a basis of two atoms per lattice point. If the stacking sequence in the first fcc lattice is $ABCABCABC\ldots$, then the stacking sequence in the second interpenetrating lattice is $abcabcabc\ldots$ Since the structure is just a doubled-up fcc lattice, the glide planes as in the fcc case, belong to the $\{111\}$ family, and the perfect dislocations have Burgers vector $\frac{1}{2}\langle 110 \rangle$. Due to the strong directional covalent bonding, the Peierls barrier in these materials is very high, and the resulting deep troughs in the Peierls barrier cause the dislocations to align primarily along the $\langle 110 \rangle$ directions, when the dislocation density is low. Thus the dislocations are primarily screw or 60° dislocations.

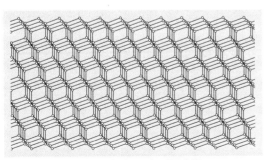

Fig. 39.3 The diamond lattice (after [39.25])

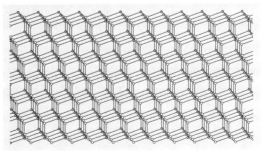

Fig. 39.4 The sphalerite lattice. *Dark* and *light* atoms show different species of atoms (after [39.25])

1344   Part F | Modeling in Crystal Growth and Defects

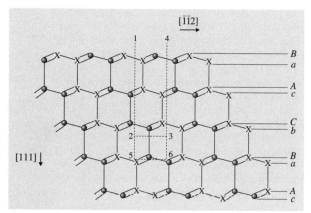

Fig. 39.5 The diamond lattice projected onto the $(1\bar{1}0)$ plane. *Bullets* represent atoms in the plane of the paper; Xs represents atoms below the plane of the paper (after [39.6])

Owing to the double-layered atomic arrangement of the diamond cubic lattice, there are two inherently different sets of dislocations, with one member of each having the same Burgers vector. These two different sets are termed the glide set and the shuffle set of dislocations. The difference is clarified by an examination of Fig. 39.5. A 60° dislocation of the glide set is formed by cutting out the material bounded by the surface 1–5–6–4, and then displacing the sides of the cut, until they are joined. The feature of these dislocations is that there is a dangling bond along surface 5–6, while all the other cut bonds on surfaces 1–5, 6–4 are made up after displacement. Thus the extra plane of atoms terminates between two layers of the different letter index, for example a and B in Fig. 39.5. Similarly, a 60° dislocation of the shuffle set is formed by cutting material along the surface 1–2–3–4 and then rejoining as before. In this case the dangling bond is along the surface 2–3, while all the bonds along surfaces 1–2 and 3–4 are filled after reconstruction. In this case the extra plane of atoms terminates between layers of the same index, such as B and b in Fig. 39.5. Three-dimensional views of the 60° glide (BC) and shuffle dislocations ($B'C'$) are given in Figs. 39.6 and 39.7, respectively. It can be seen that the shuffle and glide dislocations can transform into each other any time by the emission or absorption of point defects. The same distinction between glide and shuffle dislocations applies to dislocations of screw, edge, and intermediate orientations. To describe the dislocations the Thompson tetrahedron notation is used. Since there are two fcc lattices, there are two Thompson tetrahedrons used, one to describe the glide set (for example, dislocation BC above) and one to describe the shuffle set (for example, dislocation $B'C'$ above).

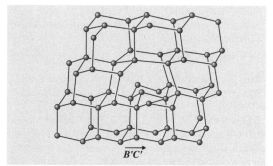

Fig. 39.7 The 60° shuffle dislocation (after [39.6])

Dislocations in the glide set can undergo dissociation into partial dislocations, in the same manner as dislocations in fcc metals. Figure 39.8 is a three-dimensional view of the glide dislocation BC dissociated into δC (30° partial) and $B\delta$ (90° partial). These partial dislocations in the glide set are glissile, and because the dislocation glides between the layers b and C, the dissociation into δC and $B\delta$ results in an intrinsic stacking fault between the two partials. The dissociation of shuffle dislocations in the set $A'B'C'D'$ is not direct, because such dissociation would produce a high-energy fault of the type $CcAaB|aBbCc$.

Hornstra [39.26], and later *Alexander* [39.27], gave an alternate description of an extended shuffle dislocation, which may be described in two completely equivalent ways: either as an association of a stacking-fault ribbon to a shuffle dislocation, or as a dissociated glide dislocation that has emitted or absorbed a line of atoms in the core of one of its partials. This is consistent with the relationship between perfect shuffle and

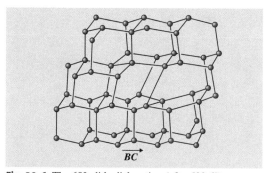

Fig. 39.6 The 60° glide dislocation (after [39.6])

glide dislocations that can be converted into each other by emission or absorption of point defects.

Dislocations of the screw, edge, and other intermediate types, of both the shuffle and glide variety, may be generated in ways equivalent to that of the 60° dislocation. For these configurations, dislocations of the glide and shuffle type may be dissociated into partial dislocations in equivalent ways. For example, a screw dislocation of the glide type may be dissociated into two 30° partial dislocations, with a stacking fault between them. These can be converted into the shuffle type by emission or absorption of point defects. A pertinent question in the literature is whether the dislocations are extended and, if so, whether they exist in the shuffle or the glide set. It was previously assumed that moving dislocations were undissociated and existed in the shuffle mode, since the number of covalent bonds to be broken is three times smaller than in the glide mode. However, it has been demonstrated that dislocations are dissociated when at rest in both the diamond and sphalerite structure [39.28,29], and remain dissociated while moving [39.30].

Since the movement of partial dislocations in the shuffle set is more difficult than those in the glide set, it is more probable that the glide set is prevalent. However, it must be pointed out that the coexistence of both sets is not ruled out in the literature, and sometimes the as yet unexplained generation of point defects during deformation of semiconductors is attributed to the existence of, and interchange between, both sets [39.31].

The situation is further complicated in the case of compound semiconductors such as InP, GaAs, InSb, and other III–V and II–VI compounds, which exist in the sphalerite or the zincblende structure (Fig. 39.4). The sphalerite structure is the same as the diamond structure, the only difference being that the two atoms at each lattice point are of two different elements in the sphalerite structure, while they are of the same element in the diamond structure. Identical to the diamond structure, the glide planes are {111}, the direction of closest packing, while the slip directions, or Burgers vector of the perfect dislocations, are again ⟨110⟩, the shortest translation vectors in the lattice. In the following paragraphs the chief differences between the two structures are illustrated.

α and β Dislocations

Apart from the complexities of the glide and shuffle sets already existing in the diamond cubic structure, there are further complexities in the sphalerite system, i.e., there are two different {111} systems. For exam-

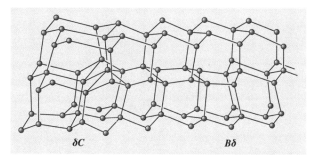

Fig. 39.8 Dissociation of a prefect 60° glide dislocation into 30° and 90° partials (after [39.6])

ple, in Fig. 39.5, the A, B, and C layers contain one atom, say In for InP, while the layers a, b, and c contain the P atoms. Opposite-sign dislocations in the sphalerite structure have the extra plane of atoms ending in different atoms. For example, in a 60° shuffle dislocation (Fig. 39.5), formed by removing material bounded by surface 1–2–3–4, the extra plane of atoms lies below the surface 2–3 and ends on the B layer in the diagram, which is an In layer. If instead a shuffle dislocation of exactly the opposite sign was formed at exactly the same place, the extra plane of atoms would lie above surface 2–3, and end on the b layer, which is a P layer. Dislocations with the edge of their extra half-planes ending on In (atoms of lower valency) layers are called β dislocations, while those with the edge of their extra half-planes ending on P (atoms of higher valency) layers are called α dislocations. This terminology, although illustrated for the shuffle set, can also be applied to glide dislocations. Since, the core structures of α and β dislocations are different, they have different properties, the most important of which is the significant difference of mobility between the two.

As pointed out before, the dislocations in compound or elemental semiconductors lie along the ⟨110⟩ directions of the glide plane, at low dislocation densities. A perfect dislocation loop would thus consist of two screw dislocations and four 60° dislocations. The geometry of a typical loop is shown in Fig. 39.9a, which shows that a glide loop possesses two types of dislocation segments: screw and 60° dislocations (60° being the angle between the dislocation line and the Burgers vector). The 60° dislocations are of two different types, α and β, since the extra half-planes of these are respectively down and up. The screw dislocations are not classified since they do not have an extra half-plane. Figure 39.9b–d show the dissociation of the α, β, and screw dislocations into partial dislocations in the glide

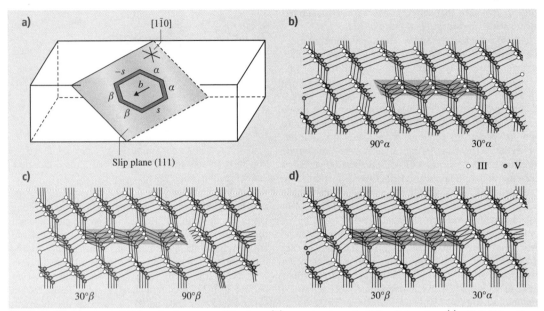

Fig. 39.9 (a) A hexagonal dislocation loop on a glide plane, (b) core structure of a 60° α dislocation, (c) 60° β dislocation, and (d) screw dislocation. The *shaded planes* represent the stacking faults (after [39.25])

set, with 60° dislocations split into 30 and 90° partials, and each screw split into a 30° partial dipole. Each of the partials can be associated with α or β character as shown.

39.6 Deformation Behavior of Semiconductors

Elemental and compound semiconductors are brittle at low temperatures, but acquire a substantial capacity for plastic deformation at temperatures above 30% of their melting point. The brittleness is a result of strong directional covalent bonds, which result in a deep Peierls potential. The plasticity of these materials is strongly strain rate and temperature dependent, again owing to the strong Peierls potential acting against dislocation movement. The dislocations glide in these materials at low speeds, in contrast to fcc metals whose plasticity is almost rate independent. The deformation behavior of these materials has been studied extensively through the stress–strain behavior of elemental and compound semiconductors such as Si, Ge, and InSb during uniaxial compression or tensile testing conducted at constant strain rate on a crystal oriented for single slip, and creep tests at constant stress for samples again oriented for single slip. The most common way of obtaining these curves is the constant-strain-rate compression test. This is done in a constant-displacement-rate machine, in which the specimen is kept between two rigid crossheads while one of them moves at a fixed speed. The axial force applied on the crosshead to compress the specimen is recorded at any instant. The force is recorded as a function of the crosshead displacement, which can then be converted into curves of shear stress versus shear strain based on the specimen dimensions. All elemental and compound semiconductors exhibit common behavior in uniaxial compression constant-strain-rate tests, oriented for single-slip conditions, when the tests are done at temperatures where the mobilities of the dislocations are similar for all materials. This commonality of behavior indicates that the underlying microscopic processes behind the deformation have a common origin. Figure 39.10 elucidates the generic features common to all elemental and III–V compound semiconductors. In this figure, the shear stress in the main slip system is plotted against the shear strain in the same system, as the deformation evolves. The striking feature of this curve is the presence of

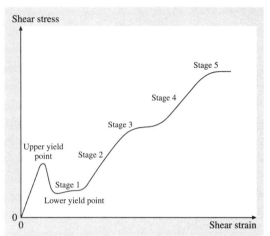

Fig. 39.10 Generic plot of shear stress versus shear strain for deformation of semiconductors in single slip

an upper and a lower yield point. The stress increases rapidly with strain at the beginning of the deformation, reaches a maximum, then decreases with increasing strain and reaches a minimum, from where the stress increases with increasing stain. The stress maximum and minimum are termed the upper and lower yield point, respectively. At low temperatures, three deformation stages following the upper and lower yield points are observed: a stage of low work hardening (I), a stage of strong work hardening (II), and a stage of increasing softening (III), where recovery processes operate in the crystals exposed to an external stress – dynamical recovery. At high temperatures, two further stages appear: a new hardening stage (IV) and a second recovery stage (V).

39.6.1 Stage of Upper and Lower Yield Points

The phenomena of upper and lower yield points was also observed in LiF. *Johnston* and *Gilman* [39.32] explained this phenomenon in terms of microscopic dislocation dynamics. In a constant-strain-rate uniaxial deformation test, the total strain rate can be split into its elastic and plastic components as

$$\dot{\varepsilon} = \dot{\varepsilon}_{pl} + \dot{\varepsilon}_{el} \,.$$

At the beginning of the deformation, the plastic strain rate is low, due to the low initial dislocation density in semiconductors. The total strain rate is then mostly elastic in origin, leading to a high initial rise in stress. This rising stress multiplies the dislocation density many-fold, leading to the plastic strain rate catching up and finally overtaking the total strain rate. As a result the stress actually falls, leading to a maximum: an upper yield point. The stress continues falling until dislocation interaction becomes important, leading to the hardening of the material, after a stress minimum: the lower yield point. *Alexander* and *Haasen* [39.33] reformulated the theory by Johnston, for semiconductors, to explain the yield phenomenon.

The central tenet of their theory is the introduction of a variable N to denote the dislocation density in the crystal. The plastic shear strain rate on the only active slip system is given by the Orowan relation

$$\frac{d\gamma_{pl}}{dt} = bNv \,, \tag{39.13}$$

where b is the Burgers vector of dislocations, v is the mean velocity of the dislocations, and γ_{pl} is the plastic shear strain on the only active slip system. Since there is only one active slip plane in the crystal, the axial component of the macroscopic plastic strain tensor, ε_{pl}, is given by $\varepsilon_{pl} = \phi\gamma_{pl}$, where ϕ is a geometrical factor relating the shear and the axial strains.

The velocity of dislocations in the slip system v is dependent on the resolved shear stress on that system and has been found experimentally [39.34] to be of the form

$$v = v_0 \left(\frac{\tau_{eff}}{\tau_0}\right)^m \exp\left(-\frac{Q}{k_B T}\right), \tag{39.14}$$

where m, τ_0, Q, and v_0 are empirical constants depending on the semiconductor material and its doping, k_B is the Boltzmann constant, τ_{eff} is the effective resolved shear stress on a dislocation and is given by

$$\tau_{eff} = \tau - A\sqrt{N} \,, \tag{39.15}$$

τ is the externally applied shear stress on the active slip system, and $A\sqrt{N}$ is the back-stress on a given dislocation due to the neighboring ones, which resist its movement through interaction. The square-root dependence of the back-stress on N is a classical one and was first derived by *Taylor* [39.35], based on an arrangement of parallel dislocations on a slip plane. Though the actual arrangement of the dislocations (the microstructure) was found to be much more complicated than that assumed by Taylor, this form of the back-stress has been found to be valid under a wide range of conditions and is widely used in work-hardening theory [39.36]. The dislocations multiply as they move, resulting in evolution of the dislocation density. It is reasonable to assume that the multiplication rate is proportional to the dislocation density and the velocity of their movement. Thus the evolution of dislocation density can be modeled by

$$\frac{dN}{dt} = \delta N v. \quad (39.16)$$

Based on experimental evidence from [39.37] δ was assumed to be directly proportional to the effective stress τ_{eff}. *Alexander* and *Haasen* [39.33] opine that the multiplication mechanism is mainly due to the action of long jogs on screw dislocations, and the number of these jogs present is proportional to the effective stress, giving rise to the factor of τ_{eff}, therefore, $\delta = k\tau_{\text{eff}}$. Combining (39.13–39.16), the following two equations are obtained

$$\frac{d\varepsilon_{\text{pl}}}{dt} = \phi b N v_0 \left(\frac{\tau - A\sqrt{N}}{\tau_0}\right)^m \exp\left(-\frac{Q}{k_B T}\right), \quad (39.17)$$

and

$$\frac{dN}{dt} = k N v_0 \left(\frac{\tau - A\sqrt{N}}{\tau_0}\right)^{m+1} \exp\left(-\frac{Q}{k_B T}\right). \quad (39.18)$$

Equations (39.17) and (39.18) form a complete set of phenomenological models, which given the externally applied shear stress τ, can model the evolution of the plastic strain and dislocation density with time. They describe the yield region in the curve shown in Fig. 39.10 rather well, and can account for the occurrence of the upper and lower yield stress. In the stress–strain curve shown in Fig. 39.10, the total strain rate, which is the sum of elastic and plastic strain rates, is maintained constant. Hence

$$\dot\varepsilon = \phi \frac{\dot\tau}{G} + \dot\varepsilon_{\text{pl}} \quad \text{or} \quad \dot\varepsilon = \phi \frac{\dot\tau}{G} + \phi b N v \quad (39.19)$$

is held constant, where G is the shear modulus.

Equations (39.17–39.19) can be solved numerically to derive the stress–strain curve for constant-strain-rate compression tests. The results of these numerical solutions have confirmed the experimental observation that the upper and lower yield stresses are very sensitive to the strain rate, the temperature, and in some cases the initial dislocation density in the material. They replicate the experimental observation of the upper and lower yield stresses of the form

$$\tau_{\text{uy}} \propto \dot\varepsilon^{\frac{1}{n}} \exp\left(\frac{U}{k_B T}\right)$$

and

$$\tau_{\text{ly}} \propto \dot\varepsilon^{\frac{1}{n}} \exp\left(\frac{U_1}{k_B T}\right),$$

where τ_{uy} is the upper yield stress and τ_{ly} is the lower yield stress. *Alexander* and *Haasen* [39.33] through some approximations arrive at the results

$$\tau_{\text{uy}} \propto \dot\varepsilon^{\frac{1}{3}} \exp\left(\frac{Q}{3k_B T}\right) \left[\ln \dot\varepsilon + \left(\frac{Q}{k_B T}\right) - \frac{3}{2}\ln N_0 + \text{const.}\right]^{1/3}$$

and

$$\tau_{\text{ly}} \propto \dot\varepsilon^{\frac{1}{3}} \exp\left(\frac{Q}{3k_B T}\right) \left(A^2 b B_1\right)^{1/3},$$

where $B_1 = v_0/\tau_0$, Q, A, v_0, and τ_0 are parameters from (39.14) and (39.15) and m has been assumed to be 1. N_0 is the initial dislocation density. From the above analysis, the following conclusions can be drawn:

1. The numerical calculations from (39.17–39.19) predict the upper and lower yield stresses for a given deformation experiment. The predictions are quite accurate for a number of materials except that the lower yield stress is slightly underpredicted.
2. A log–log plot of the lower yield stress versus $1/T$, at constant strain rate, would yield a straight line, the slope of which is related to the quantity activation energy for dislocation velocity Q from (39.14). The same quantity Q can be estimated from dislocation velocity measurements and deformation tests. Values estimated independently from both methods have been shown to coincide for a number of semiconductor materials, thus giving an indirect validation of the above formulation [39.33].
3. The upper yield stress is shown to depend on the initial dislocation density. This is true for elemental semiconductors. In compound semiconductors such as InP and GaAs, the upper yield stress is experimentally seen to be independent of the initial dislocation density [39.38, 39]. This is because the dislocations present initially are locked in their position by impurities, and the surface acts as an effective source of dislocations. Hence the initial dislocation density does not play a big role in the deformation.
4. The lower yield stress is predicted to be independent of the initial dislocation density of the material. This is found to be true experimentally, with the

lower yield stress showing only weak dependence on the initial dislocation density. *Alexander* and *Haasen* [39.33] postulated that the state of the crystal at the lower yield point is a stationary state. In this state the dislocation density is at an optimum value, which gives a minimum stress for a given strain rate. The value of this optimum dislocation density was derived to be equal to $(4/9)\tau_{ly}^2/A^2$, which is independent of the initial dislocation density. Substituting the above result into (39.17), a state of plasticity of the form $f(\dot{\varepsilon}, T, \tau_{ly}) = 0$ can be derived, independent of the state of the material at the beginning of the deformation.

5. The results of the uniaxial strain rate tests were compared with uniaxial, constant stress creep tests, with the material oriented for single slip. The generic shape of strain versus time is shown in Fig. 39.11. It can be seen that the inflexion point with strain ε_w has the largest strain rate. This again corresponds to a state of plasticity; the dislocation density at this point is again at the same optimum value, and the material is softest, corresponding to a maximum strain rate at this point. From the theory outlined above, this state of plasticity should satisfy the same relation as above: $f(\dot{\varepsilon}_w, T, \tau) = 0$, as has been verified experimentally by *Völkl* [39.40].

Thus some level of understanding has been reached for the behavior in the yield regions, though there are still limitations to the theory as will be shown later. After the stage of the yield points, the stress–strain curve is followed by five more distinct stages at higher strains. Some of these stages are found only at higher temperatures and represent alternate stages of hardening and softening of the material due to phenomena such as cross-slip and climb of dislocations.

Stage I
This is also known as the stage of easy glide. In this stage, the dislocation structure changes from a predominance of screw dislocations in the yield regions to a predominance of edge dislocations, dipoles, and multipoles. This stage is characterized by low hardening.

Stage II
This stage is characterized by a constant hardening rate $\theta = d\tau/d\varepsilon$, which is almost invariant with respect to temperature and strain rate changes. This value is also close to the hardening coefficients of stage II of fcc metals ($\approx 2.4 \times 10^{-3}G$, where G is the shear modulus of the material). The deformation in this region is characterized by inhomogeneity of dislocation distribution and a number of locally activated glide systems.

Stage III
This is characterized by the softening behavior of the crystal, during which the hardening coefficient decreases. This stage is also exhibited in fcc crystals, which might lead to assumptions about similar dislocation mechanisms operating in this range. However a detailed analysis of the temperature and strain rate dependence, together with metallographical investigations, reveals that stage III softening in semiconductors is due to climb of dislocations, while in metals it is due to cross-slip of dislocations. *Siethoff* and coworkers have extended the experimental data for this stage for Si and Ge [39.41], GaAs [39.42], and InP [39.43], and in [39.44] have fitted this data to the dislocation climb model of *Mukherjee* et al. [39.45] and the jog-dragging model of *Barret* and *Nix* [39.46]. They found the data to be more consistent with the jog-dragging model, which proposed the softening to be a consequence of interaction of point defects with dislocations when screw dislocations drag long jogs along with them. However, more light needs to be shed on this stage, before the final word can be said on deformation mechanisms of stage III.

Stages IV and V
These stages were discovered in Ge by *Brion* et al. [39.47] and occur only in a narrow tempera-

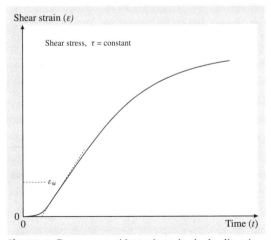

Fig. 39.11 Creep curve with specimen in single-slip orientation and constant applied stress

ture range below the melting point T_m. Stage IV, like stage II, is characterized by linear hardening, where the hardening coefficient is invariant with temperature and depends slightly on strain rate. Stage V is similar to stage III and is characterized by softening or dynamic recovery. Investigations have revealed the recovery to be due to the cross-slip of screw dislocations, and *Siethoff* [39.48] have fitted the data obtained for stage V to *Escaig*'s theory for cross-slip of screw dislocations [39.49].

An important observation from experiment is the shrinkage of the stages with a rise in temperature.

Stages IV and V do not occur at low temperatures. As the temperature is raised, each of the stages begins earlier, i.e., at lower strain, in the deformation curves. This suggests that dislocation mechanisms that lead to stages II–V begin to operate earlier at high temperatures. The deformation, due to thermal stresses, in crystal growth processes is not expected to exceed stage I, but in areas of high temperature close to the melting point, such as the crystal–melt interface, mechanisms prevalent in the later stages such as cross-slip, climb, and interaction with point defects may be important.

39.7 Application of the Haasen Model to Crystal Growth

Völkl [39.40] was the first to use the above model to study dislocation generation in the Czochralski system. In this analysis, the entire thermal and thermal–elastic stress history of a given particle in the crystal, from the particle's evolution from the melt–crystal interface to the end of the growth process, is determined. The thermal–elastic stress history so determined is used to determine the resolved shear stress history on the highest stressed slip system; an unstated assumption by Völkl is that the same slip system is the most highly stressed throughout the growth history of the particle. The resolved shear stress history is designated $\tau_t^{gs}(t)$. Völkl assumes that, as the plastic deformation proceeds, the resolved shear stress relaxes or is dissipated by the plastic strain, and the residual elastic stress $\tau_{er}(t)$ is the driving force behind plastic deformation. The relaxation process can be described by

$$\frac{d\tau_{er}(t)}{dt} + \frac{G}{\varphi}\frac{d\varepsilon_{pl}}{dt} = \frac{d\tau_t^{gs}(t)}{dt} . \quad (39.20)$$

In conjunction with (39.20), (39.17), and (39.18) with τ replaced by $\tau_{er}(t)$ this leads to

$$\frac{d\varepsilon_{pl}}{dt} = \phi b N v_0 \left(\frac{\tau_{er}(t) - A\sqrt{N}}{\tau_0}\right)^m \exp\left(-\frac{Q}{k_B T}\right), \quad (39.21)$$

$$\frac{dN}{dt} = kNv_0 \left(\frac{\tau_{er}(t) - A\sqrt{N}}{\tau_0}\right)^{m+1} \exp\left(-\frac{Q}{k_B T}\right). \quad (39.22)$$

Since the thermal–elastic history $\tau^{gs}(t)$ has already been determined from the thermal history of the crystal, through a finite element code, (39.20–39.22) can be integrated simultaneously to obtain the final dislocation density N. The dislocation density, in the volume element, after it just emerges out from the melt is taken as a fixed constant N_0, i.e., the initial condition

$$N|_{t=0} = N_0 .$$

However, according to Völkl the final values of N are reasonably independent of the N_0 chosen. It is apparent that a choice of $N_0 = 0$ will lead to a singularity. Völkl justifies his assumption of single slip by observing that an exponential dependence of plastic deformation on stress will cause the most highly loaded slip system to act and thus release the stresses on all other slip systems. *Maroudas* and *Brown* [39.50] used an isotropic generalization of the plastic strain response in the Alexander–Haasen model and performed what they called an integrated analysis of crystal growth. *Tsai* [39.51] has also applied the Alexander–Haasen model to bulk crystal growth. He generalized the Alexander–Haasen model to an isotropic plastic response instead of assuming deformation on the most highly stressed slip system. *Tsai* et al. [39.52] later formulated a multislip generalization of the Alexander–Haasen model, involving all 12 slip systems active in the semiconductor crystal. *Lambropoulos* and *Wu* [39.53] applied the model to Czochralski growth of GaAs, and investigated the effect of the shape of the interface on the final dislocation generation, while *Miyazaki* and *Kuroda* implemented this model for Czochralski (CZ) growth, using a finite element technique [39.54].

The driving force behind deformation in the growing crystal is the thermal strain. Since thermal strains are small, the deformations are likely to fall within the yield region in Fig. 39.10, thus justifying the use of the Alexander–Haasen model. However, there is no evidence that the thermal loading would favor the single-slip scenario. Moreover, as noted earlier, stages IV and V in Fig. 39.10 appear only at high temperatures. Higher temperatures are characterized by shrinkage of the stress–strain curves – the various stages occur at lower strains. This indicates that phenomena such as cross-slip and climb of dislocations, which dominate the later stages, become important at lower strains, for deformation at high temperatures. Thus, at temperatures approaching the melting point of the crystal – for which little data is available – these phenomena might be important even in the yield-point region. A multislip generalization of the Alexander–Haasen model is therefore needed, with provisions for high-temperature phenomena such as cross-slip, climb, etc. A model by *Moosbrugger* [39.55] and *Moosbrugger* and *Levy* [39.56] considers several of these features for CdTe crystals, and is the most advanced of the Alexander–Haasen variants.

39.8 An Alternative Model

The Alexander–Haasen model, though impressive, suffers from certain drawbacks for its application to compound semiconductors, and even elemental semiconductors. It is well known that the structure of the lattice in semiconductors gives rise to different types of dislocations, for example, screw, α dislocations, β dislocations, etc. All of these have different mobilities, and each one of them is expected to follow a different relation of the type (39.14). Perhaps, the adoption of a single dislocation variable N and the characterization of the dislocation velocity by a single equation of the type (39.14) is insufficient. Moreover, equations of the type (39.14) have been experimentally verified only for certain range of temperatures and stresses, which do not quite cover the entire gamut of these occurring in the crystal growth process. The specification of the initial dislocation density N_0 is also problematic. What is N_0 in the growing crystal? *Völkl* [39.40] as well as other researchers, as mentioned in the previous section, assumed a certain low N_0 along with the claim that this does not affect the final result much. This claim might be true, but such a treatment means that the final grown crystal, even for Si, always contains dislocations, since the dislocations move and multiply even at small stresses. Such a situation is incompatible with daily experience, as silicon single crystals are grown without any dislocations.

The deformation experiments looked at so far are uniaxial creep or constant-strain-rate tests, where the crystal has been oriented for single slip. In multislip deformations, or in applications such as crystal growth, dislocations in many slip systems are expected to be active. Thus what is need is a generalization of the Alexander–Haasen model, retaining the basic structure and ideas, but expanding it to include notions of multislip, different types of dislocations, modified multiplication laws, the possible effects of dopants, and some information on the initial dislocation sources in the crystal. Let us examine each of these issues to build a model.

39.8.1 Different Types of Dislocations

Semiconductors have a high Peierls energy, resulting in deep Peierls wells along the $\langle 110 \rangle$ directions. Consequently, the dislocations tend to align themselves along the $\langle 110 \rangle$ directions to minimize the energy. Thus typical dislocation loops tend to assume a hexagonal shape. Transmission electron microscopy (TEM) studies have shown that this condition holds when the dislocation density is low and the impurity concentration in the crystal is not too high [39.25, 27]. If the dislocation density is high, the local back-stresses between the dislocations will distort the hexagonal shape, and the dislocations may become curved, no longer lying in $\langle 110 \rangle$ directions. The conditions of low dislocation density might be expected to hold in crystal growth. X-ray topography by *Dudley* et al. [39.57, 58] shows long segments of screw dislocations aligned along the $\langle 110 \rangle$ directions, and hexagonal loops. The geometry

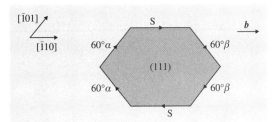

Fig. 39.12 Geometry of a hexagonal dislocation loop on the (111) plane.

of such a loop is shown in Fig. 39.12. Since there are three different dislocation types, the shear strain in this $(111)[\bar{1}10]$ slip system is given by a straightforward extension of the Orowan equation

$$\dot{\gamma}_{\text{pl}} = (N_\alpha \bar{v}_\alpha + N_\beta \bar{v}_\beta + N_s \bar{v}_s) b, \quad (39.23)$$

where N_α, N_β, and N_s are the densities of α, β, and screw dislocations, respectively, in the (111) plane with the Burgers vector in the $[\bar{1}10]$ direction. Here v_α, v_β, and v_s are the corresponding velocities, with the overbar representing an average. As the loop in the figure expands, it can be seen that the screw segments expand due to the movement of α and β segments, while these segments expand primarily by the movement of the screw dislocations. If the lengths of these segments are denoted by L_α, L_β, and L_s, their rate of change can be approximately given by

$$\dot{L}_\alpha = 2\bar{v}_s, \quad \dot{L}_\beta = 2\bar{v}_s \quad \text{and} \quad \dot{L}_s = 2(\bar{v}_\alpha + \bar{v}_\beta).$$

Since the evolution of the dislocation densities of α and β types are governed approximately by the same quantity, a good approximation would be

$$N_\alpha = N_\beta$$

and

$$\frac{N_\alpha + N_\beta}{N_s} = \frac{2\bar{v}_s}{\bar{v}_\alpha + \bar{v}_\beta}.$$

Substituting the above approximations into (39.23), we obtain the following approximation for the evolution of the shear strain

$$\dot{\gamma}_{\text{pl}} = 2 N_s \bar{v}_s b.$$

Thus the evolution of the shear strain in a slip system is governed chiefly by the velocity and density of the screw dislocations in that system. Treatment similar to the above was first proposed by *Steinhardt* and *Haasen* [39.59] to explain results of creep experiments on GaAs, and was more clearly expounded by *Yonenaga* et al. [39.60], whom the above follows closely.

Since there are 12 $\{111\}\langle\bar{1}10\rangle$ active in the lattice, the plastic shear strains on all of these have to be dyadically combined to get the total plastic strain rate, which is a second-order tensor. This can be expressed as

$$\dot{\boldsymbol{\varepsilon}}^{\text{pl}} = \sum_{i=1}^{12} N_s^i \bar{v}_s^i b (\boldsymbol{n}_i \otimes \boldsymbol{s}_i + \boldsymbol{s}_i \otimes \boldsymbol{n}_i) \text{sign}(\tau_i), \quad (39.24)$$

where index i represents the slip system, $\boldsymbol{s}_i$ represents the slip direction of the slip system i, $\boldsymbol{n}_i$ represents the normal to the slip system i, N_s^i represents the density of screw dislocations on the slip system i, and $\bar{v}_s^i$ is the average velocity of these dislocations. $\text{sign}(\tau_i)$ is the sign of the resolved shear stress along the slip system. Equation (39.24) represents a three-dimensional generalization of the Orowan equation (39.13), taking into account the geometry of a dislocation loop in compound semiconductors. The slip and slip plane normal vectors for the 12 slip systems are given in Table 39.1.

39.8.2 Dislocation Glide Velocity

The glide velocity of dislocations was investigated primarily for segments of the screw and 60° type mainly by the technique of producing dislocation half-loops through deliberate scratches on the surface of a semiconductor. Once the scratches are polished off, the material is subjected to loading (four-point bending, or tension or compression) and the glide velocities of the resulting dislocations are measured. This technique has been used for a wide range of temperature and stresses, and for various elemental and compound semiconductors. Following the work of *Chaudhuri* et al. [39.34] data has been collected for Si [39.61–63], and III–V semiconductors such as GaAs [39.59], InP [39.64–66], etc. A consensus exists regarding the results for measurements in the so-called central range of stress and temperature. *George* and *Rabier* [39.31] define this central range as

$$0.45 T_m < T < 0.65 T_m,$$
$$5 \times 10^{-5} \mu < \tau < 10^{-3} \mu,$$

where T_m is the melting point of the material and μ is the shear modulus of the material.

The temperature and shear stress dependence of the glide velocity of screw and 60° segments is as follows.

In the central range of measurements, the temperature dependence of the dislocation velocities at constant stress is given by an Arrhenius-type law of the form

$$v \propto \exp\left(-\frac{Q(\tau)}{k_B T}\right),$$

where $Q(\tau)$ is a stress-dependent activation energy, τ is the shear stress acting on that slip system, and k_B is the Boltzmann constant. The activation energy Q is of the order of 1–2 eV in most semiconductors. The stress dependence of Q has been found to fall under two different regimes. For 60° dislocations in Ge, it is weakly stress dependent for stresses above 20 MPa, and is a strongly decreasing function of stress below this stress. A similar transition stress of 10 MPa is found even for screw dislocations in Ge. In Si, the transition stress for 60° dislocations is 10 MPa, while none has been found for screw dislocations. The Arrhenius law has also been observed to breakdown in Si and Ge at some critical temperatures. *Farber* and *Nikitenko* measured 60° dislocation glide velocities up to temperatures of $0.9 T_\mathrm{m}$ in Si [39.63] and have observed some discrepancies. The velocities still depict an Arrhenius-type behavior, but the activation energy and prefactor are changed at temperatures above $0.75 T_\mathrm{m}$.

The dependence on stress of the dislocation glide velocity at constant temperature is conveniently described by a power law of the form

$$v \propto \exp(\tau^m).$$

Combining the above two dependencies, an overall dependence of the form

$$v = v_0 \left(\frac{\tau}{\tau_0}\right)^m \exp\left(-\frac{Q(\tau)}{k_\mathrm{B} T}\right) \quad (39.25)$$

is obtained. This form is applicable for higher stresses, 20–100 MPa in Ge and $>$ 30 MPa in Si, with constant m and Q. Below this range m and Q cannot be taken as constants. A double-logarithmic plot of v versus τ at constant temperature shows a bend. This is partly due to the stress dependence of $Q(\tau)$ below the critical stress mentioned above. *George* et al. [39.61] opine that, below the critical stress, Q can be expressed in the form

$$Q = Q_0 - E_i \ln\left(\frac{\tau}{\tau_0}\right).$$

Alexander [39.27] has explored the stress range above the critical stress for Si, and found that the parameter m depends not only on the type of dislocation but also on the glide system to which the dislocation belongs, and nonshear stress components etc. These complications are not well understood. However, *Sumino* [39.25] opines that the above complications are a result of impurities immobilizing the dislocations at low stresses, and also the many problem associated with the technique of measuring dislocation velocities described above. *Alexander* discusses these in reference [39.27]. *Imai* and *Sumino* [39.62] used a live x-ray topography technique to measure the velocity of dislocations in highly pure Si in the temperature range 600–800 °C and stress range 1–40 MPa. They found perfect agreement with (39.25), with $m = 1$, and $Q = 2.20$ eV for 60° dislocations and $Q = 2.35$ eV for screw dislocations.

Hence the use of (39.25) lies on shaky foundations with several opinions expressed regarding its form and range of usage for different materials. Several, theoretical models have been put forward to explain the phenomena of glide, a few of which will be reviewed, before a brief summary of literature about dislocation glide velocities in InP is presented.

Dislocation glide in diamond structured elemental and compound III–V semiconductors takes place through the formation and migration of double kinks. The theory has been formulated by *Hirth* and *Lothe* [39.6, Chap. 15]. They derived the nucleation rate of double kinks on a dislocation as

$$J = v_0 \frac{\tau b h}{k_\mathrm{B} T} \frac{\exp(-2F_\mathrm{k} - W_\mathrm{m})}{k_\mathrm{B} T},$$

where v_0 is the Debye frequency, τ is the applied shear stress, $2F_\mathrm{k}$ is the formation energy of a double kink under the action of a shear stress τ, W_m is the migration energy of the kink, b is the magnitude of the Burgers vector, and h is the kink height, or the distance between two Peierls valleys. For small kink densities or short dislocation segments, the double kinks traverse the entire length of the dislocation, without being annihilated, and the dislocation glide velocity is proportional to the length of the segment L. Hence

$$v = h L J \text{ or } v = \frac{L \tau b h^2}{k_\mathrm{B} T} \frac{v_0 \exp(-2F_\mathrm{k} - W_\mathrm{m})}{k_\mathrm{B} T}.$$

For long dislocation segments, the kinks are annihilated when they meet kinks of opposite sign, before they reach the ends of the segment. The mean free path X of the kinks in this case is given by

$$\frac{1}{\frac{X}{2} J} = \frac{X}{2 v_\mathrm{k}} \Rightarrow X = 2\sqrt{\frac{v_\mathrm{k}}{J}},$$

where v_k is the kink velocity given by $v_\mathrm{k} = (\tau b h)/(k_\mathrm{B} T) v_0 a^2 \exp(-W_\mathrm{m}/k_\mathrm{B} T)$, where W_m is kink migration energy, and a is the jump distance of a kink.

In the case of long dislocation segments, the velocity is given as

$$v = h J X = 2h(J v_\mathrm{k})^{\frac{1}{2}} = 2 \frac{v_0 \tau b h^2}{k_\mathrm{B} T} \frac{a \exp(-F_\mathrm{k} - W_\mathrm{m})}{k_\mathrm{B} T}.$$

The above expressions suggest that the activation energy of dislocation glide would vary with the dislocation length; there exists a transition region between the two regimes. *Louchet* [39.67] confirmed this. He was able to observe that the velocity of long dislocation segments was independent of length, while the velocity of short segments was proportional to their length. He determined the critical length for $\tau = 90$ MPa as $L_c = 0.4$ μm, which was interpreted as the mean free path of the kinks before annihilation. From these measurements, W_m and F_m were calculated separately, and both were of the magnitude of 1 eV. It was previously assumed that, analogously to fcc metals, the W_m in semiconductors was very small, and the presence of local obstacles along the dislocation was postulated to explain the discrepancy of a high W_m obtained. Theoretical treatments with this assumption presented by *Celli* et al. [39.68] and *Rybin* and *Orlov* [39.69] yield a very high density of obstacles, or a very high activation energy for dislocation glide velocities. The theory of local obstacles along the dislocation, opposing kink motion, is not accepted any longer, and it is now recognized that the high W_m is a result of a high secondary Peierls potential. The existence of a high Peierls means that migration of kinks is as difficult as their nucleation. This is related to the core structure of the partial dislocations, comprising the dislocation segment. It is believed that, in elemental semiconductors, the bonds on partial dislocations and kinks are reconstructed, except at certain defects along the core of the partials called solitons, or antiphase defects. It is only at these solitons that dangling bonds are present. The solitons play a direct role in the nucleation and migration of kinks, accounting for the same order of magnitude of the kink migration and nucleation energies.

Several analyses of the core structure of 90 and 30° partial dislocations, and estimates of W_m and F_k, have been performed [39.70, 71]. These analyses are based on atomistic simulations, using some empirical interatomic potentials, or ab initio methods, which are able to reproduce the quantum mechanical effects. These techniques, along with recently developed electron microscopy techniques [39.72] for the direct observation of movement and nucleation of kinks, should clarify W_m and F_k in the future. It should be pointed out that the above formalism is based on nucleation and migration of kinks in whole dislocations, while in reality the whole dislocation segments are split into partial dislocations. *Möller* [39.73] performed an analysis taking into account the presence of partial dislocations. He predicted the existence of a critical stress below which dislocation glide takes place through the correlated motion of the partials. At stresses above the critical stress, the glide took place with the movement of the two partials being independent. The correlated movement of partial dislocations has higher activation energy than the uncorrelated movement. This analysis somewhat reproduced the experimental observations in Ge and Si about the existence of two different regimes for the stress-dependence of the activation energy Q, decribed earlier in this section (Sect. 39.8.2). Möller's analysis was performed under the framework of local obstacles along the dislocation, and there is a need to adapt this to more recent ideas of a high secondary Peierls potential.

The influence of doping on the glide velocity of dislocations in semiconductors is well recorded. This is realized chiefly through electronic and metallurgical interaction. The metallurgical effect is classical and is relevant to any type of dopant or impurity in the semiconductor. It is a local effect and is due to the interaction between the impurity atom and the dislocation core. Another manifestation of this is the so-called locking stress, which certain impurities cause in certain semiconductors. At shear stresses below the locking stress, the dislocations are locked and do not glide and, once the shear stress exceeds the locking stress, the dislocations glide smoothly. Examples of this phenomenon are O, P, and S in Si. The electronic interaction is prevalent only for electrically active dopants or impurities when their concentration exceeds the intrinsic carrier concentration of the material. It has been found that III or V doping in elemental semiconductors affects the glide velocities in the same way irrespective of the size and chemical nature of the dopant. The primary influence of these electrically active dopants is that they change the Fermi energy level in the semiconductor. There are two classes of theories to explain the effect on dislocation glide velocity. *Haasen* [39.74] postulate that the line charge on a dislocation in a semiconductor depends on the position of the Fermi energy level. A charged dislocation has the formation energy of a double kink reduced, because of the electrostatic repulsion between the two kinks. This affects the dislocation glide velocity. *Hirch* [39.75] postulated that the formation of a double kink leads to the generation of acceptor E_{ka}, or donor E_{kd} levels in the band gap. Charged as well as neutral kinks can exist on a dislocation, while the concentration of the neutral kinks is constant, the concentration of charged kinks depends on the position of the Fermi level (E_F), and the charged kink's own level in the spectrum. The concentrations of negatively and positively charged kinks are given by

$$\frac{C_k^+}{C_k^0} = \exp\left(\frac{(E_{kd} + eV) - E_F}{k_B T}\right)$$

and

$$\frac{C_k^-}{C_k^0} = \exp\left(\frac{E_F - (E_{ka} + eV)}{k_B T}\right),$$

where V is the gate voltage. The above formulae are derived by the application of Fermi–Dirac statistics. The concentration of neutral kinks is constant, and hence the dislocation glide velocity is directly dependent on the concentration of charged links. If the kink's level is near the middle of the energy gap – close to the Fermi level – increasing the n-type doping would raise the Fermi level and increase the concentration of negatively charged kinks and consequently dislocation velocity. Similarly increasing p-type doping will decrease the Fermi level and raise the concentration of positively charged kinks and consequently the dislocation velocity. Consequently both n-type as well as p-type materials will have higher dislocation velocity than the intrinsic material, as in the case of silicon. On the other hand, if the kink's level is near the valence band – much below the Fermi level – the intrinsic material will have negatively charged kinks. Increasing the n-type doping will raise the Fermi level and increase the concentration of negatively charged kinks raising the dislocation velocity, while a p-type doping will lower the Fermi level and decrease the concentration of negatively charged kinks and lower the dislocation velocity. Thus in materials where the kink's level is near the valence band, the velocity should increase from p-type to n-type material, as in Ge. This gives some understanding of the doping effect.

Finally the complications introduced by the splitting into partial dislocations and the existence of two different types of atomic species in compound semiconductors should be recognized. A 60° dislocation splits into a 30 and a 90° partial, and the mobility of each partial is different. The velocity of the whole 60° dislocation depends on which partial is leading in the direction of the movement. Thus the same dislocation might show asymmetry in velocity depending on the direction of the movement. In compound semiconductors, the velocity of α and β dislocations of the same type are expected to be different because of the different core structures. Thus, a review of the theoretical and experimental aspects of dislocation glide velocity reveals that, at this stage, the theory is not yet comprehensive enough to yield concrete relations for dislocation glide velocities that can be used in a microdynamical theory of plasticity. The best that can be done currently is to adapt an expression of the type (39.14), with constant m and Q, while recognizing its inherent limitations – that the parameters m and Q are stress and temperature dependent.

There is some data available on the direct measurement of the velocity of dislocations for InP. For undoped and sulphur-doped material, the velocity of

Table 39.2 Dislocation velocity data for undoped InP (after [39.64, 76])

Dislocation	v_0	m	Q (eV)
α	40 000	1.4	1.6
β	500 000	1.8	1.7
screw	40 000	1.7	1.7

α dislocations was reported by *Nagai* [39.65] in the temperature range 250–400 °C. He fitted the velocities to a relation of the type (39.14), with $m = 2.7$, for both doped and undoped crystal. The value of Q was found to be constant, with $Q = 1.1$ eV for S-doped crystal (6.5×10^{18} cm^{-3}), and 1.0 eV for undoped InP. S doping was found to increase the velocity of α dislocations. *Maeda* and *Takeuchi* [39.66] set out to measure the effect of recombination-enhanced glide in InP crystal under the effect of minority carrier injection. In the course of their investigation, they measured the velocity dependence of β dislocations in S-doped crystal (5×10^{18} cm^{-3}) on temperature, in the range 220–410 °C. They found a Q value of 1.6 eV. *Yonenaga* and *Sumino* [39.64, 76] investigated the velocity of all three types of dislocations in InP crystals – undoped, and with a variety of dopings. The temperature range of their measurements was 300–500 °C, and stresses between 2 and 20 MPa. They fitted their data to an equation of type (39.14), with $\tau_0 = 1$ MPa. For undoped InP, they obtained the parametric fits presented in Table 39.2.

Yonnenaga and *Sumino* [39.64] have also obtained the corresponding parameters for material doped with Zn (p-doping), S (n-doping), and the isovalent impurities Ga and As. The dislocation velocity data in Table 39.2 is used in the modeling effort presented later.

39.8.3 Dislocation Multiplication

Dislocations being line defects, any movement of these will involve self-multiplication during motion and increase their lengths. If some parts of the dislocation loop are less mobile than others, the more mobile parts will move around in a spiral, increasing their length. In the loop shown in Fig. 39.12, the β segments are more mobile than the α segments, which are more mobile than the screw segments. Under the influence of stress, the β segments will move quickly through the material while screws will lag behind, resulting in an expansion of the screw segment's length. Another typical case is jogs, found on screw dislocation. These jogs are less mobile than the main dislocation; hence these lag behind,

Table 39.3 The type of interactions between dislocations of different slip systems (N: pair belong to cross-slip systems, G: pair form glissile junctions, H: Hirth locks, S: sessile junctions, C: pair belong to coplanar systems. The numbers in parentheses are f_{ij})

i \ j	1	2	3	4	5	6	7	8	9	10	11	12
1		C(0)	C(0)	S(1)	G(0)	H(1)	N(0)	G(1)	G(1)	H(1)	S(1)	G(0)
2	C(0)		C(0)	G(1)	N(0)	G(1)	G(0)	S(1)	H(1)	S(1)	H(1)	G(0)
3	C(0)	C(0)		H(1)	G(0)	S(1)	G(0)	H(1)	S(1)	G(1)	G(1)	N(0)
4	S(1)	G(0)	H(1)		C(0)	C(0)	G(1)	N(0)	G(1)	G(0)	S(1)	H(1)
5	G(1)	N(0)	G(1)	C(0)		C(0)	S(1)	G(0)	H(1)	G(0)	H(1)	S(1)
6	H(1)	G(0)	S(1)	C(0)	C(0)		H(1)	G(0)	S(1)	N(0)	G(1)	G(1)
7	N(0)	G(1)	G(1)	G(0)	S(1)	H(1)		C(0)	C(0)	H(1)	G(0)	S(1)
8	G(0)	S(1)	H(1)	N(0)	G(1)	G(1)	C(0)		C(0)	S(1)	G(0)	H(1)
9	G(0)	H(1)	S(1)	G(0)	H(1)	S(1)	C(0)	C(0)		G(1)	N(0)	G(1)
10	H(1)	S(1)	G(0)	G(1)	G(1)	N(0)	H(1)	S(1)	G(0)		C(0)	C(0)
11	S(1)	H(1)	G(0)	S(1)	H(1)	G(0)	G(1)	G(1)	N(0)	C(0)		C(0)
12	G(1)	G(1)	N(0)	H(1)	S(1)	G(0)	S(1)	H(1)	G(0)	C(0)	C(0)	

while the more mobile part of the screw dislocation spirals around the jog, multiplying its length. Jogs can be formed either by the cross-slip mechanism, or by cutting of a screw dislocation by a forest dislocation in another slip plane. An example of this was shown by live topography of a dislocation multiplication process in silicon by *Sumino* and *Harada* [39.77], where a jog was formed on a screw dislocation by a 60° dislocation on a different slip plane cutting through it. The jog subsequently acted as a spiral center for multiplication. Dislocation multiplication can also occur by the Frank–Read mechanism, as was shown by an image of a Frank–Read source obtained by *Völkl* et al. [39.40]. The evolution of the dislocation density, through the dislocation multiplication, was modeled by the equation $dN/dt = k\tau_{eff} N v$ in the Alexander–Haasen model. The basic premise of this equation was that the total length of dislocation created dN is proportional to:

1. The distance moved by the dislocation.
2. The length of already existing dislocations, since the longer the existing dislocations, the greater the probability of jogs forming on them.
3. The effective stress τ_{eff} in the equation is to represent the stress required by the spiraling segment in the parallel slip plane – which has been created by a jog – to pass the screw dislocation in the original slip plane.

Sumino and *Yonenaga* [39.76] retained the basic structure of the Haasen law and added to it features of multislip deformation and the different types of dislocations found in compound semiconductors. They introduced a dislocation multiplication law of the form

$$\frac{dN_s^i}{dt} = K N_s^i \bar{v}_f^i \tau_{eff}^i + K^* N_s^i \bar{v}_f^i \tau_{eff}^i \sum_{j \neq i} N_s^j , \quad (39.26)$$

where the index i represents one of the 12 slip systems, and the bar over the velocity denotes an average.

The above law models the evolution of screw dislocation density in all 12 possible slip systems in III–V compound semiconductors. The basic proportionality of the evolution rate to the dislocation density and velocity is retained. The average velocity of the fastest dislocation v_f^i is used, since the expansion of a spiral and a loop is controlled by the velocity of the fastest segment, which in the case of InP is the β segment. The second term models the formation of jogs on screw dislocations and the consequent expansion through spiral formation. The rate of formation of jogs is proportional to the number of times dislocations in different slip systems cut each other, which can be modeled by a product of dislocation densities on the two slip systems.

Equation (39.26) can therefore be generalized as

$$\frac{dN_s^i}{dt} = K N_s^i \bar{v}_f^i \tau_{eff}^i + K^* N_s^i \bar{v}_f^i \tau_{eff}^i \sum_{j \neq i} f_{ij} N_s^j . \quad (39.27)$$

The f_{ij} are closely related to the formation of jogs. Following reference [39.78], the interaction between the dislocations of any pair of slip systems as categorized in Table 39.3 are:

1. Pairs belonging to two cross-slip systems such as 11 and 9, denoted by "N" in Table 39.3. When the dislocations of such systems cross each other, two kinks are formed, and no jogs result.
2. Pairs forming a Hirth lock. When the dislocations of such systems cross each other, jogs are formed on both dislocations. "H" denotes such interactions in Table 39.3.
3. Pairs that are in the same plane, and whose resulting Burgers vector is therefore on the same plane as the original ones. These interactions do not result in jogs on either system, and are denoted by "C" in Table 39.3.
4. Pairs forming glissile junctions, such as systems 1 and 12 in Fig. 39.1. The interaction between such systems produces a jog on one and a kink on the other. For example, if the dislocations of the two slip systems 1 and 12 pass, a jog is formed on slip system 12, while a kink is formed on slip system 1. The interactions between systems forming glissile junctions are denoted by "G" in Table 39.3.
5. Finally, the last type of interaction is between the pairs of systems forming sessile junctions, such as 4 and 1. When the dislocations of two such systems interact, they form stable Lomer–Cottrell locks, and impede the motion of both original dislocations. Such pairs are denoted by "S" in Table 39.3.

Alternatively, if the dislocations cross each other, a jog is formed on each one. The f_{ij} are assigned values of either 1 or 0 in Table 39.3 based on whether a jog is formed on a screw dislocation of system i when cut by a forest dislocation of system j. Equations (39.24) and (39.27) form a complete viscoplastic system with 12 internal variables. The meaning of the effective stress τ_{eff}^i is clarified next.

39.8.4 Work Hardening

The flow stress can be classically split into a rate-dependent term and a hardening term, independent of the strain rate and temperature. The rate-dependent part of the applied stress is used to overcome the frictional resistance of the lattice to the glide of a dislocation, which is temperature sensitive. The rate-independent part is the resistance from interaction with other dislocations, both in the same slip system as well as threading dislocations in other slip systems, and finally impurities. In the single-slip model described before, the work-hardening term was given by $A\sqrt{N}$. Taylor [39.35] first introduced this form, drawing from

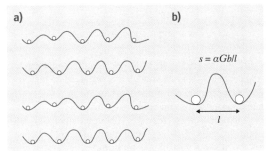

Fig. 39.13 (a) A series of dislocations facing barriers in the glide plane, and (b) a dislocation link stuck in an obstacle of strength s

theoretical treatment of the elastic interaction between parallel dislocations. The elastic theory of dislocations gives the magnitude of A as $Gb/(2\pi(1-v))$, where G is the shear modulus and v is the Poisson ratio. *Moulin* et al. [39.79] conducted a mesoscopic simulation, using discrete dislocation dynamics, for the yield point of Si, in a single-slip framework. From a fit of their data, they found the work hardening to be adequately defined by an equation of the type $A\sqrt{N}$, but the value of A was $1.48Gb/(2\pi(1-v))$. Interaction between dislocations of different slip systems gives rise to complicated hardening behavior. Apart from long-range elastic interactions, stable locks may form between dislocations of noncoplanar slip systems. *Hirth* and *Lothe* [39.6] discuss different types of locks, the most prominent of which are the Lomer–Cottrell lock and the Hirth lock.

Franciosi and *Zaoui* [39.78] have classified the different types of locks formed, and their efficiency in locking. These locks act as obstacles to other dislocations in the glide plane. Figure 39.13 shows a dislocation stuck between such obstacles. If a link of dislocation is stuck between two such obstacles of distance l (Fig. 39.13b) the shear stress s in the glide plane required to release the dislocation is of the order $s = \alpha Gb/l$, where G is the shear modulus, b is the Burgers vector, and α is some numerical factor.

Hence the resistance offered to the dislocation motion on a given glide plane is inversely proportional to $1/l$. Since l is the distance between obstacles, it is inversely proportional to the number of obstacles on the glide plane. The number of obstacles on a glide plane, due to forest dislocations on another glide system, say the α glide system, is proportional to $\sqrt{N_\alpha}$. Hence the total rate-independent resistance offered to dislocation motion in glide system β can be denoted heuristically as

$$\sum_{\alpha \neq \beta} a_{\beta\alpha} \sqrt{N_\alpha} + \frac{Gb}{2\pi(1-\nu)}\sqrt{N_\beta}, \quad (39.28)$$

where the second term is the self-hardening term of the Alexander–Haasen model. An equation of the type (39.28) has been used in the literature to justify experimental observations on latent hardening in Cu [39.78].

The coefficients $a_{\beta\alpha}$ are phenomenological and might depend on many factors including (1) the geometry of the forest dislocation configurations, which determines the distance between the intersections of these with the primary glide plane, and (2) the strength of the locks formed, which might depend on other factors such as the stacking fault energy. *Franciosi* and *Zaoui* [39.78] classified the $a_{\beta\alpha}$ based on the strength of the locks which are formed. They classify them according to whether the dislocations belong to the cross-slip systems (interaction coefficient a_1), form Hirth locks (interaction coefficient a_2), belong to coplanar systems (interaction coefficient a_3), form glissile junctions (interaction coefficient a_4), or form sessile Lomer–Cottrell locks (interaction coefficient a_5). They found the a coefficients to increase linearly with the stacking fault energy, and the anisotropy in the a coefficients to increase with decreasing stacking fault energy.

Alexander and *Crawford* [39.80] determined the interactions of different slip systems through latent hardening experiments. The procedure they used was as follows: A Ge single crystal was deformed at constant strain rate in single-slip orientation up to stage I. The lower yield stress was noted. The crystal sample was cut into smaller samples, which were then deformed at the same strain rate. Each sample was aligned for single slip, in such a way that the other secondary glide systems become the most stressed systems. The lower yield stress in each of these cases was noted, and compared with the lower yield stress in the initial sample. The results were then analyzed in the framework of the Alexander–Haasen model, with an equation similar to (39.28) adopted as the formula for the back-stress. Normalizing the $a_{\alpha\beta}$ by Gb, they found the normalized coefficients to be: 0.37 for systems forming Lomer–Cottrell locks, 0.24 for systems forming Hirth locks, 0.35 for glissile locks, and 0.64 for the cross-slip system. Thus, the cross-slip system was found to have the largest hardening coefficient. These results are at variance with those of *Franciosi* and *Zaoui* [39.78] cited above. *Alexander* and *Crawford* [39.80] explain their observation by noting that the structure in the primary glide system at the beginning of easy glide consists primarily of edge dipoles, and from the geometry it can be deduced that these intersect the cross-slip plane most effectively. In the case of crystal growth, the deformation structure is dominated by long screws, hence these coefficients should be expected to differ from those determined by *Alexander* and *Crawford* [39.80]. Also the stacking fault energy of InP is nearly four times smaller than that of Ge, hence the values might be expected to be more asymmetric.

Summarizing, an equation of the form (39.29) can be used for the hardening. Since the interaction coefficients are not available, one hopes to determine them from fitting deformation data in multislip and single-slip experiments. The effective stress in thus given by

$$\tau_{\mathrm{eff}}^i = |\tau^i| - \frac{Gb}{\beta}\left(N_s^i\right)^{1/2} - \frac{Gb}{\beta^*}\sum_{j \neq i} g_{ij}\left(N_s^j\right)^{1/2}.$$
(39.29)

The g_{ij} are set as a_1, a_2, a_3, a_4 or a_5, depending on whether the entry in Table 39.3 is N, H, C, G or S.

39.8.5 The Initial Dislocation Density

One of the main issues in the application of the Alexander–Haasen model to crystal growth is the specification of the initial dislocation densities. An important question is: what is the dislocation density in just-solidified, virgin material near the interface? Dislocations can arise at the interface due to certain phenomena at the crystal–melt interface, such as the formation of inclusions. Such dislocations are known as growth dislocations. The phenomena giving rise to them are manifestations of instability in the crystal–melt interface, and are described qualitatively in [39.81, Sect. 6.2]. Postgrowth analysis has shown that growth dislocations usually arise at inclusions formed at the growth interface, intersect the interface, and are propagated along with the movement of the interface. When the crystal is grown under stress-free conditions, the growth dislocations intersect the interface at certain specific angles, depending on their Burgers vector, in order to minimize their elastic energy. Thus growth dislocations arise from inclusions, and under stress-free conditions propagate in straight lines, intersecting the interface at a predetermined angle [39.82]. This configuration is disturbed under Czochralski growth at high temperatures, where stresses and considerable plastic deformation are prevalent. Postgrowth analysis of InP single crystals has shown the absence of growth dislocations. X-ray topographs by *Dudley* et al. [39.57, 58] show that most of the dislocations are glide or slip dislocations intersecting the sides of the crystal. They rarely intersect the crystal–melt interface. This can be interpreted to be due to the absence of the generation of dislocations due to interfacial phenomena in InP.

As discussed before, another mode for the nucleation of dislocations is the condensation of excess point defects into prismatic dislocation loops, or voids. This effect is aggravated in compounds such as GaAs, InP, and other binary compounds due to the presence of a narrow homogeneity region. In these systems, the stoichiometric compounds do not have the highest melting point; hence the first to solidify composition would be rich in one of the components. Thus, in the case of GaAs, the first to form solid is As rich, this As precipitates out during cool down, and precipitates of As are formed. These precipitates can give rise to dislocations. The modeling of point defect dynamics, and the clustering of these defects, is a complex process with many issues still unresolved.

Another important source of dislocations is surface damage. When the growing InP crystal emerges from the boric oxide encapsulant, the volatile phosphorous species can escape from the sides of the crystal, leading to extensive surface damage. This surface damage is a local distortion of the lattice and is a good site for dislocation nucleation. Under the influence of thermal stresses, the dislocations thus nucleated at the surface can move to the interior of the crystal and multiply. The rate of evaporation of the phosphorous vapor is closely related to the ambient pressure in the furnace.

Another source of dislocations is the initial seed crystal used in the CZ process. Dislocations present in the seed may move into the freshly formed material and multiply there under the influence of stresses. The necking process invented by *Dash* [39.3] prevents the movement of dislocations from the seed to the growing crystal. The term *necking* arises from the shape of the crystal being grown. The diameter of the growing crystal is decreased from the seed to a minimum value before increasing the diameter again. Thus the crystal has a neck of minimum diameter. Attempts have been made to extend the necking technique to sulfur-doped LEC-grown InP by *Dudley* et al. [39.83]. The InP crystals so grown were analyzed using synchrotron white-beam x-ray topography (SWBXT). Figure 39.14a,b presents the topographs of (110) wafers of two different as-grown InP crystals – sample A corresponding to a crystal grown with a narrow neck, and sample B corresponding to a crystal grown without necking. Figure 39.14c shows a geometric sketch of the dislocations in both the crystals. An analysis of these topographs reveals that most of the dislocations in LEC-grown InP crystals originate from the seed–crystal interface, propagate via slip planes, and

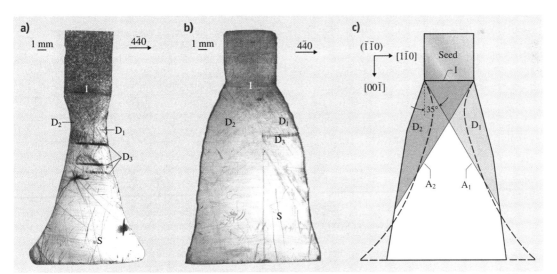

Fig. 39.14 (a) Sample A: X-ray topograph of (110) wafer of sulfur-doped InP crystal grown with a narrow neck. (b) Sample B: X-ray topograph of (110) wafer of sulfur-doped InP crystal grown without a narrow neck. (c) Schematic showing geometry and distribution of dislocations in as-grown crystals, for both shapes of crystals shown in (a,b) (after [39.83])

eventually exit the crystal through the periphery of the necked region. Based on this mechanism, dislocations exist only in the outer regions (dislocation regions) of the grown crystal separated by the four {111} planes which are extended from the four edges of the seed bottom. In the cross-sectional view of Fig. 39.14c, two of these planes, $(11\bar{1})$ and $(1\bar{1}1)$, are represented by lines A_1 and A_2, respectively. The regions of the crystal with dislocations are shaded, while those that are dislocation-free are unshaded.

As shown in Fig. 39.14c, the narrow necked portion of sample A makes the dislocation region much smaller than that of sample B. This means that the dislocations emerging from the seed–crystal interface in sample A slipped out of the neck in much shorter distances than that in sample B. A pyramid-shaped crystal, which is nearly dislocation free, was quickly formed below the dislocation regions. The above experiment and conclusions show the necessity of considering the movement of dislocations from dislocated regions to regions containing no dislocations initially. Such a formulation can be established using the continuum theory of dislocations, and would resemble the strain-gradient theories in structure. Such a formulation would involve spatial gradients of internal variables, and is not amenable to solution on the length scales involving a typical crystal wafer. Hence this line has not been pursued.

39.9 Model Summary and Numerical Implementation

39.9.1 Summary of the Model

A complete set of equations for the boundary-value problem using the constitutive model derived are summarized here, followed by a brief description of the numerical procedure.

Equilibrium of forces

$$\nabla \cdot \sigma = 0 . \tag{39.30}$$

Compatibility

$$\varepsilon = \tfrac{1}{2}\left(\nabla u + (\nabla u)^{\mathrm{T}}\right) . \tag{39.31}$$

Split of total strain

$$\varepsilon = \varepsilon^{\mathrm{el}} + \varepsilon^{\mathrm{pl}} + \varepsilon^{\mathrm{th}} . \tag{39.32}$$

Constitutive relation for thermal strain

$$\varepsilon^{\mathrm{th}} = \alpha\delta(T - T_0) . \tag{39.33}$$

Constitutive relation for elastic strain

$$\varepsilon^{\mathrm{el}} = C^{-1} : \sigma . \tag{39.34}$$

Evolution equation for plastic strain and dislocation densities

$$\dot{\varepsilon}^{\mathrm{pl}} = \sum_{i=1}^{12} N_{\mathrm{s}}^i \bar{v}_{\mathrm{s}}^i b(\boldsymbol{n}_{\mathrm{i}} \otimes \boldsymbol{s}_i + \boldsymbol{s}_i \otimes \boldsymbol{n}_i)\mathrm{sign}(\tau_i) , \tag{39.35}$$

$$\frac{\mathrm{d}N_{\mathrm{s}}^i}{\mathrm{d}t} = KN_{\mathrm{s}}^i \bar{v}_{\mathrm{f}}^i \tau_{\mathrm{eff}}^i + K^* N_{\mathrm{s}}^i \bar{v}_{\mathrm{f}}^i \tau_{\mathrm{eff}}^i \sum_{j\neq i} f_{ij} N_{\mathrm{s}}^j . \tag{39.36}$$

Auxiliary equations

$$\tau_{\mathrm{eff}}^i = |\tau^i| - \frac{Gb}{\beta}\left(N_{\mathrm{s}}^i\right)^{\tfrac{1}{2}} - \frac{Gb}{\beta^*}\sum_{j\neq i} g_{ij}\left(N_{\mathrm{s}}^j\right)^{\tfrac{1}{2}}$$

and

$$\tau^k = \boldsymbol{n}_{\mathrm{k}} \cdot \sigma \cdot \boldsymbol{s}_{\mathrm{k}} .$$

Related parameters

The values of K, K^*, β, β^*, and the interaction coefficients g_{ij} are obtained by fitting the data to results from the single-slip uniaxial compression tests at a constant strain rate. These tests were simulated using the model, and fitted to data in [39.76, 84]. The values obtained were: $K = 13 \times 10^{-6}\,\mathrm{Pa^{-1}m^{-1}}$, $K^* = 8 \times 10^{-15}\,\mathrm{m/Pa}$, $\beta = 3.3$, $\beta^* = 1.0$, $a_1 = 0.33$, $a_2 = 0.33$, $a_3 = 0.33$, $a_4 = 1.2$, and $a_5 = 1.3$. More details can be found in [39.85].

The elasticity moduli C_{ijkl} (indicial notation is used here) are based on [39.86] with the assumption that the continuum behaves as a cubic crystal with only three independent parameters. The relevant values are $C_{ijkl} = 1.02 \times 10^{11}\,\mathrm{Pa}$ when $i = j = k = l$, $C_{ijkl} = 4.4 \times 10^{10}\,\mathrm{Pa}$ when $i = k \neq j = l$ or $i = l \neq j = k$, $C_{ijkl} = 5.7 \times 10^{10}\,\mathrm{Pa}$ when $i = j \neq k = l$, and $C_{ijkl} = 0.0$ in all other cases. The coefficient of thermal expansion α is assumed to vary in quadratic fashion from $4.869 \times 10^{-6}\,\mathrm{K^{-1}}$ at 0 K and $4.410 \times 10^{-6}\,\mathrm{K^{-1}}$ at 100 K to $5.75 \times 10^{-6}\,\mathrm{K^{-1}}$ at 1000 K [39.86]. The magnitude of the Burgers vector b is $4.15 \times 10^{-10}\,\mathrm{m}$ [39.87, Table 1].

39.9.2 Numerical Implementation

A finite element technique can be adopted to obtain solutions of the boundary-value problem described above (indicial notation is used in this section). Briefly, this task involves the integration of the constitutive equations over time while ensuring equilibration of stresses in the body. Reference [39.88] gives a good description of the numerical implementation of viscoplastic equations. Assume that, at time t_n, the body is at equilibrium and that the set of displacements at each of the finite element nodes $\{u_i\}_n$, and stresses $\{\sigma_{ij}\}_n$, plastic strains $\{\varepsilon^{pl}_{ij}\}_n$, and the 12 internal variables $\{N^i_s\}_n$ at the finite element Gaussian points are all known. However between time t_n and the subsequent time t_{n+1}, the temperature has changed and hence the thermal strain has changed. The task is now to find the new displacements, stresses, plastic strain, and dislocation densities at time t_{n+1}. This is accomplished by the following procedure:

1. The finite element technique runs either a global Newton–Raphson iteration or a quasi-Newton iteration for the updated displacements $\{u_i\}_{n+1}$ at time t_{n+1}. This global iteration is based on a discretized version of (39.30).
2. A guess is made at every nodal point in the finite element domain for the set of updated displacements at the k-th global Newtonian iteration $\{u_i\}^k_{n+1}$. The updated strains, at every Gaussian point in the finite element domain, for the k-th iteration can be calculated using the discretized version of (39.31), and the thermal strain at time t_{n+1} can be calculated from (39.33), using the thermal fields at time t_{n+1}.
3. To update the stresses, plastic strains, and the dislocation densities, (39.35) and (39.36) are integrated for the time step $\Delta t = (t_{n+1} - t_n)$ with the help of (39.31–39.34) and the auxiliary equations, at every Gaussian point. The integration method can be either an implicit Euler scheme, or any first-order ordinary differential equation (ODE) integration scheme such as one of the Runge–Kutta methods. The algorithmic elasto-plastic tangent modulus $\{D_{ijmn} = \partial \sigma_{ij}/\partial \varepsilon_{mn}\}^k_{n+1}$ can also be obtained for a given integration scheme. Refer to [39.85] for a detailed development using an implicit Euler scheme and a (2)3 Fehlberg pair, which is an explicit integration scheme.
4. Once the integration is completed, one obtains the stresses $\{\sigma_{ij}\}^k_{n+1}$ and the algorithmic elasto-plastic tangent modulus at every Gaussian point in the finite element domain. These are then inserted into the global Newton–Raphson iteration, and the displacements at the next iteration $\{u_i\}^{k+1}_{n+1}$ are obtained.
5. Steps 3 and 4 are repeated until equilibrium is attained in the body to the desired tolerance level. The variables have now all been updated to time t_{n+1}.

Steps 1–5 are repeated at each new time point from the seed to the completely grown and cooled crystal. At each time point the thermal loading (strains) in the crystal has to be updated based on the temperature field prevailing in the crystal. This temperature data can be obtained either from experiment or more realistically through computational fluid dynamics (CFD) and heat-transfer-based computational models of the entire high-pressure Czochralski process.

A second complication that has been so far ignored is the addition of new material during the growth process. This addition of virgin material can be modeled as the addition of a new layer of finite elements at certain time points. The frequency of the addition of these elements depends on the general length scale of the finite element grid. If the length scale of the grid is say 'l' mm then a new layer of finite elements, with average dimension 'l' mm, can be added at time intervals corresponding to growth of 'l' mm of material. Figure 39.15 shows a layer of new elements that has just been added to the original grid. This new layer of elements corresponds to the material that has just crystallized. Now a relevant question is: what are the values of the plastic strains ε^{pl}_{ij} and the dislocation densities N^i_s in the newly crystallized material. Since the material has just crystallized, the obvious answer would be to set both the plastic strain components as well as the dislocation densities to zero. However, there are a couple of problems with this approach.

Firstly, setting the initial dislocation densities to zero would ensure that this material is always dislocation free, as the dislocation density evolution rate would always be zero according to (39.36). To resolve this, the approach of *Völkl* [39.40] is followed and the initial dis-

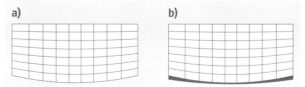

Fig. 39.15 (a) Two-dimensional projection of the grid system in the seed. (b) The grid system after 2 mm of growth. Shaded region represents the newly pulled material

location density is set to some finite but small positive value.

Secondly, setting the plastic strains to zero would give rise to huge stresses in the just-crystallized material, the reason being that nonfreshly crystallized (parent) parts of the crystal have significant plastic deformation, and the nodes on the interface between the two parts have significant displacement vectors to reflect this plastic strain. These displacements on the interface translate to large strains in the freshly crystallized material and large stresses, since this material behaves elastically due to the lack of plastic strain. This phenomenon as explained below is just a technicality. Before the crystallization of the new layer of elements, the present interface between the freshly crystallized material and the parent crystal was the crystal–melt interface, which is traction free. Hence the new layer of elements crystallize on a traction-free surface. Thus no stress from the parent crystal can propagate across this surface into the freshly crystallized material. Hence if the thin layer of freshly crystallized elements is isothermal (which is a reasonable assumption) then the new layer of elements representing the freshly crystallized material should be stress free. So what leads to the high stresses described before? The answer is that the reference state corresponding to zero plastic strain for the freshly crystallized material is different from that for the parent crystal. Hence, in order to use the same reference state for the parent crystal as well as the freshly crystallized material, some plastic strain has to be added to the freshly crystallized material.

In conclusion, the freshly crystallized elements are added in a stress-free state. To accomplish this the plastic strains at the Gaussian integration points of the freshly added elements are set such that the stresses at these Gaussian integration points are zero. This approach has been used in all the results presented in Sect. 39.10.

39.10 Numerical Results

To be of any use, the viscoplastic model needs the complete thermal history of the crystal during the growth. Such detailed information has not yet been obtained experimentally for any specific case, hence the results have been derived from the temperature field history predicted by a CFD-based model of the high-pressure Czochralski process termed MASTRAPP (multizone adaptive scheme for transport and phase change processes) [39.89]. MASTRAPP is a two-dimensional axisymmetric CFD-based algorithm applied to a simplified domain of the furnace. It simulates the temperature and flow field history during the growth process; it includes phenomena such as heat conduction in the crystal, thermal radiation, convection in the melt and gas, radiation, moving crystal–melt interface, and latent heat generation, and can handle presence of different domains such as the gas, crystal, melt, and encapsulant, and the movement of the boundaries of these different domains during the growth process. The following inputs are required in MASTRAPP as macroscopic parameters specifying the growth process.

39.10.1 Strength of Convection in the Melt and Gas

The fluid flow and temperature distributions are a result of natural convection due to buoyancy forces, and forced convection due to the rotation of the crystal and the crucible. The relevant nondimensional parameters for natural convection are the melt and gas Grashof numbers, and for forced convection are Re_s (the Reynolds number characterizing the crucible rotation) and Re_c (the Reynolds number characterizing the crystal rotation). For the Czochralski system shown in Fig. 39.2 with the diameter of the crucible being 100 mm and the height of the system – from the bottom of the crucible to the top of the heat shield – 170 mm, the Grashof numbers are of order 10^9, and the Reynolds numbers are of order 10^4. In papers dealing with numerical simulations of crystal growth, Grashof and Reynolds numbers an order of magnitude lower are used; the lower Grashof number can be justified by saying that they correspond to microgravity situations, and the lower Reynolds number to much lower revolution rates of the crystal and crucible than are typically used.

39.10.2 Temperature Boundary Condition

The reduced domain of Fig. 39.2 interacts with the other parts of the furnace through the specification of appropriate conditions at the boundaries of the reduced domain. Towards this the temperature on the crucible plus the heat shield is prescribed as a function of the distance from the base of the crucible in Fig. 39.2. The base of the crucible, and portions occupied by the melt, are maintained at 1435 K, and the temperature is linearly ramped down along the wall of the heat shield to a temperature of 427 K at the top. All the simulation results presented assume the boundary condition just described.

39.10.3 A Sample Case

A simulation of the growth of a flat-top 50 mm-diameter InP crystal from a height of 7–45 mm with a 20 mm height of encapsulation is used as a sample case. The gas and melt Grashof numbers are fixed at 10^6, and the rotational Reynolds numbers at 500. The growth axis is along the [001] direction. The results presented below correspond to the end of the growth period, before the cooling of the crystal to room temperature.

Temperature Field in the Crystal

The primary reason for the thermal stress and the consequent plastic deformation is the incompatibility of the thermal strain arising from the temperature field. The evolution of the axisymmetric temperature field and its gradients is shown in Figs. 39.16 and 39.17 respectively. At the beginning of the growth, when the seed has just thermally stabilized in the furnace, the isotherms show that heat enters the crystal from the sides and escapes from the top of the crystal. Hence the axial or vertical temperature gradient is higher in the upper parts of the crystal – of the order of 35 K/cm – than in the lower part. The radial temperature gradient is highest in the lower corners of the crystal, and is nearly zero in the central part of the crystal. As the crystal grows, the thickness of the encapsulant above the crystal decreases, thereby decreasing the thermal resistance to heat flow between the melt and the ambient gas; consequently the heat flux flowing through the crystal and the thermal gradients increase with growth.

The crystal just emerges out of the encapsulant at $t = 84$ min, and a direct pathway is established between

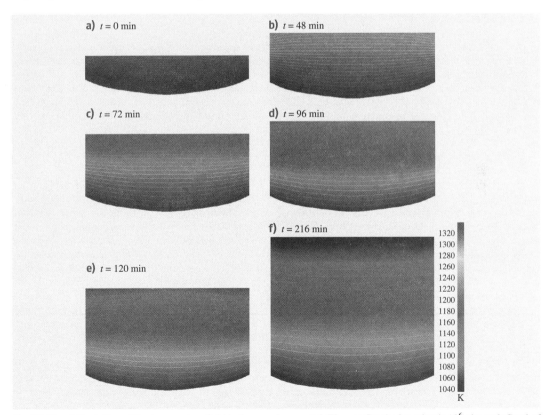

Fig. 39.16a–f Evolution of the temperature field in the growing crystal. The gas Grashof number is 10^6, the melt Grashof number is 10^6, the rotational Reynolds numbers are 500, and the encapsulant height is 20 mm

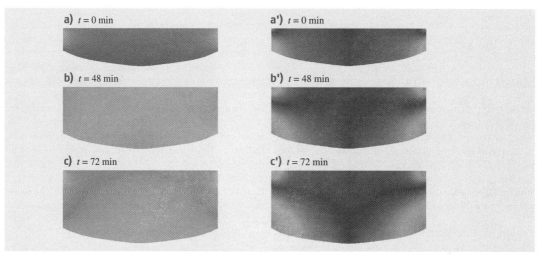

Fig. 39.17 History of thermal gradient field in the growing crystal. The gas Grashof number is 10^6, the melt Grashof number is 10^6, the rotational Reynolds numbers are 500, and the encapsulant height is 20 mm. Color scales for vertical and radial gradients are in (**f**) and (**f'**), respectively

the melt and the gas through the highly conducting crystal. Moreover, the lateral side of the part of the crystal above the encapsulant loses heat by convection and radiation to the cooler parts of the furnace, thus establishing a radially outward flow of heat, i.e., a radial temperature gradient. These give rise to a jump in the heat flux through the crystal. This is evident from comparing the frames at $t = 72$ min with the frames at $t = 96$ min in Fig. 39.17, when the top of the crystal is 1 mm below the encapsulant and 3 mm above the encapsulant respectively. In the latter frame the thermal gradients are appreciably higher. This change in the thermal profile and consequent high temperature gradients on emergence of the crystal from the encapsulant can be termed *thermal shock*.

As the crystal grows further, the length scale increases, while the temperature difference driving the heat flow between the melt and the gas remains the same; consequently vertical and radial temperature gradients decrease. The portion of the lateral surface of the crystal just above the encapsulant is hotter than the sidewalls of the furnace and loses heat by radiation and convection, leading to high radial temperature gradients in that region. Conversely, portions of the lateral surface of the crystal that are nearer to the top of the crystal are cooler than the side walls of the furnace and hence gain heat due to radiation from the sides of the furnace. *Zou* et al. report a similar behavior [39.22]. The radial temperature gradient is almost zero in most of the crystal except on the lateral surface at three specific locations: the bottom corner of the crystal near the interface due to a radially inward flow of heat from the melt into the crystal, just above the encapsulant due to cooling by convection and radiation, and finally the top of the crystal due to heating of the lateral surface by radiation.

Summarizing, the thermal history of the crystal can be divided into three stages:

1. Low thermal gradients, when the crystal is below the encapsulant and the gradients increase with growth
2. The thermal shock, when the crystal grows out from beneath the encapsulant, resulting in a jump in the thermal gradients
3. The later stages of the growth, characterized by decreasing temperature gradients and practically zero radial gradients.

Except for in three regions on the lateral surface of the crystal: one near the interface, the other just above the encapsulant, and the third at the top of the crystal (Fig. 39.17f). The crystal experiences maximum temperature gradients when it has just emerged from the encapsulant. The vertical temperature gradients obtained from the simulation, in the range of 50–100 K/cm, are of the same order as those reported in experiments with both GaAs [39.90] and InP [39.91].

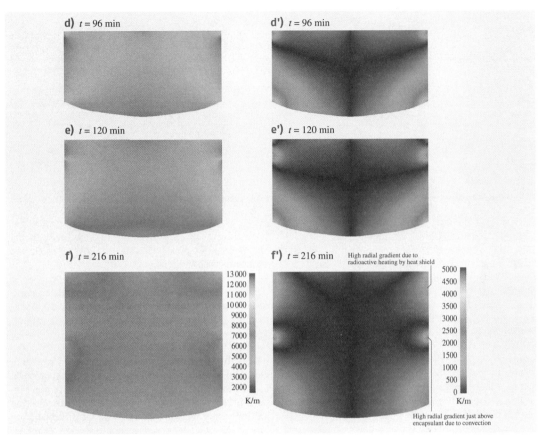

Fig. 39.17 (cont.)

Dislocation Densities after Growth Period

Figure 39.18 shows a fully grown crystal at the end of the growth period, with the colors representing the total dislocation density. The order of magnitude of the highest total dislocation density is around $1 \times 10^7 \, \text{m}^{-2}$ and occurs near the top of the crystal. For a closer inspection of the total dislocation density, Fig. 39.19 shows the distribution on (001) slices, taken at various heights of the crystal, the position of each being indicated in Fig. 39.18. On all the slices, except slice h, the dislocation density is highest near the circumference and lower towards the center. On slices a, f, and g (not shown) the dislocation density variation corresponds to the w-shaped variation mentioned often in

Fig. 39.18 Total dislocation density at the end of the growth period of a [001]-grown crystal. The vertical position of the various horizontal slices is indicated. The gas Grashof number is 10^6, the melt Grashof number is 10^6, the rotational Reynolds numbers are 500, and the encapsulant height is 20 mm ▶

x slice a	9×10^6
x slice b	8×10^6
	7×10^6
x slice c	6×10^6
x slice d	5×10^6
	4×10^6
x slice e	3×10^6
x slice f	2×10^6
x slice g	1×10^6
x slice h	0
	m^{-2}

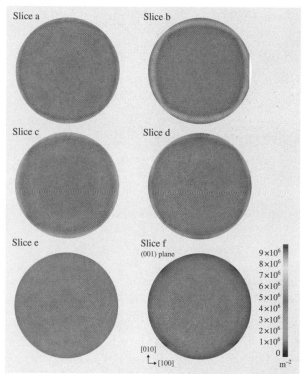

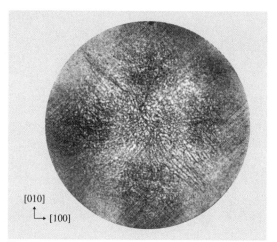

Fig. 39.19 Total dislocation densities at the end of the growth period projected on the (001) slices. The position of the various slices was indicated in Fig. 39.18. The gas Grashof number is 10^6, the melt Grashof number is 10^6, the Reynolds numbers are 500, and the encapsulant height is 20 mm

Fig. 39.20 KOH-etched (001) slice of GaAs (after [39.94], with permission)

literature [39.12, 90, 92] – the dislocation density is highest on the circumference, initially decreases as one moves radially towards the center, and then increases. The dislocation density patterns on all the slices exhibit fourfold symmetry, and reflect the symmetry of the slip systems with respect to the [001] growth axis. All the slices, except slice h, exhibit distinct areas of low dislocation density aligned along the $\langle 110 \rangle$ directions and areas of highest density along the $\langle 100 \rangle$ directions, consistent with etch-pit observations [39.12, 93]. The values for total dislocation density are in the range $2 \times 10^6 – 1 \times 10^7$ m^{-2}. It should be pointed out that the values of dislocation density are not those in the final crystal but only the fully grown crystal at the end of the growth period, with the cooling-down period still remaining. Figure 39.20 shows a KOH-etched LEC-grown GaAs(001) wafer from the work of *Buchheit* et al. [39.94]. The etch pit configuration in this figure is similar to that of slice a in Fig. 39.19. A similar configuration can be seen in the work of *Chung* et al. [39.57, 58] from x-ray topographs of LEC-grown (001)InP wafers.

Distribution of Dislocation Density on the Individual Slip Systems

Due to the symmetry of the [001] growth axis [39.95, Fig. 1], the dislocation density distributions of slip systems 1, 2, 5, 6, 7, 9, 10, and 11 on a (001) cross-section are inherently the same and can be transformed from one to the other by simple operations such as reflection and rotation by some angle. Similarly, by symmetry considerations the dislocation density distributions of slip systems 3, 4, 8, and 12, on any (001) cross-section are identical except that they are rotated with respect to each other by multiples of 90°. Hence, the dislocation density on slip systems 1 and 3 is presented, as each is representative of the other systems with which it shares symmetry.

Figure 39.21a shows the density of dislocations belonging to slip system 1, projected onto slice b. Notice the location of the maximum on the circumference. It was found by inspection that the maximum occurs on the circumference and at the same angular orientation in a majority of the slices. Similarly, Fig. 39.21b shows the density of dislocations belonging to slip system 3 on slice b. In this case too, the dislocation density maxima always occur on the circumference and at the same angular locations for most of the slices. The angular orientations of the maxima might reveal the dominant stress component responsible for the multiplication of

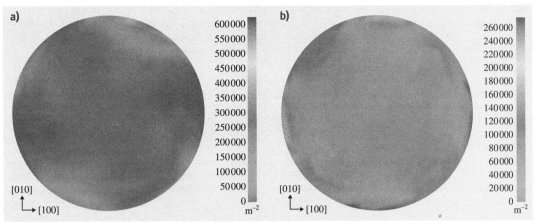

Fig. 39.21a,b Dislocation densities of slip systems 1 and 3 on slice b. The slip plane normal and slip direction for system 1 are $(11\bar{1})$ and $[101]$, and for system 3 are $(11\bar{1})$ and $[1\bar{1}0]$. (**a**) Density of dislocations belonging to slip system 1. (**b**) Density of dislocations belonging to slip system 3

dislocation densities. Since the temperature field is axisymmetric, an assumption of isotropy (for the purpose of analysis) would imply that the elastic strain field driving the deformation would be axisymmetric, and it would be appropriate to think in terms of axisymmetric stress components in cylindrical coordinates – σ_{rr}, $\sigma_{r\theta}$ σ_{rz}, $\sigma_{\theta\theta}$, $\sigma_{\theta z}$, and σ_{zz}. A useful exercise would be to isolate the effects of each of these components, by setting one of them to unity and the rest to zero, and plot the variation of the resolved shear stress on systems 1 and 3 with the polar angle. This might help explain the angular orientations of the location of the maxima for dislocation densities of slip systems 1 and 3. The results for slip system 1 are shown in Fig. 39.22a, while those for slip system 3 are shown in Fig. 39.22b. In both these figures $\sigma_{\theta\theta}$ is set to 1 and the rest to 0. A comparison of these figures with the distribution of dislocation densities on the (001) slice shown in Fig. 39.21a,b shows that the circumferential component $\sigma_{\theta\theta}$ dominates the deformation at least near the outer circumference of the slice. The reason for this is that the maxima in the resolved shear components on slip systems 1 and 3 due to $\sigma_{\theta\theta}$ occur at the same polar angles, where the dislocation density on these slip systems is high. The exercise was repeated with components other than $\sigma_{\theta\theta}$, but the polar angles of the maxima did not correspond to the polar angles of dislocation density maxima in those cases.

The deformation is multislip in some regions of the crystal. For example, consider the maximum in total dislocation density present on the circumference of slice b in the [010] direction (Fig. 39.19). In this region slip systems 1, 3, 4, and 6 are almost equally active and their dislocation densities are found to be of the order of $2.5 \times 10^6 \, \text{m}^{-2}$ (Fig. 39.21a), $2 \times 10^6 \, \text{m}^{-2}$ (Fig. 39.21b), $2 \times 10^6 \, \text{m}^{-2}$, and $2.5 \times 10^6 \, \text{m}^{-2}$, respectively. By symmetry, the other three maxima in total dislocation density present on the circumference are also composed of four, almost equally active slip systems. Similarly, multislip activity is present in the central parts of slice b. The total dislocation density here (Fig. 39.19) is of the order of $2.5 \times 10^6 \, \text{m}^{-2}$. Dislocations belonging to each of the slip systems 1, 2, 5, 6, 7, 9, 10, and 11 have an approximate density of $2 \times 10^5 \, \text{m}^{-2}$ and those belonging to each of slip systems 3, 4, 8, and 12 have a density of almost $2.5 \times 10^5 \, \text{m}^{-2}$. Thus the total dislocation den-

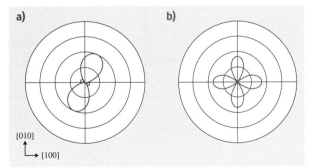

Fig. 39.22 Polar plot of resolved shear stress on slip systems 1 (**a**) and 3 (**b**), with circumferential stress component set to 1 and all others to 0. The slip-plane normal and slip direction for system 1 are $(11\bar{1})$ and $[101]$, and for system 3 are $(11\bar{1})$ and $[1\bar{1}0]$.

sity in the central parts of slice b is due to almost equal activity on all the slip systems.

39.10.4 Effect of Gas Convection and Radiation

The effect of radiative interaction of the lateral walls of the crystal with the furnace walls on the dislocation generation was isolated by artificially setting the radiative term to zero. The final total dislocation density field was observed to be practically the same as the case without radiation.

The effect of gas convection was investigated by setting the gas Grashof number to zero, while retaining radiation. Figure 39.23a shows the total dislocation density projected on a (010) surface after the end of the growth for this simulation. The total dislocation density in the case without gas convection is less than that in the case with gas convection – Grashof number 10^6 (Fig. 39.23b) – by more than a factor of 2. Thus a significant observation that can be made here is that radiation interaction of the sides of the crystal with the furnace does not significantly affect the dislocation density, while gas convection does.

Figure 39.23c shows a projection of the total dislocation density on the (010) plane at the end of the growth period (216 min) for a gas Grashof number of 10^7. The maximum dislocation density is now in the range of 3×10^7 m^{-2}. Similarly Fig. 39.23d shows the total dislocation density for the case of Grashof number 10^8. It is interesting to note that the maximum dislocation density doubled from 4×10^6 to 9×10^6 m^{-2} for the case between no gas convection and Grashof number 10^6, while a further tenfold increase in the gas Grashof number to 10^7 tripled the maximum dislocation density to 3×10^7 m^{-2}, and another tenfold increase of the Grashof number to 10^8 further increased the maximum dislocation density to the range of 2×10^8 m^{-2}. This confirms that gas convection plays a significant role in the evolution of the dislocation densities.

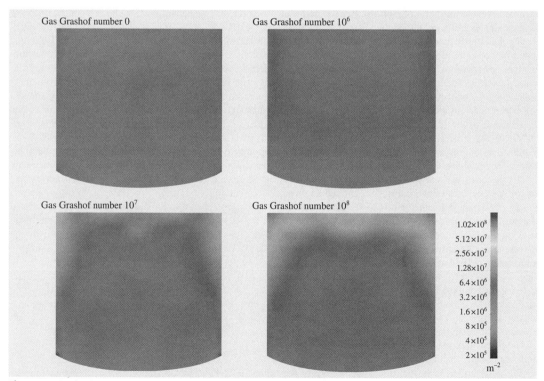

Fig. 39.23 Projection on the (010) plane passing through the growth axis, of the total dislocation density at the end of the growth period for different strengths of gas convection in the furnace. The melt Grashof number is 10^6, the Reynolds numbers are 500, and the encapsulant height is 20 mm in all cases

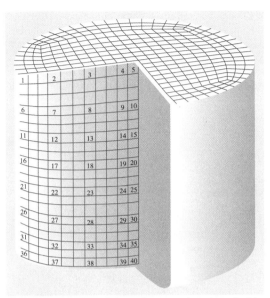

Fig. 39.24 Location of the various sample points, the history of which will be followed

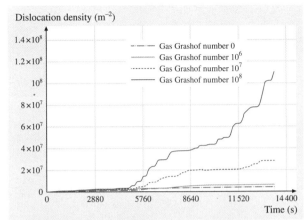

Fig. 39.25 Total dislocation density evolution for point 1, under different gas conditions. The melt Grashof number is 10^6, the rotational Reynolds numbers are 500, and the encapsulant height is 20 mm

The results for the total dislocation density at the end of the growth period for all the cases presented so far (growth with encapsulant height 20 mm with gas convection Grashof numbers of 0, 10^6 with radiation from the sides, 10^7, and 10^8) are now analyzed on the grid of points 1–40 located as shown in Fig. 39.24. The total dislocation density evolution history at point 1, for the cases of no gas convection and for gas Grashof numbers 10^6, 10^7, and 10^8, is presented in Fig. 39.25. The difference in dislocation density evolution history between the four cases again brings out the importance of gas convection. The dislocation density at point 1 is very low until a time of about 5500 s, and starts rising rapidly after this period. This timeframe corresponds to the period when the top of the crystal emerges from beneath the encapsulant (thermal shock).

39.10.5 Melt Convection and Rotation Reynolds Numbers

Melt Convection

In the calculations presented above, the Grashof number for melt convection was set at 10^6, while in real life it is closer to 10^8. Stronger melt convection might lead to increased heat flow through the crystal, and also affect the curvature of the crystal–melt interface. *Lambropou-*

los [39.15], *Lambropoulos* and *Wu* [39.53], and *Chen* et al. [39.92] have proposed that the interface curvature will have a significant effect on the evolution of the total dislocation density.

The Grashof number in the melt was increased to 10^7 and then 10^8 to study the effect on the total dislocation density. In both of these simulations, the gas Grashof number was kept at 10^8. It was found that, although the interface becomes highly convex towards the melt for a melt Grashof number of 10^8, the final dislocation density does not change appreciably from the corresponding simulation when the melt Grashof number was 10^6. This result differs from observations in literature, which suggest that a curved interface is dangerous since it leads to high stresses – the CRSS at the interface is very low and high stresses there will lead to a high dislocation density. The situation in these simulations is however more subtle. The high temperature (near the melting point) at the interface ensures that the strain (or stress) generated due to the curved interface is quickly dissipated as plastic strain due to the high velocity of the dislocations, and the resulting additional dislocation density due to the enhanced curvature is not high enough to show up in the final results at the end of the growth period, since later in its history the same material element goes through more strenuous loading at lower temperatures, thus masking the effect of the additional interface curvature.

Rotation of Crucible and Crystal

Another important factor might be the rotation of the crucible and the crystal. These are usually rotated in opposite senses, and typically the angular velocities of rotation are of the order of 5–10 rpm, which results in Re_c and Re_s of the order of 8000–16 000. The rotation of the crucible creates a swirl component in the fluid motion, and as a result centrifugal and coriolis forces arise. The rotation of the crystal creates another forced convection term, and its effect on gas convection is akin to an Ekman flow above a rotating disk. The forces due to natural convection, crucible swirl, and crystal rotation

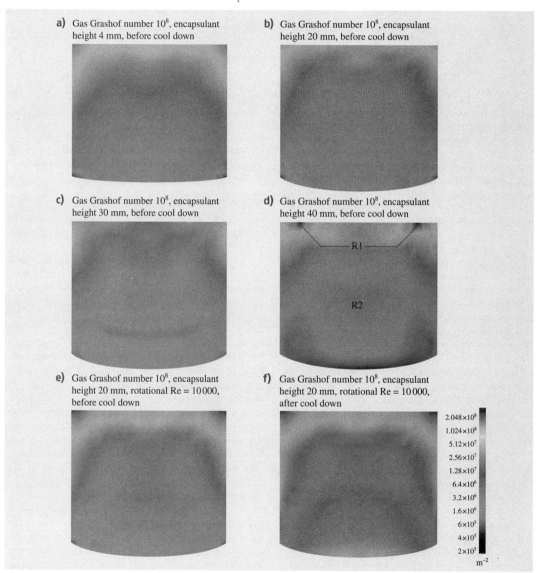

Fig. 39.26a–f Total dislocation densities projected on (010) plane passing through the growth axis. In all cases the melt Grashof Number is 10^6, and the rotational Reynolds numbers are 500 unless otherwise indicated

create a complicated pattern of gas flow, which might be oscillatory and three dimensional in nature.

An increase of the rotation Reynolds number to 2500 and 5000 did not cause much difference in the total dislocation density from the corresponding case with rotation Reynolds number of 500. However, a further increase in the rotation Reynolds number to 10 000 caused a doubling of the maximum total dislocation density to $4 \times 10^8 \, \text{m}^{-2}$, as can be seen in Fig. 39.26e. This might be consistent with the observations that at higher rotation rates the flow becomes unstable and promotes temperature and flow oscillations in the gas phase [39.96].

39.10.6 Control of Encapsulation Height

The height of encapsulation is expected to influence the total dislocation density by changing the temperature history. It will determine the instant at which the crystal will come out of the encapsulant and experience *thermal shock*, the duration of time for which the crystal is exposed to the hot gas, and the length of the crystal that is exposed to the gas. Figure 39.26a–d shows the total dislocation density projected on the (010) plane in the fully grown crystal for four different encapsulant heights (4, 20, 30, and 40 mm) with gas Grashof number of 10^8, melt Grashof number of 10^6, and rotational Reynolds numbers of 500. Though not shown in Fig. 39.26, a case for encapsulant height 50 mm under the same conditions was also simulated. This case corresponds to fully encapsulated growth, with the crystal remaining under the encapsulant throughout. The dislocation density in this case was one order of magnitude lower – less than $5 \times 10^6 \, \text{m}^{-2}$. *Elliot* et al. [39.97] report very low dislocation densities for fully encapsulated growth. The 50 mm case is to be contrasted with the case of 40 mm encapsulant height. In this case too, the crystal is under the encapsulant for 204 of the 216 min of its entire growth history, and at the end of its growth period, only the top 3 mm of the 43 mm long crystal is exposed to the gas convection. However, the dislocation levels at the top in this case are an order of magnitude higher than in the fully encapsulated case. This is further proof of the dominant role played by exposure of the crystal to gas convection in dislocation multiplication, at least as predicted by the modeling effort.

The highest dislocation density occurs in the case of encapsulant height 4 mm. The maximum occurs in the top corner of the crystal, and is of the order of $2 \times 10^8 \, \text{m}^{-2}$ in the case of 4 mm encapsulant, and decreases to around $8 \times 10^7 \, \text{m}^{-2}$ in the case of encapsulant height of 40 mm. In every case, the crystal seems to be clearly demarcated into two different zones: an upper zone where the total dislocation density is of the order of $3 \times 10^7 – 6 \times 10^7 \, \text{m}^{-2}$, and a lower region where the dislocation density is of the order of $5 \times 10^6 – 1 \times 10^7 \, \text{m}^{-2}$. The dislocation density tends to be higher towards the edges of the crystal. The area of high dislocation density is greatest in the case of encapsulant height of 4 mm.

39.10.7 The Cool-Down Period

The results presented so far are those at the end of the growth process. After the growth is the process of cooling down the fully grown crystal to room temperature. The temperature field in the crystal during the cool-down period is a function of the environment in the furnace, the rate at which the heaters in the furnace are ramped down, and other such details. These features are not available in the MASTRAPP CFD model, for a realistic simulation of the cool-down period. Therefore a reasonable approximation to the cool-down process is to ramp the temperature field in the crystal at the end of the growth period down to room temperature over a length of time. What this means is that the temperature gradients and thermal loads on the fully grown crystal are ramped down to zero.

Consequently, as the thermal loading is withdrawn, stresses develop due to the incompatibility of the plastic strain, leading to additional plastic deformation and dislocation multiplication, and finally considerable residual stresses in the cooled-down crystal. An important issue is the length of time over which the temperature ramp-down should take place. In a real Czochralski furnace, the crystal ingot is cooled down to room temperature over a few hours. Integrating the viscoplastic equations over a few hours of real time is beyond the computational resources available. Hence, the strategy adopted was to ramp down the thermal fields to room temperature over 12 min. In certain cases, we used three times this period to check for differences in the results. A clear disadvantage of the shortness of the ramp-down period is that the exaggerated unloading rate might lead to an overestimation of the final total dislocation density and residual stresses.

Figure 39.26f shows the final dislocation density after the cool-down period on a (010) slice containing the centerline, for the case with encapsulant height of 20 mm, gas Grashof number of 10^8, melt Grashof number of 10^6, and rotational Reynolds number of 10 000. To contrast with the situation the dislocation density before the cool-down period is presented in Fig. 39.26e. It can be seen that the dislocation density in the lowest part of the crystal has doubled or even tripled during the cool-down period. In fact the highest dislocation density in the crystal now occurs in the bottom and top parts of the crystal, with a distinct low-dislocation-density region in the middle part of the crystal. This seems to coincide with the experimental observation by *Chung* et al. [39.57, 58]. Though the figures are not displayed here, the cool-down calculations were performed for every case discussed previously. The results have a common feature: while the total dislocation density at the top and sides of the crystal develops during the growth period, the bottom of the crystal acquires a high dislocation density during the cool-down period. This phenomenon occurs due to the large temperature difference between the top and bottom of the crystal – roughly 300 K. Due to this, by the middle of the cooling period, the top portion of the crystal has cooled down sufficiently to be incapable of plastic deformation (since the threshold strain for plastic deformation is too high), and consequently generates considerable residual stresses. In contrast the bottom portion of the crystal is still hot (relative to the top portion) and is capable of plastic deformation. Since the crystal is a continuous medium, the residual stresses generated in the top part of the crystal generate a high loading in the lower part too, resulting in increased plastic deformation and dislocation generation there. Thus a new prediction by the modeling effort is that the dislocation density in the upper portions of the crystal develops during the growth period, while that in the lower part develops during the cool-down period. Though experimental data is scarce, this phenomenon has been reported by *Neubert* and *Rudolph* [39.98], in the context of vapor pressure controlled Czochralski (VCZ) growth of GaAs single crystals.

The cool-down period was tripled in the case of encapsulant height of 20 mm, gas Grashof number of 10^8, melt Grashof number of 10^6, and rotational Reynolds numbers of 500. A longer cool-down period means that parts of the crystal are at higher temperature for a longer time and capable of plastic deformation for a longer period. This gives rise to dislocation densities higher by a factor of 1.3 and slightly lower residual stresses at the end of the cool-down period than before.

39.10.8 The [1$\bar{1}$1] Growth Axis

All the cases presented so far have been for growth along the [001] axis. In this section, the case for growth along the [1$\bar{1}$1] axis is presented. The main difference is in the geometry of the slip systems. While in the case of [001] growth, the slip systems have fourfold symmetry with respect to the growth axis, in the case of [1$\bar{1}$1] growth, threefold symmetry is expected.

Only one case is simulated for the [1$\bar{1}$1] growth axis, with gas Grashof number of 10^8, rotation Reynolds numbers of 500, melt Grashof number of 10^6, and encapsulant height of 4 mm. Figure 39.27 presents the total dislocation density after growth and cool down in slices a–h (see Fig. 39.18 for locations of slices a–h) for the [1$\bar{1}$1] case. The total dislocation density profiles in slices d and e exhibit threefold symmetry, and those in slices a, b, and c exhibit sixfold symmetry. Both of these symmetries are common in etch pit density observations on slices [39.40, 99]. A direct comparison of the slices showed that the total dislocation density after cool-down is slightly less for the [1$\bar{1}$1] growth com-

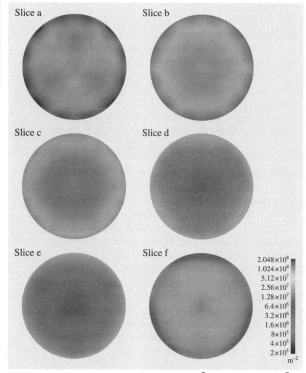

Fig. 39.27 Total dislocation density on (1$\bar{1}$1) slices of a [1$\bar{1}$1]-grown crystal after the growth and cool-down period. The gas Grashof number is 10^8, the Reynolds numbers are 500, and the melt Grashof number is 10^6. The encapsulant height is 4 mm. The position of the slices is as indicated in Fig. 39.18

pared with the [001] case. The explanation for this is not clear.

39.10.9 Summary of the Calculations and Some Comparisons

A CFD model for the high-pressure Czochralski growth of *InP* single crystals – MASTRAPP – was linked with the ABAQUS software to integrate the viscoplastic constitutive equations developed in previous sections under the thermal loading prevalent during the growth period and to predict the final total dislocation density and stress. After the growth period, the crystal temperature was ramped down from the temperature at the end of the growth period to room temperature to simulate the *cool-down* period of the crystal. A study was then conducted to delineate the effect of various parameters. The conclusions are summarized below:

1. The strength of gas convection has the most significant effect on the total dislocation density. The total dislocation density is very low – 400 cm^{-2} – for a gas Grashof number of zero, i.e., with no convection. On increasing the gas Grashof number to 10^6, the total dislocation density approximately doubles, and a further increase of the gas Grashof number to 10^7 raises the maximum total dislocation density to the order of 2000–3000 cm^{-2}; finally, an increase of the Grashof number to 10^8 raises the maximum total dislocation density to the order of 20 000 cm^{-2}. In a real furnace, the Grashof numbers are of the order of 10^9, and hence dislocation densities one order of magnitude higher can be expected, and are found in commercial wafers. In contrast, thermal radiation was found not to play a significant role in dislocation generation.
2. The phenomenon of *thermal shock* has been delineated. The thermal shock is captured clearly in both the thermal and elastic stress histories. The total dislocation density in the crystal jumps multifold after the thermal shock.
3. An important finding is that the high dislocation density at the top of the crystal is developed during the growth process after exposure to gas convection, and that at the bottom of the crystal during the cool down.
4. The dislocation density pattern on select slices for [001]-grown crystal exhibits fourfold symmetry, and the total dislocation is higher at the edges of the wafer. The dislocation density is generally highest along the $\langle 010 \rangle$ directions and lowest along the $\langle 110 \rangle$ directions, in agreement with experimental findings. Similarly, for [1$\bar{1}$1]-grown crystal, the dislocation density pattern shows sixfold and threefold symmetry.
5. The effect of the height of encapsulation on the total dislocation density at the end of the growth period has been studied. Encapsulation heights of 4, 20, 30, 40, and 50 mm are used with gas Grashof number of 10^8, melt Grashof number of 10^6, and rotational Reynolds numbers of 500. The last case corresponds to fully encapsulated growth and the total dislocation density developed is only of the order of 2×10^6–4×10^6 m^{-2}. This confirms experimental findings of fully encapsulated growth producing low-dislocation-density crystals.
6. Raising the melt Grashof number did not significantly affect the total dislocation density; however, an increase in the rotation Reynolds numbers increased the dislocation density.

If there is a common theme to the message from these calculations, it is that gas convection plays the most important role in dislocation multiplication. How does this square up with experimental observations? We report a few experimental papers which attempt to reduce the dislocation densities by adopting the same strategies explored numerically in the preceding sections – the reduction of thermal gradients – primarily achieved through certain artifices such as improving the design of the heat insulators and heaters, adjustment of the thickness of the boric oxide layer, reduction of the gas convection by reducing the ambient gas pressure as much as possible and design of appropriate baffles and convection shields, and finally stabilization of the melt flows. *Von Neida* and *Jordan* [39.100] achieved low thermal gradients in the LEC growth of GaAs by both adjusting the position of the heating source and modifying the B_2O_3 level. They noticed a reduction in dislocation densities by a factor of 10. *Elliot* et al. [39.101] achieved dislocation densities of 5000 cm^{-2}, in the LEC growth of GaAs, by reducing the thermal gradients to around 6 °C/cm. *Katagiri* et al. [39.102] used a heat shield above the heater and crucible to decrease the axial temperature gradient from 100 °C, in a conventional LEC method, to 40 °C. The crystals grown in these conditions showed large areas free of dislocations. *Hirano* and *Uchida* [39.103] and *Hirano* [39.104] developed a new thermal baffle to obtain low-dislocation-density S-doped InP crystal of 50 mm diameter. The thermal baffle was designed to suppress gas convection. However, there are serious problems associated with growth under low-thermal-gradient conditions – crystal diameter instability and instabilities at the crystal–melt interface, which might give rise to twinning. This problem is particularly severe in InP growth, which is prone to twinning because of the low stacking fault energy. These problems have been overcome by applying a vertical magnetic field to suppress the flow in the melt and stabilize the crystal–melt interface [39.105–108]. Another approach is to shield the crystal completely from gas convection by utilizing fully encapsulated growth. *Kohda* et al. [39.109] combined fully encapsulated growth and a vertical magnetic field to obtain 50 mm-diameter completely dislocation-free semi-insulating GaAs crystals.

Thus there are close experimental precedents to the modeling results. A more detailed report of these modeling results can be found in [39.110]. With the rapid increase in computing power, and the greater availability of parallel clusters, a more detailed optimization study with better temperature histories – perhaps obtained from experiment or high-fidelity CFD models – has the potential to devise growth programs that completely minimize the dislocation densities in III–V as-grown crystals.

39.11 Summary

An effort has been made to present the state of the art in the theoretical and modeling aspects of dislocation generation in III–V compound semiconductors. Sections 39.1 and 39.2 set up and motivated the discussion. Section 39.3 discussed the need for the reduction of dislocation density, and traces the origin of these, the most important being the nucleation of dislocations from point defects. Section 39.4 described models for dislocation multiplication due to thermal stress; the popular Jordan model and also a more general and computationally intensive viscoplastic framework were introduced. Section 39.5 and beyond dealt with viscoplastic models in the literature, their main features, and their origins from microscopic dislocation dynamics. Section 39.5 gave an introduction to the structural aspects of common dislocations in diamond- and zincblende-type lattices. Section 39.6 discussed the deformation behavior of semiconductors and presents the Alexander–Haasen viscoplastic model. A brief description of the various components of the model is given. Section 39.7 gave a description of existing literature on the application of Alexander–Haasen-type models to the crystal growth problem. As a thorough examination of this class of models, Sect. 39.8 and the sections that follow discussed the development and application to crystal growth of a particular viscoplastic model. Section 39.8 described in great detail the various building blocks of the model, and the many limitations and assumptions that go into such models, as well as insights into future work that can be performed. Section 39.9 described briefly the numerical implementation of the model in a finite element framework and the methodology of its coupling with computational fluid dynamics and heat-transfer-based thermal models for crystal growth processes. Section 39.10 presented the results of one such successful coupling, and various conclusions are drawn.

Finally, some open questions and avenues for research are now presented:

1. A question frequently posed is the reason for the growth of dislocation-free silicon. In view of the above conclusion, a bold hypothesis may be the absence of substantial gas convection in low-pressure Czochralski growth of silicon. It might be useful to make a viscoplastic model for silicon and investigate it further.
2. A substantial improvement can be made in the thermal models. With the advent of high-performance computing, it might be possible to develop a continuous, nonsteady, three-dimensional simulation of the entire growth process, instead of the present two-dimensional axisymmetric quasistatic series of simulations. Such a simulation would resolve instabilities in the gas flow, and the temperature and flow oscillations could be resolved with more certainty.
3. The viscoplastic model suffers from various difficulties. Firstly, the topological aspect of dislocations was neglected. Dislocations are continuous line defects and can either end at the surfaces of the material or form closed loops. Physically, this is reflected in a divergence-free dislocation density tensor. Present formulations ignore the topological aspects of dislocation density and simply model it as scalar variables. Again, it might be possible to formulate alternate models, with dislocation as line objects rather than scalar variables. This might lead to theories with a structure similar to strain gradient theories.

4. Instead of assuming an initially negligible but finite total dislocation density, a suitable alternative would be to model the initial dislocation density from condensation of point defects into Frank loops, and their subsequent transformation into glide dislocations. Such a model should necessarily be coupled with models for point defect dynamics in the crystal.
5. There is scope for a synergetic effort between experimentalists – TEM, constant-strain-rate tests – and theoreticians. There is a great deal of uncertainty in measurements of dislocation velocities and stress–strain curves. Often these experiments are done on materials from different sources; this gives rise to great variations in the results. Furthermore, the virgin state of the material before the deformation, such as the initial dislocation density, is often not recorded. Hence a consistent set of experiments should be designed to measure quantities such as the dislocation density and stress–strain curves for material from the same set of sources. At the same time quantities such as the hardening coefficients and multiplication coefficients can be resolved with a mixture of atomistic quantum and mesoscopic discrete dislocation dynamics calculations. Such an effort will involve modeling on multiple scales along with the experimental data.

References

39.1 J. Czochralski: Ein neues Verfahren zur Messung der Kristallisationsgeschwindigkeit der Metalle, Z. Phys. Chem. **92**, 219–221 (1917), in German

39.2 G.K. Teal, J.B. Little: Growth of germanium single crystals, Phys. Rev. **78**, 647 (1950)

39.3 W.C. Dash: Dislocation free silicon crystals. In: *Growth and Perfection of Crystals*, ed. by R.M. Doremus, B.W. Roberts, D. Turnbull (Wiley, New York 1958)

39.4 V. Swaminathan, A.S. Jordan: Dislocations in III/V compounds, Semicond. Semimet. **38**, 293–341 (1993)

39.5 R.J. Roedel, A.R. Von Neida, R. Caruso, L.R. Dawson: The effect of dislocations in $Ga_{1-x}Al_xAs$:Si light-emitting diodes, J. Electrochem. Soc. **126**, 637–641 (1979)

39.6 J.P. Hirth, J. Lothe: *Theory of Dislocations* (Krieger, Malabar 1992)

39.7 H. Alexander: On dislocation generation in semiconductor crystals, Radiat. Eff. Defects Solids **112**(1/2), 1–12 (1989)

39.8 B.T. Lee, R. Gronsky, E.D. Bourret: Dislocation loops and precipitates associated with excess arsenic in GaAs, J. Appl. Phys. **64**(1), 114–118 (1988)

39.9 J. Lagowski, H.C. Gatos, T. Aoyama, D.G. Lin: Fermi energy control of vacancy coalescence and dislocation density in melt-grown GaAs, Appl. Phys. Lett. **45**(6), 680–682 (1984)

39.10 W. Zulehner: Czochralski growth of silicon, J. Cryst. Growth **65**(1–3), 189–213 (1983)

39.11 M.F. Ashby, L. Johnson: On the generation of dislocations at misfitting particles in a ductile matrix, Philos. Mag. **20**, 1009–1022 (1969)

39.12 A.S. Jordan, R. Caruso, A.R. Von Neida: A thermoelastic analysis of dislocation generation in pulled GaAs crystals, Bell Syst. Technol. J. **59**(4), 593–637 (1980)

39.13 N. Kobayashi, T. Iwaki: A thermoelastic analysis of the thermal stress produced in a semi-infinite cylindrical single crystal during the Czochralski growth, J. Cryst. Growth **73**, 96–110 (1985)

39.14 M. Duseaux: Temperature profile and thermal-stress calculations in GaAs crystals growing from the melt, J. Cryst. Growth **61**(3), 576–590 (1983)

39.15 J.C. Lambropoulos: Stresses near the solid-liquid interface during the growth of a Czochralski crystal, J. Cryst. Growth **80**, 245–256 (1987)

39.16 C.E. Schvezov, I.V. Samarasekera, F. Weinberg: Calculation of the shear stress distribution in LEC gallium arsenide for different growth conditions, J. Cryst. Growth **92**, 479–488 (1988)

39.17 G.O. Meduoye, K.E. Evans, D.J. Bacon: Modelling of the growth of the LEC technique II. Thermal stress distribution and influence of interface shape, J. Cryst. Growth **97**, 709–719 (1989)

39.18 G.O. Meduoye, D.J. Bacon, K.E. Evans: Computer modelling of temperature and stress distributions in LEC-grown GaAs crystals, J. Cryst. Growth **108**, 627–636 (1991)

39.19 S. Motakef, K.W. Kelly, K. Koai: Comparison of calculated and measured dislocation density in LEC-grown GaAs crystals, J. Cryst. Growth **113**, 279–288 (1991)

39.20 F. Dupret, P. Necodeme, Y. Ryckmans: Numerical method for reducing stress level in GaAs crystals, J. Cryst. Growth **97**, 162–172 (1989)

39.21 D.E. Bornside, T.A. Kinney, R.A. Brown: Minimization of thermoelastic stresses in Czochralski grown silicon: Application of the integrated system model, J. Cryst. Growth **108**, 779–805 (1991)

39.22 Y.F. Zou, H. Zhang, V. Prasad: Dynamics of melt-crystal interface and coupled convection-stress predictions for Czochralski crystal growth processes, J. Cryst. Growth **166**, 476–482 (1996)

39.23 I. Yonenaga, K. Sumino: Impurity effects on the generation, velocity, and immobilization of dislocations in GaAs, J. Appl. Phys. **65**, 85–92 (1989)

39.24 J. Lubliner: *Plasticity Theory* (Macmillan, New York 1990)

39.25 K. Sumino: Mechanical behavior of semiconductors. In: *Handbook on Semiconductors*, Vol. 3a, ed. by S. Mahajan, T.S. Moss (Elsevier, Amsterdam 1994) pp. 73–181

39.26 J. Hornstra: Dislocations in the diamond lattice, J. Phys. Chem. Solids **5**, 129–141 (1958)

39.27 H. Alexander: Dislocations in covalent crystals. In: *Dislocations in Solids*, Vol. 7, ed. by F.R.N. Nabarro (North-Holland, Amsterdam 1986) pp. 113–234

39.28 D.J.H. Cockayne, A. Hons: Dislocations in semiconductors as studied by weak-beam electron-microscopy, J. Phys. **40**(6), 11–18 (1979)

39.29 H. Gottschalk, G. Patzer, H. Alexander: Stacking-fault energy and ionicity of cubic III–V compounds, Phys. Status Solidi (a) **45**(1), 207–217 (1978)

39.30 R. Meingast, H. Alexander: Dissociated dislocations in germanium, Phys. Status Solidi (a) **17**(1), 229–236 (1973)

39.31 A. George, J. Rabier: Dislocations and plasticity in semiconductors. I – Dislocation structures and dynamics, Rev. Phys. Appl. **22**, 941–966 (1987)

39.32 W.G. Johnston, J.J. Gilman: Dislocation velocities, dislocation densities, and plastic flow in lithium fluoride crystals, J. Appl. Phys. **30**, 129–144 (1959)

39.33 H. Alexander, P. Haasen: Dislocations and plastic flow in the diamond structure. In: *Solid State Physics*, Vol. 22, ed. by F. Seitz, D. Turnbull, H. Ehrenreich (Academic, New York 1968) pp. 28–158

39.34 A.R. Chaudhuri, J.R. Patel, L.G. Rubin: Velocities and densities of dislocations in germanium and other semiconductor crystals, J. Appl. Phys. **33**, 2736–2746 (1962)

39.35 G.I. Taylor: The mechanism of plastic deformation of crystals. Part I – Theoretical, Proc. R. Soc. Lond. Ser. A **145**, 362–387 (1934)

39.36 F.R.N. Babarro, Z.S. Basinski, D.B. Holt: The plasticity of pure single crystals, Adv. Phys. **13**, 193–323 (1964)

39.37 E. Peissker, P. Haasen, H. Alexander: Anisotropic plastic deformation of indium antimonide, Philos. Mag. **7**, 1279 (1962)

39.38 I. Yonenaga, K. Sumino: Effects of in impurity on the dynamic behavior of dislocations in GaAs, J. Appl. Phys. **62**(4), 1212–1219 (1987)

39.39 I. Yonenaga, K. Sumino: Mechanical properties and dislocation dynamics of GaP, J. Mater. Res. **4**(2), 355–360 (1989)

39.40 J. Völkl: Stress in the cooling crystal. In: *Handbook of Crystal Growth*, Vol. 2, ed. by D.T.J. Hurle (North Holland, Amsterdam 1994) pp. 823–874

39.41 H. Siethoff, W. Schröter: Work-hardening and dynamical recovery in silicon and germanium at high-temperatures and comparison with FCC metals, Scr. Metall. **17**(3), 393–398 (1983)

39.42 H. Siethoff, R. Behrensmeier: Plasticity of undoped GaAs deformed under liquid encapsulation, J. Appl. Phys. **67**(8), 3673–3680 (1990)

39.43 H. Siethoff, K. Ahlborn, H.G. Brion, J. Völkl: Dynamical recovery and self-diffusion in InP, Philos. Mag. A **57**(2), 235–244 (1988)

39.44 H. Siethoff, W. Schröeter: New phenomena in the plasticity of semiconductors and FCC metals at high temperatures, Z. Metall. **75**(7), 475–491 (1984)

39.45 A.K. Mukherjee, J.E. Bird, J.E. Dorn: Experimental correlations for high temperature creep, ASM Transactions **62**, 155–179 (1969)

39.46 C.R. Barrett, W.D. Nix: A Model for steady state creep based on the motion of jogged screw dislocations, Acta Metall. **13**, 1247–1258 (1965)

39.47 H.G. Brion, H. Siethoff, W. Schröter: New stages in stress–strain curves of germanium at high-temperatures, Philos. Mag. A **43**(6), 1505–1513 (1981)

39.48 H. Siethoff: Cross-slip in the high-temperature deformation of germanium, silicon and Indium-antimonide, Philos. Mag. A **47**(5), 657–669 (1983)

39.49 B. Escaig: Cross-slip processes in the fcc structure. In: *Dislocation Dynamics*, ed. by A.R. Rosenfield, R. Alan (McGraw-Hill, London 1968) pp. 655–677

39.50 D. Maroudas, R.A. Brown: On the prediction of dislocation formation in semiconductor crystals grown from the melt – Analysis of the Haasen model for plastic deformation dynamics, J. Cryst. Growth **108**, 399–415 (1991)

39.51 C.T. Tsai: On the finite-element modeling of dislocation dynamics during semiconductor-crystal growth, J. Cryst. Growth **113**, 499–507 (1991)

39.52 C.T. Tsai, A.N. Gulluoglu, C.S. Hertley: A crystallographic methodology for modeling dislocation dynamics in GaAs crystals grown from the melt, J. Appl. Phys. **73**, 1650–1656 (1993)

39.53 J.C. Lambropoulos, C.H. Wu: Mechanics of shaped crystal growth from the melt, J. Mater. Res. **11**, 2163–2176 (1996)

39.54 N. Miyazaki, Y. Kuroda: Dislocation density simulations for bulk single crystal growth process, Met. Mater. Int. **4**(4), 883–890 (1998)

39.55 J.C. Moosbrugger: Continuum slip viscoplasticity with the Haasen constitutive model – application to single-crystal inelasticity, Int. J. Plast. **11**, 799–826 (1995)

39.56 J.C. Moosbrugger, A. Levy: Constitutive modelling for CdTe single-crystals, Metall. Mater. Trans. A **26**(10), 2687–2697 (1995)

39.57 H. Chung, W. Si, M. Dudley, A. Anselmo, D.F. Bliss, A. Maniatty, H. Zhang, V. Prasad: Characterization of structural defects in MLEK grown InP single crystals using synchotron beam x-ray topography, J. Cryst. Growth **174**(1–4), 230–237 (1997)

39.58 H. Chung, W. Si, M. Dudley, D.F. Bliss, R. Kalan, A. Maniatty, H. Zhang, V. Prasad: Characterization

39.58 of defect structures in magnetic liquid encapsulated Kyropoulos grown InP single crystals, J. Cryst. Growth **181**(1-2), 17–25 (1997)

39.59 H. Steinhardt, P. Haasen: Creep and dislocation velocities in GaAs, Phys. Status Solidi (a) **49**, 93–101 (1978)

39.60 I. Yonenaga, U. Unose, K. Sumino: Mechanical properties of GaAs crystals, J. Mater. Res. **2**, 252–261 (1987)

39.61 A. George, C. Escaravage, G. Champier, W. Schröter: Velocities of screw and 60°-dislocations in silicon, Phys. Status Solidi (b) **53**, 483–496 (1972)

39.62 M. Imai, K. Sumino: Insitu x-ray topographic study of the dislocation mobility in high-purity and impurity-doped silicon-crystals, Philos. Mag. A **47**(4), 599–621 (1983)

39.63 B.Y. Farber, V.I. Nikitenko: Change of dislocation mobility characteristics in silicon single-crystals at elevated-temperatures, Phys. Status Solidi (a) **73**(1), K141–144 (1982)

39.64 I. Yonenaga, K. Sumino: Dislocation velocity in indium-phospide, Appl. Phys. Lett. **58**(1), 48–50 (1991)

39.65 H. Nagai: Dislocation velocities in indium phospide, Jpn. J. Appl. Phys. **20**(4), 793–794 (1981)

39.66 K. Maeda, S. Takeuchi: Recombination enhanced glide in InP single crystals, Appl. Phys. Lett. **42**(8), 664–666 (1983)

39.67 F. Louchet: On the mobility of dislocations in silicon by insitu straining in a high-voltage electron-microscope, Philos. Mag. **43**(5), 1289–1297 (1981)

39.68 V. Celli, M. Kabler, T. Ninoyama, R. Thomson: Theory of dislocation mobility in semiconductors, Phys. Rev. **131**(1), 58–72 (1963)

39.69 V.V. Rybin, A.N. Orlov: Theory of dislocation motion in low-velocity range, Sov. Phys. Solid State **11**, 2635–2641 (1970)

39.70 S. Öberg, P.K. Sitch, R. Jones, M.I. Heggie: First-principles calculations of the energy barrier to dislocation motion in Si and GaAs, Phys. Rev. B **51**(19), 13138–13145 (1995)

39.71 V.V. Bulatov, S. Yip, A.S. Argon: Atomic modes of dislocation mobility in silicon, Philos. Mag. A **72**(2), 453–496 (1995)

39.72 H.R. Kolar, J.C.H. Spencer, H. Alexander: Observation of moving dislocation kinks and unpinning, Phys. Rev. Lett. **77**(19), 4031–4034 (1996)

39.73 H.J. Möller: The movement of dissociated dislocations in the diamond–cubic structure, Acta Metall. **26**, 963–973 (1977)

39.74 P. Haasen: Kink formation in charged dislocation, Phys. Status Solidi (a) **28**(1), 145–155 (1975)

39.75 P.B. Hirsch: Mechanism for the effect of doping on dislocation mobility, J. Phys. **40**(6), 117–121 (1979)

39.76 K. Sumino, I. Yonenaga: Dislocation dynamics and mechanical behavior of elemental and compound semiconductors, Phys. Status Solidi (a) **138**, 573–581 (1993)

39.77 K. Sumino, H. Harada: In situ x-ray topographic studies of the generation and the multiplication processes of dislocations in silicon crystals at elevated temperature, Philos. Mag. A **44**(6), 1319–1334 (1981)

39.78 P. Franciosi, A. Zaoui: Multislip in fcc. crystals: A theoretical approach compared with experimental data, Acta Metall. **30**, 1627–1637 (1982)

39.79 A. Moulin, M. Condat, L.P. Kubin: Mesoscale modelling of the yield point properties of silicon crystals, Acta Metall. **47**(10), 2879–2888 (1999)

39.80 H. Alexander, J.J. Crawford: Latent hardening of germanium crystals, Phys. Status Solidi (b) **222**, 41–49 (2000)

39.81 A.A. Chernov: *Modern Crystallography III. Crystal Growth* (Springer, Berlin 1984)

39.82 H. Klapper: Generation and propagation of dislocations during crystal growth, Mater. Chem. Phys. **66**, 101–109 (2000)

39.83 G. Dhanaraj, B. Raghothamachar, J. Bai, H. Chung, M. Dudley: Synchrotron x-ray topographic characterization of defects in InP bulk crystals, Proc. Int. Conf. Indium Phosphide Relat. Mater. (2005) pp. 643–648

39.84 G.T. Brown, B. Cockayne, W.R. Macewan: Deformation behavior of single crystals of InP in uniaxial compression, J. Mater. Sci. **15**, 1469–1477 (1980)

39.85 S. Pendurti: Modeling Dislocation Generation in High Pressure Czochralski Growth of InP Single Crystals. Ph.D. Thesis (State University of New York, Stony Brook 2003)

39.86 A.S. Jordan: Some thermal and mechanical properties of InP essential to crystal growth modeling, J. Cryst. Growth **71**, 559–565 (1985)

39.87 H. Siethoff: The plasticity of elemental and compound semiconductors, Semicond. Semimet. **37**, 143–187 (1992)

39.88 J.C. Simo, T.J.R. Hughes: *Computational Inelasticity* (Springer, New York 1998)

39.89 H. Zhang, V. Prasad: A multizone adaptive process model for low and high pressure crystal growth, J. Cryst. Growth **155**, 47–65 (1995)

39.90 P. Rudolph, M. Jurisch: Bulk growth of GaAs – An overview, J. Cryst. Growth **199**(1), 325–335 (1999)

39.91 V.A. Antonov, V.G. Elsakov, T.I. Olkhovikova, V.V. Selin: Dislocations and 90°-twins in LEC-grown InP crystals, J. Cryst. Growth **235**(1-4), 35–39 (2002)

39.92 T.-C. Chen, H.-C. Wu, C.-I. Weng: The effect of interface shape on anisotropic thermal stress of bulk single crystal during Czochralski growth, J. Cryst. Growth **173**, 367–379 (1997)

39.93 J. Matsui: Study of strain variation in LEC-grown GaAs bulk crystals by synchotron radiation x-ray, Appl. Surf. Sci. **50**, 1–8 (1991)

39.94 H.M. Buchheit, A. Khoukh, M. Bejar, S.K. Krawczyk, R.C. Blanchet: Residual strain mapping in III-V materials by spectrally resolving scanning photoluminescence, Microelectron. J. **30**(7), 651–657 (1999)

39.95 S. Pendurti, V. Prasad, H. Zhang: Modelling dislocation generation in high pressure Czochralski growth of InP single crystals: Part I. Construction of a visco-plastic deformation model, Model. Simul. Mater. Sci. Eng. **13**, 249–266 (2005)

39.96 V. Prasad, H. Zhang: Transport phenomena in Czochralski crystal growth processes, Adv. Heat Transf. **30**, 313–435 (1997)

39.97 A.G. Elliot, A. Flat, D.A. Vanderwater: Silicon incorporation in LEC growth of single-crystal gallium-arsenide, J. Cryst. Growth **121**(3), 349–359 (1992)

39.98 M. Neubert, P. Rudolph: Growth of semi-insulating GaAs crystals in low temperature gradients by using the vapour pressure controlled Czochralski method (VCZ), Prog. Cryst. Growth Charact. Mater. **43**(2/3), 119–185 (2001)

39.99 G. Müller, J. Völkl, E. Tomzig: Thermal analysis of LEC InP growth, J. Cryst. Growth **64**(1), 40–47 (1983)

39.100 A.R. Von Neida, A.S. Jordan: Reducing dislocations in GaAs and InP, J. Met. **38**, 35–40 (1986)

39.101 A.G. Elliot, C.L. Wei, R. Farraro, G. Woolhouse, M. Scott, R. Hiskes: Low dislocation density, large diameter, liquid encapsulated Czochralski growth of GaAs, J. Cryst. Growth **70**, 169–178 (1984)

39.102 K. Katagiri, S. Yamazaki, A. Takagi, O. Oda, H. Araki, I. Tsuboya: LEC growth of large diameter InP single crystals doped with Sn and with S, Inst. Phys. Conf. Ser. **79**, 67–72 (1986)

39.103 R. Hirano, M. Uchida: Reduction of dislocation densities in InP single crystals by the LEC method using thermal baffles, J. Electron. Mater. **25**, 347–351 (1996)

39.104 R. Hirano: Growth of low etch pit density homogeneous 2″ InP crystals using a newly developed thermal baffle, Jpn. J. Appl. Phys. **38**(2B), 969–971 (1999)

39.105 K. Terashima, T. Fukuda: A new magnetic-field applied pulling apparatus for LEC GaAs single-crystal growth, J. Cryst. Growth **63**, 423–425 (1983)

39.106 H. Miyairi, T. Inada, M. Eguchi, T. Fukuda: Growth and properties of InP single crystals grown by the magnetic-field applied LEC method, J. Cryst. Growth **79**(1–3), 291–295 (1986)

39.107 J. Osaka, H. Kohda, T. Kobayashi, K. Hoshikawa: Homogeneity of vertical magnetic-field applied LEC GaAs crystal, Jpn. J. Appl. Phys. Part 2 – Lett. **23**(4), L194–197 (1984)

39.108 S. Ozawa, T. Kimura, J. Kobayashi, T. Fukuda: Programmed magnetic-field applied liquid encapsulated Czochralski crystal-growth, Appl. Phys. Lett. **50**(6), 329–331 (1987)

39.109 H. Kohda, K. Yamada, H. Nakanishi, T. Kobayashi, J. Osaka, K. Hoshikawa: Crystal-growth of completely dislocation-free and striation-free GaAs, J. Cryst. Growth **71**(3), 813–816 (1985)

39.110 S. Pendurti, H. Zhang, V. Prasad: Modeling dislocation generation in high pressure Czochralski growth of InP single crystals: Part II, Model. Simul. Mater. Sci. Eng. **13**, 267–297 (2005)

40. Mass and Heat Transport in BS and EFG Systems

Thomas F. George, Stefan Balint, Liliana Braescu

In this chapter several mathematical models describing processes which take place in the Bridgman–Stockbarger (BS) and edge-defined film-fed growth (EFG) systems are presented. Predictions are made concerning the impurity repartition in the crystal in the framework of each of the models. First, a short description of the real processes which are modeled is given, along with the equations, boundary conditions, and initial values defining the mathematical model. After that, numerical results obtained by computations in the framework of the model are provided, making a comparison between the computed results and those obtained in other models, and with the experimental data.

40.1 Model-Based Prediction of the Impurity Distribution – Vertical BS System 1380
    40.1.1 Burton–Prim–Slichter Uniform-Diffusion-Layer Model (UDLM) 1380
    40.1.2 Chang–Brown Quasi-Steady-State Model (QSSM) 1381
    40.1.3 Adornato–Brown Pseudo-Steady-State Model (PSSM) 1383
    40.1.4 Nonstationary Model (NSM) 1384
    40.1.5 Modified Quasi-Steady-State Model (MQSSM) and Modified Nonstationary Model (MNSM) 1386
    40.1.6 Larson–Zhang–Zheng Thermal-Diffusion Model (TDM).... 1387

40.2 Model-Based Prediction of the Impurity Distribution – EFG System 1389
    40.2.1 The Uniform-Diffusion-Layer Model (UDLM) 1389
    40.2.2 Tatarchenko Steady-State Model (TSSM) 1389
    40.2.3 Melt Replenishment Model (MRM) 1390
    40.2.4 Melt Without Replenishment Model (MWRM) 1397

References 1400

The quality of the crystal being grown depends to a considerable extent on homogeneity and the distribution of both detrimental and specially added impurities. Variations of the concentration of impurities along the crystal length and cross section lead to variations of the mechanical, electrical, and optical properties in the mass of the crystal. The impurity repartition in the crystal is determined by the processes which take place in the melt during the growth and by mass transfer. Quantitative descriptions of these processes permit the prediction of such repartition in the crystal, which is a reason to build mathematical models in order to describe processes which take place during the growth. Consequently, computer modeling of the crystal growth processes has often been described as an art as well as a science, to acknowledge the seemingly endless difficulties that arise in the application of numerical methods. Three core competencies are required for the effective use of modeling in crystal growth. One is a thorough grasp of the fundamentals of continuum transport phenomena. The second is the building of an appropriate mathematical model, and the third is a general understanding of the numerical methods necessary for solving the governing equations of transport phenomena.

The philosophy behind building the mathematical model is to provide a quantitative description of the mechanism by balancing what is wanted with what can be done. The quantitative description is made in terms of a certain number of variables (called the model variables) such that the mathematical model is a set of equations concerning these variables. The analysis of a mathematical model leads to results or predictions that can be tested against observations. It is important to realize that all models are idealizations and hence are limited in their applicability; experimental verification of the predictions can establish the authenticity of the model.

40.1 Model-Based Prediction of the Impurity Distribution – Vertical BS System

The prototype vertical BS crystal growth system (discussed here) consists of a crystal and melt contained in a cylindrical ampoule of radius R and length L pulled slowly through a vertically aligned furnace with hot and cold isothermal zones, as shown in Fig. 40.1.

The isothermal zones are separated by an adiabatic region of length L_g, designed to promote a steep axial temperature gradient and to maintain a flat solidification interface. A purpose of different mathematical models conceived to describe the processes which take place in such a system is the prediction of the distribution of detrimental and added impurities in the crystal being grown. The models discussed here describe one or several of the following processes: rejection of impurities at the melt–crystal interface, molecular diffusion, thermodiffusion, thermal convection, solutal convection, heat transfer between the furnace and ampoule walls, decrease of the melt in the ampoule, and melt–crystal interface morphology.

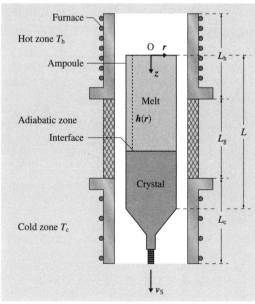

Fig. 40.1 Prototype BS crystal growth system

40.1.1 Burton–Prim–Slichter Uniform-Diffusion-Layer Model (UDLM)

For a large set of binary alloys, the rejection of impurities at the melt–crystal interface causes an unequal partitioning of the impurities between the melt and crystal during solidification. This leads to the accumulation of impurities in the melt and creates in the melt a nonuniform concentration field. A model to describe the above concentration field in the neighborhood of the melt–crystal interface (at the melt side) and the transfer of impurities in the crystal is the so-called uniform-diffusion-layer model (UDLM). The concept of an *axial boundary layer* evidently was introduced by *Burton* et al. [40.1]. Others [40.2–6] have extended this concept by introducing the notion of a *uniform diffusion layer* or *stagnant film*, without any picture for the fluid motion in the melt. Physically, the uniform diffusion layer is a thin layer masking the growing crystal which separates the crystal surface from the bulk melt [40.7].

According to UDLM, the isoconcentration lines of impurities are parallel to the interface, and the concentration decreases exponentially into the melt in this layer. The transport of impurities in the layer is molecular diffusion controlled, and the prediction is that, after an initial transient, the crystal exhibits a uniform (axial and radial) impurity distribution [40.8]. The effective partition coefficient k_{eff}, the equilibrium partition coefficient k, and the dimensionless diffusion-layer thickness $\delta = \tilde{\delta}/L$ are related by Fleming's formula

$$k_{\text{eff}} = \frac{k}{k + (1-k)\exp\left(\frac{\text{Pe Sc}}{\text{Pr}}\delta\right)}, \qquad (40.1)$$

where Pr is the Prandtl number, Pe is the Péclet number, and Sc is the Schmidt number [40.7]. The predicted impurity concentration in the crystal is given by

$$c_{\text{crys}} = k_{\text{eff}} c_0, \qquad (40.2)$$

where c_0 represents the impurity concentration in the bulk melt.

This model is appropriate for the prediction of impurities in crystals grown from a quiescent melt. Thus, a reduction in the magnitude of the buoyancy forces when processing semiconductors, such as in a low-gravity environment, has been pursued over the past several decades. The effectiveness of space processing for the growth of chemically uniform crystals is supported by experimental and theoretical studies [40.9, 10]. For example, the InSb crystal reported in [40.11] exhibits axial distribution profiles that are characteristic of diffusion-controlled mass transfer growth. In [40.7] it is shown, based on modeling studies, that the low gravity levels achieved in space are sufficient to inhibit thermal convection in small-diameter Ge and Ga-Ge melts. At the same time, it is pointed out that UDLM is an oversimplified model for accurate prediction of the impurity distribution in the crystal [40.12].

40.1.2 Chang–Brown Quasi-Steady-State Model (QSSM)

An important process in a BS system is thermal convection, the effect of which is essentially neglected in UDLM. Thermal convection is driven by buoyancy differences induced by axial and radial temperature gradients, and it may be laminar, periodic or turbulent. When the melt is below the crystal, the axial temperature gradient places the hottest melt at the bottom of the ampoule and generates convection. *Kim* et al. [40.3] used gallium-doped germanium to identify the melt length at which three forms of convection were present. In [40.13–26], the melt was oriented above the crystal in order to produce a stable axial density gradient. Because radial temperature gradients cause lateral density variations [40.27], convection still exists in this configuration. Thermal convection in a BS system contributes to the distribution of the impurities in the crystal.

The QSSM is a simplified model of the thermal convection and of the heat and mass transport in a BS system. The real process is replaced by a quasi-steady-state process viewed from a stationary reference frame and described in the cylindrical polar coordinate system shown in Fig. 40.1. The translation of the ampoule is replaced by supplying melt into the ampoule at $z = 0$ with velocity $\tilde{v}_z = v_L$ and removing the crystal at the other end of the ampoule at the rate $v_S = (\rho_S/\rho_L) v_L$, where ρ_L represents the melt density and ρ_S the crystal density. Changes in the length of the melt are neglected in QSSM. The location of the melt–solid interface is represented as $z = h(r)$, $0 \leq r \leq \Lambda$, $\Lambda = R/L$, and the unit vectors normal $\boldsymbol{n}$ and tangent $\boldsymbol{t}$ everywhere to the interface are

$$\boldsymbol{n} = \frac{\boldsymbol{e}_z - h_r \boldsymbol{e}_r}{\left(1 + h_r^2\right)^{1/2}}, \quad \boldsymbol{t} = \frac{\boldsymbol{e}_r + h_r \boldsymbol{e}_z}{\left(1 + h_r^2\right)^{1/2}}, \quad (40.3)$$

where $h_r = dh/dr$ and the set $(\boldsymbol{e}_r, \boldsymbol{e}_\theta, \boldsymbol{e}_z)$ are the unit vectors of the cylindrical polar coordinate system. According to [40.7], the equations governing the melt flow and heat and mass transport in the melt which define QSSM, in axisymmetric and dimensionless form, are

$$(\boldsymbol{v}\nabla)\boldsymbol{v} = -\nabla p + \mathrm{Pr}\nabla^2 \boldsymbol{v} + \mathrm{Ra_T Pr}\theta\boldsymbol{e}_z, \quad (40.4)$$

$$\nabla \boldsymbol{v} = 0, \quad (40.5)$$

$$\boldsymbol{v}\nabla\theta = \nabla^2 \theta, \quad (40.6)$$

$$\left(\frac{\mathrm{Sc}}{\mathrm{Pr}}\right) \boldsymbol{v}\nabla c = \nabla^2 c. \quad (40.7)$$

Here, $\nabla = \boldsymbol{e}_r \partial/\partial r + \boldsymbol{e}_z \partial/\partial z$ represents the gradient operator in cylindrical coordinates; Pr is the Prandtl number; $\mathrm{Ra_T}$ is the thermal Rayleigh number; Sc is the Schmidt number; $\boldsymbol{v}$ is the dimensionless melt flow velocity; θ is the dimensionless melt temperature; c is the impurity concentration in the melt; and $0 \leq z \leq h(r)$, $0 \leq r \leq \Lambda$, $\Lambda = R/L$.

In the crystal, the velocity field $\boldsymbol{v}$ is uniform at speed v_S

$$\boldsymbol{v} = v_S \boldsymbol{e}_z, \quad h(r) \leq z \leq 1, \quad 0 \leq r \leq \Lambda. \quad (40.8)$$

The heat transport in the crystal is described by

$$\mathrm{Pe}\,\boldsymbol{e}_z\nabla\theta = \gamma\nabla^2\theta, \quad h(r) \leq z \leq 1, \quad 0 \leq r \leq \Lambda, \quad (40.9)$$

where $\gamma = \alpha_S/\alpha_L$ represents the ratio of thermal diffusivities in crystal and melt, and $\mathrm{Pe} = v_S L/\alpha_L$ is the Péclet number for convective heat transfer. The concentration of impurities in the crystal is given by

$$c_S(r) = kc(r, h(r)), \quad 0 \leq r \leq \Lambda. \quad (40.10)$$

Boundary conditions for the melt flow field $\boldsymbol{v}$ at the interface ensure no slip tangential to the crystal and the solidification of the melt at rate v_S. The top surface and sidewall of the ampoule are assumed to be no-slip surfaces, i.e.,

$$v_r = 0, \quad v_z = \frac{\rho_S}{\rho_L}\mathrm{Pe}. \quad (40.11)$$

This last condition dictates a jump in the velocity of the ampoule wall when $\sigma = \rho_S/\rho_L \neq 1$.

The axisymmetric boundary conditions for $\boldsymbol{v}$ and θ at the axis of the cylinder are

$$v_r = \frac{\partial v_z}{\partial r} = \frac{\partial \theta}{\partial r} = 0, \quad r = 0, \quad 0 \leq z \leq 1. \quad (40.12)$$

The thermal boundary conditions incorporate the assumption that the ampoule has negligible thermal mass, the adiabatic region is a perfect insulator, and the ampoule has the temperature of the surrounding furnace; the ends of the ampoule are perfectly insulated. At the melt–crystal interface, the temperature satisfies

$$\theta(r, h(r)) = \frac{T_m - T_c}{T_h - T_c},$$
$$0 \leq r \leq \Lambda, \quad (40.13)$$

$$(n\nabla\theta)_L - K(n\nabla\theta)_S = \text{St Pe}\, n e_z;$$
$$z = h(r), \quad 0 \leq r \leq \Lambda, \quad (40.14)$$

where T_m is the melting temperature, K is the ratio of the thermal conductivities between crystal and melt, and St is the Stefan number. The boundary conditions for the impurity concentrations are

$$\frac{\partial c}{\partial z} = \left(\frac{\text{Pe Sc}}{\text{Pr}}\right)(c - 1);$$
$$z = 0, \quad 0 \leq r \leq \Lambda, \quad (40.15)$$

$$n\nabla c = \left(\frac{\text{Pe Sc}}{\text{Pr}}\right)(ne_z)(1-k)c;$$
$$z = h(r), \quad 0 \leq r \leq \Lambda, \quad (40.16)$$

$$\frac{\partial c}{\partial r} = 0; \quad 0 \leq z \leq 1,$$
$$r = 0, \quad r = \Lambda. \quad (40.17)$$

These conditions express the following: a supply melt inlet at the top of the BS system; the rejected impurities at the melt–crystal interface; and the no-flux condition which is valid at the center line and at the side wall of the ampoule.

Using finite-element numerical methods for crystals and melts with thermophysical properties similar to gallium-doped germanium, the temperature field and dopant field in the melt and in the crystal were computed [40.7, 28] for several different Ra_T numbers in QSSM. Results are presented for two Bridgman configurations: vertically stabilized (melt on the top) and vertically destabilized (melt on bottom). Steady axisymmetric flows are classified according to Ra_T and are nearly equal to the growth velocity. Three distinct types of flow patterns were observed with changing Ra_T. At low Rayleigh numbers, the streamlines were rectilinear and only slightly distorted by buoyancy forces. Increasing the Rayleigh number developed first a cellular flow, and after that a weak secondary cell for stabilized configuration. The flows in these two configurations are in opposite directions. Calculations of the transport of a dilute dopant by these flow fields reveal radial segregation levels as large as 60% of the mean concentration. Radial segregation is most severe at an intermediate value of the Rayleigh number, above which the dopant distribution along the interface levels increases. The complexity of the concentration field coupled with calculations of the effective segregation coefficient show the coarseness of the usual diffusion-layer approximation for describing the dopant distribution adjacent to the crystal.

The results reveal that UDLM is a gross oversimplification of the interactions between the complicated flow patterns and the concentration field of impurities. The diffusion-layer thickness δ considered in UDLM in the best case is an empirical fit to the effective segregation coefficient. Although the radially averaged diffusion-layer thickness computed in QSSM is of the same order of magnitude as those considered in UDLM, the actual concentration gradient adjacent to the crystal is far from being radially uniform. In the computations made in QSSM, the ratio of thermal conductivities K between the melt and the crystal and the length of the gradient zone L_g have emerged as important parameters for determining the interface shape and radial segregation. According to [40.7], the value of K sets the shape of the interface and determines the qualitative behavior of the radial segregation with increased convection, and L_g sets the degree of interaction between the cellular flow and the concentration gradient adjacent to the interface. For long enough gradient zones and moderate convection, the flow causes lower radial segregation. A critical issue not addressed in this chapter is the transition from the steady-state flow patterns predicted here to time-periodic flows observed in some experiments. Mathematically, these transitions occur as Hopf bifurcations from the steady flows and have been predicted theoretically by computer-aided bifurcation analysis for low-Prandtl-number melts in the horizontal boat geometry as studied experimentally by *Hurle* et al. [40.29]. Simulation of the crystal growth in the time-periodic regime requires a time-dependent solution of the Boussinesq equations.

The qualitative understanding of melt flow and impurities segregation offered by QSSM serves as a starting point for more refined and accurate calculations aimed at comparison with experimental data. These would require the inclusion in the model of the thermal interactions between the ampoule and the furnace, the solutal convection, the decrease of the melt in the ampoule, the thermal diffusion (Soret effect), and the effect of the precrystallization zone.

40.1.3 Adornato–Brown Pseudo-Steady-State Model (PSSM)

Thermosolutal convection which can occur in a BS system is driven by buoyancy differences induced by axial and radial temperature gradients and by concentration-dependent density differences. Convection driven by such density differences is initiated in the layer adjacent to the melt–crystal interface and is caused by the accumulation of impurities in this region. In nondilute alloy systems, the interaction of the solutal and thermal driving forces is so closely coupled through convection of the solute field that only a full numerical simulation can give a quantitative description of the flow [40.12]. PSSM is a simplified model of the real thermosolutal convection and of the heat and mass transport in a BS system which takes into account the heat transfer between the furnace and the ampoule. In PSSM, like in QSSM, the translation of the ampoule is replaced by supplying melt at the top of the ampoule with uniform velocity and composition, and removing crystal from the bottom of the ampoule at a certain growth rate. Transients in velocity, pressure, temperature, and concentration caused by the steady decrease of the length of the melt in the real ampoule, and the displacement of the ampoule in the furnace, are neglected in PSSM.

The equations (in axisymmetric and dimensionless form) governing the melt flow, and heat and mass transport in the melt, defining PSSM [40.12] are the following

$$(v\nabla) v = -\nabla p + \text{Pr} \nabla^2 v$$
$$+ \text{Pr} (\text{Ra}_T \theta - \text{Ra}_S c) e_z , \quad (40.18)$$

$$\nabla v = 0 , \quad (40.19)$$

$$v\nabla\theta = \nabla^2\theta , \quad (40.20)$$

$$\left(\frac{\text{Sc}}{\text{Pr}}\right) v\nabla c = \nabla^2 c , \quad (40.21)$$

where ∇, Pr, Ra_T, Sc, v, θ, and c have the same meanings as in (40.4–40.7), and Ra_S represents the solutal Rayleigh number ($0 \leq z \leq h(r), 0 \leq r \leq \Lambda$). The velocity field v in the crystal is uniform at speed v_S

$$v = v_S e_z ; \quad h(r) \leq z \leq 1, \quad 0 \leq r \leq \Lambda . \quad (40.22)$$

The heat transport equations in the ampoule and in the crystal are

$$\text{Pe}_a e_z \nabla\theta = \gamma_a \nabla^2 \theta ; \quad h(r) \leq z \leq 1, \quad 0 \leq r \leq \Lambda \quad (40.23)$$

$$\text{Pe}_c e_z \nabla\theta = \gamma_c \nabla^2 \theta ; \quad h(r) \leq z \leq 1, \quad 0 \leq r \leq \Lambda , \quad (40.24)$$

where $\gamma_a = \alpha_a/\alpha_L$ and $\gamma_c = \alpha_c/\alpha_L$ are the ratios between the thermal diffusivities of each phase and the melt. The Péclet number Pe is the dimensionless transition rate of the ampoule scaled by the characteristic velocity for thermal diffusion. The shape of the melt–crystal interface is set by the shape of liquids curve $\theta_L(c)$ from the binary phase diagram as

$$\theta(r, h(r)) = \theta_L(c) = \theta_L^0 + m \left(c + 1 - \frac{1}{k}\right) ,$$
$$0 \leq r \leq \Lambda , \quad (40.25)$$

where the curve is approximated by a straight line with dimensionless slope m, and θ_L^0 is the melting temperature of the alloy with concentration c_0. It is assumed that m and the partition coefficient k are independent of concentration.

The interfacial temperature and impurities field are related to the corresponding values in the bulk melt by the balances

$$(n\nabla\theta)_L - K (n\nabla\theta)_S = \text{St Pe} \, n e_z ,$$
$$0 \leq r \leq \Lambda \quad (40.26)$$

$$n\nabla c = \frac{\text{Pe Sc}}{\text{Pr}} (n e_z)(1-k) c ;$$
$$z = h(r), \quad 0 \leq r \leq \Lambda . \quad (40.27)$$

The boundary conditions for the melt flow field are the same as those presented for QSSM. The thermal boundary conditions along the ampoule wall are specified according to the heat transfer condition

$$\left.\frac{\partial \theta}{\partial r}\right|_a = Bi(z) \cdot (\theta - \theta_\infty(z)) , \quad (40.28)$$

where $Bi(z) = h_a \cdot L/k_a$ is a dimensionless heat transfer coefficient defined to include radiative, conductive, and convective transport between the furnace and the ampoule, and $\theta_\infty(z)$ is the furnace temperature distribution. Perfect thermal contact at the melt–ampoule and crystal–ampoule interfaces are assumed, so that the conductive fluxes at the boundaries are equal

$$\left.\frac{\partial \theta}{\partial r}\right|_L = K_a \left.\frac{\partial \theta}{\partial r}\right|_a , \quad K_c \left.\frac{\partial \theta}{\partial r}\right|_c = K_a \left.\frac{\partial \theta}{\partial r}\right|_a ;$$
$$r = \Lambda . \quad (40.29)$$

The top and bottom of the ampoule are assumed to be at the temperature of the hot and cold zones. The specification of the concentration field of impurities in the melt is completed by setting the diffusive flux of impurities through the inlet to be equal to Pe_c, and zero through the ampoule wall. The concentration of the impurities in the crystal is given by

$$c_S(r) = kc(r, h(r)), \quad 0 \leq r \leq \Lambda. \tag{40.30}$$

The axis of the ampoule is taken as a line of symmetry for all field variables.

Using finite-element numerical methods, computations in PSSM were performed for a binary alloy coupling heat transfer in the melt, crystal, and ampoule [40.12] with convection in the melt and segregation in the growing crystal, taking into account convection driven by both temperature and concentration gradients. Calculations presented for growth of gallium-doped germanium in ampoules of boron nitride, quartz, and graphite, along with a quartz/graphite composite ampoule, show the effects of ampoule conductivity on the flow structure and intensity. The thermal conductivity of the ampoule material and the thickness of the ampoule wall can be used as parameters to change the structure of the flow, as well as to modulate the flow intensity. The computations show that flow pattern and concentration fields have rich structures, depending on the furnace, either that constructed at Massachusetts Institute of Technology (MIT) by Wang and Witt (based on heat pipes for forming isothermal hot and cold regions separated by a nearly adiabatic zone; Fig. 40.1) or at Grenoble by Rouzaud and Favier (using a tapered heating element to establish a nearly linear temperature profile over the length of the ampoule). The vertical BS system used at MIT has a two-cell structure: one cell driven by the radial gradients at the junction of the adiabatic and hot zones of the furnace, and the other adjacent to the interface caused by the differences in the thermal conductivities of the melt, crystal, and ampoule. The constant-gradient furnace designed at Grenoble has only a single cell near the interface.

Impurities are transported mostly by diffusion between adjacent cells because the convective motion is laminar. The radial segregation is set primarily by the flow adjacent to the interface when this cell is long enough to completely contain the solute gradient for diffusion-controlled growth. The upper cell in the MIT furnace plays a small role in determining the solute distribution along the interface. Axial segregation depends on the intensity of the convective mixing through the ampoule, so that the cellular structure of the flow pattern is important. In the heat pipe system, solute transport between flow cells limits the effectiveness of this mixing and leads higher solute concentrations in the cell next to the interface. The degree of axial segregation is set mainly by this cell.

The dependence of the degree of radial segregation on Ra_T studied in both the MIT and Grenoble systems is characterized by the following: (i) segregation is lowest for either diffusion-controlled growth (small Ra_T) or for when the intense laminar mixing is present (large Ra_T); (ii) segregation reaches a maximum for intermediate values of low intensity; (iii) the ampoule material, furnace design, and stabilizing solute all affect the values of segregation. This result is crucial for understanding the effects on solute segregation of microgravity solidification and of applied magnetic fields, and answers two questions which were not clarified before: (i) the way in which the reduced convection level affects the solute segregation, and (ii) the dependence of the segregation level on the ampoule size for the same Ra_T. In both situations the body forces acting on the melt are altered in an attempt to damp convection. Both approaches are clearly effective at eliminating time-periodic and chaotic convection present in large-size systems. If the reduced gravitational field has constant magnitude and is aligned with the crystal axis, then it is modeled exactly by reducing the thermal Rayleigh number. Thus, in [40.12], it is shown that the decrease in gravitational level needed to reach diffusion-controlled growth increases with a value which depend on the cube of the ampoule radius. This proves that the segregation effect is more severe for larger ampoules.

The prediction of the melt–crystal interface shape, radial impurity distribution, and effective segregation coefficient for the growth of dilute gallium-doped germanium obtained in [40.12] agrees reasonably well with the measurements by *Wang* [40.30].

40.1.4 Nonstationary Model (NSM)

NSM is a model of the unsteady thermoconvection, heat and impurity transport in a BS system which takes into account the decrease of the melt in the ampoule. The equations that govern such phenomena can be expressed in a mobile reference frame (Fig. 40.1) in axisymmetric and dimensionless form as [40.31]

$$\frac{\partial \boldsymbol{v}}{\partial t} + (\boldsymbol{v}\nabla)\boldsymbol{v} = -\nabla p + \Pr \nabla^2 \boldsymbol{v} + \Pr \text{Ra}_T \theta \boldsymbol{e}_z, \tag{40.31}$$

$$\nabla \boldsymbol{v} = 0, \tag{40.32}$$

$$\frac{\partial \theta}{\partial t} + \boldsymbol{v}\nabla\theta = \nabla^2 \theta, \tag{40.33}$$

$$\frac{\partial c}{\partial t} + \boldsymbol{v}\nabla c = \frac{\Pr}{\text{Sc}} \nabla^2 c. \tag{40.34}$$

The meanings of the symbols which appear in the above equations correspond to those in (40.18–40.21). These equations are time dependent since the domain Ω_t occupied by the melt is time dependent. The boundary $\partial \Omega_t$ of Ω_t is decomposed as $\partial \Omega_t = (\partial \Omega_t)_{\text{top}} \cup (\partial \Omega_t)_w \cup (\partial \Omega_t)_i$, where $(\partial \Omega_t)_{\text{top}}$ and $(\partial \Omega_t)_w$ are the melt boundaries limited by the top and side walls of the ampoule, respectively, and $(\partial \Omega_t)_i$ is the melt boundary limited by the bottom of the ampoule for t satisfying $0 \leq t \leq L_g/(2v_p L)$ and by the melt–crystal interface for t satisfying $L_g/(2v_p L) \leq t \leq (1 + L_g/(2L))/v_p$, where v_p is the ampoule translation rate of the ampoule.

The boundary conditions for v, θ, and c are

$$v = 0 \quad \text{on} \quad \partial \Omega_t \,, \quad (40.35)$$

$$\frac{\partial \theta}{\partial z} = 0 \quad \text{on} \quad (\partial \Omega_t)_{\text{top}} \,;$$
$$\theta = \theta_f \quad \text{on} \quad (\partial \Omega_t)_w \,;$$
$$\theta = \theta_i \quad \text{on} \quad (\partial \Omega_t)_i \,, \quad (40.36)$$

$$\frac{\partial c}{\partial z} = 0 \quad \text{on} \quad (\partial \Omega_t)_{\text{top}} \,;$$
$$\frac{\partial c}{\partial r} = 0 \quad \text{on} \quad (\partial \Omega_t)_w \,;$$
$$\frac{\partial c}{\partial z} = \Phi_i \quad \text{on} \quad (\partial \Omega_t)_i \,. \quad (40.37)$$

Here, θ_f is the temperature of the surrounding furnace, θ_i is the temperature of the bottom of the ampoule for t satisfying $0 \leq t \leq L_g/(2v_p L)$, and temperature at the melt–crystal interface for t satisfying $L_g/(2v_p L) \leq t \leq (1 + L_g/(2L))/v_p$, $\Phi_i = 0$ for t satisfying $0 \leq t \leq L_g/(2v_p L)$; and $\Phi_i = (\text{Pe Sc})/(\text{Pr})(1-k)c$ for t satisfying $L_g/(2v_p L) \leq t \leq (1 + L_g/(2L))/v_p$. The initial conditions for the equations (40.31–40.34) are

$$v(0, r, z) = 0\,;$$
$$\theta(0, r, z) = 1\,;$$
$$c(0, r, z) = 1\,. \quad (40.38)$$

The impurity distribution in the crystal is given by

$$c_S(\xi, r) = kc\,(t, r, h(r))$$
$$= kc\left(\frac{1}{v_p}\left(\xi + \frac{L_g}{2L}\right), r, h(r)\right)\,, \quad (40.39)$$

where ξ represents the solidified fraction

$$\xi = v_p t - \frac{L_g}{2L}$$
$$\text{for} \quad \frac{L_g}{2v_p L} \leq t \leq \left(1 + \frac{L_g}{2L}\right)/v_p\,. \quad (40.40)$$

Computations in NSM were made by the finite-element method using the software Cosmos/M [40.31] in a low-gravity environment ($\text{Ra}_T = 10^3$) for gallium-doped germanium. The results reveal that, during the growth process, $|v|_{\max} = |v_z|_{\max}$ is localized at $r = 0$ at the distance $1.25 L_g/L$ from the top of the crystal, and $v_r|_{\max}$ is localized at $r = 0.125$ at the distance $L_g/(2L)$ from the top of the crystal. In the first part of the growth process, $|v|_{\max} = |v_z|_{\max}$ increases from 0.1772 to 0.1817 and $|v_r|_{\max}$ increases from 0.0598 to 0.0621. In the second part, $|v|_{\max} = v_z|_{\max}$ decreases from 0.1817 to 0.0193

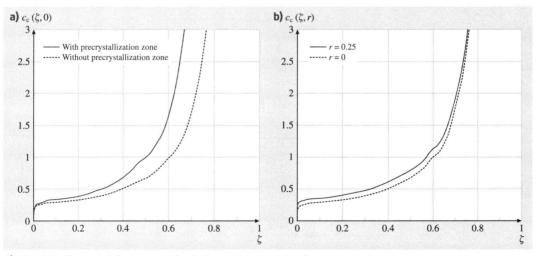

Fig. 40.2a,b Computed Ga concentration in the crystal versus ξ and r

and $|v_r|_{\max}$ decreases from 0.0621 to 0.0043. During the growth, at all points of the melt, the impurity concentration c increases but not uniformly. On the axis, parallel to the ampoule axis, c increases from the top towards the interface. In a transversal section of the ampoule, c increases from the axis towards the side wall. The impurity concentration on the axis of the crystal increases from kc_0 to c_0 in the solidified fraction $\xi \in (0; 0.6099)$, and the percent axial segregation defined by the formula $\Delta c_{\text{ax}} = |c_c(0, 0) - c_c(0.6099, 0)| \times 100/k$ is $\Delta c_{\text{ax}} = 854.5$ (Fig. 40.2a). The percent radial segregation, defined as $\Delta c_{\text{rad}} = |c_c(\xi, 0) - c_c(\xi, \Lambda)| \times 100/k$ for $\xi = 0.6099$, is $\Delta c_{\text{rad}} = 134$ (Fig. 40.2b).

Due to the decrease of the melt in the ampoule, the percent axial segregation in the crystal increases with the solidified fraction ξ. The percent radial segregation in the crystal increases for the first three-quarters and decreases thereafter. Finally, it has been shown [40.32] that, if the initial impurity distribution in the melt is not uniform like in [40.31], and if it is given by

$$c_0(r,z) = \frac{1}{\alpha(a)}\left\{1 + \frac{1-k}{k}\frac{\Lambda - r}{\Lambda}\right.$$
$$\left. \times \exp\left[-\frac{v_p \text{Sc}}{\text{Pr}}(1-z)\right] - \frac{a \cdot r}{\Lambda}\right\}, \quad (40.41)$$

then the computed impurities distribution in the crystal (in the NSM) on the solidified fraction $\xi \in (0; 0.7)$ is a constant equal to the prescribed concentration $c_c = 1$.

40.1.5 Modified Quasi-Steady-State Model (MQSSM) and Modified Nonstationary Model (MNSM)

In order to evaluate the effect of the precrystallization zone on the impurity distribution in the crystal, two different models – MQSSM and MNSM – have been reported in the literature [40.28, 33–35]. In such models, it is assumed that the melt→solid phase transition does not take place on a flat mathematical surface like in UDLM, QSSM, PSSM, and NSM, but rather in a thin region of width 10^{-9} m masking the crystal, where both phases coexists and there are periodically distributed solid inclusions (clusters) of size 10^{-10} m. This periodic microstructure is created by the crystal which, like a matrix, governs the arrangement of its own atoms from the melt into a specific crystalline lattice and creates periodically distributed clusters (solid inclusions). These clusters do not affect the melt flow but can influence the diffusion of impurities and the heat conductivity. In the isotropic case, for the impurities and heat transport, new diffusion and heat conductivity coefficients appear, and in the anisotropic case, new diffusion and heat conductivity tensors appear. These new coefficients and tensors depend on the spatial coordinates, on the geometry of clusters, and on the volumetric concentration of the clusters.

In MQSSM presented in [40.28], it is assumed that in the precrystallization zone the new diffusion coefficient decreases with respect to the spatial coordinate z (Fig. 40.1). According to [40.36, 37], the clusters do not influence the flow near the interface (i.e., the computed flow fields are identical to those obtained in QSSM model), but have a very small influence (of order 10^{-4}) on the heat transport. For example, changing the Schmidt number in the impurity transport equation from QSSM, the effect of the precrystallization zone on the axial and radial impurity distribution in the neighborhood of the melt–crystal interface was computed for different volumetric concentration of the clusters (solid inclusions). Computations performed with this model show that the influence of the precrystallization zone on the repartition of impurities is relevant. The concentration of impurities in the precrystallization zone reaches values which can be as high as ten times those obtained with QSSM. Consequently, the axial impurity distribution is nonuniform. In MNSM presented in [40.34], it is assumed that the microstructure existing in the precrystallization zone does not influence the flow and the heat transport, but that the transport of impurities is governed by the equation

$$\Pi \frac{\partial c}{\partial t} + \boldsymbol{v}\nabla c = \frac{\text{Pr}}{\text{Sc}}(\boldsymbol{I} + \boldsymbol{\Phi})\nabla^2 c, \quad (40.42)$$

where Π is porosity, $\boldsymbol{I}$ is the identity tensor, and $\boldsymbol{\Phi}$ is a tensor defined by

$$\boldsymbol{\Phi}_{kl} = -\varphi\left(\delta_{kl} - \frac{1}{|V|}\int_{\partial V} \gamma_k n_l \, dS\right). \quad (40.43)$$

In (40.43), φ represents the volumetric concentration of the clusters, $|V|$ represents the volume of a cluster, γ_k is the solution of the boundary-value problem, δ_{kl} is the Kronecker symbol, and dS represents the elementary surface [40.36]. In particular, if the clusters are spheres, the new diffusion tensor $D(\boldsymbol{I} + \boldsymbol{\Phi})$ is diagonal, and the new diffusion coefficients are given by the Maxwell–Jeffries formula [40.38]

$$\frac{D_{11}}{D} = \frac{D_{22}}{D} = \frac{D_{33}}{D} = 1 - \left(\tfrac{3}{2}\right)\varphi. \quad (40.44)$$

in these conditions, (40.42) becomes

$$\Pi \frac{\partial c}{\partial t} + \boldsymbol{v}\nabla c = \frac{\text{Pr}}{\text{Sc}}\left(1 - \tfrac{3}{2}\varphi\right)\nabla^2 c. \quad (40.45)$$

Therefore, the dispersion of impurities in the precrystallization zone is governed by (40.45), and by (40.34) in the rest of the melt.

Numerical computations of the impurities concentration have been made for the melt and crystal with thermophysical properties similar to gallium-doped germanium [40.34]: $Ra_T = 10^3$; $\varphi = 0.00985, 0.0317, 0.0491, 0.055$; $Sc = 10$; $k = 0.1$; and replacing the true unsteady process by the quasi-steady-state process. Computations show that the impurity concentration in the crystal is larger than that obtained with QSSM. The axial impurity distribution is uniform, and the radial segregation of the impurities in the crystal, computed with MQSSM, is less than that computed with QSSM.

In [40.33], it was shown rigorously (from a mathematical point of view) that the assumption of the existence of a periodic structure in the precrystallization zone permits us to deduce MNSM starting from NSM in the framework of continuum mechanics. The obtained MNSM includes the effect of the precrystallization zone. This effect gives a decrease of the molecular diffusivity D of the impurities from D to $D_{\text{eff}} = 0.917D$, and a increase of the heat conductivity K to $K_{\text{eff}} = 1.1K$ in the precrystallization zone. Numerical computations made with MNSM using the quasi-steady-state approximation [40.33, 35] show that, due to the decrease of the molecular diffusivity D, the impurity concentration in the crystal is five times larger than that computed with QSSM. Computations made in [40.31] using MNSM and those presented for a low-gravity environment ($Ra_T = 10^3$) in [40.33] show that the impurity concentration on the axis and on the walls of the crystal increases faster than that computed with NSM. The ratio of the impurity concentrations computed with MNSM and NSM at $\xi = 0.5$ (on the axis and on the walls of the crystal) is 1.45. According to [40.39], this result is in agreement with the measurements of the samples grown on the Mir space station.

Finally, in [40.40] it is shown that, if the initial impurities distribution in the melt is not uniform like in [40.31], and it is given by the formula

$$c_0(r,z) = \frac{1}{\alpha(a)}\left\{1 + \frac{1-k}{k}\frac{\Lambda - r}{\Lambda}\right. \\ \left. \times \exp\left[-\frac{v_p Sc}{Pr}(1-z)\right] - \frac{ar}{\Lambda}\right\}, \quad (40.46)$$

then the computed impurity distribution in the crystal (in the framework MNSM) on the solidified fraction $\xi \in (0; 0.7)$ is a constant equal to the prescribed concentration $c_c = 1$.

40.1.6 Larson–Zhang–Zheng Thermal-Diffusion Model (TDM)

Doped crystal grown in space in a BS system reported in [40.41, 42] exhibits significant axial compositional nonuniformities. Due to the fact that in these cases the thermosolutal convection can be neglected, the thermodiffusion of impurities is responsible for these nonuniformities. A one-dimensional mathematical model [40.43] has been developed to explain the impurity distribution in the sample presented in [40.41]. According to [40.43], for ampoules with small-diameter cross section and for a small impurity concentration, in a strictly zero-gravity environment, the mathematical model of the heat and impurities dispersion in the melt in a BS system is defined by the following dimensionless equations, boundary conditions, and initial conditions as written in a reference frame with the origin fixed at the interface

$$\frac{\partial \theta}{\partial t} = v_p \frac{\partial \theta}{\partial z} + \frac{1}{Pr}\frac{\partial^2 \theta}{\partial z^2} - \frac{2}{Pr}\frac{K_a}{K}\frac{L}{\delta}\frac{L}{R}(\theta - \theta_f),$$

for $0 < t < \frac{1}{v_p}$ and $0 < z < 1 - v_p t$, (40.47)

$$\frac{\partial c}{\partial t} = v_p \frac{\partial c}{\partial z} + \frac{1}{Sc}\frac{\partial^2 c}{\partial z^2} + \frac{1}{Sc}\frac{\partial}{\partial z}\left(\sigma \Delta T c \frac{\partial \theta}{\partial z}\right),$$

for $0 < t < \frac{1}{v_p}$ and $0 < z < 1 - v_p t$, (40.48)

$$\theta(t,z) = 0 \quad \text{for} \quad 0 < t < \frac{1}{v_p} \quad \text{and} \quad z = 0, \quad (40.49)$$

$$\frac{\partial \theta}{\partial z} = 0 \quad \text{for} \quad 0 < t < \frac{1}{v_p} \quad \text{and} \quad z = 1 - v_p t, \quad (40.50)$$

$$\frac{\partial c}{\partial z} + v_p Sc(1-k)c + \sigma c \Delta T \frac{\partial \theta}{\partial z} = 0,$$

for $0 < t < \frac{1}{v_p}$ and $z = 0$, (40.51)

$$\frac{\partial c}{\partial z} = 0 \quad \text{for} \quad 0 < t < \frac{1}{v_p} \quad \text{and} \quad z = 1 - v_p t, \quad (40.52)$$

$$\theta(0,z) = \begin{cases} 1 & \text{for } \frac{L_g}{2L} \leq z \leq 1 \\ \frac{2L}{L_g}z & \text{for } 0 \leq z \leq \frac{L_g}{2L} \end{cases}, \quad (40.53)$$

$$c(0,z) = c_0(z). \quad (40.54)$$

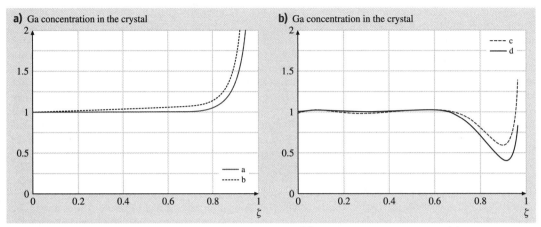

Fig. 40.3a,b Computed Ga concentration in the crystal versus ξ: (a) for the nonuniform case and (b) for the modified initial dopant repartition (after [40.44])

Equation (40.47) represents the dimensionless heat transport equation in the melt, which takes into account the heat dispersion due to the growth rate v_p, the thermal diffusivity in the melt $\lambda_L = \nu_L/\mathrm{Pr}$, and the heat loss to the environment from the sample surface. Equation (40.48) represents the dimensionless impurities transport equation in the melt which takes into account the impurity dispersion due to the growth rate v_p, the molecular diffusion $D = \nu/\mathrm{Sc}$, and the thermal diffusion σ (Soret effect). The boundary condition (40.49) expresses the melt–crystal interface as fixed at $z = 0$ and the dimensionless temperature at the interface equal to zero. In general, the shape of the melt–crystal interface is set by the conditions for the equilibrium melting temperature and the balance of conductive heat fluxes across the interface with the latent heat released there; for strictly zero gravity ($\mathrm{Ra}_T = 0$) [40.7], computations show that the interface is flat and fixed. The boundary condition (40.50) represents the no-flux condition for the heat at the top of the ampoule, and the boundary condition (40.51) gives the dopant concentration near the melt–crystal interface on the melt side. The boundary condition (40.52) is for no flux of the impurities on the top of the ampoule. The initial condition (40.53) is the temperature distribution in the melt at the time when solidification begins, and the initial condition (40.54) represents the impurity repartition in the melt when solidification begins.

The solidified fraction ξ at time t is given by

$$\xi = v_p t, \qquad (40.55)$$

and the impurity concentration in the crystal is given by

$$c_c(\xi) = kc\left(\frac{\xi}{v_p}, 0\right) \quad \text{for} \quad 0 \leq \xi \leq 1. \qquad (40.56)$$

Numerical results obtained with TDM using a second-order central difference scheme [40.43] are in agreement with the experimental results reported in [40.41]. The same numerical method with TDM gives numerical results which are qualitatively in agreement with the results reported in [40.42], as shown in [40.45]. Moreover, in [40.45] it is shown that, if the initial impurity distribution in the melt is not uniform (Fig. 40.3a) like in [40.42], but if it is given by the formula

$$c_0(x) = \frac{v_p \mathrm{Sc} L_g}{v_p \mathrm{Sc} L_g + \sigma \Delta T 2L} \left\{1 + a_1 x + a_2 x^2\right.$$
$$+ \left(\frac{1-k}{k} + \frac{\sigma \Delta T 2L}{k v_p \mathrm{Sc} L_g}\right)$$
$$\left. \times \exp\left[-\left(v_p \mathrm{Sc} + \frac{\sigma \Delta T 2L}{L_g}\right) z\right]\right\}, \qquad (40.57)$$

then the computed impurity concentration in the crystal on the solidified fraction $\xi \in (0; 0.7)$ is a constant equal to the prescribed concentration $c_c = 1$ (Fig. 40.3b).

40.2 Model-Based Prediction of the Impurity Distribution – EFG System

The prototype EFG system used in this work is presented in Fig. 40.4.

The central component of the system is a die which controls the heat and mass transfer from the crucible to the meniscus, i.e., the liquid bridge retained between the die and the crystal. Processes that take place in the melt (more precisely, in the meniscus melt) near the melt–crystal interface determine the impurity distribution in the crystal. During the past three decades, several mathematical models have been reported for predicting the impurity distribution in a crystal grown in such a system. Some of these models and the predictions obtained are presented below.

40.2.1 The Uniform-Diffusion-Layer Model (UDLM)

The UDLM presented for a BS system was extended for an EFG system considering that the effective partition coefficient k_{eff} is given by Fleming's formula [40.46]. If it is assumed that the uniform diffusion layer thickness is equal to the total height of the meniscus and the die has a long capillary channel, then the effective partition coefficient k_{eff} is equal to unity [40.47–53] for any impurity. Hence, according to UDLM, the impurity distribution in the crystal is uniform (i.e., no variations of the impurity concentration along the crystal length and cross section), and the value of the impurity concentration in the crystal is given by

$$c = k_{\text{eff}} c_\infty = c_\infty \,. \tag{40.58}$$

This prediction implies that the growth process in an EFG system has a low tolerance for rejected melt impurities, which definitively shows that UDLM is an oversimplified model. Impurities are generally trapped in the boundary layer within the meniscus melt. Interface-region convection in EFG can be manipulated through design of die capillaries to redistribute segregated impurities [40.50–53].

40.2.2 Tatarchenko Steady-State Model (TSSM)

TSSM is a model of impurity transport in the meniscus, which relates the values of k_{eff} to the parameters of capillary shaping and feeding, is presented in a book by Tatarchenko [40.46]. The stationary process is considered, and it is assumed that the meniscus and capillary melt have not been stirred, where the conditions of complete stirring are maintained in the crucible. Impurity transport in the meniscus is described by the equation

$$D\left(\frac{d^2 c}{dr^2} + \frac{1}{r}\frac{dc}{dr}\right) = -v\frac{dc}{dr} \tag{40.59}$$

with the following boundary condition at the melt–crystal interface

$$-D\left.\frac{dc}{dr}\right|_{r=r_0} = v_0(1-k)c(r_0)\,. \tag{40.60}$$

Here, c represents the impurity concentration in the melt, D is the diffusion coefficient, k is the partition

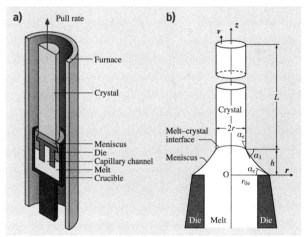

Fig. 40.4a,b Prototype EFG crystal growth system. (a) Prototype EFG system, (b) schematic EFG used in mathematical model

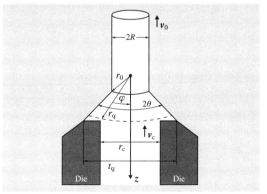

Fig. 40.5 EFG diagram for TSSM

coefficient at equilibrium, and the other notations are given in Fig. 40.5.

The melt flow rate in the meniscus is given by
$$v(r) = \frac{v_0 r_0}{r}. \quad (40.61)$$

Impurity transport in the capillary channel is described by the equation
$$D \frac{d^2 c}{dz^2} = -v_c \frac{dc}{dz} \quad (40.62)$$

with the following boundary condition
$$c \to c_\infty \quad \text{if} \quad z \to \infty. \quad (40.63)$$

Employment of these equations and boundary conditions gives the following expression describing the impurity concentration distribution in the meniscus
$$c(r) = (1 - k_{\text{eff}}) c (r_0) \left(\frac{r_0}{r}\right)^e$$
$$+ \frac{c_\infty - (1-k)\left(\frac{r_0}{r_\Phi}\right)^e (\cos\varphi - 1) c(r_0)}{1 + \left[\left(\frac{v_0 \theta r_0}{v_c \varphi r_\Phi}\right) - 1\right] \cos\varphi}, \quad (40.64)$$

where
$$\cos\varphi = \left(1 - \frac{r_c^2}{4r_\Phi^2}\right)^{1/2} \quad \text{and}$$
$$e = \frac{v_0 R}{D \sin\theta}, \quad \varphi \in [0, \Phi]. \quad (40.65)$$

Taking into account the equalities $k = c_S/c(r_0)$, $k_{\text{eff}} = c_S/c_\infty$, from (40.64) for $r = r_0$ it follows that
$$k_{\text{eff}}^{-1} = 1 + \left(\frac{v_0 \theta r_0}{v_c \varphi r_\Phi} - 1\right) \cos\varphi$$
$$+ \frac{(1-k)(1 - \cos\varphi)(r_0/r_\Phi)^e}{k}. \quad (40.66)$$

Using the relations
$$2v_0 R = v_c r_c, \quad \frac{r_0}{r_\Phi} = 2Rt_\varphi,$$
$$r_0 = \frac{R}{\sin\theta}, \quad (40.67)$$

(40.66) becomes
$$k_{\text{eff}}^{-1} = 1 + \left(\frac{v_0 r_c}{\varphi t_\varphi} - 1\right) \cos\varphi$$
$$+ \frac{(1-k)(1 - \cos\varphi)(2R/t_\varphi)^{v_0 R/D \sin\theta}}{k}. \quad (40.68)$$

Equation (40.68) relates the value of the effective coefficient of partition k_{eff} with the process parameters: v_0, $2R$, t_Φ, r_c, θ, D, and k.

When the capillary channel radius its maximum value, $r_c = t_\Phi$, (40.68) becomes
$$k_{\text{eff}} = \frac{k}{k + (1-k)(1 - \cos\varphi)\left(\frac{2R}{t_\Phi}\right)^{\frac{v_0 R}{D \sin\theta}}}. \quad (40.69)$$

According to TSSM, the impurity distribution in the crystal is uniform (i.e., no variations of the impurity concentration along the crystal length and cross section), and the value of the impurity concentration in the crystal is given by
$$c_c = k_{\text{eff}} c_\infty, \quad (40.70)$$

where k_{eff} is given by (40.68).

In order to verify the above prediction for Al impurities in Si ($k = 0.002$, $D = 5.3 \times 10^{-4}$ cm^2/s, $2R = 0.03$ cm, $t_\Phi = 0.06$ cm), the effective partition coefficient k_{eff} was computed. It was found that $k_{\text{eff}} \ll 1$, where k_{eff} increases as both the growth rate and the melt flow in the capillary channel increase [40.46]. Mass-spectrographic analysis of the impurity content in tapes grown from refined metallurgical silicon confirms that $k_{\text{eff}} \ll 1$. The sulfur distribution along the axis of a silicon tape obtained by laser emission analysis shows also that $k_{\text{eff}} \ll 1$, in agreement with the value of k_{eff} given by (40.68). According to [40.46], in order to verify the TSSM-based prediction, a series of 0.01 mass indium-doped silicon tapes was grown, where the predicted values of k_{eff} are in agreement with those obtained experimentally. The values of k_{eff} given by Fleming's formula are several orders of magnitude greater than the experimental values. For the variation of the impurity concentration observed experimentally along the crystal cross section in [40.46], the following explanation was given: For the case of a horizontal melt flow in the meniscus from the die edge towards the center, impurities with $k_{\text{eff}} < 1$ are driven off by the growing crystal and accumulate in the central part of the meniscus. Impurities may be moved away in order to improve the quality of the crystal. For this aim, the rejected impurities were studied using an axisymmetric or *displaced* die shape in an EFG system with melt replenishment [40.54, 55].

40.2.3 Melt Replenishment Model (MRM)

Recently, the influence of the die geometry and various growth conditions on the fluid flow and on the solute distribution has been studied in an EFG system with melt replenishment (MR) using numerical simulation [40.56]. The nonuniform distribution of the voids (trapped gas bubbles) along the cross section of the sapphire rod was analyzed, assuming that the voids are impurities which are present in the melt and that their transport in the capillary channel and in the meniscus is governed by the stationary convection–diffusion equation. It was also assumed that the mechanism of transfer of the voids from the melt into the crystal is similar to the transfer of impurities. Two types of die designs were investigated: one with a central capillary channel (CCC) and another with an annular capillary channel (ACC). The fluid flow is characterized by a small Reynolds number (due to the small value of the fluid velocity and of the pulling rate) and consequently the fluid flow is laminar. Computations prove that, for both die geometries, the flow field near the meniscus free surface exhibits the same aspect, with only the velocity field values being different: for the ACC die, all the velocity values are higher than for the CCC die. For small Marangoni numbers (Ma), no loop or cell structure appears in the melt flow. As the Ma number increases, the loop position slightly migrates toward the crystallization interface. The isoconcentration field and solute concentration near the solidification interface are different for the two studied shaper designs. The solute fields for the CCC and ACC dies differ in aspect and concentration values: the concentration field for CCC die shows a single maximum close to the free meniscus surface; for the ACC die, a strong maximum that is two times higher appears near the symmetry axis. Moreover, the radial segregation is higher for the ACC die than for the CCC die and increases if the Ma number increases. These numerical results with good agreement with the experiments show that the gas contained in the melt acts as a solute rejected at the interface and nucleates into bubbles at the more concentrated locations. The experimental fact that the outer bubbles are situated at a distance from the external surface of the order of the meniscus height demonstrates that Marangoni convection exists in a shaped sapphire meniscus. Finally, in order to reduce the concentration of voids, a CCC die is recommended as well as growth with low meniscus height and low pulling rate.

For optimization and control of the growth process, in order to concentrate the voids close to the crystal surface which can be removed by polishing, the dependences of the fluid flow and impurity distribution on some process parameters (e.g., the pulling rate, radius of the capillary channel, natural convection, Marangoni convection, and vertical temperature gradient) have been studied [40.44, 57–60]. First, the dependences of the melt flow and impurity distribution on the pulling rate and on the capillary channel radius were studied; the temperature variations and their consequences were completely neglected (no Marangoni convection) [40.57].

The equations which define the model are the incompressible Navier–Stokes equations and the conservative convection–diffusion equation

$$\rho_l \frac{\partial \boldsymbol{u}}{\partial t} - \eta \nabla^2 \boldsymbol{u} + \rho_l (\boldsymbol{u}\nabla)\boldsymbol{u} + \nabla p = \boldsymbol{F}\,, \qquad (40.71)$$

$$\nabla \boldsymbol{u} = 0\,, \qquad (40.72)$$

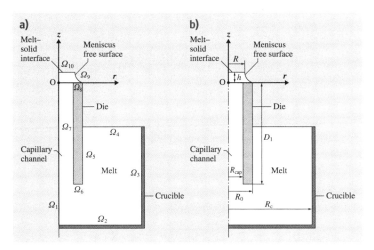

Fig. 40.6a,b EFG diagram in the case of MRM

$$\frac{\partial c}{\partial t} + \nabla(-D\nabla c + c\boldsymbol{u}) = 0, \qquad (40.73)$$

written in a cylindrical polar coordinate system as shown in Fig. 40.4. In the above equations, $\boldsymbol{u} = (u_r, u_z)$ is the velocity vector, c is the impurity concentration; $\boldsymbol{F}$ is the volume force field due to gravity, which for zero gravity is $\boldsymbol{F} = \boldsymbol{0}$ and for terrestrial conditions is $\boldsymbol{F} = \rho_l \boldsymbol{g}$ when: ρ_l is the melt density; $\boldsymbol{g}$ is the gravitational acceleration; p is the pressure; η is the dynamical viscosity; t is time; and D is the impurity diffusion. Axisymmetric solutions are sought in a two-dimensional domain with boundaries $\Omega_1 - \Omega_{10}$, as shown in Fig. 40.6.

On the boundaries $\Omega_1 - \Omega_{10}$, the following conditions are considered:

1. For the incompressible Navier–Stokes equations:

- Axial symmetry on Ω_1

$$u_r = 0. \qquad (40.74)$$

- Given inflow velocity on Ω_4

$$\boldsymbol{u} = -\frac{\rho_s}{\rho_l} \frac{R^2}{R_c^2 - R_0^2} v \boldsymbol{k}, \qquad (40.75)$$

where $\boldsymbol{k}$ represents the unit vector of the Oz-axis; this condition expresses continuous melt replenishment, i.e., the melt level in the crucible is maintained constant.

- Slip condition on Ω_9

$$\boldsymbol{u}\boldsymbol{n} = 0, \qquad (40.76)$$

where $\boldsymbol{n}$ is the unit normal vector of Ω_9.

- Given outflow velocity on Ω_{10}

$$\boldsymbol{u} = v\boldsymbol{k}. \qquad (40.77)$$

- Nonslip condition on $\Omega_2 \cup \Omega_3 \cup \Omega_5 \cup \Omega_6 \cup \Omega_7 \cup \Omega_8$

$$\boldsymbol{u} = \boldsymbol{0}. \qquad (40.78)$$

2. For the conservative convection–diffusion equation

- Axial symmetry on Ω_1

$$\frac{\partial c}{\partial r} = 0. \qquad (40.79)$$

- Impurity flux rejected into the melt on Ω_{10}

$$\frac{\partial c}{\partial \boldsymbol{n}} = -\frac{v}{D}(1 - K_0) c. \qquad (40.80)$$

- No impurity flux, i.e., insulation on $\Omega_2 \cup \Omega_3 \cup \Omega_5 \cup \Omega_6 \cup \Omega_7 \cup \Omega_8 \cup \Omega_9$

$$\frac{\partial c}{\partial \boldsymbol{n}} = \boldsymbol{n}\nabla c = 0. \qquad (40.81)$$

- Given concentration on Ω_4

$$c = C_0. \qquad (40.82)$$

The system (40.71–40.73) has a unique stationary solution, i.e., $\partial \boldsymbol{u}/\partial t = 0$ and $\partial c/\partial t = 0$, which satisfies the above boundary conditions. This describes the steady-state process.

In order to identify a nonstationary solution of the system (40.71–40.73) which describes a transient process, beside the above boundary conditions, it is necessary to specify an initial melt flow velocity

$$u_r(t_0) = 0, \quad u_z(t_0) = 0, \qquad (40.83)$$

and the initial impurity distribution

$$c(t_0) = C_0. \qquad (40.84)$$

The significance of these quantities and their values for the considered EFG system are given in Table 40.1 for Al-doped Si.

According to MRM, the value of the impurity concentration in the crystal is given by

$$c_c = k c_{\text{interface}}. \qquad (40.85)$$

Using the right-hand side of (40.85) and the stationary solution of the boundary-value problem (40.71–40.82),

Table 40.1 Nomenclature

Nomenclature		Value
c	Impurity concentration (mol/m^3)	
C_0	Initial alloy concentration (mol/m^3)	
D	Impurity diffusion (m^2/s)	5.3×10^{-8}
D_l	Die length (m)	0.04
g	Gravitational acceleration (m/s^2)	9.81
h	Meniscus height (m)	0.5×10^{-3}
K_0	Partition coefficient	0.002
η	Dynamical viscosity (kg/(m s))	7×10^{-4}
p	Pressure (Pa)	0
$Pe_{\text{mass}} = \frac{v R_{\text{cap}}}{D}$	Péclet number for mass transfer	
R	Crystal radius (m)	1.5×10^{-3}
R_{cap}	Capillary channel radius (m)	
R_c	Inner radius of the crucible (m)	23×10^{-3}
R_0	Die radius (m)	2×10^{-3}
ρ_l	Density of the melt (kg/m^3)	2500
ρ_s	Density of the crystal (kg/m^3)	2300
$\boldsymbol{u}$	Velocity vector	
v	Pulling rate (m/s)	
z	Coordinate in the pulling direction	

the following prediction can be obtained: constant impurity concentration along the crystal length and variation of the impurity concentration along the crystal cross section. The use of the right-hand side of (40.85) for a nonstationary solution of the boundary-value problem (40.71–40.84) leads to the prediction of the variation of the impurity concentration along the crystal length and along the crystal cross section during a transition period. In [40.44, 57], computations were made using COMSOL Multiphysics 3.2 software for zero gravity and for terrestrial conditions in the case of the following EFG system: crucible inner radius of $R_c = 23 \times 10^{-3}$ m, die radius of $R_0 = 2 \times 10^{-3}$ m, and die length of 40×10^{-3} m; two-thirds of the die immersed in the crucible melt; constant crucible melt height of 40×10^{-3} m; capillary channel radius of $R_{cap} = (0.5, 1, 1.5) \times 10^{-3}$ m; and pulling rate v in the range 10^{-10} m/s $\leq v \leq 10^{-6}$ m/s. The meniscus height was $h = 0.5 \times 10^{-3}$ m, and the single-crystal rod radius (being grown) was $R = 1.5 \times 10^{-3}$ m.

The computations reveal that, in the stationary case, for a given R_{cap}, there exist critical pulling rate values v' depending on R_{cap} at which the impurity concentration has a maximum. If v increases in the range $[1 \times 10^{-10}, v']$, the impurity concentration increases, and if v increases in $[v', 1 \times 10^{-6}]$, the concentration decreases slowly. Moreover, the impurity concentration on the axis and on the lateral surface of the crystal, and the absolute value of the difference between them – called the radial segregation – depend on the pulling rate and the capillary radius (Fig. 40.7). The optimal value of R_{cap} for which the crystal has the best homogeneity (radial segregation is minimum) is $R_{cap} = R = 1.5 \times 10^{-3}$ m.

Numerical results concerning a nonsteady state show that the nonstationary melt flow field and the impurity concentration field in the whole melt tend to the stationary melt flow field and to the stationary impurity concentration, respectively. The transition period, defined as that period of time after which the impurity concentration at the interface becomes equal to those obtained in the stationary case, depends on R_{cap} and v.

The impurity concentrations in the meniscus at three different instances of time $0 < t_1 < t_2 < t_3$ are presented in Fig. 40.8 for $R_{cap} = 1.5 \times 10^{-3}$ m and $v = 10^{-6}$ m/s.

These computed results are in agreement with experimental data reported by Cao et al. [40.53]: each impurity has a transient distance or crystal length until the steady-state segregation is established. The localization of the maximum of the dopant concentration on the free meniscus surface shows that the model is incomplete. More precisely, natural and Marangoni convections should be involved in order to *move* the maximum of the dopant concentration near the melt–solid interface, as was observed experimentally [40.56].

In [40.58–60] the combined effect of the buoyancy and Marangoni forces on the fluid flow and on the dopant distribution is analyzed in the stationary case. Thus, the growth process is described by the stationary incompressible Navier–Stokes equations written in the

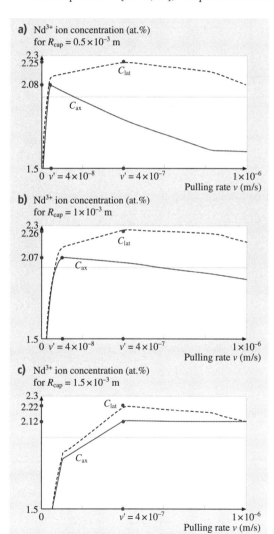

Fig. 40.7a–c Dependence of the Nd^{3+} ion concentration on the pulling rate v for (**a**) $R_{cap} = 0.5 \times 10^{-3}$ m, (**b**) $R_{cap} = 1 \times 10^{-3}$ m, and (**c**) $R_{cap} = 1.5 \times 10^{-3}$ m

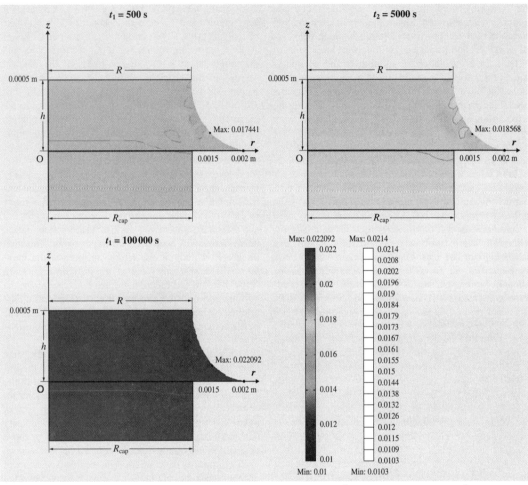

Fig. 40.8 Al concentration at three instances of time $t_1 < t_2 < t_3$, for $v = 10^{-6}$ m/s and $R_{cap} = 1.5 \times 10^{-3}$ m

Boussinesq approximation [i.e., $F = \beta \rho_l g (T - \Delta T)$, where β is heat expansion coefficient and $\Delta T = (T_0 + T_m)/2$ is the reference temperature at the free surface], and the convection–conduction and conservative convection–diffusion equations. For implementation of the Marangoni effect, the weak form of the boundary application mode is employed. The fluid density is assumed to vary with temperature as $\rho(r,z) = \rho_l [1 - \beta(T(r,z) - \Delta T)]$, and the surface tension γ in the meniscus is assumed to vary linearly with temperature as $\gamma(r,z) = \gamma_1 + (\mathrm{d}\gamma/\mathrm{d}T)(T(r,z) - \Delta T)$, where γ_1 is the surface tension at the temperature ΔT, and $\mathrm{d}\gamma/\mathrm{d}T$ is the rate of change of surface tension with the temperature.

For implementation of the Marangoni effect, the boundary condition

$$\eta \begin{pmatrix} t_r & t_z \end{pmatrix} \begin{pmatrix} 2\dfrac{\partial u}{\partial r} & \dfrac{\partial u}{\partial z} + \dfrac{\partial v}{\partial r} \\ \dfrac{\partial u}{\partial z} + \dfrac{\partial v}{\partial r} & 2\dfrac{\partial v}{\partial r} \end{pmatrix} \begin{pmatrix} n_r \\ n_z \end{pmatrix}$$
$$= \dfrac{\mathrm{d}\gamma}{\mathrm{d}T} \left(\dfrac{\partial T}{\partial r} + \dfrac{\partial T}{\partial z} \right), \quad (40.86)$$

is imposed on the free surface (meniscus). This expresses that the gradient velocity field along the meniscus is balanced by the shear stress, where $t = (t_r, t_z)$ and $n = (n_r, n_z)$. The sign of the rate $\mathrm{d}\gamma/\mathrm{d}T$, in general, depends on the material, with leading downward

($d\gamma/dT < 0$) or upward ($d\gamma/dT > 0$) flow on the free liquid surface.

The effect of the buoyancy and Marangoni forces on the fluid flow and on the impurity distribution were investigated numerically including both the cases of downward and upward flows on the liquid free surface (meniscus). Thus, two materials having negative and positive surface tension rates (Nd:Y$_3$Al$_5$O$_{12}$ (Nd:YAG) and Nd:LiNbO$_3$) were considered. The computed dopant distribution in the meniscus for different values of the Grashof and Marangoni numbers (Gr = 0.00022 and 0.02231, Ma = 0.049 and 4.913 for Nd:YAG, and Gr = 0.00252 and 0.25222, Ma = 0.671 and 67.072 for Nd:LiNbO$_3$), are illustrated in Figs. 40.9 and 40.10. Computations show that the dopant concentration increases from the initial value C_0 to a maximum value C_{max} situated at the level of the melt–crystal interface $z = h$, marked on Figs. 40.9 and 40.10. The maximum values of the Nd^{3+} ion concentration show that increases of the Marangoni and Grashof numbers lead to a decrease of C_{max} and can move it from the lateral surface of the crystal toward the axis. The effect of the Grashof number is smaller on the C_{max} localization and its values, and shows that the Marangoni convection has a dominant effect on the natural convection.

The buoyancy- and surface-driven flows perturb the forced flow: the arrows presented in Figs. 40.9 and 40.10 denote the flow of the velocity field caused by the surface-tension-driven flow, with downward flow on the meniscus surface for Nd:YAG ($d\gamma\, dT < 0$; Fig. 40.9), and upward flow on the meniscus surface for Nd:LiNbO$_3$ ($d\gamma/dT > 0$; Fig. 40.10). The maximum velocity of the fluid flow in the meniscus presents a linear dependence on the Marangoni number. More precisely, if the Marangoni number increases, then the maximum velocity of the fluid flow increases. Concerning the radial segregation, computations show that an increase of the Grashof or Marangoni number leads to a smaller radial segregation, clearly because of better mixing in these cases. The magnitude of the radial segregation is more sensitive to the Marangoni convection and less sensitive to the natural convection. The magnitude of the radial segregation changes are larger for larger Marangoni numbers.

Figure 40.10 shows a similar location of the maximum of the dopant distribution to those reported experimentally for sapphire [40.56]. The question here

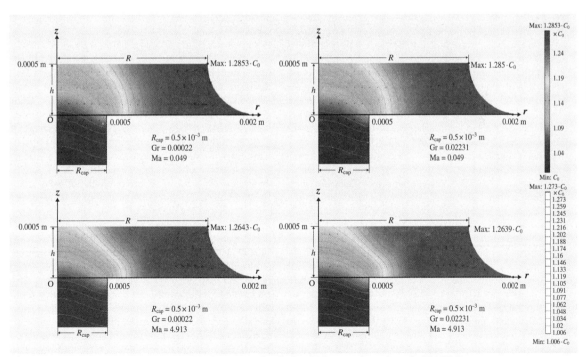

Fig. 40.9 Dependence of the dopant distribution on the Grashof and Marangoni numbers for Nd:YAG

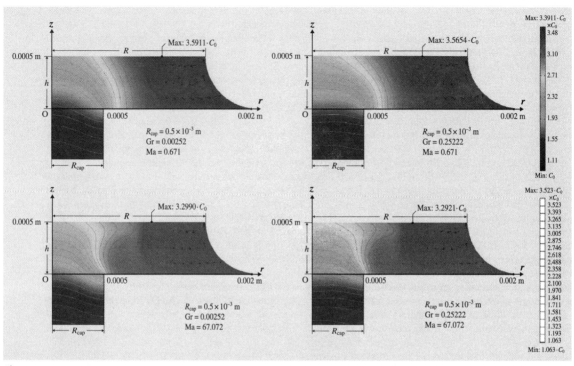

Fig. 40.10 Dependence of the dopant distribution on the Grashof and Marangoni numbers for Nd:LiNbO$_3$

is whether it is possible to determine conditions for which the maximum of the dopant distribution is located at the triple point, and at the distance from the lateral crystal surface equal to the meniscus height, respectively. There is also the question as to which case assures the best homogeneity of the crystal. The answers are found in the paper [40.60] in which the dependences of the Marangoni flow and impurity distribution on the vertical temperature gradient in sapphire (Al$_2$O$_3$) fibers are reported. Thus, for three representative values of the vertical temperature gradients k_g and for different Marangoni numbers Ma corresponding to the surface tension gradients $d\gamma/dT$ situated in the range $[-6 \times 10^{-5}, 0]$ N/(m K), the impurity distributions are computed for two cases: (i) the buoyancy is taken into account (the heat expansion coefficient has the value reported in the literature, i.e., $\beta = 3 \times 10^{-5}$ 1/K); and (ii) the buoyancy is neglected ($\beta = 0$), revealing the existence of three critical Marangoni numbers – Ma$_{c1}$, Ma$_{c2}$, Ma$_{c3}$ – as reported for two-dimensional containers, determined by the behavior of the fluid flow. For Ma in the range [0, Ma$_{c1}$], the downward flow ($d\gamma/dT < 0$) on the free liquid sur-

face leads to a steady flow, with a decrease of the maximum of the dopant concentration C_{max} located at the triple point if Ma increases. For Ma in the range [Ma$_{c1}$, Ma$_{c2}$], turbulence in the fluid flow appears, leading to an increase of C_{max}, still located at the triple point. For Ma in the range [Ma$_{c2}$, Ma$_{c3}$], this increased C_{max} is shifted inside the crystal from the melt–crystal interface at a distance on the same order as the meniscus height from the external crystal surface. If Ma is larger than Ma$_{c3}$, C_{max} is shifted into the center of the crystal.

We thus conclude that the best homogeneity of the crystal is assured in the range [0, Ma$_{c1}$] in which steady flow takes place (the maximum of the dopant distribution is located at the triple point). In this range, voids are situated at the lateral crystal surface and hence can be removed by polishing with minimum material loss. This range becomes larger if the vertical temperature gradient k_g and heat flux coefficient β decrease, which proves that a smaller vertical temperature gradient and a lower gravity, e.g., zero gravity, assures the best homogeneity of the crystal over a wide range of surface tension gradients. This suggests to practical crystal growers that a possible feedback control for delaying the Marangoni convection can be obtained by decreasing the vertical temperature gradient in the furnace.

40.2.4 Melt Without Replenishment Model (MWRM)

In [40.61], a MWRM model in which the melt flow and impurity distribution are described in an EFG system without melt replenishment (the melt level in the crucible decreases during the growth process) is performed. It is assumed that temperature variations and their consequences in the system (no Marangoni convection) are negligible. The fluid flow and impurity distribution in the crucible, in the capillary channel, and in the meniscus are considered in a time-dependent domain $\Omega(t)$, $t \in [0, T]$ by the incompressible Navier–Stokes equations in terrestrial conditions (40.71–40.72) and the conservative convection–diffusion (40.73). For these equations, axisymmetric solutions are sought in the cylindrical polar coordinate system (rOz) (Fig. 40.11).

The evolution of $\Omega(t) = \{(r(R, Z, t), z(R, Z, t)): (R, Z) \in \Omega(0)\}$ is described by the system of partial differential equations corresponding to the Laplace smoothing (Poisson equations)

$$\begin{cases} \dfrac{\partial^2}{\partial R^2}\left(\dfrac{\partial r}{\partial t}\right) + \dfrac{\partial^2}{\partial Z^2}\left(\dfrac{\partial r}{\partial t}\right) = 0, \\ \dfrac{\partial^2}{\partial R^2}\left(\dfrac{\partial z}{\partial t}\right) + \dfrac{\partial^2}{\partial Z^2}\left(\dfrac{\partial z}{\partial t}\right) = 0. \end{cases} \quad (40.87)$$

Here, R, Z represent the reference coordinates in the reference frame (ROZ), i.e., the fixed frame used for the description of $\Omega(t)$ and of the mesh velocity, as depicted in Fig. 40.11a,b. The solution $(r(R, Z, t), z(R, Z, t))$ satisfies the condition $(r(R, Z, 0), z(R, Z, 0)) = (R, Z)$.

The coupled system (40.71–40.73, 40.87) is considered in the two-dimensional domain $\Omega(t)$ with boundaries $\Omega_1 - \Omega_{12}$, and for solving it, the arbitrary Lagrangian Eulerian (ALE) technique has been used. This technique for mesh movement [40.62, 63] is an intermediate between the Lagrangian and the Eulerian methods, combining the best features of both, i.e., it allows moving boundaries without the need for mesh movement to follow the material. The moving-mesh ALE application mode solves the system of partial differential equations (PDEs) (40.87) for the mesh displacement. This system smoothly deforms the mesh given by constraints on the boundaries. By the Laplace smoothing option (which has been chosen), the software introduces deformed mesh positions as degrees of freedom in the model.

In order to solve the system (40.71–40.73, 40.87), the following boundary conditions on the boundaries $\Omega_1 - \Omega_{12}$ are considered:

1. For the incompressible Navier–Stokes (NS) equations:
   – Axial symmetry on Ω_1

   $$u_r = 0. \quad (40.88)$$

   – Given outflow velocity on Ω_3

   $$\boldsymbol{u} = v\boldsymbol{k}. \quad (40.89)$$

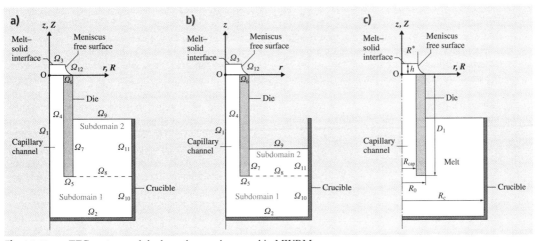

Fig. 40.11a–c EFG system and the boundary regions used in MWRM

- Nonslip condition on
  $\Omega_2 U \Omega_4 U \Omega_5 U \Omega_6 U \Omega_7 U \Omega_{10} U \Omega_{11} U \Omega_{12}$
  $$u = 0. \tag{40.90}$$
- Neutral condition on $\Omega_8 U \Omega_9$
  $$[-p\mathbf{I} + \eta(\nabla u + (\nabla u)^\top)]n = 0. \tag{40.91}$$

2. For the moving-mesh ALE:
   The domain $\Omega(t)$ is divided in two subdomains, 1 and 2 (Fig. 40.11). Within subdomain 1 there is no displacement (i.e., no motion in this subdomain), whereas within subdomain 2 there are free displacements (i.e., motion is free in this subdomain). Hence, the mesh displacement takes place only in subdomain 2 and it is constrained by the boundary conditions on the surrounding boundaries of Ω_7, Ω_8, Ω_9, and Ω_{11}. The displacement in subdomain 2 is obtained by solving the PDEs (40.87). The boundary conditions involve variables from the NS application mode. For convergence, it is important for the boundary conditions to be consistent. The usual pointwise constraints or ideal constraints for ALE cause unwanted modifications of the boundary conditions for the NS mode coupled with convection diffusion (CD). For this reason, nonideal weak constraints on the boundaries are used in the ALE application mode:
   - The mesh displacements in the r- and z-directions on Ω_9 are the following
     $$dr = 0, \quad dz = v_n t, \tag{40.92}$$
     where $v_n = -(\rho_S/\rho_l)R^{*2}/(R_c^2 - R_0^2)v$, and R^* is the crystal radius. (According to [40.62], the mesh velocity should be equal to the fluid velocity.)
   - The mesh displacement in the r-direction on $\Omega_7 U \Omega_{11}$ is $dr = 0$; the mesh displacement dz in the z-direction is not specified, i.e., the mesh follows the fluid movement;
   - The mesh displacements dr and dz in the r- and z-direction are not specified on Ω_8 (the mesh follows the fluid flow).

3. For the conservative convection–diffusion equation:
   - Axial symmetry on Ω_1, i.e., $r = 0$.
   - Impurity flux rejected into the melt on Ω_3
     $$\frac{\partial c}{\partial n} = -\frac{v}{D}(1 - K_0)c. \tag{40.93}$$
   - No impurity flux, i.e., insulation on
     $\Omega_2 U \Omega_4 U \Omega_5 U \Omega_6 U \Omega_7 U \Omega_9 U \Omega_{10} U \Omega_{11} U \Omega_{12}$
     $$\frac{\partial c}{\partial n} = n\nabla c = 0. \tag{40.94}$$

- Continuity on Ω_8
  $$n(N_1 - N_2) = 0, \quad N_i = -D_i \nabla c_i + c_i u_i, \tag{40.95}$$
  where $i = 1$ for subdomain 1 and $i = 2$ for subdomain 2. Besides the above boundary conditions, the followings initial conditions are also used:
- For the fluid flow
  $$u(t_0) = 0, \quad v(t_0) = 0 \tag{40.96}$$
  in subdomain 1, and
  $$u(t_0) = 0, \quad v(t_0) = v_n \tag{40.97}$$
  in subdomain 2 (according to [40.62] the fluid velocity should be equal to the mesh velocity).
- For the pressure
  $$p(t_0) = P_0 = -\rho_l g z \tag{40.98}$$
  in subdomain 1, and
  $$p(t_0) = 0 \tag{40.99}$$
  in subdomain 2.
- For the initial impurity distribution
  $$c(t_0) = C_0 \tag{40.100}$$
  in both subdomains.
- For the mesh displacement
  $$r(t_0) = R, \quad z(t_0) = Z. \tag{40.101}$$

Numerical investigations were carried out for an Al-doped Si rod of radius $R^* = 1.5 \times 10^{-3}$ m, grown in terrestrial conditions with a pulling rate $v = 10^{-6}$ m/s. According to [40.57], for $R_{cap} = R^* = 1.5 \times 10^{-3}$ m, the radial segregation in the Si rod grown in terrestrial conditions is minimal over the range $[10^{-7}, 5 \times 10^{-6}]$ m/s of v. In order to evaluate the way in which the resulting fluid flow and deformed geometry determine the impurity distribution in the melt and in the crystal, COMSOL Multiphysics 3.2 software was used in order to solve the coupled NS-ALE-CD application modes in the ALE frame (rOz). COMSOL Multiphysics does the mathematics necessary to manipulate, move, and deform the mesh simultaneously with the boundary movement as required by the other coupled NS-CD PDEs. The computed impurity distributions at three different instances of time are presented in Fig. 40.12.

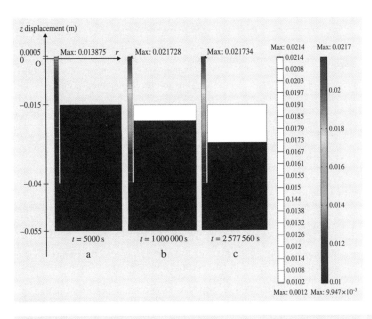

Fig. 40.12 Impurity distribution in the melt at three different instances of time $0 < t_1 < t_2 < t_3 < T$ for $R_{cap} = 1.5 \times 10^{-3}$ m and $v = 10^{-6}$ m/s for the case of no melt replenishment

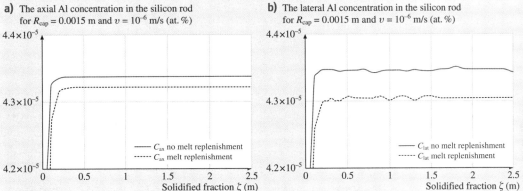

Fig. 40.13a,b Dependence of the impurity distributions on the axis (**a**) and on the lateral surface (**b**) of the crystal versus solidified fraction ξ for $R_{cap} = 1.5 \times 10^{-3}$ m and $v = 10^{-6}$ m/s, with and without melt replenishment

Figure 40.12a–c shows that, at the beginning, the concentration increases and there exists a certain time after which the impurity concentration becomes constant.

The computed impurity distributions versus solidified fraction along the crystal axis (C_{ax}) obtained with and without melt replenishment are presented in Fig. 40.13a. Similarly, the computed impurity distributions versus solidified fraction at the lateral surface (C_{lat}) of the crystal are presented in Fig. 40.13b.

Figure 40.13 shows that, after a transition period, the impurity distributions along the crystal axis (C_{ax}) and on the lateral surface (C_{lat}), computed in MRM and MWRM, become constant. In the MWRM case these constants are larger than those obtained in the MRM case: the difference is on the order of 10^{-7}–10^{-6} in the transition period, and after that, this difference is on the order of 1×10^{-7} on the crystal axis and 3×10^{-7} on the lateral surface. Moreover, Fig. 40.13 shows that the transition periods computed in MWRM are shorter. In MWRM the transition period for C_{ax} and C_{lat} is equal to 200 000 s. In MRM the transition period for C_{ax} is 300 000 s and for C_{lat} is 500 000 s.

References

40.1 J.A. Burton, R.C. Prim, W.P. Slichter: The distribution of solute on crystals grown from the melt, J. Chem. Phys. **21**, 1987–1996 (1953)

40.2 C. Wagner: Theoretical analysis of diffusion of solutes during the solidification of alloys, J. Met. **6**, 154–160 (1954)

40.3 K.M. Kim, A.F. Witt, M. Lichtensteiger, H.C. Gatos: Quantitative analysis of the thermo-hydrodynamic effects on crystal growth and segregation under destabilizing vertical thermal gradients: Ga-doped germanium, J. Electrochem. Soc. **125**, 475–480 (1974)

40.4 J.J. Favier: Macrosegregation-I: Unified analysis during non-steady state. Solidification, Acta Metall. **29**, 197–204 (1981)

40.5 J.J. Favier: Macrosegregation-II: A comparative study of theories, Acta Metall. **29**, 205–214 (1981)

40.6 D.E. Holmes, H.C. Gatos: Convective interference and "effective" diffusion-controlled segregation during directional solidification under stabilizing vertical thermal gradients; Ge, J. Electrochem. Soc. **128**, 429–437 (1981)

40.7 C.J. Chang, R.A. Brown: Radial segregation induced by natural convection and melt/solid interface shape in vertical Bridgman growth, J. Cryst. Growth **63**, 343–364 (1983)

40.8 W.A. Tiller, K.A. Jackson, J.W. Rutter, B. Chalmers: The redistribution of solute atoms during the solidification of metals, Acta Metall. **1**, 428–437 (1953)

40.9 P.R. Griffin, S. Motakef: Influence of non-steady gravity on natural convection during micro-gravity solidification of semiconductors. Part I. Time Scale Analysis, J. Appl. Microgravity II **3**, 121–127 (1989)

40.10 P.R. Griffin, S. Motakef: Influence of non-steady gravity on natural convection during micro-gravity solidification of semiconductors. Part II. Implications for crystal growth experiments, J. Appl. Microgravity II **3**, 128–132 (1989)

40.11 A.F. Witt, H.C. Gatos, M. Lichtensteiger, M.C. Lavine, C.J. Herman: Crystal growth and steady-state segregation under zero gravity, J. Electrochem. Soc. **122**, 276–283 (1975)

40.12 P.M. Adornato, R.A. Brown: Convection and segregation in directional solidification of dilute and non-dilute binary alloys, J. Cryst. Growth **80**, 155–190 (1987)

40.13 P.A. Clark, W.R. Wilcox: Influence of gravity on thermocapillary convection in floating zone melting of silicon, J. Cryst. Growth **50**, 461–469 (1980)

40.14 S.A.I. Nikitin, V. Polezhayev, A.I. Fedyushkin: Mathematical simulation of impurity distribution in crystals prepared under microgravity conditions, J. Cryst. Growth **52**, 471–477 (1981)

40.15 S.R. Coriell, R.F. Sekerka: Lateral solute segregation during unidirectional solidification of a binary alloy with a curved solid-liquid interface, J. Cryst. Growth **46**, 479–482 (1979)

40.16 S.R. Coriell, R.F. Boisvert, R.G. Rehm, R.F. Sekerka: Lateral solute segregation during unidirectional solidification of a binary alloy with a curved solid-liquid interface II. Large departures from planarity, J. Cryst. Growth **54**, 167–175 (1981)

40.17 H.M. Ettouney, R.A. Brown: Effect of heat transfer on melt/solid interface shape and solute segregation in edge-defined film-fed growth: Finite element analysis, J. Cryst. Growth **58**, 313–329 (1982)

40.18 J.P. Kalejs, L.Y. Chin, F.M. Carlson: Interface shape studies for silicon ribbon growth by the EFG technique I. Transport phenomena modeling, J. Cryst. Growth **61**, 473–484 (1983)

40.19 C.E. Chang, W.R. Wilcox: Control of interface shape in the vertical Bridgman-Stockbarger technique, J. Cryst. Growth **21**, 135–140 (1974)

40.20 T.W. Fu, W.R. Wilcox: Influence of insulation on stability of interface shape and position in the vertical Bridgman-Stockbarger technique, J. Cryst. Growth **48**, 416–424 (1980)

40.21 T.W. Clyne: Heat flow in controlled directional solidification of metals I. Experimental investigation, J. Cryst. Growth **50**, 684–690 (1980)

40.22 T.W. Clyne: Heat flow in controlled directional solidification of metals II. Mathematical model, J. Cryst. Growth **50**, 691–700 (1980)

40.23 P.C. Sukanek: Deviation of freezing rate from translation rate in the Bridgman-Stockbarger technique I. Very low translation rates, J. Cryst. Growth **58**, 208–218 (1982)

40.24 P.C. Sukanek: Deviation of freezing rate from translation rate in the Bridgman-Stockbarger technique II. Moderate translation rates, J. Cryst. Growth **58**, 219–228 (1982)

40.25 T. Jasinski, W.M. Rohsenow, A.F. Witt: Heat transfer analysis of the Bridgman-Stockbarger configuration for crystal growth I. Analytical treatment of the axial temperature profile, J. Cryst. Growth **61**, 339–354 (1983)

40.26 L.R. Morris, W.C. Winegard: The development of cells during the solidification dilute Pb-Sb alloy, J. Cryst. Growth **5**, 361–375 (1969)

40.27 A.F. Witt, H.C. Gatos, M. Lichtensteiger, C.J. Herman: Crystal growth and segregation under zero gravity, J. Electrochem. Soc. **125**, 1832–1840 (1978)

40.28 A.M. Balint, D.G. Baltean, T. Levy, M. Mihailovici, A. Neculae, S. Balint: The dopant fields in uniform-diffusion-layer, global-thermal-convection and precrystallization-zone models, Mater. Sci. Semicond. Process. **3**, 115–121 (2000)

40.29 D.T.J. Hurle, E. Jakeman, C.P. Johnson: Convective temperature oscillations in molten gallium, J. Fluid Mech. **64**, 565–576 (1974)

40.30 C.A. Wang, A.F. Witt: *Annual Report Material Processing Center* (Massachusetts Institute of Technology, Massachusetts 1984)

40.31 M.M. Mihailovici, A.M. Balint, S. Balint: The axial and radial segregation due to the thermoconvection, the decrease of the melt in the ampoule and the effect of the precrystallization-zone in the semiconductor crystals grown in a Bridgman–Stockbarger system in a low gravity environment, J. Cryst. Growth **237–239**, 1752–1756 (2002)

40.32 M.M. Mihailovici, A.M. Balint, S. Balint: On the controllability of the level of the dopant concentration and of the compositional uniformity of a doped crystal, grown in a low gravity environment by Bridgman–Stockbarger method, Int. J. Theor. Physics, Group Theory Nonlin. Opt. **10**, 425–436 (2003)

40.33 A.M. Balint, M.M. Mihailovici, D.G. Baltean, S. Balint: A modified Chang–Brown model for the determination of the dopant distribution in a Bridgman–Stockbarger semiconductor crystal growth system, J. Cryst. Growth **230**, 195–201 (2001)

40.34 A.M. Balint, M.M. Mihailovici, D.G. Baltean, S. Balint: Interface structure in the growth of semiconductor crystals using the Bridgman–Stockbarger method, Thin Solid Films **380**, 108–110 (2000)

40.35 M.M. Mihailovici, A.M. Balint, S. Balint: The dopant distribution computed in the modified Chang–Brown model using quasi-steady state approximation, Comput. Mater. Sci. **24**, 262–267 (2002)

40.36 D.G. Baltean, T. Levy, S. Balint: Transport de masse par convection et diffusion dans un milieu multi-poreux, C. r. Acad. Sci. Paris, Ser. IIB **326**, 821–826 (1998)

40.37 K. Moutsopoulos, S. Bories: Dispersion en milieux poreux hétérogènes, C. R. Acad. Sci. Paris, Ser. IIB **316**, 1667–1672 (1993), in French

40.38 J.C. Maxwell: *Electricity and Magnetism* (Clarendon, Oxford 1873)

40.39 V.I. Avetisov, Mendeleev Institute Moscow (personal communication)

40.40 M.M. Mihailovici, A.M. Balint, S. Balint: Way to improve the compositional uniformity of doped crystals grown by Bridgman–Stockbarger method in a low gravity environment, Int. J. Theor. Phys. Group Theory Nonlinear Opt. **11**, 109–119 (2004)

40.41 D.J. Larson, J. Bethin, B.S. Dressler: *Shuttle Mission 51-G, Experiment MRS77F055, Flight Sample Characterization* (Grumman Corporate Research Center, Bethpage 1988), NASA Report RE-753

40.42 P.S. Dutta, A.G. Ostrogorsky: Segregation of Ga in Ge and InSb in GaSb, J. Cryst. Growth **217**, 360–365 (2000)

40.43 L.L. Zheng, D.J. Larson Jr., H. Zhang: Role of thermotransport (Sorret effect) in macrosegregation during eutectic/off-eutectic directional solidification, J. Cryst. Growth **191**, 243–251 (1998)

40.44 L. Braescu: The dependence of the dopant distribution on the pulling rate and on the capillary channel radius in the case of a Nd:YVO$_4$ cylindrical bar grown from the melt by the EFG method, Mater. Sci. Eng. B **146**, 41–44 (2008)

40.45 E. Tulcan-Paulescu, A.M. Balint, S. Balint: The effect of the initial dopant distribution in the melt on the axial compositional uniformity of a thin doped crystal grown in strictly zero-gravity environment by Bridgman–Stockbarger method, J. Cryst. Growth **247**, 313–319 (2003)

40.46 V.A. Tatarchenko: *Shaped Crystal Growth* (Kluwer, Dordecht 1993)

40.47 H.E. LaBelle Jr., A.I. Mlavsky, B. Chalmers: Growth of controlled profile crystals from the melt: Part I – Sapphire filaments, Mater. Res. Bull **6**, 571–579 (1971)

40.48 H.E. LaBelle Jr., A.I. Mlavsky, B. Chalmers: Growth of controlled profile crystals from the melt: Part II – Edge-defined, film-fed growth (EFG), Mater. Res. Bull. **6**, 581 (1971)

40.49 B. Chalmers, H.E. LaBelle Jr., A.I. Mlavsky: Edge-defined, film-fed crystal growth, J. Cryst. Growth **13/14**, 84–87 (1972)

40.50 J.P. Kalejs: Impurity redistribution in EFG, J. Cryst. Growth **44**, 329–344 (1978)

40.51 J.P. Kalejs, G.M. Freedman, F.V. Wald: Aluminium redistribution in EFG silicon ribbon, J. Cryst. Growth **48**, 74–84 (1980)

40.52 B. Chalmers: Transient solute effects in shaped crystal growth of silicon, J. Cryst. Growth **82**, 70–73 (1987)

40.53 J. Cao, M. Prince, J.P. Kalejs: Impurity transients in multiple crystal growth from a single crucible for EFG silicon octagons, J. Cryst. Growth **174**, 170–175 (1997)

40.54 J.P. Kalejs: Interface shape studies for silicon ribbon growth by the EFG technique II. Effect of die asymmetry, J. Cryst. Growth **61**, 485–493 (1983)

40.55 J.P. Kalejs: Modeling contribution in commercialization of silicon ribbon growth from the melt, J. Cryst. Growth **230**, 10–21 (2001)

40.56 O. Bunoiu, I. Nicoara, J.L. Santailler, T. Duffar: Fluid flow and solute segregation in EFG crystal growth process, J. Cryst. Growth **275**, 799–805 (2005)

40.57 L. Braescu, S. Balint, L. Tanasie: Numerical studies concerning the dependence of the impurity distribution on the pulling rate and on the radius of the capillary channel in the case of a thin rod grown from the melt by edge-defined film-fed growth (EFG) method, J. Cryst. Growth **291**, 52–59 (2006)

40.58 L. Braescu, T.F. George, S. Balint: Evaluation and control of the dopant distribution in a Nd:LiNbO$_3$ fiber grown from the melt by the edge-defined film-fed growth (EFG) method, Photonic Fiber and Crystal Devices: Advances in Materials and Innovations in Device Applications I (Optics and Photonics 2007), SPIE Proc. **6698**, 669803:1–8 (2007)

40.59 L. Braescu, T. Duffar: Effect of buoyancy and Marangoni forces on the dopant distribution in the case of a single crystal fiber grown from the melt by the edge-defined film-fed growth (EFG) method, J. Cryst. Growth **310**, 484–489 (2008)

40.60 T.F. George, L. Braescu: Sapphire fibers grown from the melt by the EFG technique: Dependence of the impurity distribution on temperature and surface tension gradients, Photonic Fiber and Crystal Devices: Advances in Materials and Innovations in Device Applications II (Optics and Photonics 2008), SPIE Proc. **7056**, 705603-1–705603-10 (2008)

40.61 L. Braescu, T.F. George: Arbitrary Lagrangian-Eulerian method for coupled Navier-Stokes and convection-diffusion equations with moving boundaries, Proc. 12th WSEAS Int. Conf. Appl. Math. – Math'07 (2007) pp. 33–36

40.62 F. Duarte, R. Gormaz, S. Natesan: Arbitrary Lagrangian-Eulerian method for Naver-Stokes equations with moving boundaries, Comput. Methods Appl. Mech. Eng. **193**, 4819–4836 (2004)

40.63 M. Fernandez, M. Moubachir: Sensitivity analysis for an incompressible aeroelastic system, Math. Models Methods Appl. Sci. **12**, 1109–1130 (2002)

Part G Defects Characterization and Techniques

41 Crystalline Layer Structures with X-Ray Diffractometry
Paul F. Fewster, Brighton, UK

42 X-Ray Topography Techniques for Defect Characterization of Crystals
Balaji Raghothamachar, Stony Brook, USA
Michael Dudley, Stony Brook, USA
Govindhan Dhanaraj, Nashua, USA

43 Defect-Selective Etching of Semiconductors
Jan L. Weyher, Warsaw, Poland
John J. Kelly, Utrecht, The Netherlands

44 Transmission Electron Microscopy Characterization of Crystals
Jie Bai, Hillsboro, USA
Shixin Wang, Boise, USA
Lu-Min Wang, Ann Arbor, USA
Michael Dudley, Stony Brook, USA

45 Electron Paramagnetic Resonance Characterization of Point Defects
Mary E. Zvanut, Birmingham, USA

46 Defect Characterization in Semiconductors with Positron Annihilation Spectroscopy
Filip Tuomisto, Espoo, Finland

1404

41. Crystalline Layer Structures with X-Ray Diffractometry

Paul F. Fewster

X-ray scattering analysis and instrumentation has been evolving rapidly to meet the demands arising from the growth of sophisticated device structures. X-ray scattering is very sensitive to composition, thickness and defects in layered structures of typical present-day electronic device dimensions. Considerable information can be obtained from simple profiles, including an estimate of layer thickness and composition (by measuring peak separations) and a measure of the sample quality (from the peak broadening). The full simulation of the profiles takes this a stage further to interpret very complex structures and obtain more reliable parameter estimates. By obtaining two-dimensional scattering data the information becomes more extensive, including layer strain relaxation and defect analysis, quantum dot composition and shape. Generally the data is averaged over a few mm, however reducing the beam size can break this averaging to reveal inhomogeneities, isolating small regions and in some circumstances isolate individual quantum dots for analysis. This article gives an overview of the status, differentiating those methods that are easily accessible and those that require a collaborative approach because they are still being established.

41.1	X-Ray Diffractometry	1406
41.2	Basic Direct X-Ray Diffraction Analysis from Layered Structures	1407
	41.2.1 Theory	1407
	41.2.2 Interpretation of Data Collected from Planes Parallel to the Surface – the ω–2θ Scan – an Example	1409
	41.2.3 Interpretation of Data Collected from Several Reflections – The Reciprocal Space Map – An Example	1411
41.3	Instrumental and Theoretical Considerations	1412
	41.3.1 The Instrument for Collecting X-Ray Diffraction Patterns	1412
	41.3.2 Interpreting the Scattering by Simulation	1412
41.4	Examples of Analysis from Low to High Complexity	1413
	41.4.1 Established Methods	1413
	41.4.2 New Methods and New Analyses	1416
41.5	Rapid Analysis	1419
41.6	Wafer Micromapping	1420
41.7	The Future	1421
	References	1422

X-ray scattering offers a nondestructive method for determining phase composition in crystalline materials, layer thickness, interface details, shapes of crystallites and quantum dots, etc. The basic principles of the various techniques will be outlined in this chapter. Bulk materials are limited to their intrinsic properties, whereas composite materials of several thin layers extend this range, e.g., semiconductor light-emitting diodes and modern transistors. However these physical properties are influenced by the layer compositions, thickness, defect density, and other structural parameters. X-ray scattering can be used very effectively for measuring all these structural parameters, some to very significant precision. The first part of this chapter discusses the general principles of the analysis, which is applicable to achieving an initial sample characterization by direct interpretation of the diffraction pattern. The limitations of this direct interpretation are also considered and how simulation of the scattering process gives more exact results. The advantages of simulation make the analysis of complex layer structures possible and also extend the applicability of x-ray scattering to evaluating interface roughness, distortions, etc. Quantum dots are singled out as a particular area to illustrate the developments in x-ray scattering. Any analysis introduces some form of averaging, and approaches to reduce the probed volume are discussed. This has enabled detailed mapping of lateral variations, yielding a wealth of information for detailed characterization and screening, as discussed in the penultimate section.

41.1 X-Ray Diffractometry

The measured x-ray intensity scattered from a sample, in terms of magnitude and distribution, is sensitive to the arrangement and type of atoms in a crystal. If the x-ray probe is very well collimated then the resulting *high-angular-resolution* mode, or more generally phrased *high-resolution x-ray diffraction*, is very sensitive to small changes in the atomic layer spacing and deposited layer thickness. This is a consequence of the angle through which x-rays are scattered, which is inversely related to the separation of the atomic layer planes and the deposited layer thickness. So the sensitivity of x-ray scattering is effectively very high for very thin layers. Fortunately this is the region of interest for many of today's device structures; however, this is only relevant if there is sufficient scattered intensity. This aspect of the compromise between scattered intensity and high resolution will be addressed in Sect. 41.5. To indicate the bounds, it is possible to measure the thickness of layers up to $\approx 4\,\mu\text{m}$ and down to single atomic layers or a single plane of quantum dots using conventional configurations and equipment. For example, a single 1 nm $\text{In}_{0.1}\text{Ga}_{0.9}\text{As}$ layer on GaAs can be observed as an asymmetry on the main substrate profile, and determined by simulation to within ± 0.1 nm; a layer thickness above this value begins to indicate distinctive oscillations, and below this it is possible to observe this asymmetry but in practice it requires considerable care to determine the thickness. If however the layer is buried, then thinner layers of 0.1 nm can be observed by virtue of the interference between the layers above and below. The measurement of the thickness, again by simulation, appears simple and sensitive, however the thickness is correlated with the composition, and so ideally a combination of profiles are required. Any increase in thickness at constant composition increases the amplitude of the interference oscillations. The positions of the main Bragg peaks give a measure of the interplanar spacing in a crystal that can be determined to within ≈ 1 ppm, i.e., considerably less than the classical radius of an electron. However extracting dilute composition information in alloy systems requires a combination of differences in covalent radii and the concentration itself. So, to obtain the alloy composition of Al in GaAs, for example, where the difference in the covalent radius of Al to Ga is ≈ 2865 ppm, we require a method for differentiating the interatomic spacing to 28 ppm to achieve 1% precision in the concentration of Al in AlGaAs. With careful measurement and accounting for sample uncertainties this can be done routinely. For systems with larger differences in covalent radii, e.g., Ge in SiGe or In in GaN, the precision increases for the same amount of care in the measurement.

For obtaining the thickness of a layer, similar rules apply; as the thickness decreases, the width of the associated peak increases and the fringes associated with the interference of the scattering from the top and bottom surface of the layer become further apart. If the fringes are very close together and the central peak is narrow then the thickness measure becomes more problematic and requires higher-resolution instrumentation, up to the limit of the x-ray coherence length (typically $\approx 4\,\mu\text{m}$) or the layer perfection. Rough top and bottom surfaces reduce the intensity of the fringes and so can give a measure of the interface quality. As the thickness is reduced the peak broadens until eventually the scattering is indistinguishable from the inherent noise level of the instrument. If layers are stacked as in most device structures then thin layers modulate the overall pattern and appear more obvious again, e.g., interfacial layers in superlattice structures or single atomic layers separated by thicker layers. It is clear from this that x-ray scattering cannot simply be categorized in terms of what it can and cannot detect in terms of a single layer, for example, but rather the whole structure and the instrument are necessarily intertwined, and the route to extracting useful information generally comes down to simulating the scattering. However, some very useful analyses can be obtained quite directly from the measurements, although awareness of the pitfalls is important.

As far as instrument development is concerned, this can be a large and confusing subject, since as discussed above, the sample, the information to be extracted, and the instrument resolution are all linked. This has led to instruments becoming more flexible for general materials research, and as the level of understanding or the improvement in control over the crystal growth develops, the resolution often needs to be increased. This allows more detailed information, deviations from perfection, etc., to be probed more easily. Similarly the analysis goes through a cycle of approximate quick interpretation, through to more precise simulation to understand the sample more fully, until eventually the analysis becomes a control feedback method that can be routine yet sophisticated and specific to the crystal properties of interest. When the growth of the material becomes too complex, in that there are many complex parameters to assess with high degrees of correlation, then some statistical significance is required. X-ray scattering can contribute here, since as illustrated above its high sensitivity to structural parameters can give some form of measure for obtaining correlations. A few examples of this will be given in this chapter.

As mentioned above x-ray scattering is very sensitive to structural changes and generally the probe is large, usually mm and for some specialized applications down to μm. However, this does not mean that everything is averaged over that region. In fact the x-ray pattern contains the average and the fluctuations from the average for dimensions over which the x-rays are coherent, so for example, the onset of layer relaxation (the distortion around dislocations) and the existence of precipitates can be observed, both of which modify the scattering in quite characteristic ways. Examining the spatial distribution of the intensity in the scattered beam can also break this averaging effect. Chapter 42 in this volume, describe a very direct approach to obtaining defect information, and when combined with high-resolution reciprocal space mapping, the x-ray topographic contrast can be related to features in the scattering pattern [41.1, 2].

This chapter will cover the rapid direct measures that give an indication of the material's properties. It will also show how far the analysis can be taken and finally the more systematic routine analysis methods for controlled crystal growth.

41.2 Basic Direct X-Ray Diffraction Analysis from Layered Structures

41.2.1 Theory

X-rays scatter strongly from electrons, principally those that are highly localized, i.e., core electrons. The scattering amplitude from an assembly of atoms at positions r with individual scattering factors (or form factors) of f_r is given by

$$F_S = \int_r f_r e^{-2\pi r S \cdot r} \, dr \, . \quad (41.1)$$

The scattering amplitude and therefore the consequent intensity is distributed but reaches a maximum for atom positions satisfying the condition

$$|S| = \frac{1}{d} \, , \quad (41.2)$$

where d is the separation of the atomic planes. Any position in diffraction space S can be probed by manipulating the incoming (k_0) and the outgoing (k_H) x-ray beam paths, where $|k_0| = |k_H| = 1/\lambda$, such that

$$S = k_H - k_0 \, , \quad (41.3)$$

which is Bragg's relation in vector form. It can be seen from the geometry of Fig. 41.1 that

$$S = \frac{1}{d} = \frac{2\sin\theta}{\lambda} \, . \quad (41.4)$$

Therefore, to probe different regions of diffraction space, the incident beam direction with respect to the sample and the detector acceptance direction are manip-

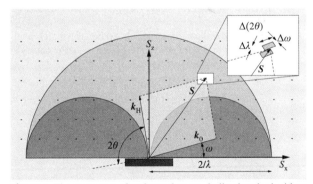

Fig. 41.1 The geometry of reciprocal space, indicating the incident and scattered beam vectors k_0 and k_H and the resultant position probed S. The regular array of reciprocal lattice points represents the atomic plane spacings, and the limiting sphere for their observation is defined by the wavelength λ. The two smaller darker grey spheres are inaccessible regions to data collection in conventional reflection mode. The *inset* gives the detail of the size of the reciprocal space probe

ulated to map the distribution of intensity (Fig. 41.1). If the incident and scattered beam directions are very well defined, as with a high-angular-resolution diffractometer, then the length d can be measured very precisely as well as small differences in d. High resolution is very useful for studying closely lattice-matched semiconductor multilayers, for example.

The scattering amplitude F_S is complex, includes aspects of absorption, and is directly related to the refractive index. For most materials the refractive index for x-rays is less than one, and therefore x-rays total-externally reflect, and the actual Bragg scattering peaks are displaced to larger angles. The refractive index is $\approx (1-10^{-5})$, which has advantages in that for most measurements only small uncertainties are introduced, e.g., in thickness measurement from fringe spacings or lattice parameter determination. On the other hand the difference in refractive indices of layers from different materials creates modulations in the specular scattering (total external scattering), which offers another method for determining thickness and composition.

So the simplest information that can be extracted from an x-ray scattering experiment is the location of the peaks; this gives the crystal plane spacing, and their intensity relates to the atomic arrangement. The width of the peak relates to the number of contributing atomic planes, so in general for thin films the peaks become broadened normal to the surface direction, but very narrow parallel to the surface direction, so the term reciprocal space describes it all. This is the general guide relevant to the early stages of analyzing x-ray diffraction profiles.

Consider the early stages of crystal growth of a new material when little is known. Suppose that it consists of some layers and the intention is to grow something epitaxially related to the thick substrate wafer. The structure could appear as in Fig. 41.2a, where some of the likely measurable features are illustrated. A reciprocal space map for a single-layer structure around a reflection close to $s_x = 0$ may appear as in Fig. 41.2b. If the substrate is a reasonable quality single crystal, then the scattering from this will be represented by a narrow peak, and if the layers are not too thick ($< 2\,\mu$m), then this could be the highest peak and can be easily identified. Its position will give a value for the interplanar spacing, and the associated lattice parameter can be obtained if the reflection index H is known, since this relates the length scale d probed to that of the material lattice parameter. The layer peak is broader and wider and is not in line with the substrate peak. If the choice of reflection is from planes parallel to the surface (i.e., $\omega \approx \theta$, $s_x = 0$; Fig. 41.1), then this difference in angle $\delta\omega$ is indicative of layer planes being tilted with respect to the substrate planes by $\delta\omega$. The width of the layer peak in the direction normal to the surface (s_z) will be related to the layer thickness L_z or some variation of strain in the layer, the microstrain $\varepsilon_\perp$, or a combination of both. This width is obtained by scanning in ω and 2θ, coupled in a 1:2 ratio. If this width is purely due to the layer thickness then $L_z \equiv 1/\Delta s_z$, and from the geometry of Fig. 41.1

$$\frac{1}{\Delta s_z} = \frac{\lambda}{[\sin\omega_1 + \sin(2\theta_1 - \omega_1)] - [\sin\omega_2 + \sin(2\theta_2 - \omega_2)]}. \tag{41.5}$$

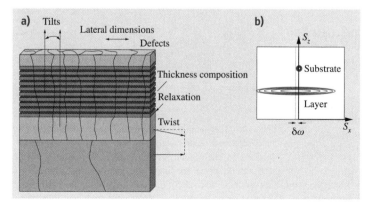

Fig. 41.2a,b The typical information that can be extracted from an x-ray scattering experiment (a), and a schematic of a reciprocal space map of an imperfect layer on a nearly perfect substrate (b)

In this case $2\theta_1 \approx 2\omega_1$ and $2\theta_2 \approx 2\omega_2$, $\Delta 2\theta = (2\theta_1 - 2\theta_2)$, and the chosen positions described by the subscripts correspond to the full-width at half-maximum (FWHM), hence

$$L_z \approx \frac{\lambda}{2\Delta\theta \cos\theta}, \quad (41.6)$$

which is the Scherrer equation [41.3]. If the peak is related to the microstrain then, from (41.2),

$$\frac{\Delta d}{d} = -\frac{\Delta S}{S}. \quad (41.7)$$

Hence for the scattering peak considered here

$$\frac{\Delta d}{d} = -\frac{\Delta s_z}{s_z} = -\frac{\lambda/(2\sin\theta)}{\lambda/(2\Delta\theta\cos\theta)} = -\cot\theta\,\Delta\theta. \quad (41.8)$$

Separating these two influences on the peak width can either be judged by the acceptable ranges for the microstrain or layer thickness, or by combining several reflections (preferably along the same zone axis, i.e., 001, 002, 003, etc.), to separate the different dependency on θ [41.4].

The broadening normal to this also has two components. These two components are composed of local tilts and some lateral dimension: for mosaic layers these can be related to the Burgers vectors of the dislocations forming the mosaic blocks. If the layer is dominated by threading dislocations then the component of tilt is small and the width will give a good estimate of the mosaic block dimension, $L_x \equiv 1/\Delta s_x$, given by

$$\frac{1}{\Delta s_x} = \frac{\lambda}{[\cos\omega_1 - \cos(2\theta_1 - \omega_1)] - [\cos\omega_2 - \cos(2\theta_2 - \omega_2)]}, \quad (41.9)$$

where in this case $2\theta_1 \approx 2\theta_2$ and the peak width $\Delta\omega = (\omega_1 - \omega_2)$, thus

$$L_x \approx \frac{\lambda}{2\Delta\omega \sin\omega}. \quad (41.10)$$

If there is a high density of misfit dislocations that have Burgers vectors out of the plane of the interface, e.g., 60° dislocations in Si, then these will have a tilt component. The distribution of tilted regions and the size of these regions start to produce a more complex picture. However if the contribution is purely due to tilted regions of large dimensions then the spread of microscopic tilts is simply given by $\Delta\omega$.

41.2.2 Interpretation of Data Collected from Planes Parallel to the Surface – the ω–2θ Scan – an Example

Consider the example of a GaN structure given in Fig. 41.3a and the reciprocal space map around the 0002 reflection (Fig. 41.3b). The reciprocal space map has many satellites associated with this periodic layer structure and two strong peaks having two very different widths along s_z and similar widths along s_x (Fig. 41.3c,d). A considerable amount of information can be extracted from this data. Firstly, the separation of the two peaks (Fig. 41.3c) relates to the difference in lattice parameter along s_z. The individual lattice parameter value along this direction is given by

$$a_z = d_z H_z = \frac{H_z}{s_z}, \quad (41.11)$$

where H_z is the index of the reflection along the surface normal direction, simply equal to 2 here. Therefore the most precise measure of the lattice parameter is obtained for larger values of s_z, as is also clear from (41.8). Referring to Fig. 41.3c, we could in the first instance assume that the underlying substrate, or a thick GaN buffer in this case, has a known lattice parameter (some database value) and therefore the layer lattice parameter $_\mathrm{L}a_z$ is simply given by combining (41.11) and (41.8) as

$$_\mathrm{L}a_z = \frac{_\mathrm{S}a_z}{_\mathrm{S}H_z}\left(1 + \frac{\Delta d}{d}\right)_\mathrm{L}H_z$$
$$\approx \frac{_\mathrm{S}a_z}{_\mathrm{S}H_z}(1 - \cot\theta\,\Delta\theta)\,_\mathrm{L}H_z, \quad (41.12)$$

where $\Delta\theta$ is the difference in Bragg angle and $_\mathrm{L}H_z$ is the reflection index in the direction normal to the surface for the layer and $_\mathrm{S}H_z$ for the substrate. For the example $_\mathrm{L}H_z = _\mathrm{S}H_z = 2$, $_\mathrm{S}a_z = 0.51851$ nm for the 0002 reflection from GaN, $\theta_\mathrm{S} = 17.284°$ for an x-ray wavelength of $\lambda = 0.1540593$ nm, using (41.4). Since the measurement gives $\Delta\theta = 0.0777°$, and if we assume that the layer has a good epitaxial relationship to the substrate or layer below, the lateral lattice parameter of the layer $_\mathrm{L}a_x$ is identical to $_\mathrm{S}a_x$ (0.31893 nm), so $_\mathrm{L}a_z = 0.52076$ nm.

Now the layer material would have a characteristic and usually unique set of lattice parameters, but in the undistorted state. The unstrained lattice parameters can be determined from taking the distorted unit cell (defined by the lattice parameters) and allowing it to relax to the expected shape. In this case InN and GaN and the phase mixture are hexagonal, such that certain unit cell parameters (a, b, c, α, β, γ) are defined, i.e., $a = b$,

$\alpha = \beta = 90°$, and $\gamma = 120°$. The layer structure in the unstrained state is then just defined by the ratio c/a, which is 1.640 for InN and 1.626 for GaN. The way in which the distorted unit cell is transformed into the unstrained unit cell depends on the elastic coefficients that relate the stress and strains. The strain normal to the surface ε_{zz} is related to strains parallel to the surface, ε_{xx} and ε_{yy}, such that they can be related to a Poisson ratio ν

$$\left(\frac{\Delta d_z}{d_z}\right)_L = \frac{_Ld_z - _Ld_{z0}}{_Ld_{z0}} = \varepsilon_{zz} = \frac{-\nu}{1-\nu}\left(\varepsilon_{xx} + \varepsilon_{yy}\right)$$
$$= \frac{-\nu}{1-\nu}\left(\frac{_Ld_x - _Ld_{x0}}{_Ld_{x0}} + \frac{_Ld_y - _Ld_{y0}}{_Ld_{y0}}\right),$$
(41.13)

where $_Ld_{x0}$, $_Ld_{y0}$ and $_Ld_{z0}$ are the lattice plane spacing for the layer parallel and normal to the surface in the unstrained state; these are parameters needed to identify the alloy composition. The Poisson ratio used in this way is dependent on the orientation, however this can be determined from the elastic coefficients or from tabulated values. Many typical semiconductors have values $\nu \approx 0.3$, whereas maintaining constant volume $\nu \approx 0.5$. In our example in the first instance we are assuming $\varepsilon_{\parallel} = \varepsilon_{xx} = \varepsilon_{yy}$, i.e., that the strain is isotropic in the interface plane. Combining (41.13) with the c/a ratio ($= K$) of the expected unstrained state for the layer, we can isolate c_L ($= H_z _Ld_{z0}$) in terms of the measurable ($_Ld_z$ or Δd_z) and known or tabulated values (H_z, K, ν, c_S, a_S) as

$$c_L = \left(\frac{2\nu}{1-\nu}\right)(Ka_S - _Ld_z H_z) + _Ld_z H_z, \quad (41.14)$$

where $_Ld_z H_z = c_S + \Delta d_z H_z$, c_S is the lattice parameter of the underlayer or substrate normal to the surface, and a_S is the lattice parameter in the plane of the interface. Within the approximations in this example, the values of K and ν can be estimated, which can be iterated since they are composition dependent. It is assumed that a_S is a known or database value, then c_L can be derived and the phase composition can be obtained by the interpolation [41.5]

$$x = \frac{c_L - c_{GaN}}{c_{InN} - c_{GaN}}. \quad (41.15)$$

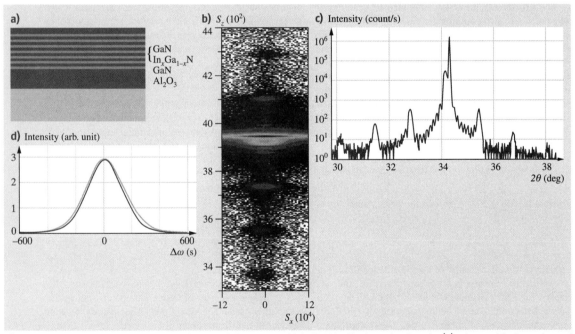

Fig. 41.3a–d The structure of an InGaN/GaN multilayer on a thick GaN buffer layer is given in (**a**), along with the reciprocal space map (**b**) and some extracted profiles along s_z (**c**) and along s_x (**d**). The intensities in (**d**) correspond to the GaN buffer (*grey*) and the superlattice (*brown*), indicating that they have similar microstructure, although the superlattice may be of better quality; the intensities are normalized for comparison

In the example discussed, $x = 2.2\%$. The purpose of the above derivations is to indicate some of the approximations required for a direct interpretation of the intensity distribution in reciprocal space. Before discussing a more rigorous approach, the influence of layer strain relaxation should be considered, since similar approaches are applicable.

41.2.3 Interpretation of Data Collected from Several Reflections – The Reciprocal Space Map – An Example

If the atom positions do not align from layer to substrate across the interface, then $_Ld_x \neq {_S}d_x$ and $a_L \neq a_S$, and hence a_L should be substituted for a_S in (41.14). This dimension can be obtained from a reciprocal space map of a reflection having a component of this in-plane spacing (Fig. 41.4, compare with Fig. 41.2b). The distance in s_x between the layer peak of Fig. 41.4 and that of Fig. 41.2 gives the lateral plane spacings in the layer $_Ld_x$ by applying (41.9) and similarly for the substrate reflection $_Sd_x$, so that $a_L = {_L}d_x \Delta H_x$. For example, when combining the $hkil$ reflections 0002 and $2\bar{1}\bar{1}4$, $\Delta H_x = \sqrt{\{4[h^2 + k^2 + hk]/3\}} = \sqrt{\{4[2^2 + 1^2 - 2]/3\}}$ for the hexagonal system with a (0001) surface plane. For the cubic system with a (001) surface, $\Delta H_x = \sqrt{\{[h^2 + k^2]\}}$. Alternatively if the substrate is unstrained and has known lattice parameters $a_L = a_S + \Delta d_x H_x$, where $\Delta d_x \equiv 1/\Delta s_x$, (41.9) the layer lattice parameter can be measured directly from the reciprocal space map of Fig. 41.4 (after accounting for any tilt). The tilt observed in the map of Fig. 41.2b changes $\omega_1 \to \omega_1 + \delta\omega$ for the layer, and ω_2 for the substrate is unchanged. This approach has been extended to the more general case, i.e., any space group, anisotropic relaxation or orientation [41.6].

From the discussion above a series of reciprocal space maps can yield a considerable wealth of information about the sample: the unit cell dimensions,

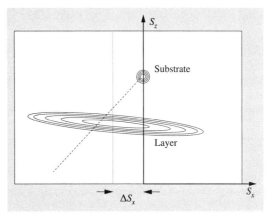

Fig. 41.4 A schematic of a reciprocal space map for a reflection from planes inclined to the surface plane. The lateral misalignment of the layer and substrate peak is indicative of differing lateral lattice parameters (layer strain relaxation) and/or relative lattice plane tilts

the distortion, and some characteristic dimensions (of mosaic blocks and thickness, for example) and their relationship to other layers or the substrate. The combination of several reflections or the shape of the peaks in reciprocal space can be used to isolate the contribution of the tilts, sizes, and strains [41.4, 6], using data as illustrated in Fig. 41.3b,d for the 0002 and 0004 reflections of GaN, for example. These direct approaches are very suitable for feedback during the early stages of crystal growth or for monitoring established growth procedures. However when there are several layers, and the material exhibits good crystal quality, the assumptions concerning the peak positions and the peak shapes become very approximate. This is because the simple interpretation assumes that each layer scatters independently, i.e., the scattering theory is represented very simply and it is assumed that the instrument and method of data collection create no artefacts.

41.3 Instrumental and Theoretical Considerations

The basics of the instrument for data collection and the appropriate scattering theories to use are discussed in this section. Many of these aspects will become clear in the examples.

41.3.1 The Instrument for Collecting X-Ray Diffraction Patterns

The reciprocal space maps of Figs. 41.2, 41.3, 41.4, etc., are a very small subset of Fig. 41.1. Referring to the expanded detail in the inset of Fig. 41.1, the representation of the angular divergence of the incident beam $\Delta\omega$, the angular acceptance of the scattered beam $\Delta 2\theta$, and the spread in wavelength of the source $\Delta\lambda$ all add to smearing of the x-ray probe [41.6]. The divergence $\Delta\omega$, the acceptance $\Delta 2\theta$, and the wavelength dispersion $\Delta\lambda$ vary throughout reciprocal space. If this capture volume is reduced too far, the intensity will be reduced significantly; however with recent advances in x-ray mirrors and lenses, and more significantly the availability of synchrotron sources, the expanding scope of x-ray scattering has not been diminished by insufficient intensity. The equipment is not significantly different from laboratory sources to synchrotron sources, but in terms of intensity differences the more exotic experiments only become feasible for synchrotron sources, especially if combinations of wavelengths are needed. However the long-term stability and convenience of laboratory sources make them the best choice for detailed analyses for most problems, and certainly for screening experiments prior to using a synchrotron source.

A good basic *workhorse* for many crystal growth studies is the instrument shown in Fig. 41.5 [41.7]. The monochromator and analyzer combination gives a very small reciprocal space probe over most of the accessible region available (defined by the wavelength used: the smaller the wavelength, the greater the accessible sphere). The intensity for laboratory sources is kept high by maintaining a reasonably large beam, $\approx 1 \times 10$ mm. The angular spread in the beam normal to Fig. 41.1 is the axial divergence, which is only restricted to $\approx 0.5°$, compared with $\approx 0.001°$ for $\Delta\omega$ and $\Delta 2\theta$, thus the measured reciprocal space map is a projection. Restricting the axial divergence creates a three-dimensional (3-D) reciprocal space map [41.8], yielding further information, but this approach will not be discussed further in this chapter. The resolution is a complex interplay between the instrument, the sample, the choice of reflection, the wavelength, and the alignment of the instrument and sample. Generally though, the sample is the determining factor and the probe is small compared with the structural features that influence the pattern, although the influence of the instrument should also be modeled.

These are only general comments on the resolution; methods of maintaining resolution and enhancing intensity will be addressed in Sect. 41.5.

41.3.2 Interpreting the Scattering by Simulation

The kinematical theory permits the basic interpretation summarized in Sect. 41.2 but is only valid for structures that are *ideally imperfect*. However, the majority of structures grown by highly controlled growth methods, e.g., molecular-beam epitaxy and metalorganic vapor-phase deposition, will result in crystal structures that are highly perfect. Also the substrate material is often Si, sapphire, MgO, GaAs, InP or $SrTiO_3$, which are nearly perfect. This level of perfection necessitates the application of dynamical theory [41.9], which includes extinction effects, the loss of incident intensity through scattering, interference with the multiply scattered wave, etc. [41.10, 11]. This can have a much more dramatic effect on the penetration depth than normal photoelectric absorption [41.12]. Also the peak positions, even for very simple structures, vary with layer thickness, creating uncertainties in the composition, thickness, and simple interpretation of periodic structures [41.13]. Perhaps the most important reason

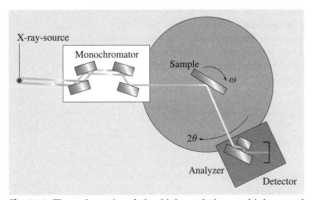

Fig. 41.5 The schematic of the high-resolution multiple-crystal diffractometer, which gives a small and well-defined reciprocal space probe over much of reciprocal space

for using dynamical theory is that, due to the increasing complexity of today's layer structures, the pattern is not a simple superposition of all the contributing layers, but a complex interference with peak shifts, beat frequencies, etc. Simulation also removes a few myths about direct interpretation, especially concerning superlattices, e.g., interfacial layers (roughening) that might appear as a loss of perfection can under some circumstances increase the satellite intensities [41.14]. Simulation has significant advantages in that the whole structure, thickness, and composition of each layer can now be analyzed simultaneously, and also all the iterative assumptions concerning the changing c/a ratio and Poisson ratio with composition, etc., are all included. The simulation of the profiles, or the reciprocal space maps, rely on the scattering amplitude (41.1), the unit cell parameters, and the elastic parameters, many of which are established database values, and of course some estimate of the structure to be simulated. The sensitivity to structural details is very high, so the need to have a good procedure for fitting the calculated to measured profiles by adjusting the model is evident. These procedures are becoming very sophisticated.

The simulation of more imperfect structures, for example, quantum dots within a perfect crystal matrix or rough interfaces in a period structure, is best achieved with the distorted-wave Born approximation (DWBA) [41.15]. This considers multiple scattering and is therefore dynamical and takes into account effects associated with the transmission of the beams. It is based on solving the scattering from a perfect structure by the incident wave. This average scattered wave defines the local electric field of the wave, which is used to calculate the disturbed wave that results from quantum dots, precipitates, interfaces, and other disturbances. The combination of the undisturbed and disturbed waves gives the resultant scattering pattern. If the disturbance becomes too large then this perturbation method is not applicable and kinematical theory becomes more appropriate and can then be considered as a single scattering theory, i.e., the Born approximation (BA) [41.16].

The analytical approaches available and discussed in the following sections therefore have theoretical models that cover: perfect to nearly perfect structures (dynamical theory), imperfections giving rise to diffuse scattering (distorted-wave Born approximation), and highly imperfect or very small perfect regions (kinematical theory). If however an experiment is carried out when $S \approx 0$, i.e., with very small incident and scattering angles (specular scattering), then we see from (41.1) that the scattering amplitude is purely the sum of the scattering strength of individual atoms, independent of their position r in the unit cell. It is then possible to consider the layers in a structure as consisting of a constant electron density, and therefore the interaction is analogous to light scattering from regions of constant refractive index; hence a much simpler, *optical* theory is valid [41.17]. Consequently specular reflectometry (close to the total external reflection condition) within a few degrees of the 000 reflection, is usually simulated with this optical theory.

41.4 Examples of Analysis from Low to High Complexity

The advances in x-ray scattering have been very rapid in recent years so this section will give a snapshot of the state of the art as well as the more established methods. Eventually the useful state-of-the-art analyses that are proven to be successful in revealing structural parameters, within a broad range of applications, will become established methods that are more accessible to more scientists. However, in the interim, these state-of-the-art methods are generally only accessible through collaboration with scientists in these fields, or through significant self-endeavor.

41.4.1 Established Methods

The interpretation of scattering profiles through simulation is now widely available for perfect and nearly perfect structures, based on dynamical theory. A typical analysis is illustrated in Fig. 41.6a, for a GaAs-based structure. The approximate thickness and composition of each layer was known from the growth and this gave the starting model for its refinement through comparison of measured and simulated profiles. This is a two-layer sample so from the start we have to determine the composition and thickness for the InGaAs and the AlGaAs layers. The peak positions give some rough starting values, however with the limited number of parameters the fitting process can cope easily with the estimated values, which were within $\approx 20\%$ of the final fit values. The fit to the data was good with just this information included, i.e., two layers; however, the intensity of the fringes between -5000 and -4000 arcsec was significantly overestimated. The inclusion of some

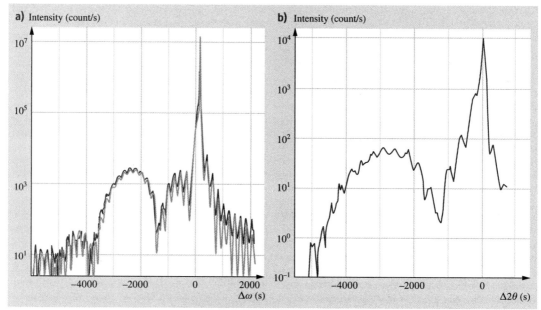

Fig. 41.6a,b The diffraction profile from an AlGaAs/ InGaAs/GaAs structure obtained with the beam-selection diffractometer (*brown*) and simulated best fit profile (*light brown*) (**a**), and the static diffractometer, discussed in Sect. 41.5 (**b**). The data collected in (**a**) and (**b**) were collected in 7.3 min and 1 s, respectively

grading at the InGaAs to AlGaAs interface had a strong influence on this broad InGaAs layer fringe (with the AlGaAs thickness fringes superimposed). This grading reduced the intensity to achieve an improved fit (Fig. 41.6a). This additional graded layer is only observable with very high intensities or longer data collection times. This data was obtained with the beam-selection diffractometer, which is composed of an x-ray mirror, scattering from the 004 reflection of the sample to a 004 Ge analyzer in front of the detector [41.18]. This gives a substrate peak intensity of ≈ 30 Mcps, so the ω–2θ scan data was collected in just over 7 min with a 0.005° step size. The final parameters were 0.0130 μm of $In_{0.145}Ga_{0.855}As$, a 0.0059 μm graded layer of InAlGaAs, and 0.0773 μm of $Al_{0.258}Ga_{0.742}As$. These values compare with the best fit of 0.0157 μm of $In_{0.145}Ga_{0.855}As$ and 0.0808 μm of $Al_{0.258}Ga_{0.742}As$ when no interfacial grading was included, which would be the case if the broad InGaAs layer fringes were not observed.

For very complex samples the fitting process requires more persistence and time, mainly because of the high degree of correlation between parameters and the complicated interference in the scattering from individual layers. Typical structures that fall into this category are vertical-cavity surface-emitting lasers (VCSEL). This example is composed of quaternary AlGaInP layers flanking a ternary GaInP quantum well, which are surrounded by Bragg stacks of AlGaAs/AlGaAs periodic layers; the laser wavelength of the radiation generated in the quantum well is defined by its thickness and composition, and the efficiency of the cavity relies on the Bragg-stack composition modulation and thickness. There will be some interfacial grading and further layers making a total in excess of 450. This is quite a challenge, but possible to solve [41.19]. To add to the difficulty the surface is offset from the (001) crystal plane by 10°. Careful control of the fitting process and appropriate choice of reflections gives a very good indication of the sensitivity of certain parameters and some clear bounds that fit the x-ray data. The final fit compared with the experimental profile for the 006 reflection is shown in Fig. 41.7.

The sensitivity to composition and thickness is a clear advantage of x-ray scattering, and this is enhanced at increasing scattering angles (41.8). When the scattering angle is very small, the composition sensitivity from strain is negligible, but the scattering is still

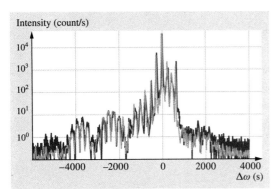

Fig. 41.7 The complex scattering from a VCSEL structure (*brown*) is shown along with the final fit to this profile (*grey*) (data courtesy of P. Kidd). The result gave bounds for the layer composition and thickness throughout this structure consisting of > 450 layers

sensitive to the composition because of the differences in scattering strength (41.1). This can be very useful for isolating the influence of compensating strain in structures such as $Si_{1-x-y}Ge_xC_y$ [41.20]. C has a very small covalent radius, but scatters very weakly, whereas Ge has a covalent radius closer to Si and scatters strongly. Since in this case Ge scatters strongly, the composition can be obtained from the reflectometry profile to within $\approx 0.1\%$ in x, provided the C concentration is estimated to within $\approx 1\%$ in y. Since the typical C concentration is < 1%, the Ge concentration can be determined to good precision; the reflectometry curve is given in Fig. 41.8a. The C concentration is then extracted from the 004 profile (Fig. 41.8b); however, this relies heavily on the assumptions concerning the C covalent radius. For the result to be consistent with secondary-ion mass spectrometry (SIMS), the covalent radius probably matches that in cubic SiC or the relationship between the lattice parameter and composition is far from linear.

From the above analyses it is clear that composition, thickness, and grading due to interdiffusion can be obtained quite directly. The interface roughness in the plane can also be estimated, although the theory does become a little more problematic. As with all the examples above, the structural model is derived via an iterative procedure by comparing simulated with measured profiles. Thickness, composition, and even interfacial grading are reasonably predictable, but lateral roughness is less clearly described. There are several models of interfaces, ranging from a truncated fractal [41.21], staircase for vicinal surfaces [41.22], and castellated surfaces, Holý within [41.6]. The resulting profiles are quite different and require estimates of the extent of the roughness replication layer to layer, i.e., none, partial, and full replication. The truncated fractal

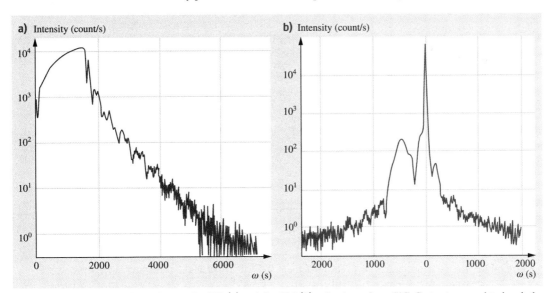

Fig. 41.8a,b The two profiles, close to the 000 (**a**) and the 004 (**b**) reflections, for a SiGeC structure used to break the correlation between parameters. By fitting the two profiles the composition of the Ge and C could be determined, because of their different influences on the two profiles

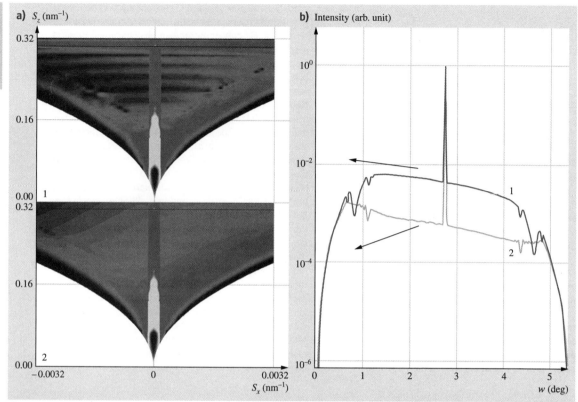

Fig. 41.9a,b The change in diffuse scattering close to the specular profile for various interface types: (1) castellated fully replicated and (2) unreplicated layer to layer for a periodic superlattice. The structure simulated is composed of (AlAs 50 nm + GaAs 50 nm) × 10 on GaAs, all with raised regions of width 50 ± 50 nm and height 1 ± 0.2 nm

description has a maximum length that is often termed the lateral correlation length; however, this is not an easy parameter to visualize. The nature of the jaggedness is also described by a single parameter and the scattering is derived from the correlation between different heights across the interface. The staircase and castellated surfaces require a distribution of heights and widths, with the former having an inclination angle. The castellated surface will almost equate to a fractal surface at one extreme. It should be clear from this that a good physical model for the likely interface shape is required to obtain some significance in the results. If the roughness is not replicated from layer to layer then the distributed intensity around the specular profile is rather uniform and weak, whereas replicated roughness brings the diffuse scattering into bands (Fig. 41.9). The most appropriate theory for modeling the scattering is DWBA, but this is only applicable for small roughness and a limited s_z range. If the roughness becomes significant or the simulation is taken to large s_z, then the kinematic approach (BA) is more applicable. Making a decision on the cross-over of DWBA to BA depends on several aspects [41.23], but certainly interfaces that are comparable in width to the layer thickness itself should use the BA, although this will not reproduce the region close to the critical angle or Yoneda wings [41.24] associated with surface roughness and transmission factors.

41.4.2 New Methods and New Analyses

All the methods and analyses described above are generally available. However, because the science in this field is progressing rapidly, some of the most exciting work requires considerable effort to reproduce. In many instances though, the equipment to undertake the work is either available commercially or for highly specialized experiments a synchrotron may be the only route. This section will consider some methods that extend the analysis beyond those discussed above; it does not pretend to be comprehensive, but rather picks out a few advances.

Dislocations are one of the significant defects that can destroy devices, and they influence the scattering, as will be clear from Chap. 42 on x-ray topography. Their influence is clear in reciprocal space maps, even at the very early stages of layer relaxation [41.25]. These misfit dislocations have a clear influence since they cause lattice plane rotation and strain that can be modeled [41.26]. The influence of threading dislocations is less clear since the Burgers vectors are not necessarily additive, but do contribute to a twist that is often interpreted purely in terms of peak broadening, although the influence of strain must also be considered [41.27]. Generally misfit dislocations, in GaN for example, are accompanied with threading dislocations, and obtaining the proportions of each requires analysis of the diffraction peak tails as well as the peak widths. A crucial aspect in this analysis is the correlation between dislocations, as this can have a profound influence on the FWHM [41.27, 28]. *Holý* et al. [41.29] are now taking this to the next stage, by reducing the probe volume to 1 μm or less, such that it is closer to the dislocation separation. It may then be possible to determine the threading dislocation densities in GaN epitaxial-lateral-overgrown structures, by averaging the diffraction profiles over several positions. However the simulation of these structures is not yet fully resolved.

The development in the analysis of quantum dots gives a good indication of what can be achieved by x-ray scattering experiments. The interpretation of scattering of buried dots can be based on DWBA. This is most appropriate in semiconductor systems where the dots form a region with a different lattice parameter and scattering strength. The distortion will then extend to the surface where the stress is relieved. The distribution of the dots can be measured by the lateral satellites associated with the Bragg peaks. For ordered arrays of dots these produce very strong effects [41.30]. However, for the more typical self-assembled dots, there may be only one broad satellite. This lateral spacing (41.9) between the satellite and the layer peak will give some measure of the dot spacing and its variation [41.31]. Combining spacings from planes inclined to the surface at several azimuths will give the average arrangement in the plane. For dots on the surface, the best approach is to use grazing-incidence small-angle scattering [41.32] to give satellites either side of the specular scattered beam. To extract details of the shape of buried dots, in-plane scattering appears to be the most unambiguous route [41.33], by modeling the strain, which must be anisotropic [41.34]. Generally though, these approaches have been entirely concentrated at synchrotron sources, although the analysis of buried dots [41.34] was carried out with a standard laboratory instrument with a sealed x-ray source (Fig. 41.5) and has even been extended to the analysis of a single layer of quantum dots buried 100 nm below the surface [41.35].

As discussed in Sect. 41.4.1, the shape (thickness) and not the strain is obtained from scattering close to the 000 *reflection* condition, and therefore the grazing-incidence small-angle scattering configuration will only give the shape of the dots. However to obtain the strain state of surface dots one uses in-plane scattering, i.e., grazing-incidence and grazing-exit scattering close to Bragg peaks, as in the analysis of buried dots. If the dots are known to have a regular shape or the shape can be deduced, some estimate of the strain state (composition or strain relaxation) can be obtained [41.36], although this can lead to complications with uncertainties about the shape [41.37]. Extracting the dot shape is clearly an iterative method based on some initial guess, and this averaging of the diffraction process will result in a weighted average of the information. If the x-ray source is coherent and an individual dot can be isolated, then the resulting diffraction pattern can be directly inverted to reconstruct the dot shape [41.38]. In practice isolating one dot is difficult, as is having perfect coherence, however incorporating a few dots with a partially coherent beam gives adequate data to reconstruct the dot shapes [41.39]. The procedure uses the measured amplitudes and iteratively solves for the phase information using methods developed in optics. The scattering volume is so small that the kinematical (BA) scattering theory is sufficient. The typical data collected from an experiment with a sealed laboratory source, averaging over $\approx 10^7$ dots [41.35], compared with that collected from a coherent synchrotron source for a line of $\approx 10^1$ dots [41.39], is presented in Fig. 41.10. The former obtains the dot shape and composition indirectly from the strain distribution created by the dots, whereas the coherent diffraction experiment gives the scattered amplitudes, i.e., the Fourier coefficients for the shape of surface dots.

The arrangements of quantum dots can be either regular or self-assembled, and both can be studied

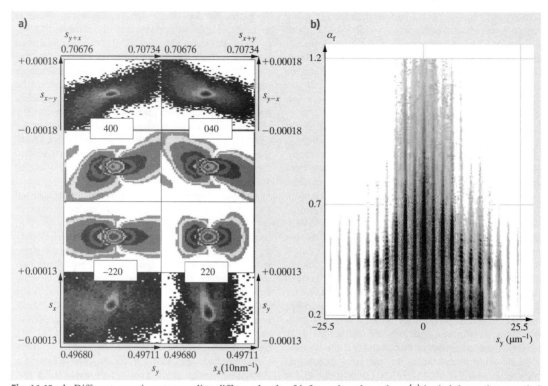

Fig. 41.10a,b Different experiments revealing different levels of information about dots: (**a**) buried dots using a sealed laboratory source, in-plane reciprocal space maps obtained just above the critical angle, for various reflections with their best-fit simulations, (**b**) the forward scattering, close to 000, from surface dots obtained with a coherent synchrotron source where the incident beam is below the critical angle for the substrate and the resultant scattering is purely from the dots (data courtesy of I. Vartanyants)

with x-ray scattering to yield useful information. Quantum wires on the other hand are nearly always regular since their growth is governed by vicinal wafers or patterning. These structures give impressive scattering patterns (Fig. 41.11) that are easily observable with laboratory sources [41.40]. They also change depending on whether the quantum wires or Bragg gratings are parallel or perpendicular to the incident x-ray beam. A rather pragmatic approach to the analysis can be applied. Effectively the Fourier transform based on the BA (kinematical) theory of a trial model shape and period can be used to obtain some very useful shape parameters. This will not give reliable intensities or account for

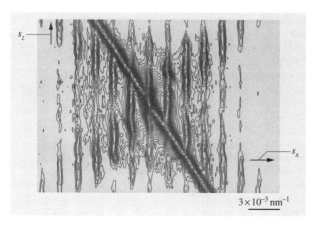

Fig. 41.11 The scattering close to the 004 reflection for a laterally periodic grating on a GaAs substrate. The data was obtained with a high-resolution diffractometer (data courtesy of M. Gailhanou). The *long streak* comes from the analyzer that had insufficient channel bounces before reaching the detector ◄

the true shapes of the scattering peaks for detailed interpretation. All combinations of the beam paths need to be considered [those scattered directly from the grating, those transmitted (refracted) through the grating, etc.] and are coherently brought together to account for all the cross-hatching of intensity, etc. based on the DWBA [41.41]. Surface features, dots or wires, do have an added complication to the analysis in that the strain influence is unlikely to be homogeneous and requires finite-element analysis to predict the strain distribution. However obtaining a good strain model can result in a high level of detail. Surface quantum wires studied with a detailed strain model suggest that SiGe/Si superlattice wires have an oxide coating [41.42]. For buried gratings the strains dominate the scattering [41.43], as with buried quantum dots.

41.5 Rapid Analysis

The methods discussed so far in this chapter have concentrated on analysis, both direct and simulation, and some specialized applications, with some discussion on the instrumentation. However it is largely the development in instrumentation that has made x-ray scattering so much more accessible. The instrumentation at synchrotron x-ray sources is very similar to how x-ray laboratories used to be, i.e., components are built specifically for the type of analysis required, so the set up time can be a significant factor. This does have the advantage in that some of the latest technology can be incorporated very rapidly, but the instruments do require highly trained specialists to set up the experiments, and it all becomes a significant team effort. Commercial laboratory machines are, on the other hand, rather versatile, with x-ray mirrors, lenses, and a plethora of monochromators, analyzers, and detectors. State-of-the-art laboratory equipment has exchangeable components that are precision-aligned so that several experimental configurations can be used without removing the sample. So in many ways the components, through significant cross-fertilization of ideas, are rather similar, although the intensity and wavelength choice at synchrotrons outstrips the laboratory source. However, with the use of mirrors, lenses, etc., the laboratory source gives sufficient intensity for the majority of applications. Some of the most interesting advances have been made in detector technology, which until recently were saturated by the high count rates. New developments in solid-state detectors, which are pixel-array single-photon counting detectors, will be a fascinating area to watch to create new opportunities.

Since the push is for increased speed and feedback in crystal growth there have been some interesting developments. In the quest for higher speed the intensity needs to be increased, or the data can be collected in parallel. Higher intensity usually comes at a cost to the resolution, although this can be overcome by making the experiment much more specific to the analysis. The first example of this was given in Sect. 41.4.1, where the more specific beam-selection diffractometer gives a 30× increase in intensity compared with conventional instrumentation with an x-ray mirror (Fig. 41.5). This is still a scanning method, which cannot compete with parallel data collection. The discussion (Sect. 41.3.1) on the influence of the instrument considered only resolution in diffraction space. However, when this is combined with the *real* space resolution, it is possible to achieve high-resolution profiles instantaneously [41.44]. Consider the profile in Fig. 41.6a, which already has 30× the intensity of a conventional instrument (Fig. 41.5); it took ≈ 7 min to collect. However the profile given in Fig. 41.6b was collected in 1 s. This static diffractometer has rather simple components: an x-ray source, slit, and a linear position-sensitive detector. The simulation of the scattering profile requires modeling of the beam paths to account accurately for all the features. This method can be applied to all wafer-based samples, by finding a suitable reflection, or if that is not possible by choosing an alternative wavelength. The highest resolution relies on a high-quality sample; however, useful information can still be extracted from imperfect materials, as will be shown in Sect. 41.6.

There are several growth chambers at synchrotrons for in situ characterization. Some of the early work concentrated on analyzing the surface reconstruction [41.45], and the investigation of dislocations by topography [41.46]. Some of the methods discussed in Sect. 41.4.2 have been applied in situ at synchrotrons to study the evolving crystal growth, and it is possible to see features even when 0.5 monolayer has been deposited [41.47]. A laboratory source, with the advantage of bringing the x-ray experiment to the metalorganic chemical vapor deposition (MOCVD) growth chamber, has been used for studying the growth of InGaN on GaN [41.48]. The source is used in combination with a Johansson monochromator to focus the beam onto the rotating sample, and despite the complications associated with rotation and wobbling, the evolution of the InGaAs on a GaN buffer layer is very clear. The integration time is $\approx 17\%$ of the rotation period. This is not high-resolution data, but perfectly sufficient to monitor the growing layer. This instrument is again relatively simple with a source and focusing Johansson crystal (which captures the divergent beam from the source and focuses it onto the sample); the scattered wave is collected by a position-sensitive detector.

These rapid analyses could be very useful for either quick feedback during the initial stages of crystal growth or controlling an established process. However, there are some interesting possibilities associated with rapid data collection and high intensity that will be addressed in the next section. Although these examples are specific for the material of interest, they are easily adapted; the concentration at this stage is on composition and thickness measurement and some estimate of *quality*.

41.6 Wafer Micromapping

Lateral changes in the composition, thickness, and *quality* across a wafer can be very revealing to the crystal grower. Generally the x-ray probing beam is of the order of mm, and therefore can suffer from significant averaging, adding uncertainty to the interpretation. As an example, the influence of peak widths and peak shifts contribute to the measured peak widths. The benefit of rapid data collection is that the probing beam can be reduced and still achieve reasonable measurement times. The typical sizes of the beam of the beam-selection diffractometer and the static diffractometer are $100\,\mu\text{m} \times 10\,\text{mm}$ to achieve these high data collection rates, so reducing this to $100\,\mu\text{m}$ double-pinholes reduces the intensity considerably ($> 100\times$). The intensity achieved is still very usable, although requiring longer count times or a compromise on counting statistics. As an example consider this far-from-perfect GaN sample, which is first probed with the beam-selection diffractometer using the $100\,\mu\text{m}$ double-pinholes, to obtain an estimate of quality; this is achieved by centering on the 0002 GaN peak and scanning in ω (Fig. 41.12a). The width of this peak is related to small tilted regions or some finite lateral dimension, e.g., the distance between defects or mosaic block size. The resulting map over the $2.5 \times 5\,\text{mm}$ sample at $50\,\mu\text{m}$ intervals is given in Figs. 41.12b,c, where the peak positions and peak widths across the wafer are shown. Clearly the peak width varies on this $100\,\mu\text{m}$ scale (Fig. 41.12b) and is most pronounced in one region that is between two areas that have quite different peak positions (Fig. 41.12c). These may be large mosaic blocks or regions of large curvature. If the measurement of the peak width was obtained on the mm scale then all these contributions would be combined and may have led to a different interpretation.

A similar analysis has been performed with the static diffractometer, but for speed obtained with a $200\,\mu\text{m}$ probe and stepping at $100\,\mu\text{m}$ intervals. This analysis gave the variation in composition and period across the wafer. The satellites are very weak (Fig. 41.12d) from this structure (the In concentration was rather low), but still an indication of the period variations could be obtained and did not indicate any significant fluctuations. The composition, though, does vary significantly over the wafer, with the center-right region being 10% below that of the lower region of the wafer.

The variation in the composition and the period can probably be linked quite clearly to the growth of the structure. However, the material *quality* can help in screening wafers. Reducing the probe clearly begins to separate the peak width and the peak position, and since the width, rather than the position, is generally related to distorted regions this may correlate with the local defect density and hence optical emission in GaN lasers and light-emitting diodes (LEDs), for example. The reduction in the probe size has two effects, not only the isolation of small regions but also the variation in peak broadening effects is likely to become larger as the averaging effects are reduced [41.49]. In the extreme case when the probe is close to a micron, the effects of individual dislocations will be observed, as discussed in Sect. 41.4.2. *Baumbach* et al. [41.50] have extended micromapping towards this extreme, using an extension of their *rocking-curve imaging* technique [41.51], which requires a well-collimated monochromatic beam from a synchrotron source; the spatial resolution is

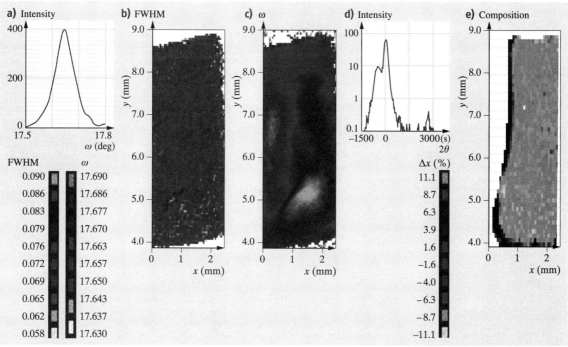

Fig. 41.12a–e Examples of wafer mapping on the 100 μm scale on a GaN wafer with a InGaN/GaN superlattice. Data collected with the beam-selection diffractometer: (**a**) a typical profile, (**b**) the distribution of the FWHM $\Delta\omega$, (**c**) the distribution of the peak position ω. The data collected with the static diffractometer; (**d**) a typical profile; and (**e**) the variation in composition across the wafer

defined by the pixel size of an area detector. By rocking the sample an array of profiles is achieved. An interesting aspect with this is the cross-over between topography and conventional high-resolution scattering, where the latter will not only break the averaging effect of larger probes but could provide ways of interpreting the contrast in topography that can arise from several complicating aspects. This combination of microdiffraction or topography (real space) and reciprocal space has many advantages in unraveling the structural information that is not revealed on the mm scale.

41.7 The Future

There is little doubt that x-ray scattering is very sensitive to structural details, and these methods are pushing the technological, theoretical, and instrumental developments. The intensity and quality of x-ray sources is no longer a restriction, with many synchrotrons existing around the world. The possibilities from coherent x-ray sources will become clearer as these facilities become more widely available, e.g., not only at synchrotrons but at x-ray free-electron laser facilities. The improvements in detectors will also create many opportunities, and are very necessary to cater for the large intensities now available.

Instrument development is benefiting from the cross-fertilization of laboratory and synchrotron-based work, with sophisticated optics being of use for both types of x-ray sources. Theoretical developments will also be a challenge as more scientists struggle to interpret the scattering from ever more exotic materials and structures. Enhanced computing power will benefit this field, for example, when modeling the diffuse scattering or optimization for extracting parameters of very complex materials.

It is very difficult to pinpoint specific areas where the most significant method developments will occur, since there is a large range of possibilities: coherent scattering, microtomography, micromapping, etc., but the most significant will be determined by the materials requiring analysis. The main aspect though, is that this whole field is moving very rapidly, creating an impressive toolset for materials characterization.

References

41.1 P.F. Fewster, N.L. Andrew: Interpretation of the diffuse scattering close to Bragg peaks by x-ray topography, J. Appl. Cryst. **26**, 812–819 (1993)

41.2 P.F. Fewster: Multicrystal x-ray diffraction of heteroepitaxial structures, Appl. Surf. Sci. **50**, 9–18 (1991)

41.3 P. Scherrer: Bestimmung der Grösse und der inneren Struktur von Kolloidteilchen mittels Röntgenstrahlen, Nachr. Ges. Wiss. Göttingen **2**, 98–100 (1918), in German

41.4 G.K. Williamson, W.H. Hall: X-ray line broadening from filed aluminium and wolfram, Acta Metall. **1**, 22–31 (1953)

41.5 L. Vegard: Die Konstitution der Mischkristalle und die Raumfüllung der Atome, Z. Phys. **5**, 17–26 (1921), in German

41.6 P.F. Fewster: *X-ray Scattering from Semiconductors*, 2nd edn. (Imperial College Press, London 2003)

41.7 P.F. Fewster: A high-resolution multiple-crystal multiple-reflection diffractometer, J. Appl. Cryst. **22**, 64–69 (1989)

41.8 P.F. Fewster: Reciprocal space mapping, Crit. Rev. Solid State Mater. Sci. **22**, 69–110 (1997)

41.9 P.P. Ewald: Zur Theorie der Interferenzen der Röntgenstrahlen in Kristallen, Phys. Z. **14**, 465–472 (1913), in German

41.10 C.G. Darwin: The theory of x-ray reflextion, Philos. Mag. **27**, 315–333 (1914)

41.11 C.G. Darwin: The theory of x-ray reflextion, Philos. Mag. **27**, 675–690 (1914)

41.12 J.A. Prins: Die Reflexion von Röntgenstrahlen an absorbierenden idealen Kristallen, Z. Phys. **63**, 477–493 (1930), in German

41.13 P.F. Fewster: X-ray diffraction from low dimensional structures, Semicond. Sci. Technol. **8**, 1915–1934 (1993), review article

41.14 J.M. Vandenberg, M.B. Panish, H. Temkin, R.A. Hamm: Intrinsic strain at lattice-matched $Ga_{0.47}In_{0.53}As/InP$ interfaces as studied with high-resolution x-ray diffraction, Appl. Phys. Lett. **53**, 1920–1922 (1988)

41.15 V. Holý, U. Pietsch, T. Baumbach: *High-Resolution X-ray Scattering from Thin Films and Multilayers* (Springer, Berlin, Heidelberg 1999)

41.16 M. Born, E. Wolf: *Principles of Optics*, 6th edn. (Cambridge Univ. Press, Cambridge 1980)

41.17 L.G. Parrat: Surface studies of solids by total reflection of x-rays, Phys. Rev. **95**, 359–369 (1954)

41.18 P.F. Fewster: A 'beam-selection' high-resolution x-ray diffractometer, J. Appl. Cryst. **37**, 565–574 (2004)

41.19 P. Kidd: Investigation of the precision in x-ray diffraction analysis of VCSEL structures, J. Mater. Sci. Mater. Electron. **14**, 541–550 (2003)

41.20 J. Zhang, J.H. Neave, X.B. Li, P.F. Fewster, H.A.W. El Mubarek, P. Ashburn, I.Z. Mitrovic, O. Buiu, S. Hall: Growth of SiGeC layers by GSMBE and their characterization by x-ray techniques, J. Cryst. Growth **278**, 505–511 (2005)

41.21 S.K. Sinha, E.B. Sirota, S. Garoff, H.B. Stanley: X-ray and neutron scattering from rough surfaces, Phys. Rev. B **38**, 2297–2311 (1988)

41.22 E.A. Kondrashkina, S.A. Stepanov, R. Opitz, M. Schmidbauer, R. Köhler, R. Hey, M. Wassermeier, D.V. Novikov: Grazing-incidence x-ray scattering from stepped interfaces in AlAs/GaAs superlattices, Phys. Rev. B **56**, 10469–10482 (1997)

41.23 I.D. Feranchuk, S.I. Feranchuk, A.P. Ulyanenkov: Self-consitent description of x-ray reflection and diffuse scattering from rough surfaces, unpublished work presented at XTOP 2006, Baden-Baden (2006)

41.24 Y. Yoneda: Anomalous surface reflection of x-rays, Phys. Rev. **131**, 2010–2013 (1963)

41.25 P. Kidd, P.F. Fewster, N.L. Andrew: Interpretation of the diffraction profile resulting from strain relaxation in epilayers, J. Phys. D Appl. Phys. **28**, A133–A138 (1995)

41.26 V. Kaganer, R. Köhler, M. Schmidbauer, R. Opitz, B. Jenichen: X-ray diffraction peaks due to misfit dislocations in heteroepitaxial structures, Phys. Rev. B **55**, 1793–1810 (1997)

41.27 V. Kaganer, O. Brandt, A. Trampert, K.H. Ploog: X-ray diffraction peak profiles from threading dislocations in GaN epitaxial films, Phys. Rev. B **72**, 045423–045434 (2005)

41.28 S. Daniš, V. Holý: Diffuse x-ray scattering from misfit and threading dislocations in $PbTe/BaF_2/Si(111)$ thin layers, Phys. Rev. B **73**, 014102–014107 (2006)

41.29 V. Holý, T. Baumbach, D. Lübbert, M. Elyyan, P. Mikulík: Diffuse x-ray scattering from dislocations in epitaxial layers – Beyond the ensemble

41.30 averaging, unpublished work presented at XTOP 2006, Baden-Baden (2006)

41.30 V. Holý, J. Strangl, G. Springholz, M. Pinczolits, G. Bauer, I. Kegel, T.H. Metzger: Lateral and vertical ordering of self-assembled PbSe quantum dots studied by high-resolution x-ray diffraction, Physica B **283**, 65–68 (2000)

41.31 A.A. Darhuber, P. Schittenhelm, V. Holý, J. Strangl, G. Bauer, G. Abstreiter: High-resolution x-ray diffraction from multilayered self-assembled Ge dots, Phys. Rev. B **55**, 15652–15663 (1997)

41.32 M. Schmidbauer, T. Wiebach, H. Raidt, M. Hanke, R. Köhler, H. Wawre: Self-organised ordering of $Si_{1-x}Ge_x$ nanoscale islands studied by grazing incidence small-angle x-ray scattering, J. Phys. D Appl. Phys. **32**, A230–A233 (1999)

41.33 J. Stangl, V. Holý, T. Roch, A. Daniel, G. Bauer, J. Zhu, K. Brunner, G. Abstreiter: Grazing incidence small-angle x-ray scattering study of buried and free-standing SiGe islands in a SiGe/Si superlattice, Phys. Rev. B **62**, 7229–7236 (2000)

41.34 P.F. Fewster, V. Holý, N.L. Andrew: Detailed structural analysis of semiconductors with x-ray scattering, Mater. Sci. Semicond. Process. **4**, 475–481 (2001)

41.35 P.F. Fewster, V. Holý, D. Zhi: Composition determination in quantum dots with in-plane scattering compared with STEM and EDX analysis, J. Phys. D Appl. Phys. **36**, A217–A221 (2003)

41.36 T.H. Metzger, I. Kegel, R. Paniago, J. Piesl: Grazing incidence x-ray scattering: an ideal tool to study the structure of quantum dots, J. Phys. D Appl. Phys. **32**, A202–A207 (1999)

41.37 D. Grigoriev, M. Hanke, M. Schmidbauer, P. Schäfer, O. Konovalov, R. Köhler: Grazing incidence x-ray diffraction at free-standing nanoscale islands: fine structure of diffuse scattering, J. Phys. D Appl. Phys. **36**, A225–A230 (2003)

41.38 I.A. Vartanyants, I.K. Robinson: Imaging of quantum array structures with coherent and partially coherent diffraction, J. Synchrotron Rad. **10**, 409–415 (2003)

41.39 I.A. Vartanyants, I.K. Robinson, J.D. Onken, M.A. Pfeifer, G.J. Williams, F. Pfeiffer, H. Metzger, Z. Zhong, G. Bauer: Coherent x-ray diffraction from quantum dots, Phys. Rev. B **71**, 245302.1–245302.9 (2005)

41.40 P. van der Sluis, J.J.M. Binsma, T. van Dingen: High resolution x-ray diffraction of periodic surface gratings, Appl. Phys. Lett. **62**, 3186–3188 (1993)

41.41 T. Baumbach, M. Gailhanhou: X-ray diffraction from epitaxial multilayered surface gratings, J. Phys. D Appl. Phys. **28**, 2321–2327 (1995)

41.42 Y. Zhang, V. Holý, J. Strangl, A.A. Darhuber, P. Mikulík, S. Zerlauth, F. Schäffler, G. Bauer, N. Darowski, D. Lübbert, U. Pietsch: Strain relaxation in periodic arrays of Si/SiGe quantum wires determined by coplanar high-resolution x-ray diffraction and grazing incidence diffraction, J. Phys. D Appl. Phys. **32**, A224–A229 (1999)

41.43 A. Ulyanenkov, N. Darowski, J. Grenzer, U. Pietsch, H. Wang, A. Forchel: Evaluation of strain distribution in freestanding and buried lateral nanostructures, Phys. Rev. B **60**, 16701–16714 (1999)

41.44 P.F. Fewster: A 'static' high-resolution x-ray diffractometer, J. Appl. Cryst. **38**, 62–68 (2005)

41.45 M. Sauvage-Simkin, Y. Garreau, R. Pinchaux, M.B. Véron, J.P. Landesman, J. Nagle: Commensurate and incommensurate phases at reconstructed (In,Ga)As(001) surfaces: x-ray diffraction evidence for a composition lock-in, Phys. Rev. Lett. **75**, 3485–3488 (1988)

41.46 S.J. Barnett, C.R. Whitehouse, A.M. Keir, G.F. Clark, B. Usher, B.K. Tanner, M.T. Emeny, A.D. Johnson: X-ray topography of lattice relaxation in strained layer semiconductors: Post growth studies and a new facility fo in-situ topography during MBE growth, J. Phys. D Appl. Phys. **26**, A45–A49 (1993)

41.47 D.K. Satapathy, B. Jenichen, W. Braun, V. Kaganer, L. Däweritz, K.H. Ploog: In situ grazing incidence x-ray diffraction study of strain evolution during growth and post growth annealing of MnAs on GaAs(113)A, J. Phys. D Appl. Phys. **38**, A164–A168 (2005)

41.48 C. Simbrunner, A. Navarro-Quezada, K. Schmidegg, A. Bonanni, A. Kharchenko, J. Bethke, K. Lishka, H. Sitter: In-situ x-ray diffraction during MOCVD of III-nitrides, Phys. Status Solidi (a) **204**, 2798–2803 (2007)

41.49 P.F. Fewster, G.A. Tye, N.L. Andrew, P. Kidd: Using the static diffractometer for micro-high-resolution x-ray scattering, unpublished work presented at XTOP 2006, Baden-Baden (2006)

41.50 T. Baumbach, D. Lübbert, V. Holý, P. Mikulík, L. Helfen, P. Pernot, M. Elyyan, S. Keller, B. Heskell, J. Speck: Epitaxial lateral overgrowth of GaN studied by x-ray micro-diffraction imaging, unpublished work presented at XTOP 2006, Baden-Baden (2006)

41.51 D. Lübbert, C. Ferrari, P. Mikulík, P. Pernot, L. Helfen, N. Verdi, D. Korytár, T. Baumbach: Distribution and Burgers vectors of dislocations in semiconductor wafers investigated by rocking-curve imaging, J. Appl. Cryst. **38**, 91–96 (2005)

1424

42. X-Ray Topography Techniques for Defect Characterization of Crystals

Balaji Raghothamachar, Michael Dudley, Govindhan Dhanaraj

X-ray topography is the general term for a family of x-ray diffraction imaging techniques capable of providing information on the nature and distribution of structural defects such as dislocations, inclusions/precipitates, stacking faults, growth sector boundaries, twins, and low-angle grain boundaries in single-crystal materials. From the first x-ray diffraction image, recorded by Berg in 1931, to the double-crystal technique developed by Bond and Andrus in 1952 and the transmission technique developed by Lang in 1958 through to present-day synchrotron-radiation-based techniques, x-ray topography has evolved into a powerful, nondestructive method for the rapid characterization of large single crystals of a wide range of chemical compositions and physical properties, such as semiconductors, oxides, metals, and organic materials. Different defects are readily identified through interpretation of contrast using well-established kinematical and dynamical theories of x-ray diffraction. This method is capable of imaging extended defects in the entire volume of the crystal and in some cases in wafers with devices fabricated on them. It is well established as an indispensable tool for the development of growth techniques for highly perfect crystals (for, e.g., Czochralski growth of silicon) for semiconductor and electronic applications. The capability of in situ characterization during crystal growth, heat treatment, stress application, device operation, etc. to study the generation, interaction, and propagation of defects makes it a versatile technique to study many materials processes.

- 42.1 Basic Principles of X-Ray Topography 1426
  - 42.1.1 Contrast 1426
  - 42.1.2 Resolution 1427
- 42.2 Historical Development of the X-Ray Topography Technique 1428
- 42.3 X-Ray Topography Techniques and Geometry 1430
  - 42.3.1 Conventional X-Ray Topography Techniques 1430
  - 42.3.2 Synchrotron-Radiation-Based X-Ray Topography Techniques 1431
  - 42.3.3 Recording Geometries 1435
- 42.4 Theoretical Background for X-Ray Topography 1435
  - 42.4.1 Limitation of Kinematical Theory of X-Ray Diffraction 1436
  - 42.4.2 Dynamical Theory of X-Ray Diffraction 1436
- 42.5 Mechanisms for Contrast on X-Ray Topographs 1440
  - 42.5.1 Orientation Contrast from Subgrains and Twins 1440
  - 42.5.2 Extinction Contrast 1441
- 42.6 Analysis of Defects on X-Ray Topographs 1445
  - 42.6.1 Basic Dislocation Analysis 1445
  - 42.6.2 Contrast from Inclusions 1446
  - 42.6.3 Contrast Associated with Cracks 1448
- 42.7 Current Application Status and Development 1449
- References 1450

The industrial demand for high-quality single crystals has increased manyfold following the inventions of the transistor and the laser, which led to a wide range of applications [42.1]. In semiconductor technology the increasing density of devices on individual integrated circuits requires a high level of homogeneity in terms of the chemical composition and structural perfection of crystals [42.2]. The presence of imperfections in crystals used in solid-state lasers, including semiconductor lasers, and nonlinear and electrooptical devices, negatively impacts on their reliable operation [42.3]. The study of such imperfections or defects is also important from the point of view of understanding the influence of imperfections on crystal growth processes, and conversely feedback can be used to develop higher-quality crystals [42.4]. Hence it is essential for process engineers and crystal growers to characterize and have knowledge of the amount, distribution, and nature of defects present in crystals. Even though there are several techniques based on optical, electron, and atomic force microscopy or x-ray imaging to reveal dislocations and other defects, none of these methods are ideal. These characterization methods have different ranges of capabilities and limitations, and hence should normally be used complementarily. However, compared with all other techniques, x-ray topography is powerful enough to image extended defects in the entire volume of the crystal and in some cases in wafers with devices fabricated on them. The capability of this technique to image both defects intersecting the wafer surface as well as those located in the bulk combines features specific to chemical etching and transmission electron microscopy (TEM), respectively. Using synchrotron radiation, the x-ray topography technique can be applied to study large wafers up to 300 mm diameter [42.5], crystal plates, and even as-grown boules weighing several kilograms [42.6]. Presently, x-ray topography has become a very important tool in fundamental research as well as in industrial applications, as described in detail in recent literature [42.7, 8].

X-ray topography is the general term for a family of x-ray diffraction imaging techniques capable of providing information on the nature and distribution of structural defects in single crystal materials. The complete name *x-ray diffraction topography* is slightly more informative, indicating that the technique is concerned with the topography of the internal diffracting planes, i.e., local changes in the spacing and rotations of these planes, rather than with external surface topography. Even though the name *x-ray topography* does not articulate the correct meaning of the technique, it is commonly used by researchers, perhaps for convenience or as a continuation from its first usage in early literature [42.9]. This technique is usually nondestructive and suitable for studying single crystals of large cross-section with thickness ranging from hundreds of micrometers to several millimeters. It is superior to the complementary TEM technique in that it enables imaging of the entire crystal and displaying defects particularly at dislocation densities lower than $10^4\,\mathrm{mm}^{-2}$. In x-ray diffraction topography, a collimated area-filling ribbon of x-rays is incident on the single-crystal sample, at a set Bragg angle, and the corresponding area-filling diffracted beam is projected onto a high-resolution x-ray film or detector. The two-dimensional diffraction spot thus obtained constitutes an x-ray topograph and precisely displays the variation of the diffracted intensity as a function of position depending upon the local diffracting power as well as the prevailing overall diffraction conditions. Local diffracting power is affected by the distorted regions surrounding a defect, leading to differences in intensities between these regions and the surrounding more perfect regions. This intensity variation gives rise to contrast, and different defect types can be characterized from the specific contrast produced by the way they distort the local crystal lattice and thereby the local diffracting power. The absence of magnification enables the correlation of the relative position of the image of a defect with its location inside the crystal. Quantitative information such as the line direction and Burgers vector of a dislocation can be obtained by detailed interpretation of the variations in contrast obtained under different diffraction conditions. Such interpretation requires an understanding of the mechanisms of contrast formation. These mechanisms are sensitive functions of the diffraction conditions and are derived from the kinematical and dynamical theories of x-ray diffraction [42.10].

42.1 Basic Principles of X-Ray Topography

42.1.1 Contrast

In general, the individual spots obtained from Laue diffraction patterns from crystals do not have uniform contrast. The localized variations in the intensity within any individual diffracted spot arise from the deviation in structural uniformity in the lattice planes causing the spot, and this forms the basis for the x-ray topographic technique. The real information available in an x-ray topograph is manifested in the form of contrast within the projected diffraction spot. This topographic contrast arises from differences in the intensity of the diffracted beam as a function of spatial coordinate inside the crystal. The diffracted intensity is a sensitive function of local crystal perfection. For example, under the correct diffraction conditions, highly mosaic regions of a crystal (i. e., those regions comprising small subgrains which are slightly tilted with respect to the perfect crystal) will diffract kinematically, whereas nearly perfect regions of the crystal will diffract dynamically. Kinematical diffraction is characterized by an absence of the primary extinction effects, which are the essence of dynamical diffraction. These primary extinction effects, created by the interference between diffracted waves and incident waves inside the crystal, lead to an overall reduction in diffracted intensity for a perfect crystal over the mosaic crystal. Conversely, there is an increase in diffracted intensity as the degree of mosaicity increases, eventually reaching the limit of the *ideally imperfect crystal* [42.11]. The difference between the intensities diffracted from one region of the crystal which diffracts kinematically and another which diffracts dynamically is one of the ways in which dislocations can be rendered visible in topography. However, the situation is not as simple as stated here. Even for a crystal which diffracts dynamically, the diffracted intensity is a sensitive function of the local distortion in the crystal. In addition, the type of contrast that arises from a particular type of distortion is also a function of the absorption conditions in the crystal. The effects of such conditions on contrast mechanism will be discussed in Sect. 42.5 with emphasis on mechanisms which give rise to dislocation images.

42.1.2 Resolution

The contrast from individual defects will only be clearly discernible if the spatial resolution is adequate. There is no magnification involved in topography and spatial resolution is controlled solely by geometrical factors. The Bragg law is defined as

$$\lambda = 2d \sin\theta_B , \quad (42.1)$$

where λ is the x-ray wavelength, d is the spacing between the diffracting planes, and θ_B is the Bragg angle, which is the angle between the incident rays and the diffracting planes. When the Bragg condition is satisfied, possible incident and diffracted beam directions lie, diametrically opposed, on the surface of a cone with semi-apex angle $90° - \theta_B$, the axis of which is the active reciprocal lattice vector g (i. e., the normal to the diffracting planes). Therefore if an x-ray source has finite size, which it always does, the possibility arises that at a given point in the crystal the diffracting planes will receive radiation, at the correct Bragg angle, from several different points located at different positions on the source (which produce rays lying on the surface of the Bragg cone). The locus of these points on the source is the arc defined by the intersection between the operative Bragg cone and the source surface. This acceptance of rays with a finite vertical divergence imparts a finite vertical divergence to the diffracted beam, which emanates from the point of interest in the crystal, and hence gives rise to a blurring effect in the resultant *image* of the point. In fact the diffracted rays emanating from the point of interest in the crystal will lie on the arc defined by the intersection between the same Bragg cone and the detector plane (the length of this arc will, of course, be defined by the divergence angle of the accepted rays from the arc on the source). This is illustrated in Fig. 42.1, which shows how an image of a point P in the crystal, recorded with rays which emanate from points lying on the arc abc on the source, is spread over the arc $a'b'c'$ on the detector. Therefore, the angle subtended by the longest dimension of the source in the direction perpendicular to the plane of incidence (defined by the incident and diffracted beam directions, and the active reciprocal lattice vector) is the crucial parameter, which should be minimized. The amount of blurring is proportional to the distance between the crystal and detector, and can be approximately written as

$$R = \frac{SD}{C} , \quad (42.2)$$

where R is the blurring (or the effective resolution), S is the maximum source dimension in the direction perpendicular to the plane of incidence, D is the specimen–film distance, and C is the source–specimen distance. In the plane of incidence, diffraction occurs over a finite range of angles determined by the acceptance angle of the crystal in that plane or the rocking curve width. This finite acceptance angle means that a given point in the crystal receives radiation from a finite width on the source (for the case of a horizontal plane of incidence). This width can be determined by back-projecting the fan of incident rays accepted at a point in the crystal to the source itself. Again a finite divergence, this time in the horizontal plane, can be attributed to the diffracted beam, and again a blurring effect will be evident. However, as can be verified by simple calculation, the limiting factor in determining spatial resolution is

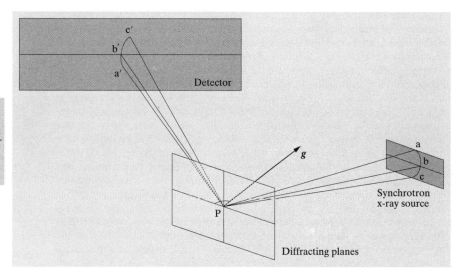

Fig. 42.1 Schematic diagram showing the effect of finite source dimension on resolution; g is the active reciprocal lattice vector. The angle bPg, which is the semi-apex angle of the Bragg cone, is $90° - \theta_B$, where θ_B is the Bragg angle

the finite source height (i.e., the source dimension perpendicular to the incidence plane), rather than the finite effective source width (i.e., the dimension parallel to the incidence plane).

Since the spatial resolution is proportional to the distance between the crystal and detector, this distance should be minimized. However, the optimum spatial resolution of the photographic detectors typically utilized in topography is limited, by photoelectron tracking between adjacent grains in the emulsion of the film, to around 1 μm. The specimen–film distance is usually set to yield a calculated spatial resolution which approximately coincides with this. It should be noted that, while this spatial resolution is greatly inferior to that of electron microscopy, it is more than sufficient for high-quality crystals of many materials currently grown.

42.2 Historical Development of the X-Ray Topography Technique

Even though the first topographic image of a single crystal was recorded as early as 1931 [42.12], the real potential of the technique was understood only in 1958 when *Lang* [42.13] demonstrated the imaging of individual dislocations in a silicon crystal. Different topographic geometries were developed independently during this period, namely, the *Berg–Barrett* reflection technique [42.12, 14], the double-crystal technique [42.15], and the Lang technique [42.13] and its variant, the scanning oscillator technique [42.16].

In 1931, *Berg* carried out x-ray diffraction imaging on crystal surfaces using characteristic radiation falling at a very low angle to the crystal surface and obtained point-to-point variation of the reflected intensity resulting in a striated image on a photographic plate placed near the crystal. The striated images produced were attributed to plastic deformation in the crystals. Even though it was realized that x-rays could be used as a powerful tool for studying inhomogeneities in crystals, no further attempt was made to use this tool until *Barrett* [42.14] recorded defect structure images from single crystals of silicon ferrite. *Barrett* improved *Berg's* reflection technique by minimizing the distance between the crystal and the photographic emulsion plate and also by using fine-grain high-resolution x-ray film. This technique is commonly known as Berg–Barrett reflection topography and is used to study large-size crystals. *Wooster* and *Wooster* [42.17] obtained topographic images revealing the defect structure from diamond surfaces using characteristic copper radiation. With further improvement in this technique, *Newkirk* [42.18] showed that individual dislocations could be resolved and their Burgers vectors could be experimentally determined. The *Schulz* technique [42.19] is another reflection topographic technique that uses white radiation from a point-focus x-ray generator.

The divergent x-ray beam diffracted by the crystal is recorded on a film. The white beam ensures diffraction from all the misoriented regions in the crystal. This simple technique can be used to reveal misorientations quickly in the form of separations and overlaps in the image.

Bond and *Andrus* [42.15] developed a high-resolution double-crystal technique for studying structural imperfections on the habit faces of natural quartz crystals. In this technique, x-rays from a line-focus source are Bragg-reflected from a highly perfect monochromator crystal and then diffracted from the specimen crystal in either reflection or transmission mode. High sensitivity is achieved in this technique because the first crystal further narrows the divergence of the slit-collimated beam, which subsequently probes the specimen crystal. This technique is highly sensitive to lattice misorientations: lattice tilts and lattice parameter changes down to 10^{-8} can be detected. The sensitivity can be increased further by introducing another monochromator or beam-conditioner crystal, and rocking curve measurements can also be carried out to assess the crystal quality.

Ramachandran [42.9] studied cleaved plates of diamond crystals using a white x-ray beam from a tungsten target in the transmission Laue geometry. He was the first to use the term *x-ray topograph* to describe the image contrast of full-size diffracted spots. The Laue spots were recognized as a topographic map of the crystal plate exhibiting variations in structure. He felt that the x-ray topographic technique might provide information on lattice perfection and compared the topographs with luminescence images. The term *x-ray topograph* became established in the literature when *Wooster* and *Wooster* [42.17] used the same term in describing diffraction images obtained from diamond surfaces revealing the defect structure. *Guinier* and *Tennevin* [42.20] studied both orientation and extinction contrast effects in aluminum in transmission mode using a polychromatic x-ray beam. *Tuomi* et al. [42.21] used a similar geometry and performed the first x-ray topography experiments using synchrotron radiation on silicon samples.

Important developments in x-ray topography took place in 1957–1958 when *Lang* developed a transmission technique to image the defects in crystals [42.13, 22]. He examined crystal sections using a narrow collimated characteristic radiation beam, a technique referred to as *section topography*. Linear and planar defects could be mapped with this technique by translating the specimen to known positions and taking a series of section topographs. Using this technique, low-angle boundaries in melt-grown metallic crystals were imaged [42.22]. By incorporating a linear traverse mechanism to translate the crystal plate and film cassette synchronously, *Lang* was able to image individual dislocations in an entire silicon crystal sample using projection topography, a milestone in the field of x-ray topography. In the Lang technique, an incident beam of narrow width and low horizontal divergence (about 4–5 arcmin) is obtained using a slit. The secondary slit placed on the other side of the sample blocks the direct beam while allowing the diffracted beam to pass through. It is possible to obtain an image on the film using K_{α_1} only. This technique can permit up to 1 μm resolution under optimized conditions. Coincidentally, during the same year, individual dislocations in silicon as well as germanium crystals were imaged by *Newkirk* [42.18] using the Berg–Barrett technique, and by *Bonse* and *Kappler* [42.23] using double-crystal topography, respectively. The tremendous development in x-ray topography during this period leading to resolving individual dislocations was also due to the fact that better-quality semiconductor crystals such as silicon and germanium were becoming available due to improved crystal growth technology. After *Lang*'s work [42.13, 22] on imaging of individual dislocations, x-ray topography has become an important quality control tool for assessment of semiconductor wafers before and after device fabrication. However, using this technique it is difficult to record topographs of crystals having elastic strains. Using the scanning oscillator technique developed by *Schwuttke* [42.16], it is possible to record transmission topographs of large-size wafers up to 150 mm in diameter, containing appreciable amount of elastic and/or frozen-in strain. This technique is based on the Lang technique with provision to oscillate the crystal and film simultaneously when the crystal is scanned. The oscillation is chosen to cover the whole reflecting range of the crystal to obtain a complete image of the crystal. This technique was widely used to image dislocations in transmission mode in large-size silicon wafers.

While laboratory-based x-ray topography techniques continue to be used, in recent decades the availability of numerous synchrotron radiation facilities providing intense, low-divergence x-ray beams of wide spectral range has allowed the development of synchrotron x-ray topography techniques. This is discussed in detail in Sect. 42.3.2.

42.3 X-Ray Topography Techniques and Geometry

X-ray topographic techniques can be broadly classified into two types: conventional x-ray topographic techniques based on laboratory x-ray sources such as target tubes, rotating anodes, and similar sources; and synchrotron x-ray topographic techniques based on synchrotron radiation.

42.3.1 Conventional X-Ray Topography Techniques

Berg–Barrett Topography

This is one of the oldest x-ray topography technique and is usually based on Bragg (reflection) geometry. The basic experimental setup (Fig. 42.2a) uses an extended x-ray source and the crystal is aligned such that the diffracting conditions are satisfied for the characteristic K_α lines from a set of Bragg planes. The crystal is cut such that the incident beam makes a small angle to the specimen surface and the diffracted beam emerges almost normal to the specimen surface. The photographic film can be placed very close to the specimen surface (as near as 1 mm). This technique is simple and uses low-cost equipment. The use of reflection geometry permits only imaging up to the x-ray penetration depth below the surface of the crystal, as determined by the extinction distance or absorption distance. This allows for the study of high-dislocation-density materials ($\approx 10^6$ cm^{-2}). Limitations of this technique include image doubling due to the K_α doublet as well as significant loss of spatial resolution with increasing specimen–film distance. These limitations can be overcome by appropriate adjustments to the recording geometry. This technique is often used for initial assessment of crystals of new materials [42.24]. The transmission Berg–Barrett method [42.12, 14] (Fig. 42.2b) (also know as the Barth–Hosemann geometry) is similar to the Lang technique except that it suffers from high background scattered radiation that limits its use for studying dynamical images.

Lang X-Ray Topography

The Lang x-ray topography technique [42.13, 22] is the most widely used laboratory technique and is based on the transmission geometry. Figure 42.3 shows the basic experimental setup, where the x-ray source is collimated to allow diffraction from one K_α line (usually K_{α_1}). The diffracting planes are typically nearly perpendicular to the crystal surface and the diffracted beam passes through a secondary slit that blocks the direct beam before striking the photographic plate. When the incident beam width is small compared with the base of the Borrmann fan formed by the extremes of the diffracted and transmitted beams with the crystal surface (typically about 10 μm), an image of a section through the crystal is obtained. This is the section topograph which allows the study of the three-dimensional defect distribution of defects. By translating both the crystal and film across the stationary beam, an image of the whole crystal can be obtained and this is the projection topograph. However, this results in loss of information available in the section topograph. The section topograph is therefore considered to be more fundamental. Usually, a much wider slit than that used in section topography is used when recording the projection topograph. Commercial Lang cameras consisting of a two-circle goniometer with precision translation stage and adjustable beam slits are available and widely used for laboratory-based x-ray topography [42.24].

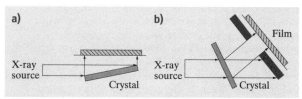

Fig. 42.2a,b Schematic diagrams of (**a**) reflection and (**b**) transmission Berg–Barrett techniques of topography

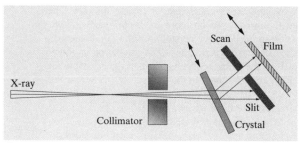

Fig. 42.3 Schematic diagram of the Lang projection technique. The topograph image due to K_{α_1} alone is recorded on the film. The secondary slit blocks the direct beam while allowing the diffracted beam to pass through. The crystal and film are translated synchronously and the whole image of the crystal is recorded

42.3.2 Synchrotron-Radiation-Based X-Ray Topography Techniques

The advent of dedicated sources of synchrotron radiation has enabled the development of a new realm of topography known as synchrotron topography. It has numerous advantages over conventional radiation techniques. These advantages derive from the high brightness, tunability, and natural collimation of synchrotron radiation.

White-Beam X-Ray Topography

One of the most important synchrotron topographic techniques developed is white-radiation topography [42.21, 25]. This technique is basically analogous to the Laue transmission technique, except with greatly enhanced capabilities which derive from the aforementioned natural collimation and high intensity of the synchrotron beam. The natural collimation (typically $\approx 2 \times 10^{-4}$ rad in the vertical plane coupled with an acceptance angle of typically a few milliradians in the horizontal plane) allows the use of very long beamlines (≈ 25 m) to maximize the area of the beam delivered at the sample without incurring significant losses in the total intensity originally available at the tangent point. This situation contrasts with that which would hold for isotropic emission from a conventional source. The large beam area delivered at the sample location allows studies to be carried out on relatively large-scale single crystals, and crystals as large as 150 mm or even 300 mm in diameter can be imaged by using precision translation stages similar to those used in the Lang technique (although exposure times are much shorter). Long beamlines also lead to small angles subtended by the source at points in the specimen, which in turn leads to excellent geometrical resolution capabilities. For example at the Stony Brook Topography Station, on beamline X-19C at the National Synchrotron Light Source (NSLS), the optimum theoretically attainable spatial resolution is $\approx 0.04\,\mu\mathrm{m/cm}$ of specimen–film distance.

If a single crystal is oriented in the beam, and the diffracted beams are recorded on a photographic detector, each diffraction spot on the resultant Laue pattern will constitute a map of the diffracting power from a particular set of planes as a function of position in the crystal with excellent point-to-point resolution (typically on the order of less than 1 µm). In other words each diffraction spot will be an x-ray topograph (Fig. 42.4a,b). The excellent geometrical resolution capability has another important consequence since it relaxes to a certain extent the requirement of having very small specimen–film distances (< 1 cm) in order to achieve good resolution. Thus, crystals can be surrounded with elaborate environmental chambers, necessitating considerable increases in specimen–detector distances (which would lead to intolerable resolution losses in conventional systems) without significant loss of resolution. The high intensity over the wide spectral range of the radiation emitted has several important advantages. Not only are the exposure times necessary to record a topograph drastically reduced from the order of days on conventional systems to seconds on a synchrotron, but since we have a white beam, a multiplicity of images is recorded simultaneously (Fig. 42.4c). This leads to great enhancement in the rate of data acquisition. The multiplicity of images also enables extensive characterization of strain fields present in the crystal. For example, instantaneous dislocation Burgers vector analysis can become possible by comparison of dislocation images obtained from several different Laue spots. Similarly, lattice rotation in the specimen can be characterized by analysis of the asterism (the spreading of a Laue spot) observed from several different Laue spots. In addition the good signal-to-noise ratios obtained associated with the high intensity also open up the possibility of direct imaging of topographs. Thus it becomes possible to conduct truly dynamic, quasi-real-time studies of crystals subjected to some kind of external stimulus such as applied fields, applied stress, heating, cooling, etc.

Synchrotron topography can also be used to image the surfaces of as-grown boules or large-size crystal plates in reflection geometry. This can reveal the overall distribution of defects and distortion around the cylindrical surface of these crystals. Investigating as-grown boules enables observation of the true microstructures and striations developed during growth and can substantially reduce the time and processing costs in cutting and polishing. Figure 42.5 shows reflection topographic images recorded from an as-grown boule grown by the Bridgman–Stockbarger method in microgravity. Topographs could be recorded covering the entire length of the boule in strips using the synchrotron beam. This boule revealed different defect structures depending on the amount of contact with the crucible wall.

Synchrotron x-ray topography in the reflection geometry can also be used to examine substrate/epilayer systems that have devices fabricated upon them. The features that make up the device topology typically provide contrast on x-ray topographs. The contrast usually originates from the strain experienced by the crystal at the edges of growth mesas, or metallization layers, although some absorption contrast may also be superimposed on this. Topographs recorded from such structures provide an image of the defect microstructure superimposed on the backdrop of the device topology. Direct comparisons can be drawn between the performance of specific devices and the distribution of defects within their active regions. This has made it possible to determine the influence of threading screw dislocations (closed and hollow core) on device performance [42.26, 27]. The back-reflection geometry is particularly useful here, since it gives a clear image of the distribution of screw dislocations on the background device topology that is imaged with sufficient clarity to unambiguously identify the device. An example of

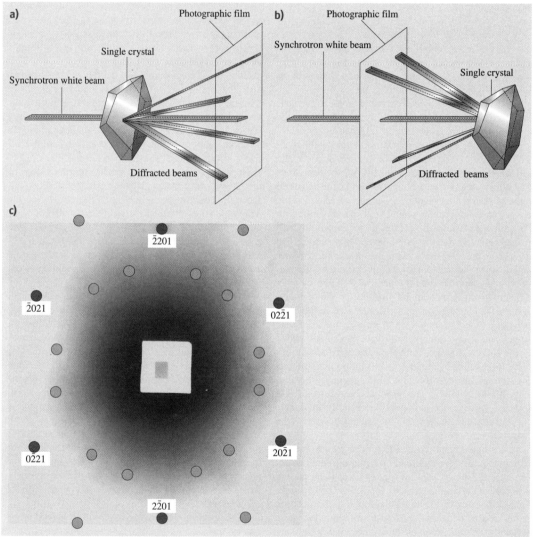

Fig. 42.4a–c Schematic diagram of the white-beam diffraction pattern recorded in (**a**) transmission geometry and (**b**) back-reflection geometry. (**c**) Actual transmission x-ray diffraction pattern recorded from an (0001) AlN single crystal on a SR-45 $8'' \times 10''$ x-ray film

a back-reflection image recorded from a crystal with devices fabricated on it is shown in Fig. 42.6.

Another important advantage of the white-beam topographic technique over conventional topographic techniques is its enhanced tolerance of lattice distortion. The wide spectrum available in the synchrotron beam allows crystals which either exhibit a uniform range of lattice orientation (for example, uniformly bent by a small amount) or which contain several regions of distinctly different orientation (for example, containing subgrains, grains or twinned regions) or regions of different lattice parameter (for example, containing more than one phase or polytype) to be imaged in a single exposure. Analysis of Laue spot shape or Laue spot asterism (the deviation from the shape expected from an undistorted crystal) enables quantitative analysis of lattice rotation. Simultaneous measurement of the variation in diffracted wavelength as a function of position in the Laue spot, for example, using a solid-state detector, also enables determination of any variation in the spacing d of the particular diffracting planes. For those crystals that contain several regions of distinctly different orientations, so-called orientation contrast becomes evident whereby the two neighboring regions of crystal separated by the boundary (a twin boundary or grain boundary, for example) each give rise to diffracted beams which travel in different directions in space, leading to image shifts on the detector. Analysis of the directions of these diffracted beams from the measured image shifts can enable the orientation relationships between the two regions of crystal to be established, leading to, for example, determination of twin laws. An example of a diffraction pattern recorded from a nominally (111) CdZnTe single crystal containing twins is shown in Fig. 42.7. Detailed analysis of the orientation relationships between the segments of the various diffraction spots enables the twin operation to be defined as a 180° rotation about [111].

By comparison, in the conventional topographic case, the maximum tolerable range of misorientation is defined by the convolution of the characteristic line width with the beam divergence. Since both of these quantities are small (typical line widths are of the order of 10^{-3} Å, equivalent to an angular divergence of ≈ 100 arcsec, and typically, beam divergences must be less than ≈ 20 arcsec in order to obtain the necessary angular resolution in the plane of incidence), misorientations greater than a few arcsec lead to the situation

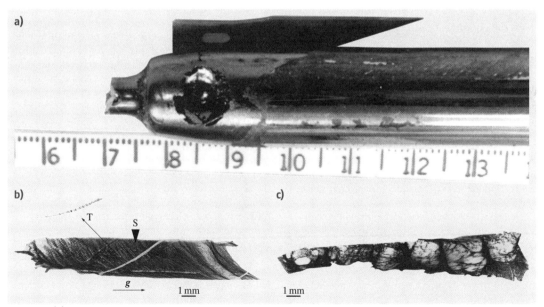

Fig. 42.5 (a) Optical picture of an as-grown CdZnTe boule grown in microgravity showing regions with different extent of wall contact. Reflection topographs recorded from the surface of as-grown CdZnTe boule grown in microgravity showing the defect structures. (b) Wall-contact region showing dense slip band networks and twins (S: slip bands, T: twin); (c) free-surface region showing a lower defect density and no twins

Fig. 42.6 Back-reflection white-beam x-ray topograph recorded from a 6H-SiC single crystal with thyristors fabricated upon it. The *small white spots* distributed over the image are 1c and larger screw dislocations. The location of these dislocations with respect to the device topology can be clearly discerned, enabling the influence of the defects on device performance to be determined. The *large white feature* corresponds to damage inflicted by a probe

where only part of the crystal fulfills the diffracting condition at a given time. In other words, Bragg contours are produced. These contours delineate those regions of the crystal which are in the diffracting condition from those which are not. Since their presence means that there are regions not set for diffraction, effectively there are blind spots in the crystal. Clearly this is a very undesirable situation if dynamic-type studies in large single crystals are to be conducted. Similar problems are encountered in monochromatic synchrotron-radiation topography.

Contrast mechanisms operative in white-beam topography are largely similar to those which are operative in Lang topography, with some notable subtle differences [42.28]. However, it has recently been demonstrated that many of these differences are attributable to the large difference in typical specimen–film distances for the two techniques (tens of centimeters for white-beam topography and ≈ 1 cm for Lang topography, see [42.29] for details). Some differences are also attributable to the added complication of harmonic contamination (the presence of several orders of

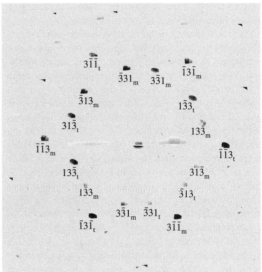

Fig. 42.7 Transmission x-ray diffraction pattern recorded from a CdZnTe crystal showing the presence of a twin. Detailed analysis of the orientation relationships between the segments of the various diffraction spots enables the twin operation to be defined as a 180° rotation about [111]

diffraction in a single diffraction spot), which becomes apparent in white-beam topography, potentially leading to a situation where more than one contrast mechanism operates in a single topograph. Harmonic contamination can, however, be avoided by judicious choice of diffraction geometry. In Sect. 42.5, a review of basic contrast mechanisms, pertinent to both Lang and white-beam topography, is presented.

Monochromatic-Beam X-Ray Topography

When a white synchrotron beam is passed through a monochromator, an x-ray topograph is obtained when the crystal is set to the Bragg angle for a specific set of lattice planes for the selected x-ray energy. Images from different atomic planes are acquired by orienting the sample to satisfy the Bragg condition for those planes and orienting the detector to the new scattering angle ($2\theta_B$) to record the image. With monochromatic radiation [42.30] only one topograph is recorded at a time, but the experimenter controls the energy or wavelength of the x-ray beam, the x-ray collimation, the energy or wavelength spread of the x-ray beam, and the size of the incident beam on the sample crystal. The synchrotron beam suffers significant loss of intensity on passing through the monochromator setup. This reduces

Fig. 42.8 Transmission x-ray topograph recorded from a CdZnTe crystal using a monochromatic synchrotron beam

the probability of radiation damage to the crystal. In addition to wavelength selection, the monochromator also improves angular collimation, and the plane-wave approximation is applicable. Monochromators used at synchrotron-radiation facilities can be either single- or multiple-crystal designs which can condition the x-ray beam to achieve optimal spatial and angular resolutions. An x-ray topograph (Fig. 42.8) recorded in transmission geometry from a CdZnTe single crystal shows dislocations, inclusions, and twinned regions.

42.3.3 Recording Geometries

X-ray topographs are acquired utilizing recording geometries based on either reflection (Bragg) from the surface, or transmission (Laue) through the bulk, of the sample crystal. In general, the reflection topograph geometry is employed for thick crystals or when absorption conditions and/or defect densities are too high to permit the use of transmission geometry. Owing to its surface sensitivity, topography in reflection geometry is also useful for characterization of surface defect structures within semiconductor heterostructures and epitaxial thin films. In all reflection-type geometries, defect information can be obtained from the volume defined by the effective area of the incident beam on the crystal and the penetration depth of the x-ray beam. The penetration depth of x-rays is determined either by the kinematical penetration depth (in imperfect crystals) or by the dynamical penetration depth (in highly perfect single crystals). The kinematical penetration depth (t_p^k) can be determined simply by geometrical relations between the incident and diffracted beams and the sample surface, and is given by

$$t_p^k = \frac{\mu_0(\lambda)}{(\mathrm{cosec}\,\Phi_0 + \mathrm{cosec}\,\Phi_H)}, \tag{42.3}$$

where $\mu_0(\lambda)$ is the linear absorption coefficient and Φ_0 and Φ_H are the angles of the incident and diffracted beams with respect to the surface, respectively. The dynamical penetration depth (t_p^d) is defined to be equal to half the extinction distance ξ_g (42.15). In transmission topography, all the defects within the crystal volume are recorded, provided that the absorption is low enough to permit sufficient transmission through the crystal. Since the x-rays pass through the entire thickness of the sample, this technique is used to characterize the overall bulk defect content of a crystal, such as dislocation networks and inclusions. The high intensities from the synchrotron source enable one to record images from relatively thick specimens using this technique. Laue transmission technique finds limited application in studying crystals of low perfection and high absorption coefficients. The back-reflection technique, commonly used for orienting single crystals, can also be used to record x-ray topographs of crystals containing specific defects such as superscrew dislocations (micropipes) in SiC [42.31]. The grazing Bragg–Laue [42.32] and the grazing-incidence reflection geometries allow precise tuning of the penetration depth for depth profiling studies of epitaxial thin films.

42.4 Theoretical Background for X-Ray Topography

X-ray topography is an imaging technique based on x-ray diffraction. Understanding contrast formation on x-ray topographs requires knowledge of the theory of x-ray diffraction in solids. In this section, the principles of the kinematical and dynamical theories of x-ray diffraction are briefly discussed with their implications for x-ray topography.

42.4.1 Limitation of Kinematical Theory of X-Ray Diffraction

In the kinematical theory of x-ray diffraction, initially developed to account for the intensities observed in x-ray diffraction studies, the amplitudes of the scattered waves are considered to be at all times small compared with the incident wave amplitude, and scattering from each volume element in the sample is treated as being independent of that from other volume elements. For small crystals, of dimensions less than a micrometer in diameter, and in heavily deformed crystals where the dislocations act to divide the crystal into a mosaic structure of independently diffracting cells, the kinematical theory may be employed satisfactorily to predict diffracted intensities. However, for large single crystals that are also highly perfect, significant discrepancies are found to exist between the measured and theoretically predicted intensities of diffracted beams. The diffracted intensity is predicted to increase continuously with increasing size of crystals and the diffracted intensity is actually predicted to become larger than the incident intensity beyond a certain size (about $1\,\mu\mathrm{m}$). This is evidently incorrect since it violates the principle of conservation of energy. Under such conditions, the kinematical theory breaks down and the volume elements can no longer be treated as independent of one another. As one can recall from wave theory, x-rays diffracted once from an atomic plane experience a phase change of $\frac{\pi}{2}$. When these waves are scattered again by the backside of the diffracting planes, they propagate in the same direction as the incident beam but are 180° out of phase. This gives rise to an attenuation of the incident intensity due to destructive interference between the primary incident and the secondary scattered beams, which in turn leads to a reduction in the total diffracted beam intensity. This is the so-called primary extinction effect, shown schematically in Fig. 42.9.

42.4.2 Dynamical Theory of X-Ray Diffraction

Breakdown of the kinematical theory in large, nearly perfect, single crystals is clearly demonstrated by the phenomenon of primary extinction. For a better understanding, it is necessary to take into account all the wave interactions within the crystal. The dynamical theory of x-ray diffraction considers the total wavefield inside a crystal while diffraction is taking place as a single entity.

The fundamental problem in the dynamical theory is to find solutions to Maxwell's equations in a periodic medium (i.e., the crystal) matched to solutions which are plane waves (the incident k_0 and diffracted k_h x-ray beams). These solutions must reflect the periodicity of the crystal, and such functions are knows as Bloch or lattice functions and may be represented by a Fourier series with appropriate Fourier coefficients.

Maxwell's equations are

$$\nabla \times \boldsymbol{E} = -\frac{1}{c}\frac{\partial \boldsymbol{B}}{\partial t} \quad \text{and} \quad \nabla \times \boldsymbol{H} = -\frac{1}{c}\frac{\partial \boldsymbol{D}}{\partial t}\,, \quad (42.4)$$

where $\boldsymbol{E}$ is the electric field, $\boldsymbol{B}$ the magnetic induction, $\boldsymbol{D}$ the electric displacement, $\boldsymbol{H}$ the magnetic field, and c the velocity of light in vacuum, assuming that the electric conductivity is zero at x-ray frequencies. Properties of the crystal are introduced by representing the crystal as a medium with a periodic, anisotropic, complex dielectric susceptibility χ. Since χ is a Bloch function, it can be expressed as a Fourier sum over all the reciprocal lattice vectors $\boldsymbol{R}_h$ as

$$\chi(r) = \sum_h \chi_h \exp[-2\pi\mathrm{i}(\boldsymbol{R}_h \cdot \boldsymbol{r})]\,, \quad (42.5)$$

where the Fourier coefficients are given by

$$\chi_h = -\frac{r_e \lambda^2}{\pi V} F_h\,,$$

where r_e is the classical electron radius, λ is the wavelength of the x-rays, V is the volume of the unit cell,

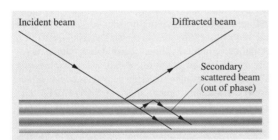

Fig. 42.9 Schematic diagram demonstrating the phenomenon of primary extinction in a perfect crystal. Diffracted beam (90° out of phase) is scattered by the backside of the diffracting planes to produce secondary scattered beam that is 180° out of phase with the incident beam, resulting in attenuation of the intensity of the incident beam, which in turn reduces the total diffracted beam intensity

R_h is the reciprocal lattice vector, and F_h is the structure factor of the unit cell.

Inside the crystal, the allowed wavevectors should satisfy the Laue equation

$$K_0 + R_h = K_h, \quad (42.6)$$

where K_0 and K_h are the incident and diffracted wavevectors inside the crystal.

Solutions for the electric displacement D are Bloch or lattice functions of the type

$$D = \sum D_n \exp[2\pi i(\nu t - K_h \cdot r)]. \quad (42.7)$$

Using the expressions for D (42.7) and χ (42.5) in Maxwell's equations, we obtain the fundamental equations of the dynamical theory that give the amplitudes of an infinite set of plane waves which together constitute a wavefield that satisfies the wave equation. In x-ray diffraction, only very rarely does more than one reciprocal lattice point provide a diffracted wave of appreciable amplitude. Thus, we need only consider two waves to have appreciable amplitude in the crystal: that associated with the incident wave and that associated with the diffracted wave from a reciprocal lattice vector h. The solution of Maxwell's equations inside the crystal is then expressed in terms of the amplitudes D_0 and D_h of these two waves by

$$\left. \begin{array}{l} [k^2(1+\chi_0) - K_0 \cdot K_0] D_0 + k^2 C \chi_{\bar{h}} D_h = 0 \\ k^2 C \chi_h D_0 + [k^2(1+\chi_0) - K_h \cdot K_h] D_0 = 0 \end{array} \right\}, \quad (42.8)$$

where C is the polarization factor and is unity for σ polarization (where vectors D_0 and D_h are perpendicular to the plane containing K_0 and K_h, and so are parallel) and $\cos(2\theta_B)$ for π polarization (where vectors D_0 and D_h lie in the plane containing K_0 and K_h, perpendicular to K_0 and K_h, respectively, but are not parallel to one another). Introducing the deviation parameters α_0 and α_h that express the deviation of the incident and diffracted wavevectors from the kinematic assumption

$$\left. \begin{array}{l} \alpha_0 = 1/2k \left[K_0 \cdot K_0 - k^2(1+\chi_0) \right] \\ \alpha_h = 1/2k \left[K_h \cdot K_h - k^2(1+\chi_0) \right] \end{array} \right\}. \quad (42.9)$$

We can write the solution as

$$\alpha_0 \alpha_h = k^2 C^2 \chi_h \chi_{\bar{h}}. \quad (42.10)$$

The amplitude ratio R can be written as

$$R = \frac{D_h}{D_0} = \frac{2\alpha_0}{C \chi_{\bar{h}} k} = \frac{C \chi_h k}{2\alpha_h} = \left(\frac{\alpha_0 \chi_h}{\alpha_h \chi_{\bar{h}}} \right)^{1/2}. \quad (42.11)$$

Equations (42.10) and (42.11) are the fundamental equations of the two-beam dynamical theory that allow us to predict the wavefields and their intensities inside (and outside) the crystal.

Dispersion Surface

The wave equations satisfying Maxwell's equations can be represented geometrically by a construction known as the dispersion surface, illustrated in Fig. 42.10. In the kinematical condition, the center of the Ewald sphere is at the Laue point L which is at the intersection of spheres of radius k about the origin O and reciprocal lattice point H in the dispersion plane. In the dynamical condition, the wavevector inside the crystal is corrected for the mean refractive index resulting in a shorter wavevector and therefore the loci of these wavevectors are represented by spheres of diameter $k(1 + \chi/2)$ about O and H. This results in a shift of the intersection of these spheres and the dispersion plane from the Laue point L to the Lorentz point L_0 (Fig. 42.10a). Figure 42.10b shows the region around the Laue point at very high magnification ($\approx 10^6$). On the scale of the picture, the spherical sections O'O'' and H'H'' of the projections of the spheres can be approximated as planes. The deviation parameters α_0 and α_h are measured perpendicularly from the planes O'O'' and H'H'', respectively, and denote the *tie point* A at which the tails of wavevectors K_0 from O and K_h from H intersect and diffraction occurs.

Thus, the dispersion equation is the loci of all such tie points and is an equation of a hyperboloid of revolution with OH as its axis. At the diameter points $\alpha_0 = \alpha_h$ so that the semidiameter of the hyperbola is given by

$$d_h = \frac{1}{2} kC \left(\chi_h \chi_{\bar{h}} \right)^{1/2} \sec \theta. \quad (42.12)$$

There are two independent dispersion surfaces for the two polarizations σ and π. The dispersion surface has two branches, the upper one denoted as branch 1 and the lower one as branch 2. Waves from the two branches are in antiphase. The direction of energy flow is described by the Poynting vector, parallel to $E \times H$, and it has been shown that this is perpendicular to the dispersion surface at the tie point [42.33, 34].

The boundary conditions at the crystal surface require that the tangential components of both E and H of the wavevectors should be continuous across the surface. The waves must be matched in amplitude at the crystal surface and in phase velocity parallel to the surface. The wavevectors inside the crystal differ from that

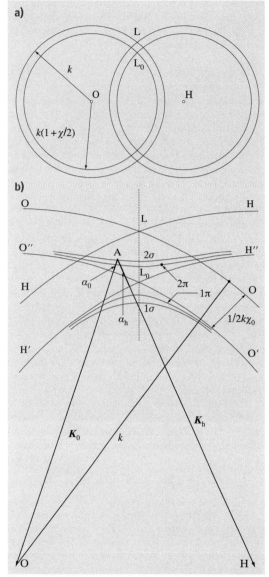

Fig. 42.10a,b Construction of the dispersion surface: (a) spheres of radius k and $k(1+\chi/2)$ about the origin O and reciprocal lattice point H in reciprocal space showing the position of the Laue point L and Lorentz point L_0; (b) dispersion surface for σ and π polarization states

outside only by a vector normal to the crystal surface

$$K_0 - k_0 = \delta n \,, \quad (42.13)$$

where n is a unit vector normal to the surface and δ is a scalar variable. A normal drawn line from the tip of the incident wavevector k_0 intersects the dispersion surface at the excited tie point and determines the tail of the wavevector K_0. In the Laue case (Fig. 42.11a), there are two points excited, one on each branch, labeled A and B. From each tie point, wavevectors directed towards O and H can be generated. There are thus four wavevectors generated in the crystal for each polarization, eight in all. At the exit surface of the crystal, the waves split up into diffracted and forward-diffracted beam and the boundary condition can be similarly determined. In the Bragg case (Fig. 42.11b), the normal from the surface intersects either two tie points on the same branch of the dispersion surface or none at all. The Poynting vectors associated with the two tie points are different; the energy flow from one point is directed into the crystal, but that from the other is directed outwards. The latter therefore does not generate any wavefields inside the crystal and can be ignored. Thus a single wavefield is generated for each polarization. When no tie points are selected, no wavefields are generated inside the crystal and total reflection occurs.

Borrmann Effect

The amplitudes of the wavefields are Bloch functions and modulated with the periodicity corresponding to the Bragg planes. The intensity I of each of this wavefields can be shown to be

$$I = D^2 = D_0^2[1 + R^2 + 2RC\cos(2\pi \mathbf{h} \cdot \mathbf{r})] \,. \quad (42.14)$$

The intensity is modulated by the factor $\cos(2\pi \mathbf{h} \cdot \mathbf{r})\cos(2\pi h \cdot r)$, which has a maxima at $\mathbf{h} \cdot \mathbf{r} = n$ and minima at $\mathbf{h} \cdot \mathbf{r} = (2n+1)/2$ for integer n ($\mathbf{h}$ is the diffraction vector). That is, maxima and minima of the standing wavefield occur either at, or halfway between, atomic planes (Fig. 42.12). Analysis of (42.14) reveals that the sign of R is opposite for wavefields with tie points on opposite branches of the dispersion surface, and intensity maxima occurs at the atomic plane when R is positive and minima at the atomic plane when R is negative. Maximum modulation occurs when both R and C equal unity, i.e., in centrosymmetric crystal ($\chi_h = \chi_{\bar{h}}$) for the σ polarized wave at the exact Bragg condition; i.e., when tie points excited on the two branches are at the diameter of the dispersion surface. Under such conditions, the wavefield with intensity maxima at the atomic planes (say branch 2) will suffer greater photoelectric absorption because the electron density is maximum at the atomic planes whereas the branch 1 wavefield, with intensity

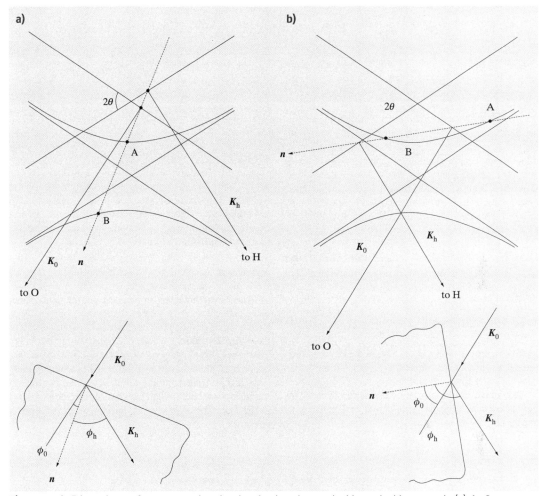

Fig. 42.11a,b Dispersion surface construction showing the tie points excited by an incident wave in (**a**) the Laue geometry: one tie point in each branch (A on branch 1 and B on branch 2) is excited (**b**) the Bragg geometry: two tie points (A and B) on the same branch are excited

maxima between the atomic planes, suffers minimum absorption. This effect is known as anomalous transmission or the Borrmann effect and was discovered by *Borrmann* in calcite [42.35, 36]. Presence of the Borrmann effect is indicative of crystal perfection.

Pendellösung Effect

An incident plane wave excites two tie points on the dispersion surface and generates two Bloch wavefields. The wavevectors associated with these tie points differ. The difference in wavevector leads to a difference in the propagation velocity, and interference effects can occur between the Bloch waves. This gives rise to the production of beats, a phenomenon referred to as Pendellösung. The period of the beats is given by the extinction distance ξ_g, which is the reciprocal of the dispersion surface diameter d_h, for the case of exact fulfillment of the Bragg condition

$$\xi_g = d_h^{-1} = \frac{\cos\theta_B}{Ck(\chi_h \chi_{\bar{h}})^{1/2}} = \frac{\pi V \cos\theta_B}{r_e \lambda C(F_h F_{\bar{h}})^{1/2}},$$
(42.15)

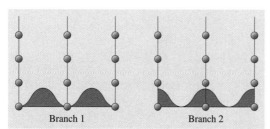

Fig. 42.12 Standing wavefields with a period corresponding to the spacing between the Bragg planes produced at the exact Bragg condition. Branch 1 waves, which have a minimum intensity at the atomic positions, suffer minimal absorption while branch 2 waves, which have a maximum intensity at the atomic positions, are strongly absorbed because of maximum electron density at the atomic planes

where F_h and $F_{\bar{h}}$ are the structure factors of (hkl) and $(\bar{h}\bar{k}\bar{l})$, respectively. When the Bragg condition is not exactly met, the extinction distance is given by

$$\xi'_g = \frac{\xi_g}{(1+\eta^2)^{1/2}}, \qquad (42.16)$$

where $\eta = \sin 2\theta_B/(C|\chi_h\chi_{\bar{h}}|)^{1/2}\Delta\theta$ is the deviation parameter expressing the departure $\Delta\theta$ of the tie points from the exact Bragg condition.

The above results can be extended to cover asymmetric reflections by the introduction of the terms γ_0 and γ_h, the cosines of the angles between the surface

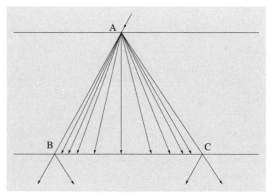

Fig. 42.13 Borrmann fan bounded by the incident (AB) and diffracted (AC) beams showing the distribution of energy for an incident spherical wave that excites all tie points along the dispersion surface

normal and the incident and diffracted beams, respectively, in the appropriate places.

The above analysis applies to an incident plane wave. However, in practice, laboratory sources used have a significant angular divergence because of which the entire dispersion surface is excited simultaneously. This gives rise to energy flow within the Borrmann fan bounded by the incident and diffracted beam directions (Fig. 42.13). Defects at any point within the Borrmann fan may contribute to the change of diffracted intensity at the exit surface of the crystal, and hence to image contrast.

42.5 Mechanisms for Contrast on X-Ray Topographs

Information from an x-ray diffraction topograph is obtained in the form of topographic contrast. As described earlier, local changes in orientation and spacing of crystal lattice planes give rise to local differences in either diffracted beam direction or intensity, which under the appropriate experimental conditions, are manifested as observable contrast on an x-ray topograph. There are two fundamental mechanisms for contrast in x-ray topographs:

1. Orientation contrast
2. Extinction contrast.

Extinction contrast can be further classified into three types:

1. Direct image contrast
2. Dynamical contrast
3. Intermediary contrast.

42.5.1 Orientation Contrast from Subgrains and Twins

Orientation contrast is explicitly defined as resulting from inhomogeneous intensity distributions arising purely from the overlap and/or separation of diffracted x-rays with varying directions. Orientation contrast is observed in crystals containing regions of different orientations such as grains, subgrains, and twins. Misorientations caused by dilations as well as rotations of the lattice can also lead to orientation contrast. Occur-

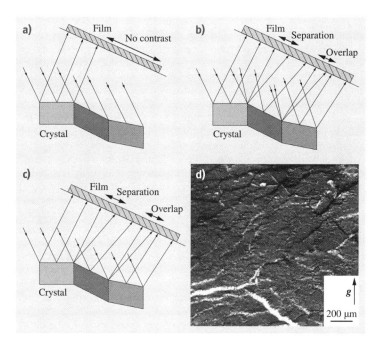

Fig. 42.14a–d Orientation contrast arising from misoriented regions: (**a**) monochromatic radiation (beam divergence < misorientation); (**b**) monochromatic radiation (beam divergence > misorientation); and (**c**) continuous radiation. (**d**) Reflection topograph from an HgCdTe single crystal. The *white bands* correspond to separation between images of adjacent subgrains, while the *dark bands* correspond to image overlap. The tilt angle is estimated at 1–4 arcsec

rence of orientation contrast depends on the nature of x-rays as well as on the nature of the misorientation, as illustrated in Fig. 42.14. If either the incident beam divergence or the range of wavelengths available in the incident x-ray spectrum is smaller than the misorientation between the blocks projected onto the incidence plane (Fig. 42.14a), then only one of the blocks can diffract at a given time. This produces one form of orientation contrast, i.e., the presence or absence of diffracted intensity. Under such conditions, one can adjust the diffraction geometry to bring the other block into the diffraction condition and thus obtain information on the mutual misorientation. However, if either the incident beam divergence or the range of wavelengths available in the incident spectrum allows each block to diffract independently (Fig. 42.14b or c), then the diffracted beams emanating from the individual blocks will travel in slightly different directions and give rise to image overlap if they converge, or separation if they diverge. This produces another form of orientation contrast. The degree of misorientation can be determined by measuring the image shifts (the amount of overlap or separation) as a function of specimen–detector distance on several different reflections. A white-beam x-ray topograph (Fig. 42.14d), recorded from a HgCdTe crystal that shows the overlap and separation of the subgrain images, clearly demonstrates orientation contrast.

42.5.2 Extinction Contrast

Extinction contrast arises when the scattering power around the defects differs from that in the rest of the crystal. Interpretation of this contrast requires understanding of the dynamical diffraction effects as discussed in the previous section (see also [42.37]). In this chapter, extinction contrast formation mechanisms are addressed with respect to images of dislocations on topographs since dislocations are among the most common defects studied by x-ray topography. Topographic contrast of dislocations (as well as other defects) consists of *direct, dynamic, and intermediary images* corresponding to the three different types of extinction contrast. In transmission geometry, which of the three types of image can be observed is determined by the absorption conditions. Absorption conditions are usually defined by the product of the linear absorption coefficient μ and the thickness of the crystal, t, traversed by the x-ray beam, i.e., μt. For topographs recorded under low absorption conditions ($\mu t < 1$), the dislocation image is dominated by the direct image contribution. Under intermediate absorption conditions, i.e., $5 > \mu t > 1$, all three components can contribute, while for high absorption cases ($\mu t > 6$), the dynamical contribution (in this case known as the Borrmann image) dominates.

The Direct Dislocation Image

The direct dislocation image is formed when the angular divergence or wavelength bandwidth of the incident beam is larger than the angular or wavelength acceptance of the perfect crystal [42.38]. Under this condition, only a small proportion of the given incident beam will actually undergo diffraction, with most of the incident beam passing straight through the crystal and simply undergoing normal photoelectric absorption. However, it is possible that the deformed regions around structural defects, such as dislocations and precipitates, present inside the crystal are set at the correct orientation for diffraction provided that their misorientation is larger than the perfect crystal rocking curve width and not greater than the incident beam divergence. The effective misorientation $\delta\theta$ around a defect is the sum of the tilt component in the incidence plane $\delta\varphi$ and the change in the Bragg angle θ_B due to dilation δd and is given by

$$\delta\theta = -\tan\theta_B \frac{\delta d}{d} \pm \delta\varphi . \qquad (42.17)$$

Therefore, the distorted region will give rise to a new diffracted beam. Further, if the distorted region is small in size then this region will diffract kinematically and will not suffer the effective enhanced absorption associated with extinction effects to which the diffracted beams from the perfect regions of the crystal are subjected. The enhanced diffracted intensity from the distorted regions compared with the rest of the crystal gives rise to topographic contrast. This is known as direct or kinematical image formation. This form of contrast dominates under low absorption conditions ($\mu t < 1-2$). Generally, the direct dislocation image formed by this extinction contrast model has been used to explain observed contrast features. Although the intensity increase for the direct image was in most cases qualitatively interpreted in a correct way [42.38], detailed measurements of dislocation image width made previously sometimes do not strictly coincide with the predictions of this theory [42.39].

From studies of the direct dislocation images of growth dislocations with large Burgers vectors (superscrew dislocations or micropipes) in x-ray topographs recorded from SiC single crystals, it was recently shown by *Huang* et al. [42.31, 40–42] that the extinction contrast theory alone is incapable of explaining the contrast features associated with superscrew dislocations on synchrotron topographs. By using a simple ray-tracing simulation method, it was demonstrated that the direct images of superscrew dislocations consist mainly of orientation contrast. Moreover, it was shown that this method is also applicable to elementary dislocations, indicating that it is a general phenomenon that orientation contrast makes a significant contribution to the direct dislocation image.

In the orientation contrast model, the mosaic region around the dislocation is divided into a large number of cubic diffraction units with their local misorientations coinciding with the long-range displacement field of the dislocation. These units diffract x-rays kinematically according to their local lattice orientation. Traces of the inhomogeneously diffracted x-rays are projected onto the recording plate to obtain the direct

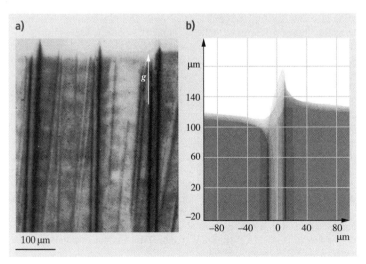

Fig. 42.15 (a) Synchrotron transmission topograph of superscrew dislocations in 6H-SiC ($g = 0006$, $\theta_B = 8.5°$, sample–film distance $d = 20$ cm); (b) simulation of pure orientation contrast of a $5c$ superscrew dislocation. The simulation parameters are chosen to coincide with the experimental conditions of (a)

image. Thus the direct image here is due to orientation contrast arising from the overlap and separation of inhomogeneously diffracted x-rays with continuously varying directions. Figure 42.15a shows several images of giant screw dislocations in SiC recorded in transmission geometry. The Burgers vector of dislocation 1 was independently measured [using optical phase-contrast microscopy, atomic force microscopy (AFM), and other x-ray topography techniques] to be $5c$ ($|b| = 75.85$ Å). The full-width w of the image is around $40\,\mu$m, and the separation between the maximum intensity peaks L_0 is around $24\,\mu$m. The conventional misorientation contour model predicts an image width of several hundred micrometers, which is clearly not the case. On the other hand, a good correlation is clearly evident between the observed image and the image simulated using the orientation contrast model (Fig. 42.15b). The simulated image consists of pure orientation contrast resulting from the separation or overlap of the inhomogeneously diffracted x-rays. In a similar way, images of superscrew dislocations recorded in back-reflection geometry can be successfully simulated, as shown in Fig. 42.16. It has also been shown that, under low absorption conditions, the above orientation contrast formation mechanism also applies to ordinary dislocations with Burgers vectors smaller than that of micropipes [42.7].

Origins of Dynamical Contrast from Dislocations

As discussed in Sect. 42.4, the dynamical theory of x-ray diffraction obtains solutions to Maxwell's equations in a periodic medium (i. e., the crystal) matched to solutions that are plane waves (the incident and diffracted x-ray beams). The wave equations satisfying Maxwell's equations can be represented geometrically by the dispersion surface [42.43], and a wavefield propagating in the crystal is represented by a tie point on the dispersion surface and comprises two waves corresponding to the incident and diffracted x-ray beams (Figs. 42.9 and 42.10). Dynamical contrast arises from the interaction of this wavefield with the dislocation distortion field (for a review see [42.37]). Under low absorption conditions this dynamical contribution to the dislocation contrast is mostly unobservable due to the fact that the image is dominated by the *direct image* contribution. Dynamical contrast becomes more observable as absorption increases, thereby attenuat-

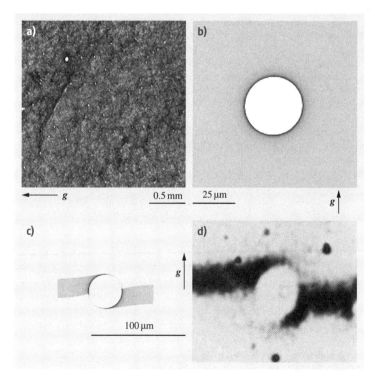

Fig. 42.16 (a) Back-reflection synchrotron topograph of a (0001) 4H-SiC wafer showing the circular images of superscrew dislocations ($g = 00016$, $\lambda = 1.25$ Å, sample–film distance $d = 20$ cm); (b) enlargement of one of the circular images in (a); (c) computer-simulated white-beam back-reflection x-ray section topograph of a screw dislocation (Burgers vector magnitude $b = 3c$) simulated under the diffraction conditions of (a); (d) back-reflection x-ray section topograph of a screw dislocation ($b = 3c$)

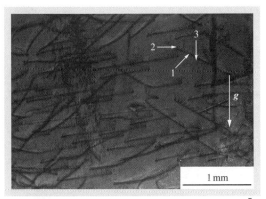

Fig. 42.17 SWBXT transmission topograph ($g = 10\bar{1}0$, $\lambda = 0.75$ Å) recorded from an AlN single crystal under intermediate absorption conditions ($\mu t = 8$) showing the direct (1), dynamical (2), and intermediary (3) images of a dislocation

ing the *direct image* contribution (the kinematically diffracted rays contributing to this image do not survive the absorption). Under these conditions, only wavefields associated with one branch of the dispersion surface (usually branch 1) which are close to the exact Bragg condition survive due to the Borrmann effect [42.35]. These wavefields can survive absorption even when the value of μt is significantly greater than unity.

For a perfect crystal, the incident boundary conditions determine the position of the tie point on the dispersion surface. However for an imperfect crystal, the local lattice distortion can modify the position of the tie point of a wavefield as it passes through the crystal. This can occur by two mechanisms depending on the nature of the distortion field. Tie point migration along the dispersion surface occurs when the wavefield encounters a shallow misorientation gradient upon passing through the crystal (e.g., regions away from the dislocation line). The variation of the misorientation should be less than the rocking curve width over an extinction distance, i.e.,

$$R_c \leq g\xi_g^2 \,, \tag{42.18}$$

where R_c is the radius of the curvature of the reflecting plane and ξ_g is the extinction distance. Under these conditions, both the direction of propagation of the wavefield and the ratio of the amplitudes of its components are changed. This phenomenon can be treated using the so-called eikonal theory [42.44], analogous to its counterpart in geometrical optics, formulated for the case of light traveling through a region of varying refractive index. As a wavefield approaches the long-range distortion field of a dislocation, rays will bend in opposite directions on either side of the core, potentially producing opposite contrast. However, if the dislocation is not located close to the crystal surface, the ray bending experienced above the defect may be compensated for by that experienced below the defect, suggesting that the contrast effects should cancel out. Net contrast is nevertheless observed, since any deviation of the wavefields from the direction of propagation corresponding to the perfect crystal region forces them to experience enhanced absorption, thereby producing a loss of intensity. Under these conditions, the dislocation image will appear white and diffuse. In cases where the dislocation is close to the exit surface of the crystal, the lattice curvature above and below the defect is asymmetric due to the requirements of surface relaxation. This means that the ray bending experienced above the defect is no longer compensated by that experienced below, with the result that opposite contrast is observed from regions either side of the defect.

On the other hand, when the wavefield encounters a sharp misorientation gradient upon passing through the crystal (e.g., regions close to a dislocation core), the strain field would completely destroy the conditions for propagation and force the wavefield to decouple into its component waves. When these component waves reach the perfect crystal on the other side of the defect, they will excite new wavefields (the so-called phenomenon of interbranch scattering). These newly created wavefields will have tie points which are distributed across the dispersion surface, and since only those wavefields with tie points close to the exact Bragg condition survive, they will be heavily attenuated, leading to a loss of intensity from the region surrounding the dislocation. Such images, known as Borrmann images, appear white on a dark background. An example is shown in Fig. 42.17, which is a detail from a white-beam x-ray topograph recorded from an AlN single crystal.

The Intermediary Image

The intermediary image arises from interference effects at the exit surface between the new wavefields created below the defect (as described in Sect. 42.4.2) and the undeviated original wavefield propagating in the perfect regions of crystal. Usually, these images often appear as a bead-like contrast along *direct* dislocation images on projection topographs. Under moderate absorption conditions when the defect (e.g., dislocation line) is inclined to the surface, the intermediary image forms a fan within the intersections of the exit and entrance surface of the dislocation and has an oscillatory contrast with depth periodicity over an extinction distance ξ_g. Again, this is illustrated in Fig. 42.17.

42.6 Analysis of Defects on X-Ray Topographs

Topography, both synchrotron and conventional, is well suited for analysis of low densities ($< 10^6$ cm^{-2}) of dislocations in crystals. The restriction to low densities arises from the fact that topographic dislocation images can be anywhere from around 5 to around 15 μm wide, so that greater densities would lead to image overlap and therefore loss of information.

42.6.1 Basic Dislocation Analysis

Determination of Line Direction
For dislocations created by slip processes, knowledge of the line direction as well as detailed information on the Burgers vector of the dislocations is required to fully assign the active slip system. Knowledge of both line direction and Burgers vector of crystal-growth-induced dislocations is also very important in understanding their origin and for the development of strategies to reduce the density of such dislocations. The projected directions of direct images of growth dislocations have also been used very successfully to compare with line energy calculations designed to determine why particular line directions are preferred by such dislocations in crystals (for a review see [42.45]). The line direction of a dislocation can be obtained by analyzing its direction of projection on two or more topographs recorded with different reciprocal lattice vectors. The use of analytical geometry enables the line direction to be determined either directly from the measured direction of projection [42.46] or indirectly by comparing calculated projected directions of expected dislocation line directions for the material of interest with the measured projection directions [42.47]. Such analysis is most readily carried out on direct images of dislocations since they are the most well defined, although similar analysis can be performed on dynamical or intermediary images.

Determination of Burgers Vector Direction
For sufficiently low dislocation densities, standard Burgers vector analysis, which enables the determination of the direction of the Burgers vector, is readily carried out in the low absorption regime, using the $\mathbf{g} \cdot \mathbf{b} = 0$ criterion for invisibility of screw dislocations, and the combination of $\mathbf{g} \cdot \mathbf{b} = 0$ and $\mathbf{g} \cdot \mathbf{b} \times \mathbf{l} = 0$ criterion for invisibility of edge or mixed dislocations (where $\mathbf{b}$ is the dislocation Burgers vector and $\mathbf{l}$ is the dislocation line direction). These criteria are also used in the analysis

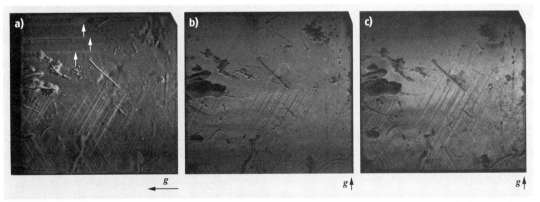

Fig. 42.18a–c Synchrotron white-beam topographs recorded in transmission geometry from a ZnO single crystal containing slip dislocations: (a) $\mathbf{g} = 11\bar{2}0$; (b) $\mathbf{g} = 1\bar{1}00$; (c) $\mathbf{g} = 2\bar{2}01$. Note the disappearance of dislocation segments (indicated by *arrows* on (a)) on (b) and the weak contrast on (c). The Burgers vector of these dislocations is determined to be $\frac{1}{3}[11\bar{2}0]$ and, from the line directions, it is determined that these are screw dislocations

of TEM images of dislocations. An example of dislocation analysis in a single crystal of zinc oxide (ZnO) is presented in Fig. 42.18. These dislocations were likely formed by deformation process during postgrowth cooling. These dislocations lie in the basal plane of the 2H crystal structure of ZnO and are visible in Fig. 42.18a ($g = 11\bar{2}0$) and invisible on the $1\bar{1}00$ (Fig. 42.18b) and the $2\bar{2}01$ (Fig. 42.18c) reflections, although weak contrast is observed. Application of the $g \cdot b = 0$ criterion to possible $\frac{1}{3}\langle 11\bar{2}0 \rangle$ Burgers vectors which lie in the basal plane shows that the dislocations have Burgers vector $\frac{1}{3}[11\bar{2}0]$.

Determination of Burgers Vector Sense and Magnitude

The determination of the actual sense and magnitude of the Burgers vector of a dislocation, once the Burgers vector direction has been determined by $g \cdot b$ analysis, requires more detailed analysis. *Chikawa* [42.48] developed conventional-radiation divergent-beam techniques for the determination of the sense of both edge and screw dislocations through measurement of the sense of tilt of lattice planes surrounding the dislocation core. For the case of screw dislocations, *Mardix* et al. [42.49] subsequently further developed Chikawa's divergent-beam technique to enable the magnitude of this tilt to be measured as a function of distance from the core. This could then be fitted to the corresponding theoretical expression, enabling the Burgers vector magnitude to be determined. *Si* et al. [42.50] developed two methods, one section and one projection, which are analogous to those of *Chikawa* [42.48], and *Mardix* et al. [42.49], but which make use of synchrotron white radiation. In these techniques the lattice tilt surrounding dislocations of mainly screw character was measured with an accuracy that was significantly improved due to the relaxation of the requirement for short specimen–film distances which is inevitable in conventional radiation techniques. *Chen*, *Dudley*, and co-workers [42.51, 52] have similarly determined the dislocation sense of micropipes and closed-core threading screw dislocations in 4H-SiC by comparing grazing-incidence synchrotron white-beam x-ray topographic images of the dislocations with corresponding images simulated by the ray-tracing method.

42.6.2 Contrast from Inclusions

Individual point defects are not visible on x-ray topographs, but when such defects cluster to form a precipitate or inclusion, contrast can be observed. Under low absorption conditions, *direct* or kinematical images of precipitates are formed on x-ray topographs and these typically consist of two dark half-circles separated by a line of no contrast perpendicular to the projection of the diffracted vector. This is simply due to the fact that distortions parallel to a given set of atomic planes are not discernible. An example is shown in Fig. 42.19, which is a transmission synchrotron white beam x-ray topography (SWBXT) image recorded from an AlN crystal. Under higher absorption conditions, dynamical contrast can be observed. When the precipitate is close to the x-ray exit surface, opposite contrast either side of the defect can usually be observed. This contrast will usually reverse with reversal of the sign of the reflection vector. The contrast is produced by tie point migration in the region above the defect. Since the defect is close to the exit surface, the reflecting planes rotate very sharply in order to meet the surface at the preferred angle, and so the curvature becomes too large for the eikonal theory to handle. Consequently, the contrast developed above the defect is *frozen-in*. The black–white contrast not only reverses with the sense of the reflection vector but also with the sense of the strain in the lattice. This can be used to determine the nature of the precipitate. If the contrast is enhanced on the side of positive g, then the lattice is under compression, and if reduced it is under tension. This empirical rule was first determined by *Meieran* and *Blech* [42.53].

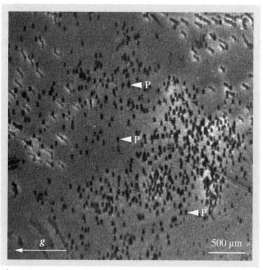

Fig. 42.19 Detail from a synchrotron white-beam topograph recorded in transmission from a AlN single crystal showing precipitate contrast (P) under low absorption conditions

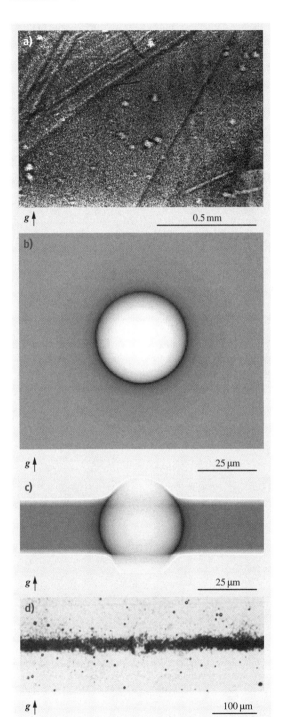

Precipitates/inclusions may also be imaged using the back-reflection geometry, as shown in Fig. 42.20a for a 3C-SiC platelet containing inclusions near the surface [42.54]. The circular spots of inclusions are similar to the images of superscrew (micropipes) in 4H-SiC (Fig. 42.16) except the screw dislocation images have more distinct white contrast than the inclusion images, there being some diffracted intensity inside the inclusion spots making them appear gray rather than white.

Images of inclusions in back-reflection topographs can be understood by considering ray-tracing computer simulations based on the displacement field of a spherical inclusion. The ray-tracing concept is the same as the orientation contrast model (see Sect. 42.5.2). Figure 42.20b shows a simulated image of an inclusion with a diameter comparable to the white spots in Fig. 42.20a. The simulation a $3c$ axial screw dislocation in Fig. 42.16b and that of an inclusion in Fig. 42.20b are similar, in that intensity reflected from the area centered on their defects is concentrated into a dark ring. This effect is complete for the case of the screw dislocation of Fig. 42.16b with intensity being absent from the ring, whereas for the case of the inclusion the intensity within its dark ring is merely depleted.

The difference between the mechanisms of image formation in the back-reflection topography of axial screw dislocations and inclusions can be analyzed more intimately with section topographs. Simulated section topographs may be obtained by utilizing only the diffracting elements of the crystal that fall within the projected path of a narrow strip of incident rays. For a screw dislocation (Fig. 42.15c), on the section topograph, the strip of diffracted intensity is displaced downward and upward at either side of the white circular center of the simulated screw dislocation image to break into two tails. The upper tail originates from the left half of the illuminated dislocation core, while the lower tail corresponds to the right half. The presence of the tails stretching in directions out of their illuminated sides indicates that the diffracted rays are

Fig. 42.20 (a) Synchrotron white-beam back-reflection x-ray topographs from a 3C-SiC platelet ($g = 12\,12\,12$, $\lambda = 1.24$ Å) showing inclusions; (b) computer-simulated back-reflection white-beam x-ray topograph ($\theta_B = 80°$) of an inclusion 14 μm beneath the surface of a 3C-SiC sample; (c) computer-simulated white-beam back-reflection x-ray section topograph of an inclusion; (d) recorded synchrotron white-beam back-reflection x-ray section topograph of an inclusion 14 μm beneath the surface ($g = 12\,12\,12$, $\lambda = 1.24$ Å) ◄

twisted from the incident beam direction. The twisting direction of the rays is opposite to the dislocation sense. Back-reflection section topography is then capable of discerning the senses of axial screw dislocations. The section topograph of an inclusion (Fig. 42.20d) lacks this two-tailedness.

Like the traverse topograph of Fig. 42.20a, the defect image again shows a gray center of depleted intensity bracketed by a dark perimeter. The diffracting planes in the vicinity of a spherical inclusion lack the helical twist that those in the core of a screw dislocation have. A spherical inclusion's strain field bulges the diffracting planes above it into convex curvatures, tilting them radially outward from the defect's center. In the simulated section topograph of an inclusion in Fig. 42.20c, the absence of twist is evident as the image is symmetrical about a line drawn through the center of the defect parallel to the g vector. Diffracted intensity is depleted from the central area of the defect image and concentrated at its edges, forming two curved bars of dark contrast. This is roughly what is seen in the section topograph of an inclusion in Fig. 42.20b.

42.6.3 Contrast Associated with Cracks

X-ray topography is also capable of discerning the deformation fields associated with cracks in single crystals. *George* et al. [42.55] have used x-ray topography to measure the extent of the plastic zone associated with the stress concentrations that can arise at the tip of a crack. *Raghothamachar* et al. [42.56] reported on deformation initiated at cracks in AlN crystals. This was achieved both in static mode using Lang topography [42.55] and synchrotron white-beam x-ray topography [42.56] and in dynamic mode using monochromatic synchrotron-radiation topography [42.57]. In a similar way, the elastic deformation field associated with crack tips can also be characterized. For example, for the case of *mud cracks*, which are often observed in GaN films grown on either SiC or sapphire substrates, the elastic field of the cracks can penetrate down into the substrate and be observable by x-ray topography. *Itoh* et al. [42.58] published an early report of such topographic observations in GaN films grown on sapphire. The presence of these mud cracks is evidence for the existence of tensile stresses in the films. For the case of SiC substrates, these tensile stresses may arise from thermal expansion mismatch, although this cannot explain their existence for the case of sapphire substrates. For the latter case, the tensile stresses are believed to be associated with grain coalescence [42.59]. TEM observation of the interface regions in the GaN/sapphire system indicates the absence of plastic deformation associated with the crack tips, rather just the presence of an elastic deformation field. This elastic deformation field (for both SiC and sapphire cases) behaves with respect to x-ray topographic observation like an edge dislocation, i.e., one can define a displacement vec-

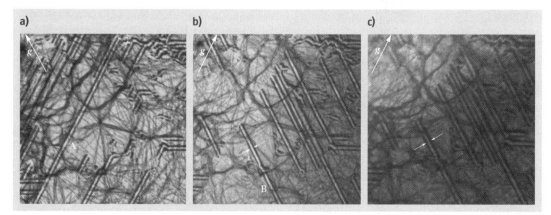

Fig. 42.21a–c SWBXT images recorded from a SiC substrate with GaN epilayers. (a) $g = \bar{2}110$, $d = 10$ cm; (b) $g = \bar{1}2\bar{1}0$, $d = 10$ cm; (c) $g = \bar{1}2\bar{1}0$, $d = 24.7$ cm. The pairs of *dark parallel lines* arise from the elastic deformation associated with *mud cracks* in the GaN film which penetrate into the SiC substrate (intensity from the GaN film is too low to be visible on this image). Crack images A are visible in (a) but invisible in (b), while the reverse is true for crack images B. This demonstrates the applicability of the $g \cdot u = 0$ and $g \cdot u \times l = 0$ criteria. The *arrows* on (b) and (c) highlight the increase in the separation between the two parallel components of the crack images as the specimen–film distance is increased

tor (u) associated with the crack which is oriented in the plane of the interface and normal to the long axis (l) of the crack. This means that the crack image disappears when the conditions $g \cdot u = 0$ and $g \cdot u \times l = 0$ are simultaneously satisfied. Figure 42.21 shows topographs of a cracked GaN/SiC sample recorded under various diffraction conditions: $g = \bar{2}110$ and $d = 10$ cm, where d is the sample–film distance (Fig. 42.21a), $g = \bar{1}2\bar{1}0$ and $d = 10$ cm (Fig. 42.21b), and $g = \bar{1}2\bar{1}0$ and $d = 24.7$ cm (Fig. 42.21c). All the topographs were taken with the GaN film on the x-ray exit surface. From these figures we can see that all the cracks running along [$\bar{2}110$] which have u along [$01\bar{1}0$] are invisible in Fig. 42.21a while all the cracks along [$\bar{1}2\bar{1}0$] which have u along [$\bar{1}010$] are invisible in Fig. 42.21b, due to the aforementioned invisibility criteria. Ray tracing, carried out by recording SWBXT images of the crack deformation fields at different specimen–film distances, enables determination of the *sense* of the lattice rotation (i.e., on which side of the interface the *extra half-plane* would lie) as well as the exact location of the misoriented volume with respect to the core of the crack. This is demonstrated in Fig. 42.21b,c. The separation between the parallel components of the images in Fig. 42.21c is larger than that in Fig. 42.21b, indicating that the cracks open up towards the GaN film surface. Further quantitative ray-tracing analysis reveals that the misoriented volumes are located at a distance of 30 μm from the crack core and that the lattice rotations are of the order of 20 arcsec.

42.7 Current Application Status and Development

Photographic films continue to be the detectors of choice for recording x-ray topographs, although large-area charge-coupled device (CCD) detectors can be used when high resolution is not paramount. Currently, holographic films are capable of recording x-ray topographs with submicrometer resolution but require long exposure times even for an intense synchrotron beam. Nuclear emulsions (Ilford plates) which have a grain size below 1 μm are also suitable to record high-resolution x-ray topographs but they are expensive and require special handling and developing procedures. Single-side-coated high-resolution x-ray films (Agfa D3-SC, Kodak SR-45, Fuji IX20, etc.) which have a grain size on the order of 1 μm are usually adequate for recording most x-ray topographs. These have sufficient contrast and resolution and can be developed fast enough to allow real-time observation and feedback for recording multiple topographs. CCD detectors are used instead of films when quick imaging is needed in some dynamic studies; however, the results are somewhat limited due to the poor resolution (25–40 μm).

As an efficient technique to explore the microstructure of various materials, x-ray topography has been extensively applied in both the research and industrial fields. Through observation and analysis of structural features and defects, it plays a key role in the evaluation of materials quality, the investigation of defect formation mechanisms, and the processing–microstructure–properties correlation. For the development of the crystal growth process of any material, x-ray topography is a critical tool that goes hand in hand with the crystal growth technique employed. X-ray topography has also been used to characterize thin films and heterostructures. While conventional x-ray topography can be used as a laboratory tool for quick evaluation of materials, synchrotron-based techniques have the capability for a wide range of experimental modifications, including in situ studies of crystal growth, phase transformations, device performance, etc. Further advances in digitization and processing of images can allow for rapid and automated analysis of x-ray topographs.

Modifications to the x-ray topography technique have also been developed in recent years. One example is the technique of x-ray reticulography [42.60]. Based on this technique, a novel stress measurement procedure has been developed that is capable of noninvasively measuring the complete stress tensor of any single-crystal material. This technique can be applied to a wide range of materials, such as those used in the semiconductor industry for developing or improving the packaging processes for the ever-reducing dimensions of microprocessors, or to measure stresses in as-grown single-crystal materials for comparison with performance and simulations.

References

42.1 H.J. Scheel: The development of crystal growth technology. In: *Crystal Growth Technology*, ed. by H.J. Scheel, T. Fukuda (Wiley, New York 2003) pp. 1–14

42.2 C. Claeys, L. Deferm: Trends and challenges for advanced silicon technologies, Solid State Phenom. **47-48**, 1–16 (1993)

42.3 D.T.J. Hurle: *Handbook of Crystal Growth: Thin Films and Epitaxy* (Elsevier, New York 1994)

42.4 G. Dhanaraj, T. Shripathy, H.L. Bhat: Growth and defect characterization of L-arginine phosphate monohydrate, J. Cryst. Growth **113**, 456–464 (1991)

42.5 S. Kawado, S. Iida, Y. Yamaguchi, S. Kimura, Y. Hirose, K. Kajiwara, Y. Chikaura, M. Umeno: Synchrotron-radiation x-ray topography of surface strain in large-diameter silicon wafers, J. Synchrotron. Radiat. **9**, 166–168 (2002)

42.6 B. Raghothamachar, H. Chen, M. Dudley: Unpublished results (1998)

42.7 M. Dudley, X. Huang: X-ray topography. In: *Microprobe Characterization of Optoelectronic Materials*, ed. by J. Jimenez (Gordon Breach/Harwood Academic, Amsterdam 2003) pp. 531–594

42.8 G. Dhanaraj, X.R. Huang, M. Dudley, V. Prasad, R.H. Ma: Silicon carbide crystals – Part I: Crystal growth and characterization. In: *Crystal Growth Technology*, ed. by K. Byrappa, T. Ohachi (William Andrew/Springer, New York 2003) pp. 181–232

42.9 G.N. Ramachandran: X-ray topographs from diamond, Proc. Indian Acad. Sci. A **19**, 280–294 (1944)

42.10 A. Authier: Contrast of images in x-ray topography. In: *Diffraction and Imaging Techniques in Materials Science*, ed. by S. Amelinckx, R. Gevers, J. van Landuyt (North Holland, Amsterdam 1978) pp. 715–757

42.11 J.E. White: X-ray diffraction by elastically deformed crystals, J. Appl. Phys. **21**, 855–859 (1950)

42.12 V.W. Berg: Über eine röntgenographische Methode zur Untersuchung von Gitterstörungen an Kristallen, Naturwissenschaften **19**, 391–396 (1931), in German

42.13 A.R. Lang: Direct observation of individual dislocations by x-ray diffraction, J. Appl. Phys. **29**, 597–598 (1958)

42.14 C.S. Barrett: A new microscopy and its potentialities, AIME Transactions **161**, 15–65 (1945)

42.15 W.L. Bond, J. Andrus: Structural imperfections in quartz crystals, Am. Mineral. **37**, 622–632 (1952)

42.16 G.H. Schwuttke: New x-ray diffraction microscopy technique for study of imperfections in semiconductor crystals, J. Appl. Phys. **36**, 2712–2714 (1965)

42.17 W. Wooster, W.A. Wooster: X-ray topographs, Nature **155**, 786–787 (1945)

42.18 J.B. Newkirk: Method for the detection of dislocations in silicon by x-ray extinction contrast, Phys. Rev. **110**, 1465–1466 (1958)

42.19 L.G. Schulz: Method of using a fine-focus x-ray tube for examining the surfaces of single crystals, J. Met. **6**, 1082–1083 (1954)

42.20 A. Guinier, J. Tennevin: Sur deux variantes de la methode de luae et leurs applications, Acta Cryst. **2**, 133–138 (1949), in French

42.21 T. Tuomi, K. Naukkarinen, P. Rabe: Use of synchrotron radiation in x-ray-diffraction topography, Phys. Status Solidi (a) **25**, 93–106 (1974)

42.22 A.R. Lang: Point-by-point x-ray diffraction studies of Imperfections in melt-grown crystals, Acta Cryst. **10**, 839 (1957)

42.23 U. Bonse, E. Kappler: Röntgenographische Abbildung des Verzerrungsfeldes einzelner Versetzungen in Germanium-Einkristallen, Z. Naturforsch. A **13**, 348 (1958), in German

42.24 D.K. Bowen, B.K. Tanner: *High Resolution X-ray Diffractometry and Topography* (Taylor Francis, London 1998) p. 174

42.25 J. Miltat: White beam synchrotron radiation. In: *Characterization of Crystal Growth Defects by X-ray Methods*, ed. by B.K. Tanner, D.K. Bowen (Plenum, New York 1980) pp. 401–420

42.26 P.G. Neudeck: Electrical impact of SiC structural defects on high electric field devices, Mater. Sci. Forum **338-342**, 1161–1166 (2000)

42.27 P.G. Neudeck, W. Huang, M. Dudley: Breakdown degradation associated with elementary screw dislocations in 4H-SiC p+n junction rectifiers. In: *Power Semiconductor Materials and Devices*, ed. by S.J. Pearton, R.J. Shul, E. Wolfgang, F. Ren, S. Tenconi (Materials Research Society, Warrendale 1998) pp. 285–294

42.28 M. Hart: Synchrotron radiation – Its application to high speed, high resolution x-ray diffraction topography, J. Appl. Crystallogr. **8**, 436–444 (1975)

42.29 X.R. Huang, M. Dudley, J.Y. Zhao, B. Raghothamachar: Dependence of the direct dislocation image on sample-to-film distance in x-ray topography, Philos. Trans. R. Soc. Ser. A **357**, 2659–2670 (1999)

42.30 D.R. Black, G.G. Long: X-ray Topography, Special Publication 0960-10 (National Institute of Standards and Technology (NIST), Washington 2004)

42.31 M. Dudley, S. Wang, W. Huang, C.H. Carter Jr., V.F. Tsvetkov, C. Fazi: White beam synchrotron topographic studies of defects in 6H-SiC single crystals, J. Phys. D Appl. Phys. **28**, A63–A68 (1995)

42.32 M. Dudley, J. Wu, G.-D. Yao: Determination of penetration depths and analysis of strains in single crystals by white beam synchrotron x-ray topography in grazing Bragg-Laue geometries, Nucl. Instrum. Methods B **40/41**, 388–392 (1989)

42.33 N. Kato: The flow of x-rays and materials waves in ideally perfect single crystals, Acta Crystallogr. **11**, 885–887 (1958)

42.34 N. Kato: The energy flow of x-rays in an ideally perfect crystal: comparison between theory and experiments, Acta Crystallogr. **13**, 349–356 (1960)

42.35 G. Borrmann: The extinction diagram of quartz, Phys. Z. **42**, 157–162 (1941)

42.36 G. Borrmann: Absorption of Röntgen rays in the case of interference, Phys. Z. **127**, 297–323 (1950)

42.37 B.K. Tanner: *X-ray Diffraction Topography* (Pergamon, Oxford 1976)

42.38 B.K. Tanner: Contrast of defects in x-ray diffraction topographs. In: *X-ray and Neutron Dynamical Diffraction: Theory and Applications*, ed. by A. Authier, S. Lagomarsino, B.K. Tanner (Plenum, New York 1996) pp. 147–166

42.39 J.E.A. Miltat, D.K. Bowen: On the widths of dislocation images in x-ray topography under low-absorption conditions, J. Appl. Cryst. **8**, 657–669 (1975)

42.40 X.R. Huang, M. Dudley, W.M. Vetter, W. Huang, C.H. Carter Jr.: Contrast mechanism in superscrew dislocation images on synchrotron back reflection topographs. In: *Applications of Synchrotron Radiation Techniques to Materials Science IV*, ed. by S.M. Mini, D.L. Perry, S.R. Stock, L.J. Terminello (Materials Research Society, Warrendale 1998) pp. 71–76

42.41 X.R. Huang, M. Dudley, W.M. Vetter, W. Huang, W. Si, C.H. Carter Jr.: Superscrew dislocation contrast on synchrotron white-beam topographs: An accurate description of the direct dislocation image, J. Appl. Cryst. **32**, 516–524 (1999)

42.42 X.R. Huang, M. Dudley, W.M. Vetter, W. Huang, S. Wang, C.H. Carter Jr.: Direct evidence of micropipe-related pure superscrew dislocations in SiC, Appl. Phys. Lett. **74**, 353–355 (1999)

42.43 B.W. Batterman, H. Cole: Dynamical diffraction of x-rays by perfect crystals, Rev. Mod. Phys. **36**, 681–717 (1964)

42.44 P. Penning, D. Polder: Anomalous transmission of x-rays in elastically deformed crystals, Philips Res. Rep. **16**, 419–440 (1961)

42.45 H. Klapper: Defects in non-metal crystals. In: *Characterization of Crystal Growth Defects by X-ray Methods*, ed. by B.K. Tanner, D.K. Bowen (Plenum, New York 1980) pp. 133–160

42.46 D. Yuan, M. Dudley: Dislocation line direction determination in pyrene single crystals, Mol. Cryst. Liq. Cryst. **211**, 51–58 (1992)

42.47 J. Miltat, M. Dudley: Projective properties of Laue topographs, J. Appl. Cryst. **13**, 555–562 (1980)

42.48 J.I. Chikawa: X-ray topographic observation of dislocation contrast in thin CdS crystals, J. Appl. Phys. **36**, 3496–3502 (1965)

42.49 S. Mardix, A.R. Lang, I. Blech: On giant screw dislocations in ZnS polytype crystals, Philos. Mag. **24**, 683–693 (1971)

42.50 W. Si, M. Dudley, C.H. Carter Jr., R. Glass, V.F. Tsvetkov: Determination of Burgers vectors of screw dislocations in 6H-SiC single crystals by synchrotron white beam x-ray topography. In: *Applications of Synchrotron Radiation to Materials Science*, ed. by L. Terminello, S. Mini, D.L. Perry, H. Ade (Materials Research Society, Warrendale 1996) pp. 129–134

42.51 Y. Chen, M. Dudley: Direct determination of dislocation sense of closed-core threading screw dislocations using synchrotron white beam x-ray topography in 4H silicon carbide, Appl. Phys. Lett. **91**, 141918 (2007)

42.52 Y. Chen, G. Dhanaraj, M. Dudley, E.K. Sanchez, M.F. MacMillan: sense determination of micropipes via grazing-incidence synchrotron white beam x-ray topography in 4H silicon carbide, Appl. Phys. Lett. **91**, 071917 (2007)

42.53 E.S. Meieran, I.A. Blech: X-ray extinction contrast topography of silicon strained by thin surface films, J. Appl. Phys. **36**, 3162 (1965)

42.54 W.M. Vetter, M. Dudley: The contrast of inclusions compared with that of micropipes in back-reflection synchrotron white-beam topographs of SiC, J. Appl. Cryst. **37**, 200–203 (2004)

42.55 G. Michot, K. Badawi, A.R. Halim, A. George: X-ray topographic study of crack-tip dislocation patterns in silicon, Philos. Mag. A **42**, 195–214 (1980)

42.56 B. Raghothamachar, M. Dudley, J.C. Rojo, K. Morgan, L.J. Schowalter: X-ray characterization of bulk AlN single crystals grown by the sublimation technique, J. Cryst. Growth **250**, 244–250 (2003)

42.57 G. Michot, A. George: In situ observation by x-ray synchrotron topography of the growth of plastically deformed regions around crack tips in silicon under creep conditions, Scr. Metall. **16**, 519–524 (1982)

42.58 N. Itoh, J.C. Rhee: Study of cracking mechanism in GaN/α-Al$_2$O$_3$ structure, J. Appl. Phys. **58**, 1828–1837 (1985)

42.59 E.V. Etzkorn, D.R. Clarke: Cracking of GaN films, J. Appl. Phys. **89**, 1025–1034 (2001)

42.60 A.R. Lang, A.P.W. Makepeace: Synchrotron x-ray reticulography: principles and applications, J. Phys. D Appl. Phys. A **32**, 97–103 (1999)

1452

43. Defect-Selective Etching of Semiconductors

Jan L. Weyher, John J. Kelly

In the present chapter we first briefly consider mechanisms for the etching of semiconductors (Sect. 43.1) and relate these principles to methods for controlling surface morphology and revealing defects (Sect. 43.2). Section 43.3 describes in some detail defect-sensitive etching methods. Results are presented for the classical (orthodox) method used for revealing dislocations in Sect. 43.3.1. More recently developed open-circuit (photo)etching approaches, sensitive to both crystallographic and electrically active inhomogeneities in semiconductors, are reviewed in Sect. 43.3.2. In particular, attention will focus on newly introduced etchants and etching procedures for wide-bandgap semiconductors.

43.1	**Wet Etching of Semiconductors: Mechanisms** 1454
	43.1.1 Chemical Etching 1454
	43.1.2 Electrochemical Etching 1454
	43.1.3 Electroless Etching 1456
	43.1.4 Photogalvanic Etching 1458
43.2	**Wet Etching of Semiconductors: Morphology and Defect Selectivity** ... 1459
	43.2.1 Chemical Etching 1459
	43.2.2 Electrochemical Etching 1459
	43.2.3 Electroless Etching 1460
	43.2.4 Photogalvanic Etching 1461
43.3	**Defect-Selective Etching Methods** ... 1461
	43.3.1 Orthodox Etching for Revealing Dislocations 1461
	43.3.2 Electroless Etching for Revealing Defects 1469
References 1473	

Wet etching processes are widely used in the fabrication of semiconductors. There are three main application fields for this technique: device pattern formation, polishing, and visualization of damage or defects. This chapter is restricted to the description of defect-selective etching (DSE). Among the methods used for revealing and analyzing defects in semiconductors, such as x-ray topography and diffraction, transmission electron microscopy (TEM), cathodoluminescence (CL), electron-beam-induced current (EBIC), and laser scattering tomography (LST), etching has several favorable features that make the technique attractive for assessment of the quality of single crystals and device structures. These are: simple and relatively low-cost equipment, no limits to the size of samples to be examined, very quick data acquisition, and the broad range of defects that can be revealed and analyzed.

The rapid development of various etching systems and methods for different semiconductors began after the discovery in the 1950s of a direct correspondence between the outcrops of dislocations and etch pits. Numerous review papers appeared in regular journals and books summarizing the current developments in this field [43.1–12]. Although for elemental and *classical* compound semiconductors different etching systems are well known, challenging problems arise due to the technological developments of the new generation of wide-bandgap compound semiconductors (group III nitrides and SiC). These materials are characterized by very high chemical resistance to the majority of the known acid-based etching systems, show strong polarity-related anisotropy of properties, and contain a very high density and diversity of defects, not occurring in classical compound semiconductors (e.g., inversion domains, macro- and nanopipes, pinholes, and extended stacking faults). New etching systems and more sophisticated etching methods as well as their calibration are therefore essential.

43.1 Wet Etching of Semiconductors: Mechanisms

For etching of semiconductors in solution (or in a melt) one can distinguish two mechanisms [43.7, 13]. In *chemical etching* valence electrons are exchanged *locally* between surface bonds of the semiconductor and an active agent in the etchant. When all back bonds are broken, the surface atom passes into solution. The other type of mechanism involves *free* charge carriers (valence-band holes or conduction-band electrons) which, when localized at the surface, cause rupture of surface bonds. If an electrochemical cell, with a voltage source and a counterelectrode, is used to supply charge carriers to the semiconductor, we speak of *electrochemical etching*. Both electrons and holes may cause decomposition of the solid. Alternatively, free charge carriers can be supplied from a chemical species in solution under open-circuit conditions: this is referred to as *electroless etching*. In addition, if minority carriers are required for etching, the semiconductor may be illuminated. In this case we can distinguish *photoelectrochemical etching* and *electroless photoetching*. Finally, there is a hybrid form of photoetching that requires a counterelectrode but no voltage source. This *photogalvanic etching* method has been used extensively for wide-bandgap semiconductors.

In this section we first briefly review the various etching mechanisms. This forms the basis for a consideration of defect-selective etching in Sect. 43.2, a theme that will be further developed in Sect. 43.3.

43.1.1 Chemical Etching

A typical example of a chemical etching reaction is that of InP in concentrated HCl solution [43.14]. By electron exchange In-P and H-Cl bonds are broken and $InCl_3$ and PH_3 are formed as products

$$InP + 3HCl \rightarrow InCl_3 + PH_3 . \tag{43.1}$$

The products are subsequently hydrolyzed in the aqueous solution. Etching of InP in HBr follows a similar mechanism [43.7]. The chemical etching of silicon in alkaline solution is another important reaction [43.15, 16]. These are complex, multistep processes. A general feature is that their rate is controlled by surface kinetics and, as a consequence, etching is strongly anisotropic. For example, the etch rate of the Si (111) face in KOH solution and of the (111) In face of InP in concentrated HCl solution is much lower than that of other crystallographic faces. Unlike electrochemical or electroless etching, the kinetics of chemical etching is not dependent on the electrochemical potential of the sample.

For revealing defects in many semiconductors (Si, SiC, GaN) a NaOH/KOH eutectic is often used [43.12]. Despite the long and widespread application of this system, surprisingly little is known about the etching mechanism. In all these cases the semiconductor is oxidized, i.e., loses valence electrons, during etching. For example, GaN will be converted to Ga^{3+} ($Ga(OH)_3$ or Ga_2O_3) in the melt. OH^- ions cannot act as a reducing agent, i.e., electron acceptor. We suggest that trace water in the eutectic is responsible for bond rupture and enhanced etching at defects. For example, in the case of GaN a possible reaction could be

$$GaN + 3H_2O \rightarrow Ga(OH)_3 + NH_3 . \tag{43.2}$$

Water, strongly bound in the solvation shell of the cations and anions, is available for bond rupture. The product $Ga(OH)_3$ will dissolve in the melt.

43.1.2 Electrochemical Etching

For the kinetics of electrochemical etching one has to distinguish between reactions based on majority and minority carriers [43.7, 13]. In this section we first consider majority-carrier (holes in a p-type or electrons in an n-type semiconductor) and then minority-carrier reactions. To guide the reader we show in Fig. 43.1 a simple electrochemical cell with three electrodes: the semiconductor working electrode (WE), a platinum counterelectrode (CE), and a reference electrode (RE). With the aid of a voltage source (e.g., a potentiostat) the electrochemical potential V of the semiconductor can be varied with respect to that of the reference electrode. As a result, a current flows between the working electrode and the counterelectrode. This current j is measured as a function of V. The semiconductor can be illuminated to give photocurrent.

Majority-Carrier Reactions

If, in a p-type semiconductor, holes are localized at the surface (i.e., the bonding electrons are removed) surface bonds will be broken. In the case of GaN, three holes are required to dissolve one formula unit of the solid in acidic solution [43.17].

$$GaN + 3h^+ \rightarrow Ga^{3+} + \tfrac{1}{2}N_2 . \tag{43.3}$$

The solid is *oxidized*; the Ga^{3+} ions dissolve and molecular nitrogen escapes to the gas phase. From (43.3)

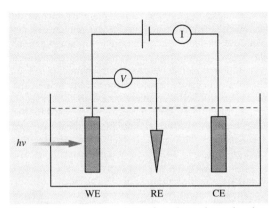

Fig. 43.1 A schematic view of a three-electrode electrochemical cell. The potential V of the semiconductor working electrode (WE) is regulated with respect to that of the reference electrode (RE) using a voltage source. As a result, a current j can flow between the working electrode and a counterelectrode (CE). In a typical experiment j is measured as a function of V either in the dark or under illumination. If electrons flow from WE to CE the current is *anodic* (j_a); an oxidation reaction occurs at the semiconductor. If electrons flow in the opposite direction (from CE to WE) the current is *cathodic* (j_c) and a reduction reaction occurs at the semiconductor

it follows that the rate of the reaction should depend on the surface hole concentration p_s, which is related to the hole concentration in the bulk p_b by

$$p_s = p_b \exp\left(\frac{eV_{sc}}{k_B T}\right), \qquad (43.4)$$

where eV_{sc} is the band bending in the semiconductor at the interface with the solution. In an electrochemical experiment the band bending, and thus the reaction rate, can be varied by means of the applied potential V. A change in the applied potential can lead to a potential drop across the space-charge layer of the semiconductor (Fig. 43.2) and across the double layer (Helmholtz layer) in solution. For simplicity, we neglect the latter; this will not markedly affect the argument in this section [43.7, 13]. In that case the band bending is given by

$$eV_{sc} = e(V - V_{fb}), \qquad (43.5)$$

where V_{fb} is the flat-band potential. At this potential, an important reference point in semiconductor electrochemistry, there is no space-charge layer or electric field present at the semiconductor surface (see the center

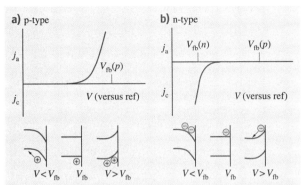

Fig. 43.2a,b Current–potential characteristics for majority-carrier reactions at (a) a p-type and (b) an n-type semiconductor. In case (a) accumulation of valence-band holes at the surface leads to oxidation of the electrode and an anodic current in the external circuit. The flat-band potential ($V_{fb}(p)$) is denoted, as well as energy-band schemes of the semiconductor at the interface with solution. In case (b) accumulation of electrons at the surface causes the reduction of H^+ ions or H_2O to hydrogen or, in some cases, the reduction of the semiconductor. Note: the difference in flat-band potential for n- and p-type forms of the same semiconductor corresponds to the difference in Fermi levels of the two types

picture at the bottom of Fig. 43.2a). At potentials considerably negative with respect to the flat-band potential of a p-type semiconductor ($V < V_{fb}(p)$) the surface hole concentration is very low ($p_s \ll p_b$): this corresponds to depletion. Oxidation reactions requiring holes cannot take place and no current flows in the external circuit. At potentials approaching $V_{fb}(p)$ an *anodic* current is measured which increases exponentially with increasing potential (Fig. 43.2a). For almost all semiconductors this process involves oxidation and dissolution of the solid: anodic etching (as in (43.3)). A reduction reaction at the counterelectrode (e.g., hydrogen evolution) provides the charge carriers required to oxidize the semiconductor. The anodic current density can be converted to an etch rate by applying Faraday's law [43.7]. It is clear that the etch rate can be simply regulated via the applied potential.

Because of the absence of holes in the valence band, anodic oxidation is not possible with an n-type semiconductor (in the dark). Instead reactions involving electrons via the conduction band can be expected. In aqueous solution this is generally the reduction of protons at low pH [43.18]

$$2H^+ + 2e^- \rightarrow H_2, \qquad (43.6a)$$

or of water at higher pH

$$2H_2O + 2e^- \rightarrow H_2 + 2OH^-, \quad (43.6b)$$

giving hydrogen evolution. Since the surface electron concentration increases exponentially with decreasing potential, i.e., $n_s = n_b \exp(-eV_{sc}/k_B T)$, the rate of the cathodic current associated with these reactions also increases markedly (Fig. 43.2b). (In this case oxidation of water at the counterelectrode provides the electrons required at the semiconductor working electrode.) For a few semiconductors the presence of electrons at the surface can give rise to decomposition of the solid. For example, at negative potential the n-type semiconductor ZnO is electrochemically reduced

$$ZnO + 2H^+ + 2e^- \rightarrow Zn + H_2O, \quad (43.7)$$

a reaction that competes with hydrogen evolution [43.18]. Another example is n-type InP [43.19]

$$InP + 3H^+ + 3e^- \rightarrow In + PH_3. \quad (43.8)$$

Zinc and indium are reactive metals that dissolve readily in acidic and alkaline solution.

Minority-Carrier Reactions

Reactions requiring minority carriers (electrons in a p-type or holes in an n-type semiconductor) can be observed when the electrode is illuminated with

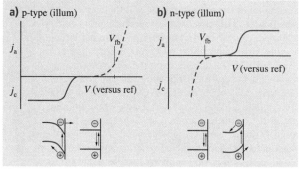

Fig. 43.3a,b The influence of suprabandgap illumination (illum) on the current–potential characteristics of (**a**) a p-type and (**b**) an n-type semiconductor. In case (**a**) the electric field of the space-charge layer drives the photogenerated minority carriers (electrons) to the surface, where a reduction reaction takes place. The holes are registered as a cathodic photocurrent in the external circuit. At positive potentials, electrons and holes recombine. In case (**b**) holes migrate to the surface, causing oxidation of the semiconductor: an anodic photocurrent is measured. In both cases the majority-carrier processes (Fig. 43.2) are shown as *dashed lines*

suprabandgap light. What is important for etching is the effective spatial separation of the photogenerated electrons and holes (the prevention of recombination) allowing the minority carriers to react at the surface.

In a p-type semiconductor electrons and holes generated by light are separated by the electric field of the space-charge layer at negative potentials (see inset in Fig. 43.3a). At the surface the electrons are used for the reduction of H^+ ions or H_2O (43.6a,b). In the case of p-type InP the semiconductor itself is decomposed under illumination (43.8). The photogenerated holes are swept to the back contact and registered as a cathodic photocurrent in the external circuit. The constant limiting photocurrent at negative potential generally depends on the light intensity. At more positive potentials going towards the flat-band condition the electric field at the surface decreases and electron–hole recombination competes with the electrochemical reaction of the electrons. Because of the importance of surface recombination at semiconductor electrodes in solution and the poor kinetics of reactions such as hydrogen evolution, the onset of photocurrent is generally at a potential considerably negative with respect to the flat-band value (up to 0.6 V), in contrast to a Schottky diode.

Holes created by suprabandgap light in an n-type semiconductor can give rise to photoelectrochemical oxidation of the solid (43.3). As in the case of the p-type electrode, electron–hole recombination generally dominates at potentials close to V_{fb} (Fig. 43.3b). The onset of photocurrent, anodic in this case, is at positive potentials where the electric field moves the holes to the surface and the electrons to the back contact. For both cases, n- and p-type, the etch rate depends not only on the applied potential but also on the light intensity.

43.1.3 Electroless Etching

In electrochemical etching, the charge carriers are supplied to the semiconductor–solution interface via an external circuit. The disadvantage of this approach is the need for a counterelectrode, a voltage source, and, in well-defined experiments, a reference electrode (Fig. 43.1). Etching of semiconductors under open-circuit conditions is possible, in principle, if the charge carriers can be supplied from a redox species in solution [43.13]. While electron injection from solution into the conduction band of semiconductors can occur, this generally does not lead to decomposition. On the other hand, extraction of electrons from the valence band (i.e., *injection of holes* into the band) can be used for etching many semiconductors of both p- and n-type (see the next paragraph). There is another form of electroless etching in which minority carriers are generated by suprabandgap light. This will be dealt with in the paragraph *Electroless Photoetching*.

Electroless Etching in the Dark

The principle of electroless etching is shown in Fig. 43.4 for a p-type semiconductor, e.g., GaAs, in an alkaline solution containing an oxidizing agent (a strong electron acceptor) such as the $Fe(CN)_6^{3-}$ ion [43.20]. Two electrochemical reactions occur: the oxidation of the semiconductor which, in the case of GaAs, requires six valence-band holes

$$GaAs + 10OH^- + 6h^+ \rightarrow GaO_2^- + AsO_3^{3-} + 5H_2O\,,$$
(43.9)

and the reduction of the oxidizing agent by hole injection into the valence band (VB)

$$Fe(CN)_6^{3-} \rightarrow Fe(CN)_6^{4-} + h^+(VB)\,.$$
(43.10)

The potential dependence of the rates of these two reactions is shown in Fig. 43.4. The form of the anodic current–potential curve (a) for reaction (43.9) has been introduced before. Reduction of the oxidizing agent (43.10) starts at the equilibrium (Nernst) potential of the

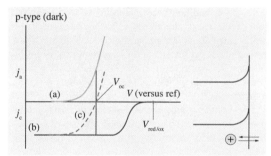

Fig. 43.4 The principle of electroless etching of a p-type semiconductor. Curve a is the anodic current–potential curve for oxidation/dissolution of the electrode. Curve b shows the cathodic current–potential curve for reduction of the oxidizing agent by hole injection into the valence band. This reaction starts at the equilibrium potential $V_{red/ox}$ and its rate becomes diffusion limited at negative potentials. The current measured with such a system, i.e., the sum of the anodic and cathodic contributions, is shown as a *dotted line*. At the open-circuit potential V_{oc} the two *partial* currents are equal ($j_a = j_c$). This is shown in the schematic diagram on the *right*. It is clear that at V_{oc} the semiconductor is etched

$Fe(CN)_6^{3-}/Fe(CN)_6^{4-}$ redox couple. The rate increases as the potential is made negative (curve b, Fig. 43.4). In the present case, the rate finally becomes diffusion controlled, i.e., the cathodic current is independent of the applied potential. In an electrochemical experiment, the measured current–potential curve (c) is the sum of the two *partial* curves a and b.

From Fig. 43.4 it is clear that at the open-circuit potential, indicated by V_{oc}, the two electrochemical reactions (43.9) and (43.10) occur at a significant rate; the anodic and cathodic currents are equal ($j_a = j_c$). For this system the electroless etch rate is determined by mass-transport-controlled hole injection (curve b). From this it follows that the etch rate can be enhanced by either increasing the concentration of the oxidizing agent in solution or by improving the hydrodynamics of the system.

In a similar way holes injected by $Fe(CN)_6^{3-}$ into the valence band of n-type GaAs will give rise to etching at open-circuit potential. From this discussion it is clear that, for electroless etching of n-type and p-type materials, the oxidizing agent must be capable of extracting electrons from the valence band of the solid; the electron acceptor levels of the oxidizing agent must correspond in energy to the valence band of the solid. For wide-bandgap semiconductors such as GaN and SiC with an energetically low-lying valence band this is no longer possible. Even the strongest oxidizing agents fail to inject holes under normal conditions. Consequently, electroless etching is not possible for these materials.

Electroless Photoetching

As for photoelectrochemical etching that depends on minority carriers, electroless etching may also be promoted by illumination with suprabandgap light [43.21–23]. The principle is shown schematically in Fig. 43.5a. At open-circuit potential a photon generates an electron–hole pair. An electron acceptor in solution (Ox^+) captures the electrons very effectively from the conduction band, allowing the holes to break surface bonds and thus cause etching. The electron acceptor in this case is different from that required for electroless etching in the dark: its acceptor levels must correspond to the conduction band of the solid and thus be higher in energy. For electroless photoetching the rates of reaction of electrons and holes at the surface are equal and must be higher than the rate of electron–hole recombination.

Electroless photoetching can be understood in electrochemical terms on the basis of Fig. 43.5b for an n-type semiconductor. As in Fig. 43.4, the two partial

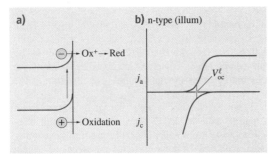

Fig. 43.5a,b The principle of electroless photoetching of an n-type semiconductor. Light generates electron–hole pairs (a). The electrons are captured by an electron acceptor (Ox^+) in solution. The holes cause oxidation and dissolution of the semiconductor. These reactions compete with electron–hole recombination. (b) Shows the current–potential characteristics of the two processes: photoanodic oxidation of the semiconductor (by minority carriers) and cathodic reduction of the oxidizing agent (by majority carriers). At the open-circuit potential under illumination (illum) V_{oc}^ℓ both reactions occur at the same rate ($j_a = j_c$) and the semiconductor is photoetched

current–potential curves are shown: (a) for the photoanodic oxidation of the semiconductor via the valence

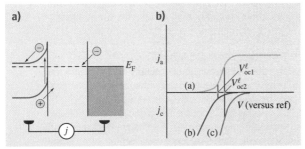

Fig. 43.6a,b The principle of photogalvanic etching. An n-type semiconductor is short-circuited to a metal in solution (without a voltage source). Electrons and holes generated by light (a) are separated by the electric field at the semiconductor surface. The holes cause etching. The electrons pass to the metal, where an oxidizing agent is reduced. Since the two electrodes are short-circuited, the Fermi level E_F is constant in the system. (b) shows the corresponding electrochemistry. Curve a gives the potential dependence of the minority-carrier reaction (photoanodic oxidation). Curve b describes reduction of the oxidizing agent (via majority carriers). The open-circuit potential is V_{oc1}^ℓ. If the surface area or the catalytic activity of the metal is increased (curve c) the open-circuit potential shifts to V_{oc2}^ℓ and the photoetch rate is significantly increased

band and (b) for the reduction of the electron acceptor via the conduction band (this reaction, involving majority carriers, can occur in the dark). At the open-circuit potential under illumination V_{oc}^ℓ both reactions occur and the semiconductor is photoetched. From this description it is clear that the systems shown in Fig. 43.3 will not give rise to photoetching. For example, the photocathodic reduction reaction at the p-type electrode (Fig. 43.3a) is only observed at negative potentials, a range in which the semiconductor cannot be oxidized (see anodic curve). In these cases the kinetics of the minority-carrier reaction is poor and electron–hole recombination dominates in the vicinity of the flat-band potential.

43.1.4 Photogalvanic Etching

For the case in which the rate of the majority-carrier reaction (e.g., electron transfer in the example of Fig. 43.5) is hindered by unfavorable kinetics, photoetching under open-circuit conditions may be promoted by short-circuiting the semiconductor to a noble metal in solution [43.24, 25]. A metal can be chosen that *catalyzes* the majority-carrier reaction (this may be the reduction of oxygen, naturally present in solution, or an added oxidizing agent). The principle is shown in Fig. 43.6a. Illumination creates in the semiconductor electrons and holes that are separated by the electric field of the space-charge layer. The holes react at the surface, causing oxidation and etching. The electrons, instead of having to react at the semiconductor–solution interface, now pass to the metal via the external circuit. Reduction of the oxidizing agent from solution occurs at the metal.

The corresponding electrochemical picture is shown in Fig. 43.6b. Curve a represents the potential dependence of the current of the illuminated n-type semiconductor electrode, while curve b is the cathodic current–potential curve measured in a separate experiment at the metal. Under open-circuit conditions (V_{oc1}^ℓ) the two currents must be equal ($j_a = j_c$). From Fig. 43.6 it is clear that photoetching will occur in this case. The ratio of the surface areas of semiconductor and metal exposed to solution is important for the photoetch rate. An increase in the area of the metal leads to an increase in the rate of reduction of the oxidizing agent (curve c). This causes a shift in the open-circuit potential to V_{oc2}^ℓ ($> V_{oc1}^\ell$) and an increase in the photoetch rate [43.25].

The principle shown in Fig. 43.6 is that of a galvanic cell (the semiconductor is the anode, the metal the cathode) that operates under illumination. It is, in fact, a solar cell that could supply electrical energy. This type of etching we refer to as *photogalvanic*.

Photogalvanic etching has an advantage over photoelectrochemical etching in that a voltage source is not required. However, the experiment is less well defined.

The open-circuit potential, and thus the photoetch rate, depend on what is happening at the metal, i.e., its active area, (catalytic) activity etc. An Ohmic resistance in the system, e.g., a poor contact to the semiconductor, complicates the picture [43.25] and can give rise to a considerable reduction in etch rate.

43.2 Wet Etching of Semiconductors: Morphology and Defect Selectivity

In the kinetics of wet etching of solids two types of limitation of the etch rate can be distinguished:

- Surface kinetics: an interfacial chemical step is rate limiting. In this case the crystallographic orientation of the surface is generally important (etching may be anisotropic). The activation energy, due to a chemical process, is high ($> 40\,\text{kJ/mol}$). These are clearly conditions which may favor defect revealing.
- Mass transport: the rate-limiting step is convective diffusion of active etching species from solution to the surface, or of etching products away from the surface. The activation energy is considerably lower ($< 20\,\text{kJ/mol}$), being due to the temperature dependence of the diffusion coefficient (i.e., of the solution viscosity). Etching in this case has a *polishing* effect: the concentration gradient responsible for dissolution is higher at protrusions and lower at recessed areas of the surface, thereby giving rise to surface leveling. This effect is enhanced if a soluble surface film is involved in the etching process.

In this section we consider briefly the surface morphology and the basis for defect selectivity in the etching systems described above.

43.2.1 Chemical Etching

The chemical etchants considered in Sect. 43.1.1 (concentrated HCl or HBr for InP, concentrated alkaline solution for Si) are kinetically controlled systems: they show strong anisotropy. The atomic structure of the surface is important in determining etching morphology; this is very clear in the Si case [43.15, 16]. Such etchants may be highly defect selective, e.g., HBr for InP [43.26]. Eutectic etching corresponds to this class. Since the etch rate at *disturbed* areas of the surface (e.g., at dislocations) is high, etch pits result in this case. In Sect. 43.3.1 defect revealing in these chemical etchants is considered in more detail.

Chemical etching, on the other hand, may also be mass-transport controlled. Etching of III–V semiconductors (GaAs, InP) in aqueous solutions of halogens (chlorine, bromine, iodine) follows a chemical mechanism. The etch rate may be controlled by either kinetics or diffusion; this is also the case for etchants based on hydrogen peroxide [43.7]. Kinetic control gives defect selectivity [43.26]. Under mass-transport-controlled conditions these etchants are used for surface preparation and polishing. In the case of the halogens, water may be replaced as a solvent by an alcohol; bromine in methanol is a widely used etchant for GaAs.

43.2.2 Electrochemical Etching

The degree of defect selectivity in electrochemical etching depends on whether majority or minority carriers are responsible for etching.

Majority-Carrier Reactions

The kinetics of electrochemical etching of *p-type* semiconductors can be sensitive to surface structure: the rate constants of reactions such as those given by (43.3) may depend on surface properties, e.g., on whether the surface is Ga or N terminated. The defect selectivity of such anodic etching of p-type materials has not been studied much; it seems, however, in most cases to be inferior to that of other methods described in this chapter.

Anodic etching may be mass-transport limited. For example, for GaAs in alkaline solution the exponentially increasing current observed at positive potentials (Fig. 43.2a) levels off, becoming potential independent. This is due to limitation of the dissolution process by mass transport of OH^- ions in solution (43.9). The anodic etch rate depends on both the OH^- concentration and the hydrodynamics [43.20]. Another example is the anodic etching of p-type Si in acidic fluoride solution [43.27]. In both cases the semiconductor is *electropolished*.

Cathodic decomposition of n-type semiconductors via conduction-band electrons (as in (43.7)) is clearly not an attractive approach to etching. However, *de Wit* et al. [43.18] have shown that electrochemical reduction of polycrystalline ZnO films reveals very clearly the grain boundaries. This indicates that, in certain cases, this approach could perhaps be interesting for defect revealing.

Minority-Carrier Reactions

Crystallographic and other defects usually give rise to electronic states in the bandgap of semiconductors. Such states can act as centers for very effective recombination of electrons and holes. In the onset of the anodic photocurrent–potential curve (Fig. 43.3b) surface recombination competes with the electrochemical reaction: holes required for dissolution of the solid are lost by recombination. This gives a significantly reduced etch rate at the surface defects, which show up as hillocks [43.7, 21, 28–31]. At positive potentials the photocurrent becomes independent of potential: the band bending at the surface is sufficient to separate effectively the photogenerated electrons and holes, thus preventing recombination. The limiting photocurrent can be determined by the light intensity or by mass transport in solution [43.25]. In the latter case electropolishing can be achieved.

43.2.3 Electroless Etching

As in Sect. 43.1.2 we must make a distinction here between electroless etching in the dark and under illumination. In this section we also consider an exceptional class of electroless system: Sirtl and adapted-Sirtl etchants.

Electroless Etching in the Dark

Since electroless etching requires strong oxidizing agents whose acceptor levels show a significant overlap with the valence band of the semiconductor, the rate of hole injection (e.g., (43.10)) is generally high and the etch rate is mass-transport controlled. In such a case one would not expect defect selectivity. However, in previous work [43.7, 20] we have shown that, since defects represent highly reactive areas on the surface, a galvanic cell can be formed between such areas and the more *noble* perfect surface: holes injected over the whole surface are used to etch preferentially defective areas, thus revealing defects as etch pits. This effect is expected when the etch rate is determined by the hole injection reaction (e.g., (43.10)). If, on the other hand, the oxidation reaction (43.9) is rate limiting and diffusion controlled, defects are not revealed. Instead polishing is observed. This subtle difference between the two cases can be understood on the basis of the electrochemistry of the systems [43.20]. In chemical etching systems free charge carriers (holes in the valence band) are not involved. Consequently, galvanic effects are not observed with chemical etchants.

Electroless Photoetching

As in the case of photoelectrochemical etching, electron–hole recombination competes with open-circuit photoetching. Etchants with H_2O_2 as electron acceptor have been used successfully to reveal defects in III–V semiconductors such as GaAs, GaP, and GaAsP [43.7, 21, 32].

Sirtl-Type Etchants

In 1961 *Sirtl* and *Adler* developed an etchant for Si based on aqueous solutions of CrO_3 and HF [43.33]. This *Sirtl etchant* proved very effective in revealing crystallographic defects in various orientations of Si. In addition, the system has been used for the delineation of junctions between layers of different dopant concentration. Subsequently, a modified form of the Sirtl etch was proposed for GaAs [43.34]. This system, termed diluted Sirtl with light (DSL), was highly successful in the characterization of III–V semiconductors. With the appropriate choice of solution composition, all crystallographic defects could be revealed in n-type, p-type, and semi-insulating GaAs with high sensitivity. A special feature of this system is the exposure of defects as hillocks during etching, not only under illumination but also in the dark.

The electrochemistry of Si in Sirtl and of GaAs in DSL etchants has been studied thoroughly [43.35–39] and far-reaching conclusions could be drawn with regard to etching mechanisms. It was shown that, except at very high HF concentrations, the dissolution of both semiconductors follows an electroless mechanism: the holes required for oxidation of the semiconductor are supplied by a Cr VI complex in solution.

In the case of GaAs [43.35–37] three ranges of etchant composition could be defined on the basis of the concentration ratio of HF to CrO_3. For $[HF]/[CrO_3]$ ratios below ≈ 10 and $[HF] < 10\,M$, the dissolution process is kinetically controlled (range A). In the dark p- and n-type etching kinetics are the same. Defects are revealed as hillocks and the defect sensitivity of n-type GaAs in higher. Illumination with suprabandgap light enhances both the etch rate and defect sensitivity of n-type GaAs but has no effect on p-type. For $[HF]/[CrO_3] > 20$ and $[HF] < 10$ (range B), etching is controlled by mass transport of the Cr VI complex in solution. Defect sensitivity is low for both p- and n-type crystals. For solutions containing a HF concentration above 10 M (range C) a purely chemical attack on Si by HF species very likely occurs; this is analogous to the reaction between InP and HCl in (43.1).

An important feature of the electroless system is the formation of a passivating surface layer that contains both Cr VI and Cr III species. In range A the coverage by the surface film is high. Etching kinetics are determined by film formation (due to hole injection from Cr VI) and film removal (via intermediates of the semiconductor oxidation reaction with the aid of HF). In range B at relatively high HF concentration, the surface coverage decreases to low values. Hole injection by Cr VI, and thus GaAs oxidation/dissolution, are limited by mass transport of the oxidizing agent.

Defects are revealed as hillocks or ridges on the surface of p- and n-type GaAs crystals, both in the dark and under illumination. Defect sensitivity in the dark can be explained by assuming a reduced bond strength at the defect. This results in a local increase in surface coverage by the passivating film and, consequently, a reduction in etch rate. For n-type GaAs under illumination locally enhanced recombination at defects leads to a further decrease in etch rate and improved defect sensitivity. Other morphological features of the system which were explained include surface roughening and growth striations.

A subsequent study of the Si/Sirtl system [43.38, 39] showed features very similar to those observed in DSL, including the importance of the formation of an adsorbed mixed-valence chromium complex during etching.

43.2.4 Photogalvanic Etching

Photogalvanic dissolution of an n-type semiconductor is, in many respects, similar to that of photoelectrochemical dissolution: in the latter, the photocurrent at fixed light intensity is regulated via the applied potential; in the former, the photocurrent is determined by the rate of reaction of the oxidizing agent at the metal electrode, which depends on concentration, and the electrode area and (catalytic) activity. In photogalvanic etching, conditions can be chosen to correspond to the rising part of the photocurrent–potential curve (Fig. 43.3b); in this case competition between dissolution and recombination gives defect selectivity in many systems. Alternatively, a photocurrent limited by mass transport in solution can be achieved by ensuring that the the hole flux to the surface (dependent on light intensity) is much higher than the diffusion of a rate-determining species (e.g., OH^- ions) in solution. In this case photogalvanic polishing is obtained [43.25].

43.3 Defect-Selective Etching Methods

In this section we consider in more detail methods for defect-selective etching (DSE). We distinguish two types of system, namely:

- Orthodox, also called classical or preferential, etchants (strong acids, molten bases); these are, in fact, the chemical systems described in Sects. 43.1.1 and 43.2.1.
- Newer systems based on electroless etching in the dark and under illumination, including photogalvanic etching.

43.3.1 Orthodox Etching for Revealing Dislocations

In this subsection, factors determining the kinetics of formation of pits and their morphology are described, as is the role of thermodynamics, i.e., the elastic energy of dislocations, as well as the calibration of these orthodox etchants and special applications.

Kinetics-Related Conditions

The mechanism of pit formation in semiconductors is governed by kinetically controlled, surface chemical reactions in both acidic and molten-salt etches. The details and some examples of etching systems have been described in Sects. 43.1.1 and 43.2.1. This etching method was discovered and used as early as the 19th century for disclosing the crystallographic symmetry of natural crystals, which was read from the shape of etch pits on the crystal faces. When the association between the pits and dislocations was found, the etch pit density (EPD) became a standard parameter describing the structural quality of semiconductor single crystals and substrates for device producers. Apart from the density of dislocations, much information can be obtained from the morphology of the etch pits, provided the basic requirements for formation of well-developed pits are fulfilled. These criteria were discussed in detail in earlier reviews [43.5, 6, 8] and are summarized in the following. The formation of pits proceeds via repeated nucleation

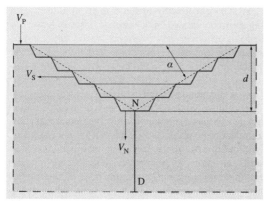

Fig. 43.7 Schematic drawing of an etch pit originating at a dislocation (D)

events at the emerging point of a dislocation on the etched surface, as shown in Fig. 43.7. Once the stable nucleus (N) is formed it grows further by horizontal step movement.

For practical applications, i.e., for obtaining clearly visible etch pits, the rate of formation of nuclei (V_N) must be larger than the step propagation velocity (V_S), while the rate of etching of the perfect surface (V_P) should be as low as possible (preferably zero). These criteria can be achieved by optimizing the etching temperature and the composition of the etchants, but the actual morphology of the etch pits, particularly the inclination angle of the side walls α, also depends on other factors such as the degree of decoration of dislocations (the presence of Cottrell atmospheres) and the type of dislocation. As an example, Fig. 43.8a shows the result of DSE of Ga-polar GaN in molten salts: deep pits are formed with $\alpha = 25-50°$ as derived from atomic force microscopy (AFM) section analysis in Fig. 43.8b. Etching of the same material in hot phosphoric acid or in a mixture of phosphoric and sulfuric acids (HH etch) yields very shallow pits with α below $10°$. In the latter etch large shallow pits *sweep away* smaller neighboring pits, which might result in the erroneous estimation of the total density of dislocations. In order to avoid underestimation of the EPD in these acidic etches, the temperature and time have to be carefully optimized [43.40, 41]. Such optimization may allow edge and mixed/screw dislocations from the diverse size of pits to be distinguished [43.41], but from our experience it can be concluded that the procedure has to be repeated for samples grown in different conditions.

Factors Influencing Pit Morphology

There are a number of factors which may influence the morphology and relative size of etch pits formed on dislocations. These are:

- Crystallographic orientation (symmetry) of the etched surface
- Polarity of surfaces in compound semiconductors
- Composition of etching medium
- Changes of chemical composition around dislocations (Cottrell atmospheres) and in the matrix
- Position of dislocations with respect to the surface
- Elastic energy of dislocations represented by their Burgers vectors.

Symmetry. The correspondence between crystallographic symmetry and the shape of etch pits in single crystals is probably one of the best recognized and most

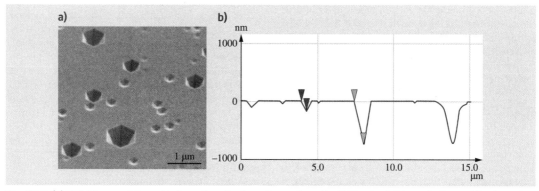

Fig. 43.8 (a) SEM image of deep pits formed on the Ga-polar surface of a GaN epitaxial layer during etching in molten KOH + NaOH eutectic. Sample tilt $45°$. (b) AFM section profile across the pits from (a) (courtesy of G. Nowak)

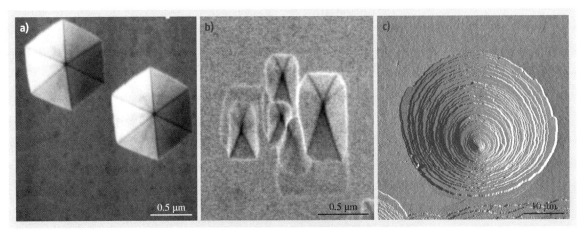

Fig. 43.9 (a,b) SEM and (c) AFM images of GaN single crystals etched on (a) (0001), (b) (10$\bar{1}$0), and (c) (000$\bar{1}$) surfaces in molten salts (E-etch) (AFM image courtesy of S. Müller)

used advantages of the preferential etching method. Numerous etches are known for different semiconductors that show this characteristic, e.g., Sirtl etch for Si [43.33], molten KOH for GaAs [43.42], and several etching systems for InP [43.26, 43, 44]. A similar correlation has recently been shown for GaN in molten KOH + NaOH eutectic (E-etch after [43.12]) and is demonstrated in Fig. 43.9a,b: well-defined hexagonal and rectangular etch pits are formed during etching of (0001) Ga and (10$\bar{1}$0) surfaces, respectively. It can be concluded that a properly calibrated etching method constitutes a fast and unambiguous tool for establishing the crystallographic orientation of semiconductor bulk crystals and epitaxial layers.

Polarity. Preferential etching is also sensitive to the bonding direction between atoms forming compound semiconductors. As a rule, polar surfaces show markedly different growth and etching behavior. In zincblende III–V materials (e.g., GaAs and InP) polar {111}B surfaces are very reactive because the presence of the group V atom at this face gives a higher electron density than that of the {111}A face. (In the extreme case the B face would have an electron pair in the dangling bond while this would be missing at the A face.) As a result of this difference it is easy to recognize the polar {111} faces using well-known etches. The sensitivity of etching methods to the atomic configurations of the etched surfaces has an important practical application for recognition of nonequivalent ⟨110⟩ and ⟨$\bar{1}$10⟩ directions on the nonpolar {001} faces of III–V compound semiconductors. For instance, the etch pits formed on the (001) surface of GaAs during etching in molten KOH are elongated in the [0$\bar{1}$1] direction, as was unequivocally established by calibration with the x-ray Lang technique [43.45]. Consequently, the definition of the orientation and identification flats on the zincblende III–V substrates, given in standard commercial specifications, is based on the shape of the etch pits, as shown in Fig. 43.10.

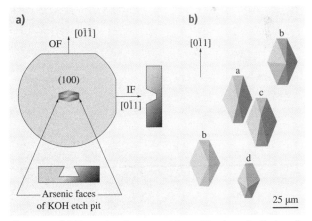

Fig. 43.10 (a) Example of commercial specification of a GaAs substrate with the description of orientation (OF) and identification flats (IF) on the basis of the shape of KOH-related etch pits. (b) Pits formed in molten KOH on the (001) surface of GaAs at dislocations differently inclined with respect to the surface

III–V materials with a wurtzite lattice (GaN, AlN) have very strong polarity-related anisotropy of properties, which shows up, for example, in markedly different chemical reactivity of (0001) Ga and (000$\bar{1}$) N surfaces. Following standard assignment based on comparative hemispherically scanned x-ray photoelectron diffraction (HSXPD), TEM, and etching studies [43.46–48], it was shown that N-polar surfaces of GaN can be dissolved in dilute aqueous solution of KOH/NaOH even at room temperature, while Ga-polar surfaces remain intact in such etching conditions. Fast dissolution of the N-polar surface of GaN in KOH was explained by adsorption of hydroxide ions at the negatively charged dangling bond of each nitrogen atom at this surface and subsequent formation and dissolution of gallium oxide [43.49]. As a result, numerous pyramids are formed on the (000$\bar{1}$) N surface and, once the whole surface is covered by them, the etching process is terminated. As for Si (111) surfaces etched in concentrated KOH solution [43.15, 16], such pyramids are bounded by chemically stable low-index {1$\bar{1}$01} planes and are not related to any specific defects (though for GaN this conclusion is still tentative). Similar behavior was reported for the nitrogen-polar surface of AlN [43.11] and this makes DSE of these surfaces more demanding. Dislocations on the N-polar surface of GaN single crystals can be revealed in molten eutectic [43.12] at relatively low temperatures as compared with the Ga-polar surfaces. In contrast to the Ga-polar surface, the resultant etch pits are circular, with a very irregular terraced morphology (Fig. 43.9a,c). The most convincing example of the difference of the polarity-dependent pit morphology was obtained by revealing dislocations in the vicinity of inversion domains (IDs) in Mg-doped homoepitaxial GaN [43.50]. The upper part of the image in Fig. 43.11a shows hexagonal pits formed on the Ga-polar surface while in the lower N-polar area circular pits are visible.

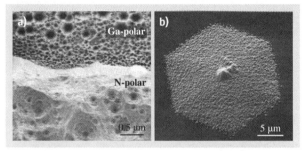

Fig. 43.11a,b SEM images of (**a**) GaN epitaxial layer with the neighboring Ga- and N-polar areas etched in molten E+M etch [43.45] (courtesy of G. Kamler) and (**b**) GaN N-polar epitaxial layer after etching in hot HH [43.46]

DSE of N-polar heteroepitaxial GaN layers is more difficult because, in addition to the high reactivity of this side, a higher density and variety of defects (e.g., inversion domains of different diameters, higher level of impurities, i. e., carrier concentration [43.51–53]) are inherent to these materials. Etching in molten salts results in fast dissolution of N-polar heteroepitaxial layers and overall roughness, without the possibility of attributing the tiny etch features to particular defects. It seems more promising to etch in hot acids, which reveals large and small inversion domains (IDs) in the form of protruding pyramidal etch features, as shown in Fig. 43.11b. However, evaluation of dislocation density in this material using orthodox etching methods does not seem to be viable.

Etchant Composition. The morphology of etch pits can be tailored by changing the ratio of constituents of the etching solution, etching conditions (e.g., temperature) or by adding so-called inhibitors. Theoretical considerations and rules were discussed and variable pit morphologies in different etching systems were demonstrated in earlier reviews [43.5,6]. A very clear example of the dependence of the pit morphology on the etchant composition was found in the orthodox HBr-$K_2Cr_2O_7$ (BCA) etching system developed for InP [43.26]. In

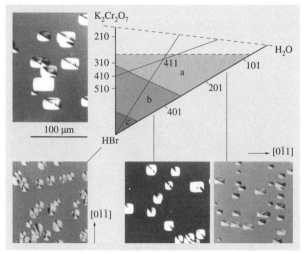

Fig. 43.12 HBr-rich corner of the BCA ternary etch system for InP. The differential interference contrast (DIC) optical images show the morphology of etch pits as a function of composition of etchants (with permission from [43.26])

the HBr-rich corner of the ternary BCA diagram there are three composition fields, denoted a, b, and c in Fig. 43.12. Etchants from these fields result in the formation of pits with different morphology: the pits formed in region b are square, while etchants from regions a and c yield rectangular pits, elongated in the [0$\bar{1}$1] and [0$\bar{1}$1] directions, respectively.

The kinetics of DSE can be influenced by adding traces of ions (inhibitors) to the solution. Their role is to *poison* kinks and ledges inside the pits, which slows down horizontal movement of steps during etching (decrease of V_S in Fig. 43.7). This effect can have practical application when kinetics-related conditions of pit formation are not fulfilled, i.e., when the pits are too shallow as in etching of GaN in hot acids. It was recently shown that, by adding Al^{3+} and Fe^{3+} ions to H_3PO_4, merging of pits on GaN heteroepitaxial layers can, indeed, be avoided and well-defined deep pits are formed [43.54].

Effect of Decoration and Composition. In semiconductors, foreign atoms (dopants, impurities) are attracted by dislocations due to the presence of a strain field and charge (Coulomb interaction). The resultant impurity (Cottrell) atmosphere may have a dual influence on the formation of etch pits:

1. By releasing the strain, the elastic energy of the dislocation is diminished, which should decrease the rate of nucleation (V_N in Fig. 43.7).
2. The chemical potential is changed and this may locally change the chemical reactivity in the given etching solution.

As was already pointed out by *Amelinckx* [43.2] it is difficult to predict which of these two factors will prevail. GaAs etching in molten KOH resulted in the formation of larger etch pits on so-called grown-in (i.e., strongly decorated) dislocations than on stress-induced dislocations [43.8]. A similar effect was observed in bulk GaN single crystals on which the size of pits formed on grown-in and indentation-induced dislocations was compared [43.55]. Morphology of pits can also be strongly influenced by heavy doping or alloying, as was shown for In-doped GaAs: the elongated hexagonal pits formed in molten KOH on nondoped GaAs (similar to these shown in Fig. 43.10b) were gradually transformed into regular hexagonal pits with increasing content of In [43.56].

Geometrical Position of Dislocations with Respect to the Surface. This effect seems to be obvious and was recognized already in the early 1960s [43.2] and subsequently well substantiated in numerous papers and reviews, e.g., [43.5, 8, 57, 58]. Most impressive was the work distinguishing five groups of etch pits and thereby five directions of dislocation lines in GaAs epitaxial layers after molten KOH etching [43.59]. These are: $a = \langle 001 \rangle$, $b = \langle 011 \rangle$ or $\langle 101 \rangle$, $c = \langle 112 \rangle$, $d = \langle 211 \rangle$, and $e = \langle 121 \rangle$ directions. Similar results were later obtained on bulk liquid-encapsulated Czochralski (LEC)-grown GaAs crystals and described in a review paper [43.8]. In Fig. 43.10b etch pits are shown in a GaAs substrate demonstrating the presence of four differently inclined dislocations (*a–d* types) in one place.

The majority of dislocations in GaN heteroepitaxial layers are perpendicular to the (0001) surfaces because

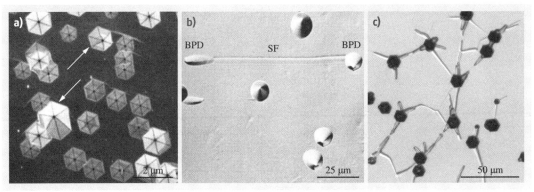

Fig. 43.13 (a) SEM image of thick HVPE-grown GaN layer after etching in molten E at 450 °C for 5 min. (b,c) DIC images of SiC substrate after etching in molten E at 520 °C for 5 min. The sample in (b) is 8° off axis

these are threading dislocations. The etch pits formed on them are, therefore, center-symmetrical, as shown in Fig. 43.9a. The same holds for bulk GaN single crystals and thick hydride vapor phase epitaxy (HVPE)-grown layers, though some inclined dislocations could be found (see two the non-center-symmetrical pits marked by arrows in Fig. 43.13a and the pit on the N-polar surface in Fig. 43.9c). In bulk and epitaxial SiC, so-called basal-plane dislocations are easily recognized from the characteristic morphology of shell-like etch pits. They are frequently observed both in SiC substrates and in 8° off-axis epitaxial layers [43.58, 60, 61]. The characteristic appearance of basal-plane dislocations (BPDs) and a bounding stacking fault (SF) in misoriented SiC is shown in Fig. 43.13b. In Fig. 43.13c, for comparison, pits are shown on an exactly (0001)-oriented SiC surface: hexagonal etch pits are on dislocations perpendicular to the surface (screw dislocations (SDs)), and grooves on dislocations almost parallel to the surface, i.e., basal plane dislocations (BPDs). In Fig. 43.13c the BPDs form dislocation nodes with SDs and a three-dimensional (3-D) network in the bulk of the material.

Thermodynamic Factors: Elastic Energy of Dislocations

From Cabrera's theory on the thermodynamics of pit formation [43.6] it follows that the critical value of the chemical potential difference ($\Delta \mu$) of a stable nucleus of a pit at the outcrop of a dislocation (N in Fig. 43.7) depends inversely on the elastic energy (E_{el}) of the dislocation

$$\Delta \mu = 2\pi^2 \Omega \frac{\gamma^2}{E_{el}}, \qquad (43.11)$$

where γ is the edge free energy and Ω the molecular volume.

The elastic energy value differs for different types of dislocations [43.63]

$$E_s = Gb^2 \alpha, \qquad (43.12)$$

$$E_e = Gb^2 \alpha \left(\frac{1}{1-\nu} \right), \quad \text{and} \qquad (43.13)$$

$$E_m = Gb^2 \alpha \left(1 - \nu \cos^2 \frac{\theta}{1-\nu} \right), \qquad (43.14)$$

for screw, edge, and mixed dislocations, respectively (where G is the shear modulus, b the Burgers vector, α a geometrical factor, ν Poisson's constant, and θ the angle between screw and edge components of the Burgers vector of mixed dislocations). Heteroepitaxial layers of GaN are the best material for considering the influence of elastic energy of dislocations on the formation of pits during orthodox etching: in this material all three types of dislocations, including nanopipes, usually coexist. In addition, numerous TEM studies of cross-sectional specimens have showed that the vast majority of dislocations in epitaxial GaN layers are of the threading type and are perpendicular to the surface. As a result, the influence of the tilt of a dislocation line on the size and/or morphology of pits can be excluded. Figure 43.14a shows the typical surface morphology of a heteroepitaxial GaN sample after

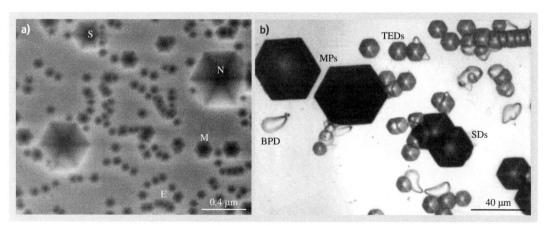

Fig. 43.14 (a) SEM image of heteroepitaxial GaN layer after etching in molten E+M. (b) Optical image of etch pits formed on different defects in 6H-SiC wafer during etching in molten KOH (with permission from [43.62])

etching in molten E+M orthodox etch (eutectic of KOH + NaOH + 10%MgO [43.64]). Four size grades (denoted N, S, M, and E) can be discerned in this material, representing nanopipes and screw-, mixed-, and edge-type dislocations, respectively. The assignment of the pit size to the type of dislocations was done on the basis of recent direct calibration of orthodox etching by TEM [43.64–66].

In the wurtzite lattice of GaN the Burgers vectors of dislocations are

$$\boldsymbol{b}_e = \frac{1}{3}\langle 11\bar{2}0 \rangle, \qquad (b_e = a, b_e^2 = a^2),$$

$$\boldsymbol{b}_m = \frac{1}{3}\langle 11\bar{2}3 \rangle, \qquad (b_m = \sqrt{c^2 + a^2},$$

$$b_m^2 = 3.66a^2),$$

$$\boldsymbol{b}_s = [0001], \qquad (b_s = c, b_s^2 = 2.66a^2),$$

$\boldsymbol{b}_{\text{nano}} = n \times \boldsymbol{b}_s$ where $n = 1, 2, \ldots$, and a, c are the lattice parameters.

In the ideal lattice structure of an etched GaN sample, the size of pits should depend on the magnitude of the Burgers vector of dislocations, i.e., the largest pits should be formed on nanopipes and the smallest on the edge dislocations. This pit size sequence is indeed valid, with the exception of screw dislocations, on which the pits are usually larger than on the mixed ones. The arguments for explaining this seeming discrepancy have been discussed in recent work [43.66] and are based on the fact that both edge and mixed dislocations are characterized by a larger deformation of the lattice, i.e., the contribution of the Poisson-constant-related term in (43.12–43.14), than the screw dislocations. The resultant higher attractive forces may be responsible for more effective decoration of edge and mixed dislocations and, in this way, for a release of strain around these defects. This, in turn, would result in a less favorable energetic condition for the formation of etch pits on edge and mixed dislocations.

The energy of defects has an even more pronounced effect on the size of pits in SiC. Very large differences between the size of pits formed on micropipes, screw, threading, and basal plane dislocations were found after an optimized etching procedure in molten KOH [43.62], as is obvious from Fig. 43.14b.

Calibration of Etching

Each new DSE system requires confirmation of its reliability in revealing all dislocations. Since the size and morphology of pits may vary depending on the type and status of dislocations, as was shown in the previous section, calibration by a direct method may allow one to identify the types of dislocation and provide an unequivocal interpretation of the etch features. Different approaches are used for this purpose, e.g., x-ray topography of etched samples [43.67] (suitable for dislocation density below 10^5 cm^{-2}), comparison with cathodoluminescence (CL) [43.68] or another calibrated etching method [43.40], sequential etching and calibration by TEM (e.g., [43.69, 70]). The latter is the most attractive because, apart from the direct association of the etch pit with the underlying defect, it yields information on the exact type of dislocation and on any additional characteristic features, such as decoration. The method became very popular especially after the introduction of the focused ion beam (FIB) technique, though the conventional cross-sectional approach is still used [43.71, 72]. Figure 43.15 shows a typical set of TEM images (specimen prepared by FIB) of the etch

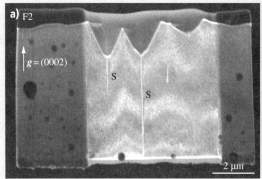 

Fig. 43.15a,b Bright-field TEM images taken with different diffraction conditions showing the association of large etch pits with screw dislocations (with permission from [43.59])

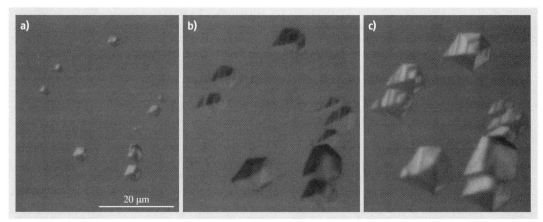

Fig. 43.16a–c DIC images of ZnO single crystal after sequential etching in 0.7% aqueous solution of HCl for (**a**) 30 s, (**b**) 60 s, and (**c**) 90 s

pits and underlying dislocations in thick HVPE-grown GaN. The method is particularly useful for samples with a moderate dislocation density, for which the conventional cross-sectional specimen preparation method is not effective because of the low probability of finding any defects in the thin foil.

The simplest way to verify the reliability of an orthodox etch for revealing dislocations is sequential etching: the pits formed on dislocations develop in size after each subsequent etching step and always have a point bottom indicating the position of the outcrop of the linear defect, as demonstrated by the set of images (made with the same magnification) in Fig. 43.16. Equally effective and simple is simultaneous etching of a two-sided polished sample. This method is, however, limited to relatively thin samples (e.g., substrates for epitaxy) containing threading dislocations.

Special Applications of Orthodox Etching

Well-controlled orthodox etching has frequently been used for revealing the origin of defects in epitaxial layers and device structures. For this purpose multiple (sequential) etching and photography are employed, which allows in-depth tracing of the defects in the sample. The results presented in [43.73] illustrate well the potential of this method: the clusters of screw dislocations revealed on the top of thick SiC epitaxial layers (up to several micrometer) were shown to be formed as a result of the closing of micropipes from the substrate. A similar approach was used for tracking the origin of pairs of dislocations in SiC epitaxial layers [43.74] and conversion of BPDs from the substrates into threading edge dislocations (TEDs) in SiC epitaxial layers [43.75]. Application of orthodox etching for examination of device structures is also very attractive since it permits one to establish at which interface in multilayered samples the dislocations are nucleated. The principle of the method is based on the fact that the point-bottomed etch pits formed on dislocations are transformed into flat-bottomed pits when the interface at which nucleation occurred is reached [43.76]. The method was recently used for studying GaN-based laser structures [43.77] and was calibrated by TEM [43.78]. Figure 43.17a shows the result of etching in molten E of the laser structure similar to that from [43.78], in which the dislocations were nucleated at different interfaces (the depth of the flat-bottomed pits is different),

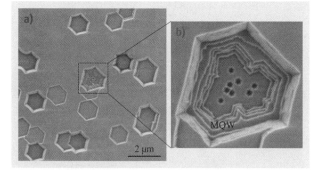

Fig. 43.17a,b SEM images of (**a**) MQW GaN-based laser structure after deep molten E+M etching. In (**b**) an enlarged fragment of image (**a**) is shown (etching and images: courtesy of G. Kamler)

with the exception of a cluster of dislocations propagating from the layers beneath the multiple quantum well (MQW) (see the enlarged central complex pit in Fig. 43.17b).

43.3.2 Electroless Etching for Revealing Defects

In another approach to etching, described in Sects. 43.1.2–43.1.4, free charge carriers (holes) localized at the surface cause the breaking of bonds. Holes can be supplied by an external voltage source (electrochemical etching), by an electron acceptor in solution (electroless etching) or by illumination (photoetching). Electrochemical etching has not been widely used for defect studies. Electroless etching, which operates under open-circuit conditions, has proved very successful. One of the most versatile electroless etchants for GaAs is the $CrO_3/HF/H_2O$ (DSL) system [43.37, 79]. This can be used for revealing dislocations both in the dark and under illumination. For wide-bandgap semiconductors such as GaN and SiC electroless etching in the dark is not possible (Sect. 43.1.3). In this case (photo)electrochemical or photogalvanic etching is an option. The latter is also described in the *Photogalvanic Etching* paragraph for revealing dislocations. These etching approaches can also be used for studying electrically active inhomogeneities in semiconductors. Special applications of electroless etching are dealt with in the final section.

Dislocations

We first consider the $CrO_3/HF/H_2O$ system in the dark and then two open-circuit photoetching approaches.

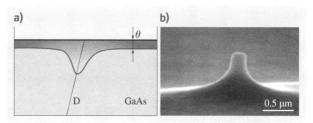

Fig. 43.18 (a) Hypothetical section profile of the thickness of the passivating layer θ formed on GaAs surface with an emerging dislocation D, immersed in $CrO_3/HF/H_2O$ etching solution. (b) SEM image of the protruding etch feature formed on a dislocation in GaAs during dark etching in the same solution

$CrO_3/HF/H_2O$ *in the Dark.* In this DS system (without light) the holes required for oxidation and dissolution of the semiconductor are injected into the valence band by hexavalent chromium ions in solution (Sect. 43.2.3). It has been shown that defect-selective etching in this case is determined by a surface passivating layer consisting of a Cr mixed-valence complex. The thickness of the passivating layer θ (or its coverage) determines the final morphological characteristics of the dislocation-related etch features [43.37]. The deformation field around dislocations reduces the Ga–As bond strength and, in this way, locally increases the thickness θ of the passivating layer (Fig. 43.18a). As a result during dark etching the outcrops of dislocations are revealed in the form of nanometer-size hillocks (tips), as shown in Fig. 43.18b. The surrounding Cottrell atmosphere has only a weak influence on the final shape of the etch features at the dislocation; it may contribute by diminishing the dislocation-related deformation field due to

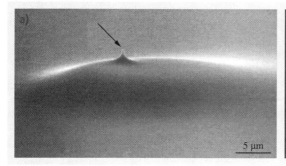
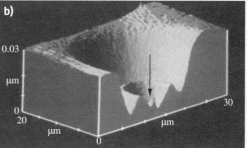

Fig. 43.19 (a) SEM and (b) phase-stepping microscopy (courtesy of P. Montgomery) images of dislocations and related Cottrell atmospheres, revealed by photoetching of n-type GaAs grown from: (a) slightly As-rich and (b) Ga-rich melt. The *arrows* indicate the etch features formed on dislocations

Table 43.1 Schematic representation of the surface profiles across the etch features formed on dislocations in GaAs after dark etching (DS) and photoetching (DSL) (with permission from [43.9]) (D: dislocations, dashed line: extent of electrically active zone with properties different from those of the dislocation-free matrix)

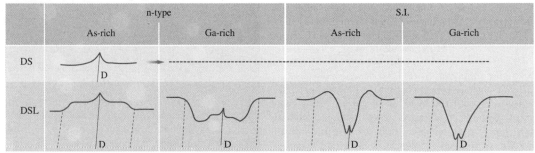

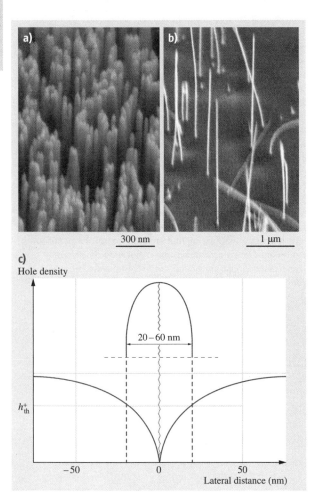

Fig. 43.20a–c SEM images of whisker-like features formed on dislocations during open-circuit photoetching of (**a**) high-dislocation-density MOCVD-grown and (**b**) low-dislocation-density HVPE-grown GaN layers. (**c**) Model of the formation of a whisker on a single dislocation during open-circuit photoetching of n-type GaN (with permission from [43.80]) ◄

the local increased concentration of foreign atoms or native point-type defects.

Photoetching

DSL System. When the surface of the semiconductor is illuminated with suprabandgap light during etching, complex etch features are formed on the dislocation sites. The final morphology of the etch features now depends on the electrical properties of both the dislocation and the surrounding atmosphere. Similar to etching in the dark, submicron protrusions are formed at the outcrops of dislocations (Fig. 43.19a), but now they reflect a cylindrical region depleted of carriers due to the recombination of electrons and holes at the dislocation. The dislocation-related atmospheres are revealed as either hillocks or depressions depending on their electronic nature: when the local recombination is stronger than in the matrix, the etch rate is locally decreased and large hillocks or ridges are formed [43.81] as can be recognized in Fig. 43.19a. On the other hand, the atmosphere containing excessive holes (as in GaAs grown from Ga-rich melt, in which Ga_{As}^{++} double acceptors are present) locally increases the etch rate and depressions are formed around the dislocation-related protrusions (Fig. 43.19b). Such complex etch features are formed in As- and Ga-rich GaAs during DSL photoetching and the electronic nature of defects responsible for the influence of atmospheres was disclosed by photoluminescence and EBIC studies [43.82, 83]. It was also shown that the section profile across the complex etch features around dislocations constitutes a fingerprint of the type of GaAs, as was discussed in [43.9] and demonstrated by the data in Table 43.1.

Photogalvanic Etching. Electroless etching of wide-bandgap semiconductors (GaN, SiC) can be performed only with the help of ultraviolet (UV) light and a supporting electrode (photogalvanic etching) for the reasons discussed in Sect. 43.1.4. The only widespread etching system used to date is based on aqueous solutions of KOH with dissolved oxygen as electron acceptor [43.24, 25], although more complex etching solutions containing a strongly oxidizing component ($K_2S_2O_8$) are emerging [43.85]. Since dilute KOH solutions (in the range 0.002–0.01 M KOH) are suitable for revealing dislocations, it is necessary to employ stirring during photoetching in order to ensure the supply of OH^- ions for the oxidation reaction at the surface. Other technical details and the limitations of this method have been discussed in several recent papers [43.11, 51, 52, 84, 86]. During etching of GaN in this system the recombination of electrons and holes at the dislocations is very effective; this leads to formation of whisker-like etch features with almost unlimited length (Fig. 43.20a,b). The diameter of the whiskers remains in the tens of nanometer range and represents a tube of the material from which holes are depleted due to recombination of photocarriers at the dislocation (see the model in Fig. 43.20c, in which h_{th}^+ describes the critical number of holes required for dissolution of a GaN molecule as follows from (43.3)). The fact that each whisker contains a dislocation was confirmed by direct TEM calibration of the photoetched samples [43.51, 53, 87–89].

Photogalvanic etching is also effective for revealing different defects in SiC. It was shown that dislocations, stacking faults, macropipes, and chemical inhomogeneities could be visualized in aqueous KOH solutions used with UV light [43.58]. Figure 43.21 shows a characteristic image of protruding features formed on dislocations parallel to the surface and pinned by two micropipes.

Electrically Active Inhomogeneities

Electroless photoetching constitutes an attractive tool for examination of extended electrically active inhomogeneities inherent to compound semiconductors, e.g., growth striations [43.90]. It was shown that well-

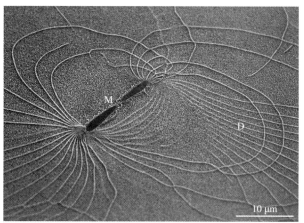

Fig. 43.21 SEM image of etch features formed on micropipes (M) and pinning dislocations (D) parallel to the (0001) Si surface of SiC substrate revealed by photoetching

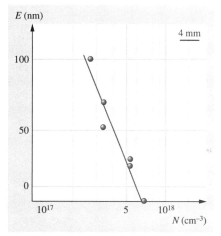

Fig. 43.22 Plot of relative etch depth E versus carrier concentration N for n-type GaAs photoetched in DSL (with permission from [43.84])

controlled photoetching is ultrasensitive to very small differences in carrier concentration both in n-type and in semi-insulating (SI) semiconductors and, after calibration by an appropriate method, can be used for quantitative evaluation of local differences in carrier concentration. Figure 43.22 shows a plot of the relative photoetch depth E of Si-doped GaAs as a function of the carrier concentration n; the latter was determined by the EBIC method. This strong inverse dependence of E on n can be understood on the basis of the illuminated Schottky diode. Electron–hole pairs are generated by light to a depth determined by the absorption coefficient of the semiconductor. For photoetching to occur, the valence-band holes must reach the surface. Two processes contribute. Holes created within the space-charge (depletion) layer are driven to the surface by the electric field. Since the thickness of the depletion is inversely proportional to the square root of the electron density ($W_{sc} \propto 1/n^{1/2}$), this contribution will decrease as the dopant density increases. Holes may also reach the edge of the space-charge layer by diffusion and then migrate to the surface. Photocurrent measurements on GaP Schottky diodes and p–n junctions [43.91] showed a strong inverse dependence of the apparent diffusion length of the minority carriers on the carrier concentration in the range of Fig. 43.22. As a result of these two effects, diffusion and migration, the hole flux to the surface of an illuminated SI or n-type semiconductor may be expected to decreases as the electron density increases. This, of course, will lead to a drop in photoetch rate, as shown for GaAs in Fig. 43.22.

A similar result was subsequently demonstrated for n-type GaN [43.92, 93]. The local photoetch rate again showed an inverse dependence on carrier concentration, determined in this case by Raman measurements. Very convincing experimental evidence of the general validity of this relationship has been found during etching of the N-polar heteroepitaxial GaN layer containing Ga-polar inversion domains (IDs). During simultaneous growth of areas of different polarities more impurities and doping element are incorporated into the N-polar material; it becomes more heavily doped, say n+, while Ga-polar IDs have lower carrier concentration, say n. During photoetching the latter areas are etched more rapidly, i.e., deep craters are formed at IDs, as shown in Fig. 43.23a and explained by the model in Fig. 43.23b. The experimental details of such peculiar etching behavior have been discussed in recent papers [43.51–53].

Photoelectrochemical measurements on n-type GaN electrodes in the same KOH solution as used for electroless photoetching confirm the explanation given above. As the dopant density of the electrode is increased the onset for photocurrent is shifted to positive potential, i.e., the hole flux to the surface decreases. This trend can explain why bulk GaN single crystals, unintentionally doped (contaminated) with oxygen and carbon to the level of carrier concentration close to $10^{20}\,\text{cm}^{-3}$, cannot be photoetched in aqueous KOH solutions.

The relationship between the etch depth and carrier concentration established for GaAs in the DSL etching system has recently been confirmed for photogalvanic etching of GaN in KOH solutions [43.92]. A quantitative correlation of the etch rate and carrier concentration was obtained by micro-Raman determination of the carrier concentration [43.93].

Special Application of Electroless Etching

There are some features of this method which allow analysis of specific defects occurring in compound semiconductors. It was shown that the popular Abra-

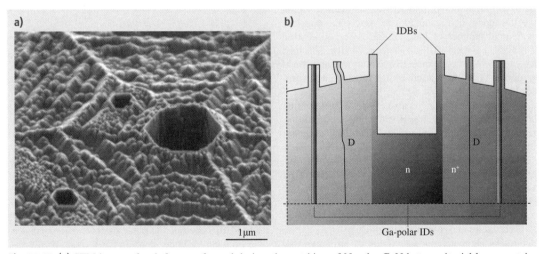

Fig. 43.23 (a) SEM image of etch features formed during photoetching of N-polar GaN heteroepitaxial layer containing inversion domains. (b) Schematic representation of a cross-section of the photoetched GaN layer from (a) (with permission from [43.53])

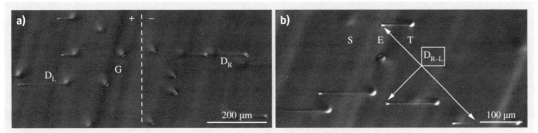

Fig. 43.24a,b DIC optical images of n-type GaAs after open-circuit photoetching

hams and Burocchi (AB) etch for GaAs is characterized by a so-called memory effect: the hillocks and ridges formed on defects remain intact even after very deep etching, even though the defects are no longer at the initial position [43.94]. On the basis of this result a projective etching method was developed and used for analysis of dislocations which moved in GaAs single crystals due to thermal stress during cooling after growth (G–S dislocations after [43.79]). This three-step etching procedure permits one to define the glide/climb system (glide/climb direction and plane) of any displaced dislocation. Numerous examples of these types of dislocations were shown and analyzed previously [43.8–10]. Figure 43.24 shows a set of images which illustrates the possibility offered by the photoetching method in analyzing the behavior of G–S dislocations. In Fig. 43.24a some dislocations moved to the left (denoted D_L) while others moved to the right (denoted D_R): most probably the strain was of the opposite sign, as indicated by "+" and "−" on the left and right side of the dashed line. The fact that in the central part of the image there are some dislocations that did not move [so-called grown-in dislocations (G), after [43.79]] supports this interpretation. In Fig. 43.24b some dislocations (denoted D_{R-L}) clearly *changed their mind* during the movement: first they moved from the starting position (S) to the right, stopped for some time (this can be recognized from the presence of the Cottrell atmosphere at the turning point (T) where the change of the direction of glide/climb occurred) and later moved to the left of the image, where they finally stopped at position E. Worth noting is the different size of the Cottrell atmospheres at positions S, T, and E, which is most probably the result of slower diffusion of decorating point defects due to decreasing temperature between the subsequent stop events of the D_{R-L} dislocations.

References

43.1 H.C. Gatos, M.C. Lavine: Characteristics of the {111} surfaces of the III–V intermetallic compounds, J. Electrochem. Soc. **107**, 427–433 (1960)

43.2 S. Amelinckx: *Direct Observation of Dislocations*, Supplement to Solid State Physics, Vol. 6 (Academic, New York 1964) pp. 15–50

43.3 D.J. Stirland, B.W. Straughan: A review of etching and defect characterization of gallium arsenide substrate material, Thin Solid Films **31**, 139–170 (1976)

43.4 D.C. Miller, G.A. Rozgonyi: Defect characterization by etching, optical microscopy and x-ray topography. In: *Handbook on Semiconductors*, Vol. 3, ed. by S.B. Keller (North-Holland, Amsterdam 1980) pp. 218–246

43.5 R.B. Heimann: Principles of chemical etching – the art and science of etching crystals. In: *Crystals: Growth, Properties and Applications*, Vol. 8, ed. by J. Grabmaier (Springer, Berlin, Heidelberg 1982) pp. 173–224

43.6 K. Sangwal: *Etching of Crystals* (North-Holland, Amsterdam 1987)

43.7 P.H.L. Notten, J.E.A.M. van den Meerakker, J.J. Kelly: *Etching of III–V Semiconductors: an Electrochemical Approach* (Elsevier, Oxford 1991)

43.8 J.L. Weyher: Characterization of compound semiconductors by etching. In: *Handbook on Semiconductors*, Vol. 3, ed. by S. Mahajan (Elsevier, Amsterdam 1994) pp. 995–1031

43.9 J.L. Weyher: Characterization of bulk as-grown and annealed III-Vs by photo-etching and complementary methods, Inst. Phys. Conf. Ser. **146**, 399–408 (1995)

43.10 J.L. Weyher, C. Frigeri, S. Müller: Selective etching and complementary microprobe techniques (SFM, EBIC). In: *Microprobe Characterization of Optoelectronic Materials*, Vol. 17, ed. by J. Jimenez (Taylor & Francis, New York 2003) pp. 595–689, Chap. 8

43.11 D. Zhuang, J.H. Edgar: Wet etching of GaN, AlN and SiC: a review, Mater. Sci. Eng. **R 48**, 1–46 (2005)

43.12 J.L. Weyher: Characterization of wide-band-gap semiconductors (GaN, SiC) by defect-selective etching and complementary methods, Superlattices Microstruct. **40**, 279–288 (2006)

43.13 J.J. Kelly, D. Vanmaekelbergh: Chemical and electrochemical etching of semiconductors. In: *Semiconductor Micromachining*, Fundamental Electrochemistry and Physics, Vol. 1, ed. by S.A. Campbell, H.J. Lewerenz (Wiley, Chichester 1997), Chap. 2

43.14 P.H.L. Notten: The etching of InP in HCl solutions: a chemical mechanism, J. Electrochem. Soc. **131**, 2641–2644 (1984)

43.15 J.J. Kelly, H.G.G. Philipsen: Anisotropy in the wet-etching of semiconductors, Curr. Opin. Solid State Mater. Sci. **9**, 84–90 (2005)

43.16 R.A. Wind, M.A. Hines: Macroscopic etch anisotropies and microscopic reaction mechanisms: using microfabrication to unravel the complicated chemistry of KOH/Si etching, J. Phys. Chem. B **106**, 1557–1569 (2002)

43.17 I.M. Huygens, K. Strubbe, W.P. Gomes: Electrochemistry and photoetching of n-GaN, J. Electrochem. Soc. **147**, 1797–1802 (2000)

43.18 A.R. de Wit, M.D. Janssen, J.J. Kelly: Electrochemical characterization of polycrystalline ZnO layers, Appl. Surf. Sci. **45**, 21–27 (1990)

43.19 A. Gagnaire, J. Joseph, A. Etcheberry, J. Gautron: An ellipsometric study of the electrochemical surface modifications of n-InP, J. Electrochem. Soc. **132**, 1655–1658 (1985)

43.20 P.H.L. Notten, J.J. Kelly: Evidence for cathodic protection of crystallographic facets from GaAs etching profiles, J. Electrochem. Soc. **134**, 444–448 (1987)

43.21 J.J. Kelly, J.E.A.M. van den Meerakker, P.H.L. Notten: Electrochemistry of photoetching and defect-revealing in III-V materials, Dechema-Monogr. **102**, 453–464 (1986)

43.22 J. van de Ven, H.J.P. Nabben: Anisotropic photoetching of III-V semiconductors I. Electrochemistry, J. Electrochem. Soc. **137**, 1603–1610 (1990)

43.23 J. van de Ven, H.J.P. Nabben: Anisotropic photoetching of III-V semiconductors II. Kinetics and structural factors, J. Electrochem. Soc. **138**, 144–152 (1991)

43.24 C. Youtsey, I. Adesida, G. Bulman: Highly anisotropic photoenhanced wet etching of n-type GaN, Appl. Phys. Lett. **71**, 2151–2153 (1997)

43.25 L. Macht, J.J. Kelly, J.L. Weyher, A. Grzegorczyk, P.K. Larsen: An electrochemical study of photoetching of heteroepitaxial GaN: kinetics and morphology, J. Cryst. Growth **273**, 347–356 (2005)

43.26 J.L. Weyher, R. Fornari, T. Görög, J.J. Kelly, B. Erné: HBr-$K_2Cr_2O_7$-H_2O etching system for indium phosphide, J. Cryst. Growth **141**, 57–67 (1994)

43.27 V. Lehmann: *Electrochemistry of Silicon. Instrumentation, Science, Materials and Applications* (Wiley-VCH, Weinheim 2002)

43.28 M.M. Faktor, J.L. Stevenson: The detection of structural defects in GaAs by electrochemical etching, J. Electrochem. Soc. **125**, 621–629 (1978)

43.29 C.R. Elliott, J.C. Regnault: The detection of strucutural defects in indium phosphide by electrochemical etching, J. Electrochem. Soc. **128**, 113–116 (1981)

43.30 A. Yamamoto, S. Tohno, C. Uemura: Detection of structural defects in n-type InP crystals by electrochemical etching under illumination, J. Electrochem. Soc. **128**, 1095–1100 (1981)

43.31 R. Bhat: Photoelectrochemical defect delineation in GaAs using hydrochloric acid, J. Electrochem. Soc. **132**, 2284–2285 (1985)

43.32 L. Blok: Characterization of vapour grown (001) $GaAs_{1-x}P_x$ layers by selective photo-etching, J. Cryst. Growth **31**, 250–255 (1975)

43.33 E. Sirtl, A. Adler: Cromasäure-Flußsäure als spezifisches System zur Ätzgrubenentwicklung auf Silizium, Z. Metallkd. **52**, 529–531 (1961), in German

43.34 J. Weyher, J. van de Ven: Selective etching and photoetching of {100} Gallium Arsenide in CrO_3-HF aqueous solutions. Part I: Influence of composition on etching behaviour, J. Cryst. Growth **63**, 285–291 (1983)

43.35 J. van de Ven, J.E.A.M. van den Meerakker, J.J. Kelly: The mechanism of GaAs etching in CrO_3-HF solutions I. Experimental results, J. Electrochem. Soc. **132**, 3020–3026 (1985)

43.36 J.J. Kelly, J. van de Ven, J.E.A.M. van den Meerakker: The mechanism of GaAs etching in CrO_3-HF solutions II. Model and discussion, J. Electrochem. Soc. **132**, 3026–3033 (1985)

43.37 J. van de Ven, J.L. Weyher, J.E.A.M. van den Meerakker, J.J. Kelly: Kinetics and morphology of GaAs etching in aqueous CrO_3-HF solutions, J. Electrochem. Soc. **133**, 799–806 (1986)

43.38 J.E.A.M. van den Meerakker, J.H.C. van Vegchel: Silicon etching in CrO_3-HF solutions. I: High [HF]/[CrO_3] ratios, J. Electrochem. Soc. **136**, 1949–1953 (1989)

43.39 J.E.A.M. van den Meerakker, J.H.C. van Vegchel: Silicon etching in CrO_3-HF solutions. II: low [HF]/[CrO_3] ratios, J. Electrochem. Soc. **136**, 1954–1957 (1989)

43.40 P. Visconti, D. Huang, M.A. Reshchikov, F. Yun, R. Cingolani, D.J. Smith, J. Jasinski, W. Swider, Z. Liliental-Weber, H. Markoç: Investigation of defects and surface polarity in GaN using hot wet etching together with microscopy and diffraction techniques, Mater. Sci. Eng. B **93**, 229–233 (2002)

43.41 J. Chen, J.F. Wang, H. Wang, J.J. Zhu, S.M. Zgang, D.G. Zhao, D.S. Jiang, H. Yang, K.H. Ploog: Measurement of threading dislocation densities in GaN by wet chemical etching, Semicond. Sci. Technol. **21**, 1229–1235 (2006)

43.42 J. Angilello, R.M. Potemski, G.R. Woolhouse: Etch pits and dislocations in {100} GaAs, J. Appl. Phys. **46**, 2315–2316 (1975)

43.43 K. Akita, T. Kusunoki, S. Komiya, T. Kotani: Observation of etch pits produced in InP by new etchants, J. Cryst. Growth **46**, 783–787 (1979)

43.44 S.N.G. Chu, C.M. Jodluk, A.A. Ballman: New dislocation etchant for InP, J. Electrochem. Soc. **129**, 352–354 (1982)

43.45 S. Komiya, T. Kotani: Direct observation of dislocations in GaAlAs-GaAs grown by the LPE method, J. Electrochem. Soc. **125**, 2019–2024 (1978)

43.46 E.S. Hellman: The polarity of GaN: a critical review, MRS Internet J. Nitride Semicond. Res. **3**, 1–11 (1998)

43.47 M. Seelmann-Eggebert, J.L. Weyher, H. Obloh, H. Zimmermann, A. Rar, S. Porowski: Polarity of (00.1) GaN epilayers grown on a (00.1) sapphire, Appl. Phys. Lett. **71**, 2635–2637 (1997)

43.48 J.L. Rouvière, J.L. Weyher, M. Seelmann-Eggebert, S. Porowski: Polarity determination for GaN films grown on (0001) sapphire and high-pressure grown GaN single crystals, Appl. Phys. Lett. **73**, 668–670 (1998)

43.49 D. Li, M. Sumiya, S. Fuke, D. Yang, D. Que, Y. Suzuki, Y. Fukuda: Selective etching of GaN polar surface in potassium hydroxide solution studied by x-ray photoelectron spectroscopy, J. Appl. Phys. **90**, 4219–4223 (2001)

43.50 G. Kamler, J. Borysiuk, J.L. Weyher, R. Czarnecki, M. Leszczynski, I. Grzegory: Selective etching and TEM study of inversion domains in Mg-doped GaN epitaxial layers, J. Cryst. Growth **282**, 45–48 (2005)

43.51 J.L. Weyher, F.D. Tichelaar, H.W. Zandbergen, L. Macht, P.R. Hageman: Selective photoetching and transmission electron microscopy studies of defects in heteroepitaxial GaN, J. Appl. Phys. **90**, 6105–6109 (2001)

43.52 L. Macht, J.L. Weyher, P.R. Hageman, M. Zielinski, P.K. Larsen: Direct influence of polarity on structural and electro-optical properties of heteroepitaxial GaN, J. Phys. Condens. Matter **14**, 13345–13350 (2002)

43.53 J.L. Weyher, L. Macht, F.D. Tichelaar, H.W. Zandbergen, P.R. Hageman, P.K. Larsen: Complementary study of defects in GaN by photoetching and TEM, Mater. Sci. Eng. B **91/92**, 280–284 (2002)

43.54 M.G. Mynbaeva, Y.V. Melnik, A.K. Kryganovskii, K.D. Mynbaev: Wet chemical etching of GaN in H_3PO_4 with Al ions, Electrochem. Sol.-State Lett. **2**, 404–406 (1999)

43.55 J.L. Weyher, M. Albrecht, T. Wosinski, G. Nowak, H.P. Strunk, S. Porowski: Study of individual grown-in and indentation-induced dislocations in GaN by defect-selective etching and transmission electron microscopy, Mater. Sci. Eng. B **80**, 318–321 (2001)

43.56 H. Ono, J. Matsui: Influence of In atoms on the shape of dislocation etch pits in LEC In-doped GaAs crystals, Jpn. J. Appl. Phys. **25**, 1481–1484 (1986)

43.57 V. Gottschalch, W. Heinig, E. Butter, H. Rosin, G. Freydank: H_3PO_4-etching of {001}-faces of InP, (GaIn)P, GaP, and Ga(AsP), Krist. Tech. **14**, 563–569 (1979)

43.58 J.L. Weyher, S. Lazar, J. Borysiuk, J. Pernot: Defect-selective etching of SiC, Phys. Status Solidi (a) **202**, 578–583 (2005)

43.59 T. Takenaka, H. Hayashi, K. Murata, T. Inoguchi: Various dislocation etch pits revealed on LPE GaAs{001} layer by molten KOH, Jpn. J. Phys. **17**, 1145–1146 (1978)

43.60 T. Ohno, H. Yamaguchi, S. Kuroda, K. Kojima, T. Suzuki, K. Arai: Direct observation of dislocations propagated from 4H-SiC substrate to epitaxial layer by x-ray topography, J. Cryst. Growth **260**, 209–216 (2004)

43.61 D. Siche, D. Klimm, T. Hölzel, A. Wohlfart: Reproducible defect etching of SiC single crystals, J. Cryst. Growth **270**, 1–6 (2004)

43.62 S.A. Sakwe, R. Müller, P.J. Wellmann: Optimization of KOH etching parameters for quantitative defect recognition in n- and p-type doped SiC, J. Cryst. Growth **289**, 520–526 (2006)

43.63 D. Hull, D.J. Bacon (Eds.): *Introduction to Dislocations* (Pergamon, Oxford 1984)

43.64 G. Kamler, J.L. Weyher, I. Grzegory, E. Jezierska, T. Wosinski: Defect-selective etching of GaN in a modified molten bases system, J. Cryst. Growth **246**, 21–24 (2002)

43.65 J.L. Weyher, P.D. Brown, J.L. Rouvière, T. Wosinski, A.R.A. Zauner, I. Grzegory: Recent advances in defect-selective etching of GaN, J. Cryst. Growth **210**, 151–156 (2000)

43.66 J.L. Weyher, S. Lazar, L. Macht, Z. Liliental-Weber, R.J. Molnar, S. Müller, V.G.M. Sivel, G. Nowak, I. Grzegory: Orthodox etching of HVPE-grown GaN, J. Cryst. Growth **305**, 384–392 (2007)

43.67 F. Secco d'Aragona: Dislocation etch for (100) planes in silicon, J. Electrochem. Soc. **119**, 948–951 (1972)

43.68 K. Motoki, T. Okahisa, S. Nakahata, N. Matsumoto, H. Kimura, H. Kasai, K. Takemoto, K. Uematsu, M. Ueno, Y. Kumagai, A. Koukitu, H. Seki: Growth and characterization of freestanding GaN substrates, J. Cryst. Growth **237–239**, 912–921 (2002)

43.69 J.L. Weyher, P.D. Brown, J.L. Rouvière, T. Wosinski, A.R.A. Zauner, I. Grzegory: Recent advances in defect-selective etching of GaN, J. Cryst. Growth **210**, 151–156 (2000)

43.70 M. Albrecht, H.P. Strunk, J.L. Weyher, I. Grzegory, S. Porowski, T. Wosinski: Carrier recombination at single dislocations in GaN measured by cathodoluminescence in a transmission electron microscope, J. Appl. Phys. **92**, 2000–2005 (2002)

43.71 K. Shiojima: Atomic force microscopy and transmission electron microscopy observations of KOH-etched GaN surfaces, J. Vac. Sci. Technol. B **18**, 37–40 (2000)

43.72 K. Engl, M. Beer, N. Gmeinwieser, U.T. Schwarz, J. Zweck, W. Wegscheider, S. Miller, A. Miler, H.-J. Lugauer, G. Brüderl, A. Lell, V. Härle: Influence of an in situ-deposited SiN_x intermediate layer inside GaN and AlGaN layers on SiC substrates, J. Cryst. Growth **289**, 6–13 (2006)

43.73 I. Kamata, H. Tsuchida, T. Jikimoto, K. Izumi: Structural transformation of screw dislocations via thick 4H-SiC epitaxial growth, Jpn. J. Appl. Phys. **39**, 6496–6500 (2000)

43.74 S. Ha, H.J. Chung, N.T. Nuhfer, M. Skowronski: Dislocation nucleation in 4H silicon carbide epitaxy, J. Cryst. Growth **262**, 130–138 (2004)

43.75 Z. Zhang, T.S. Sudarshan: Evolution of basal plane dislocations during 4H-silicon carbide homoepitaxy, Appl. Phys. Lett. **87**, 161917-1–161917-3 (2005)

43.76 G. Kamler, J. Borysiuk, J.L. Weyher, A. Presz, M. Wozniak, I. Grzegory: Application of orthodox defect-selective etching for studying GaN single crystals, epitaxial layers and device structures, Eur. Phys. J. Appl. Phys. **27**, 247–249 (2004)

43.77 J.L. Weyher, G. Kamler, G. Nowak, J. Borysiuk, B. Lucznik, M. Krysko, I. Grzegory, S. Porowski: Defects in GaN single crystals and homo-epitaxial structures, J. Cryst. Growth **281**, 135–142 (2005)

43.78 G. Kamler, J. Smalc, M. Wozniak, J.L. Weyher, R. Czarnecki, G. Targowski, M. Leszczynski, I. Grzegory, S. Porowski: Selective etching of dislocations in violet-laser diode structures, J. Cryst. Growth **293**, 18–21 (2006)

43.79 J.L. Weyher, J. van de Ven: Selective etching and photoetching of GaAs in CrO_3-HF aqueous solutions. Part III: Interpretation of defect-related etch figures, J. Cryst. Growth **78**, 191–217 (1986)

43.80 J.L. Weyher, L. Macht: Defects in wide band-gap semiconductors: selective etching and calibration by complementary methods, Eur. Phys. J. Appl. Phys. **27**, 37–41 (2004)

43.81 C. Frigeri, J.L. Weyher: Electron beam induced current and photoetching investigations of dislocations and impurity atmospheres in n-type LEC GaAs, J. Appl. Phys. **65**, 4646–4653 (1989)

43.82 J.L. Weyher, C. Frigeri, P.J. van der Wel: Complementary DSL, EBIC and PL study of grown-in defects in Si-doped GaAs crystals grown under Ga- and As-rich conditions by LEC method, J. Cryst. Growth **103**, 46–53 (1990)

43.83 J.L. Weyher, P.J. van der Wel, C. Frigeri: Spatially resolved study of dislocations in Si-doped LEC GaAs by DSL, PL and EBIC, Semicond. Sci. Technol. **7**, A294–A299 (1992)

43.84 J.L. Weyher, L. Macht, G. Kamler, J. Borysiuk, I. Grzegory: Characterization of GaN single crystals by defect-selective etching, Phys. Status Solidi (c) **0**(3), 821–826 (2003)

43.85 J.A. Bardwell, J.B. Webb, H. Tang, J. Fraser, S. Moisa: Ultraviolet photoenhanced wet etching of GaN in $K_2S_2O_8$ solution, J. Appl. Phys. **89**, 4142–4149 (2001)

43.86 B. Yang, P. Fay: Etch rate and surface morphology controle in photoelectrochemical etching of GaN, J. Vac. Sci. Technol. B **22**, 1750–1754 (2004)

43.87 C. Youtsey, L.T. Romano, I. Adesida: Gallium nitride whiskers formation by selective photoenhanced wet etching of dislocations, Appl. Phys. Lett. **73**, 797–799 (1998)

43.88 C. Youtsey, L.T. Romano, R.J. Molnar, I. Adesida: Rapid evaluation of dislocation densities in n-type GaN films using photoenhanced wet etching, Appl. Phys. Lett. **74**, 3537–3539 (1999)

43.89 S. Lazar, J.L. Weyher, L. Macht, F.D. Tichelaar, H.W. Zandbergen: Nanopipes in GaN: photoetching and TEM study, Eur. Phys. J. Appl. Phys. **27**, 275–278 (2004)

43.90 C. Frigeri, J.L. Weyher, L. Zanotti: Study of segregation inhomogeneities in GaAs by means of DSL photoetching and EBIC measurements, J. Electrochem. Soc. **136**, 262–266 (1989)

43.91 M.L. Young, D.R. Wight: Concentration dependence of the minority carrier diffusion length and lifetime in GaP, J. Phys. D Appl. Phys. **7**, 1824–1837 (1974)

43.92 J.L. Weyher, R. Lewandowska, L. Macht, B. Lucznik, I. Grzegory: Etching, Raman and PL study of thick HVPE-grown GaN, Mater. Sci. Semicond. Process. **9**, 175–179 (2006)

43.93 R. Lewandowska, J.L. Weyher, J.J. Kelly, L. Konczewicz, B. Lucznik: Calibration of the PEC etching of GaN by Raman spectroscopy, J. Cryst. Growth **307**, 298–301 (2007)

43.94 D.J. Stirland, R. Ogden: A dislocation "etch-memory" effect in gallium arsenide, Phys. Status Solidi (a) **17**, K1–K4 (1973)

44. Transmission Electron Microscopy Characterization of Crystals

Jie Bai, Shixin Wang, Lu-Min Wang, Michael Dudley

Since the first observation of dislocations published in 1956, transmission electron microscopy (TEM) has become an indispensable technique for materials research. TEM not only provides very high spatial resolution for the characterization of microstructure and microchemistry but also elucidating the mechanisms controlling materials properties. The results of TEM analyses can also shed light on possible ways for improving the crystal quality. With the recent development of the electron exit wave reconstruction technique, the resolution of TEM has exceeded the typical Scherzer point resolution of ≈ 0.18 nm and observation of dislocation cores with an accuracy of 10 pm has been achieved. Most TEM studies are carried out in a static status; however, dynamic studies using in situ heating, in situ stressing, and even in situ growth can be conducted to study the development, interaction, and multiplication of defects.

44.1 Theoretical Basis
  of TEM Characterization of Defects 1477
  44.1.1 Imaging of Crystal Defects
    Using Diffraction Contrast............. 1478
  44.1.2 Phase-Contrast High-Resolution
    Transmission Electron Microscopy
    (HRTEM)....................................... 1482
  44.1.3 Diffraction Techniques 1484
  44.1.4 STEM, EELS, and EFTEM
    in Microanalysis 1489
  44.1.5 FIB for TEM Sample Preparation..... 1493
44.2 Selected Examples of Application
  of TEM to Semiconductor Systems........... 1493
  44.2.1 Studies
    of Conventional Heteroepitaxial
    Semiconductor Systems................. 1494
  44.2.2 TEM Studies of Large-Mismatch
    Heteroepitaxial Systems............... 1500
  44.2.3 Application
    of STEM, EELS, and EFTEM.............. 1509
44.3 Concluding Remarks: Current Application
  Status and Development 1514
References 1515

44.1 Theoretical Basis of TEM Characterization of Defects

Modern TEM has three basic operation modes: imaging, diffraction, and spectroscopy analyses. In a transmission electron microscope, the objective lens takes the electrons emerging from the exit surface of the specimen, disperses them to create a diffraction pattern in the back focal plane, and recombines them to form an image in the image plane. When the lenses of the imaging system are adjusted so that the back focal plane of the objective lens coincides with the object plane of the intermediate lens, the diffraction pattern is projected onto the viewing screen. Alternatively, if the image plane of the objective lens works as the object plane of the intermediate lens, an image will be projected onto the viewing screen.

The contrast in the images recorded in imaging mode can be formed by several mechanisms:

1. Amplitude contrast (generally called diffraction contrast). The most commonly used bright field (BF) and dark field (DF) imaging techniques form images of dislocations with either the transmitted or diffracted beam under a two-beam condition. By excluding other beams, such images are formed by amplitude contrast.
2. Phase contrast and Z-contrast. When the transmitted and diffracted beams are made to combine, preserving both amplitude and phase information, a lattice image of the planes that are diffracting or even struc-

tural images of the individual atom columns may be resolved directly (high-resolution TEM, HRTEM).
3. Mass thickness contrast. Incoherent (Rutherford) elastic scatter of electrons can form mass thickness contrast. Such contrast is generally weaker and overshadowed by the stronger effects of electron diffraction, except in cases where there are large differences in atomic number or when diffraction is weak. The investigation of microstructures of crystals mainly includes defects and interfaces. The imaging and analysis of defects are generally carried out with diffraction contrast under two-beam conditions while studies of interfaces are generally conducted under multibeam phase-contrast mode (HRTEM).

In diffraction mode, patterns such as selected-area diffraction patterns, Kikuchi patterns, and convergent beam diffraction patterns can provide crystallographic information such as the orientation, crystallographic symmetry, phase, strain, etc. When a nearly parallel electron beam illuminates the specimen and a particular area in the first image is selected by an aperture, a selected-area diffraction (SAD) pattern is formed; otherwise, if the beam converges onto a small area of the specimen, convergent-beam electron diffraction (CBED) consisting of diffraction discs will occur. On the other hand, large-angle convergent-beam diffraction (LACBED) is a relatively newly established technique which is extensively applied to the analysis of various defects. A LACBED image is formed by moving the specimen out of the object plane of a CBED setting and forming a combined image from both real space (the shadowed specimen image) and reciprocal space (diffracted Bragg lines).

The spectroscopy analyses in a modern analytical TEM include energy-dispersive x-ray spectroscopy (EDS) and electron energy-loss spectroscopy (EELS) that allow quick analysis of material chemistry on the nanoscale. With the newly developed energy-filtered TEM (EFTEM), quick elemental mapping not relying on the scanning technique is possible.

44.1.1 Imaging of Crystal Defects Using Diffraction Contrast

Defects in crystals can be described in terms of translational vectors which represent displacements of atoms from their regular positions in the lattice. Assuming the general displacement vector is R, the Howie–Whelan equations describing the change in amplitude of the direct beam ϕ_0 and the amplitude of the diffracted beam ϕ_g can be written as [44.1]

$$\frac{d\psi_0}{dz} = \frac{i\pi}{\xi_0}\phi_0 + \frac{i\pi}{\xi_g}\phi_g \exp(2\pi i s z + 2\pi i \mathbf{g}\cdot\mathbf{R}),$$

$$\frac{d\psi_g}{dz} = \frac{i\pi}{\xi_g}\psi_0 \exp(-2\pi i s z - 2\pi i \mathbf{g}\cdot\mathbf{R}) + \frac{i\pi}{\xi_g}\psi_g,$$

where ξ_0 and ξ_g are the extinction coefficients for the direct and diffracted beam, s is the deviation vector, z is the depth of the defect in the specimen, and $\mathbf{g}$ is the reflecting vector.

When an objective aperture is used to exclude either the diffracted electrons or transmitted electrons under a two-beam condition, a bright-field (BF) or dark-field (DF) image is formed. By excluding other beams, such images are formed by amplitude contrast. The contrast can be used to determine the displacement field of the defect. Diffraction contrast and the appearance of features in BF and DF images depend sensitively on the deviation from the Bragg condition (deviation parameter s). Maximum transmission occurs in BF when s is small and positive, and it is under these conditions that most BF images are usually obtained [44.2].

Defects Characterized with Diffraction Contrast

Dislocations. In practical Burgers vector analysis of dislocations, the sample is tilted to a particular two-beam position with the deviation parameter s being set to a positive value. The presence of the dislocation bends the planes on one side into Bragg orientation and forms bright–dark line pair contrast in BF or DF images. A conventional technique to study the Burgers vector is $\mathbf{g}\cdot\mathbf{b}$ analysis ($\mathbf{g}$ is the reflecting plane while $\mathbf{b}$ is the Burgers vector of the dislocation) using their null-contrast properties. For dislocations with different nature, the $\mathbf{g}\cdot\mathbf{b}$ null-contrast rule works differently. Pure screw dislocations fully lose their contrast when the reflection plane satisfies $\mathbf{g}\cdot\mathbf{b} = 0$. For pure edge dislocations, $\mathbf{g}\cdot\mathbf{b} = 0$ and $\mathbf{g}\cdot(\mathbf{b}{\wedge}\mathbf{u}) = 0$ should be simultaneously satisfied to make the dislocation invisible. For mixed-type dislocations, $\mathbf{g}\cdot\mathbf{b}$, $\mathbf{g}\cdot\mathbf{b}_e$, and $\mathbf{g}\cdot\mathbf{b}_e{\wedge}\mathbf{u}$ are all required to be zero to minimize contrast (it is not possible to make it completely invisible). In practice, dislocations generally show faint contrast when $\mathbf{g}\cdot\mathbf{b} = 0$ even if $\mathbf{g}\cdot\mathbf{b}_e$ and $\mathbf{g}\cdot\mathbf{b}_e{\wedge}\mathbf{u}$ are not zero. Thus, the conventional method to determine the Burgers vector of dislocations is by finding two reflections $\mathbf{g}_1$ and $\mathbf{g}_2$ for which the invisibility criterion holds, and the Burgers vector may be determined using: $(\mathbf{g}_1{\wedge}\mathbf{g}_2) \parallel \mathbf{b}$. It is recommended that at least three consistent cases of effective invisibility are found with $w < 1.0$ (where

w is the parameter describing the deviation from the Bragg condition $w = \xi_g s$) and using low-index reflections to avoid confusion. A detailed description of dislocation analysis is presented by *Edington* [44.3]. For strongly anisotropic materials, computer simulation must be used to determine both the direction and magnitude of the Burgers vector by employing variation of $\boldsymbol{g}, s, \boldsymbol{u}, \boldsymbol{b}$ and the foil normal.

Stacking Faults. A stacking fault is a planar defect at which the regular stacking sequence of the crystal is locally interrupted and a relative shift $\boldsymbol{R}$ of the top part of the crystal with respect to the lower part is introduced. For a column passing through the faults, a phase shift $2\pi \boldsymbol{g} \cdot \boldsymbol{R}$ is produced between the waves emitted by the two parts of the crystal located on each side of the fault. Therefore, a phase factor $\alpha = 2\pi \boldsymbol{g} \cdot \boldsymbol{R}$ is introduced into the main and diffracted beam amplitudes. When the fault is inclined to the specimen surface, contrast takes the form of light and dark fringes parallel to the line of intersection of the fault plane with the surface. Thus the observation and analysis of fringes formed by stacking faults inclined to the TEM foil normal can be used to determine the fault vector. This is the so-called $\boldsymbol{g} \cdot \boldsymbol{R}$ analysis used to characterize stacking faults. When $\boldsymbol{g} \cdot \boldsymbol{R}$ is zero or an integer, no contrast is presented and the fault is invisible. For other values of $\boldsymbol{g} \cdot \boldsymbol{R}$, however, the fault produces contrast. The precise form of the fringe contrast depends on the diffraction conditions employed, and this enables fault vector and type to be determined [44.5].

Polarity. Polarity is a consequence of the noncentrosymmetrical structure frequently encountered in compound semiconductors. Polarity reversal and symmetry of both (0001) wurtzite-type and (111) zincblende-type structures have been studied extensively [44.6]. One of the techniques to characterize polarity is the multiple dark-field TEM technique. *Serneels* et al. [44.7] conducted the theoretical calculations and predicted that, under multiple-beam conditions along a noncentrosymmetric zone axis, the inverted region of the crystal should be different in brightness from the surrounding matrix. The difference arises from the violation of Friedel's law. Taking GaN/AlN as an example, when multiple dark-field images are taken along a nonsymmetrical zone axis (say, $\langle 11\bar{2}0 \rangle$ or $\langle 10\bar{1}0 \rangle$) with $\boldsymbol{g} = 0002$ and $\boldsymbol{g} = 000\bar{2}$, an inversion domain will show either brighter or darker contrast than the matrix material [44.8]. *Jasinski* et al. applied this method to characterize the V-like inversion domains in InN films [44.9]. Dark-field images recorded with (0002) and (000$\bar{2}$) reflections under multiple diffraction conditions show a reversal defect-matrix contrast, as illustrated in Fig. 44.1. This method does not determine if the domain has Ga or N-polarity, but it may prove the presence of inversion.

High-Resolution Diffraction Contrast Imaging
For the characterization of crystal defects, the geometry and the character of the individual defects are two critical features of interest. In many studies, such as those involving the dissociation of dislocations, defects of high density without strain field overlap, defects with very small size etc., examination of images with high resolution is required. Apart from instrumental limits, the resolution is also determined by factors such as diffraction contrast. Two ways of achieving high-resolution diffraction contrast are described below.

Weak-Beam Dark-Field (WBDF) Technique. Since its first report by *Cockayne* et al. [44.4], the weak-beam dark-field (WBDF) technique has become a convenient and important method to achieve high-resolution diffraction contrast images of crystal defects. In par-

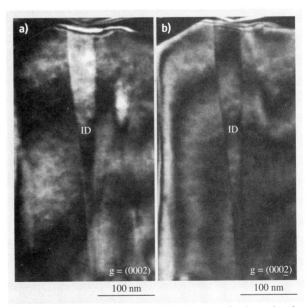

Fig. 44.1a,b Two-beam dark-field TEM images of an inversion domain in InN film ((**a**) recorded with $\boldsymbol{g} = 0002$; (**b**) recorded with $\boldsymbol{g} = 000\bar{2}$). A domain-matrix contrast reversal can be noticed (after [44.4], © AIP 1969)

ticular, it is widely employed to get sharp images of dislocation lines, resolving pairs of dislocations, and in precisely locating the positions of dislocation cores combined with simulation. The major advantage of the weak-beam approach over the two-beam dynamical technique lies in its improved resolution and the high contrast of the image. With the WBDF technique, only the diffraction contrast from the core of the defect contributes to the image. Consequently, systematic analysis of a given crystal defect is possible at a resolution approaching the limit of the microscope. Though the resolution of WBDF is not as high as the direct lattice resolution technique, it does have additional advantages such as allowing thicker specimens, reduced requirement of lens aberration, better contrast, and the validity of $\bm{g}\cdot\bm{b}$ analysis. In this technique, the TEM sample is tilted to a large, positive value of s, so that only the crystal planes close to the dislocation core are bent into a diffraction condition with $s \approx 0$ while neighboring lattice planes are away from the Bragg condition. By doing this, a sharp image of the near-core region of the dislocation can be recorded.

High-Order Reflection Method. The image width of a defect is approximately $\zeta_g/3$, where ζ_g is the extinction distance for the reflection $\bm{g}$ [44.10]. Thus the image can be made narrower and the resolution increased by using high-order reflection with long extinction length. This technique was introduced by *Bell* and *Thomas* and is especially applicable at high voltages [44.11]. In this technique the specimen is tilted to put the high-order reflection into strong diffracting condition and form the image with transmitted beam. Compared with WBDF, this technique allows shorter exposure time and hence minimizes the risk of losing resolution because of mechanical or other instabilities. The disadvantage is that the image resolution is improved at the cost of contrast. If reflections of too high an order are employed, the defect will no longer be visible.

Quantitative Determination of the Indices of Line Features

Determination of the crystallographic direction of linear defects such as straight dislocation lines, needle-shaped precipitates, etc. can be very important when trying to understand their formation mechanisms. In a similar way, knowing the direction of the lines of intersection between planar defects or between inclined planar defects and the sample surface can also help in understanding their origins. For the case of a dislocation, in order to fully characterize this defect, both its line direction and Burgers vector (magnitude and direction) need to be determined. TEM can be conveniently used to determine the line direction by tilting the specimen to at least two different orientations and measuring the angle, on the given micrograph, between the direction normal to the dislocation line direction and the trace of the reflecting plane (normal to $\bm{g}$). Stereographic projection analysis can then be used to deduce the line direction $\bm{u}$ from these measurements as described by *Edington* [44.12]. This simply involves plotting the two (or three) normals to the dislocation line direction on a stereographic projection and the subsequent drawing of a great circle through these normals. The pole of this great circle is then the line direction of the dislocation in the crystal. Figure 44.2 is a schematic showing this stereographic analysis. The various points are plotted on a standard projection.

The dashed line shows the great circle corresponding to the trace of the plane perpendicular to the incident beam direction, $\bm{B}_1 \cdot \bm{g}_1$ is the $\bm{g}$ vector used for this particular image, $\bm{g}_1 \times \bm{B}_1$ is the intersection of the trace of the reflecting plane on the image plane, and $\bm{D}_1$ is

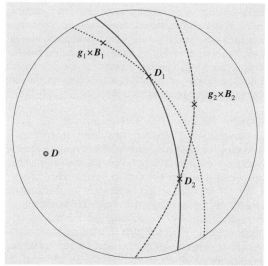

Fig. 44.2 Schematic showing the stereographic analysis. $\bm{g}_1 \times \bm{B}_1$ and $\bm{g}_2 \times \bm{B}_2$ are the intersections between the traces of the $\bm{g}$ vectors utilized with the respective image planes (planes perpendicular to $\bm{B}_1$ and $\bm{B}_2$); $\bm{D}_1$ and $\bm{D}_2$ are the intersections between the traces of the planes perpendicular to the dislocation line direction observed on the two images. $\bm{D}$ is the pole of the great circle containing $\bm{D}_1$ and $\bm{D}_2$

the intersection of the trace on the image plane of the plane normal with the dislocation line. δ_1 is the angle between $g_1 \times B_1$ and D_1 measured from the TEM image and plotted on the great circle corresponding to the trace of the plane normal to B_1. The dotted line is the great circle corresponding to the trace of the plane normal to another incident beam direction B_2, with g_2 being the reference g vector and D_2 being another observed direction normal to the dislocation line which is oriented at the measured angle δ_2 from $g_2 \times B_2$. The solid line is the great circle that runs through D_1 and D_2; the pole of this great circle is the dislocation line direction D.

Unfortunately, such graphical techniques can be tedious, inconvenient, and imprecise. Bai [44.13] introduced an analytical, vector version of this technique wherein the trace of the g vector on a given image is expressed as the vector $g \times B$. With suitably designed coordinate systems being defined, measurement of the angle (δ) between ($g \times B$) and the normal to the dislocation line direction enables it to be expressed as a vector D_n. Repeating this measurement for a second image with different g vector and beam direction B generates another vector D_n and the cross-product of two such vectors enables the line direction of the dislocation in the crystal to be determined.

The three coordinate systems utilized in the calculation (shown in Fig. 44.3) are defined as follows [44.14]:

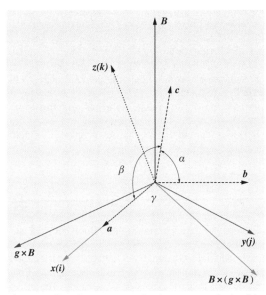

Fig. 44.3 Coordinate systems for the vector analysis of the determination of the line direction of a linear feature

1. A Cartesian coordinate system consisting of the beam direction (B), $g \times B$, and $B \times (g \times B)$
2. The crystal coordinate system comprising the three basis vectors a, b, c and three angular parameters α, β, γ of the unit cell for the crystal being studied
3. A Cartesian coordinate system x, y, and z, in which the axis x is along a, b lies in the xy plane, and c is in the upper half-space (since a, b, c roughly form a right-handed relationship).

The normal of the line feature which lies at an angle δ to the trace of the g vector can be represented as $D_n = (g \times B/|g \times B|)\cos\delta + [B \times (g \times B)/|B \times (g \times B)|]\sin\delta$. B and g are expressed as vectors in crystal system and therefore are the vectors $g \times B$ and $B \times (g \times B)$. All of these vectors then have to be expressed as vectors referred to the Cartesian coordinate system xyz in order to carry out the necessary vector calculations.

The expressions enabling the basis vectors a, b, c to be referred to the Cartesian coordinate system xyz are

$$a = a_x i + a_y j + a_z k = ai,$$
$$a_x = a, \quad a_y = a_z = 0,$$
$$b = b_x i + b_y j + b_z k = b(\cos\gamma i + \sin\gamma j),$$
$$b_x = b\cos\gamma, \quad b_y = b\sin\gamma, \quad b_z = 0,$$
$$c = c_x i + c_y j + c_z k,$$
$$c_x = c\cos\beta, \quad c_y = c\frac{(\cos\alpha - \cos\beta\cos\gamma)}{\sin\gamma},$$
$$c_z = \sqrt{c^2 - c_x^2 - c_y^2},$$
$$V_c = (a \times b) \cdot c = abc_z \sin\gamma,$$
$$a_x^* = \frac{b_y c_z}{V_c}, \quad a_y^* = -\frac{b_x c_z}{V_c}, \quad a_z^* = (b_x c_y - b_y c_x),$$
$$b_y^* = \frac{a_x c_z}{V_c}, \quad b_z^* = -\frac{a_x c_y}{V_c},$$
$$c_z^* = \frac{a_x b_y}{V_c},$$
$$b_x^* = c_x^* = c_y^* = 0.$$

$B_1(U_1, V_1, W_1)$ and $g_1(h_1, k_1, l_1)$ can be expressed in the xyz system as

$$B_1 = B_x^1 i + B_y^1 j + B_z^1 k,$$
$$B_x^1 = U_1 a_x + V_1 b_x + W_1 c_x, \quad B_y^1 = V_1 b_y + W_1 c_y,$$
$$B_z^1 = W_1 c_z,$$
$$g_1 = g_x^1 i + g_y^1 j + g_z^1 k,$$
$$g_x^1 = h_1 a_x^*, \quad g_y^1 = h_1 a_y^* + k_1 b^* y,$$
$$g_z^1 = h_1 a_z^* + k_1 b_z^* + l_1 c_z^*.$$

Then $g_1 \times B_1$ and $B_1 \times (g_1 \times B_1)$ can be written as

$$(g_1 \times B_1)_x = g_y^1 B_z^1 - g_z^1 B_y^1 ,$$
$$(g_1 \times B_1)_y = g_z^1 B_x^1 - g_x^1 B_z^1 ,$$
$$(g_1 \times B_1)_z = g_x^1 B_y^1 - g_y^1 B_x^1 ,$$
$$[B_1 \times (g_1 \times B_1)]_x = B_y^1 (g_1 \times B_1)_z - B_z^1 (g_1 \times B_1)_y ,$$
$$[B_1 \times (g_1 \times B_1)]_y = B_z^1 (g_1 \times B_1)_x - B_x^1 (g_1 \times B_1)_z ,$$
$$[B_1 \times (g_1 \times B_1)]_x = B_x^1 (g_1 \times B_1)_y - B_y^1 (g_1 \times B_1)_x .$$

The normal of the dislocation line D_1 in this case can be written as

$$D_1 = D_x^1 i + D_y^1 j + D_z^1 k ,$$
$$D_x^1 = \cos \delta_1 (g_1 \times B_1)_x + \sin \delta_1 [B_1 \times (g_1 \times B_1)]_x ,$$
$$D_y^1 = \cos \delta_1 (g_1 \times B_1)_y + \sin \delta_1 [B_1 \times (g_1 \times B_1)]_y ,$$
$$D_x^1 = \cos \delta_1 (g_1 \times B_1)_z + \sin \delta_1 [B_1 \times (g_1 \times B_1)]_z .$$

Similarly, we can get $D_2 = D_x^2 i + D_y^2 j + D_z^2 k$. Then the dislocation line $D = D_1 \times D_2$ can be written as

$$D = D_x i + D_y j + D_z k ,$$
$$D_x = D_y^1 D_z^2 - D_z^1 D_y^2 ,$$
$$D_y = D_z^1 D_x^2 - D_x^1 D_z^2 ,$$
$$D_z = D_x^1 D_y^2 - D_y^1 D_x^2 .$$

Now the index of $D(D_x, D_y, D_z)$ is in the Cartesian coordinate system xyz. We can convert this to a vector referred to the crystal system using the following relationships

$$U = D_x p + D_y p' + D_z p'' ,$$
$$V = D_y q' + D_z q'' ,$$
$$W = D_z r'' ,$$

where

$$p = \frac{1}{a} ,$$
$$p' = -\frac{b_x}{a b_y} ,$$
$$p'' = -\frac{1}{a}\left[-\frac{c_y b_x}{b_y c_z} + \frac{c_y}{c_z}\right] ,$$
$$q' = \frac{1}{b_y} ,$$
$$q'' = -\frac{c_y}{b_y c_z} ,$$
$$r'' = \frac{1}{c_z} .$$

The main source of error in this calculation relates to the accuracy of the determination of B, the measurement of the angle δ, and the integralization of the decimal indices.

Generally, a g vector in a low-index selected-area diffraction pattern with B as zone axis is used in this process. The error in determination of B and g and hence δ can be avoided by carefully adjusting the zone axis B along the electron beam. The 180° ambiguity does not create a problem because the coordinates $g \times B$ and $B \times (g \times B)$ go with the assigned g. The critical point is the sense of δ. If D_n is anticlockwise from $g \times B$, care must be taken regarding the sense of rotation sense of δ with respect to the trace of the known g. An accurate calibration of the rotation between the image and the diffraction pattern is required for the application of this technique.

44.1.2 Phase-Contrast High-Resolution Transmission Electron Microscopy (HRTEM)

For dislocations in semiconductor materials, the individual dislocations exhibit significant deviations from average predictions due to various reasons and exhibit an unusual large data scattering that is commonly of unknown origin. Therefore, it is desirable to characterize individual dislocations with truly atomic resolution and to compare the results with theoretical calculations. Phase-contrast high-resolution TEM to date has broken the 0.1 nm barrier and it is possible to combine theory and experiment to accurately identify atomic column positions with better than 10 pm precision [44.15].

As opposed to the amplitude caused contrast observed in conventional TEM (where the image is formed by one beam), HRTEM images are formed with combined transmission and diffracted beams and the contrast is therefore a composite composed of both amplitude and phase information. There is no relative displacement between the location of a defect and the contrast variation caused by the defect in the HRTEM image, which is an advantage compared with diffraction imaging. However, there are limitations in direct interpretation of the HRTEM image in terms of the sample structure and composition. Since HRTEM images are sensitive to factors such as specimen thickness and orientation, objective lens defocus, spherical and chromatic aberration, etc., precise interpretation of the images may require extensive simulations. The effect of the interference between the beams can be predicted with the phase contrast transfer function (CTF) of the objective lens and therefore a HRTEM image can be simulated. Since the CTF is focusing dependent, it is crucial to choose the optimum defocus to fully exploit the capabilities of electron microscopy in HRTEM mode. However, there is no simple optimized solution. Better experimental results can be achieved by conducting a specimen exit wave reconstruction with a series of through-focal images [44.16]. In this technique, a series of about 20 pictures is shot under the same imaging conditions with the exception of the focus, which is incremented between each take. Together with exact knowledge of the CTF the series allows for computation of the phase change. As mentioned earlier, amplitude change also contributes to image contrast. Consequently, sample thickness plays a critical rule in the appearance of a lattice image. When the sample is thin enough that amplitude variations do not contribute to the image, the HRTEM image is formed purely by phase contrast and it shows directly a two-dimensional projection (down some low-index direction) of the crystal with defects and all.

A serious limitation to the interpretation of high-resolution images is the limited signal-to-noise ratio. By noise we refer to any contribution to image intensity which is not essentially a part of the signal, e.g., lattice image. Major contributions to noise are disordered surface layers, due to either contamination or specimen thinning damage, and shot noise in the electron beam or recording media. High noise-to-signal ratio could lead to errors in interpretation of crystal/amorphous interfaces.

There are two resolution limits in axial HRTEM which apply to imaging of the simple weak-phase object. The point-to-point resolution, attainable near the Scherzer defocus condition, is $d_{pp} \approx 0.66(C_s \lambda^3)^{1/4}$ (where C_s is the spherical aberration coefficient and λ is the electron wavelength). The information limit is $d_{in} \approx 2\sqrt{E/(C_c \lambda \Delta E)}$, where C_c is the coefficient of chromatic aberrations. For a totally unknown weak-phase object one can identify detail to d_{pp} simply and to d_{in} by careful image simulation or reconstruction from a through-focal series [44.17].

Dislocation Core Observation by HRTEM

Recent developments in electron microscopy have enabled microscopists to resolve the dislocation core structure down to the atomic level. For diamond cubic and zincblende semiconductors resolving dumbbell atom columns along the dislocation line requires resolution better than the typical Scherzer point resolution of ≈ 0.18 nm of most high-resolution transmission electron microscopes. By reconstructing an electron exit wave through-focal series, *Xu* et al. resolved a 30° partial dislocation in GaAs:Be with accuracy of the atomic position at the dislocation core region within 10 pm [44.18]. A modern aberration-corrected TEM can easily achieve point-to-point resolution of 0.1 nm.

Interfaces of Epitaxial Systems

HRTEM plays a valuable role in the characterization of interfaces in epitaxial systems. The investigation generally includes studies of the misfit dislocations, measurement of rigid shifts, and identification of the roughness and extent of the interface. The major requirement for characterization of interfaces is that both crystals should be well aligned on their appropriate zone axes and the interface should be aligned so that it is edge-on to the incident beam. The microscope should be well aligned and operate near the Scherzer focus and the thickness on both sides of the interface should be considerably less than half an extinction thickness (if possible).

For conventional epitaxial structures, e.g., SiGe on Si, low-index zones such as [100] and [110] are typically used, but [111] and [112] can provide further useful projections when three-dimensional views are required in order to specify the interface structures fully. For heterostructures of III–V compounds such as GaAs/Al$_x$Ga$_{1-x}$As, it is extremely difficult to determine the location of the interface under the imaging conditions traditionally required for high-resolution imaging, namely, thin crystals and optimum defocus, and hence impossible to determine the interface sharpness. In order to highlight the compositional variations it is necessary to find thickness and defocus combinations which accentuate differences in the images from the two constituent materials. To get distinct interface, dark-field images formed by the 200 diffraction spot can be used to precisely determine the interface. This is due to the fact that, in the kinematical approximation, the intensity in the 200 diffraction beam is proportional to $(f_{III} - f_V)^2$, where f_{III} and f_V are the atomic scattering amplitudes of the group IIIA and group V elements, respectively. When this diffracted beam is strongly excited, the Al$_x$Ga$_{1-x}$As layers appear much brighter than the GaAs layers. Such images provide an accurate measure of the layer thickness and of the flatness of the interface [44.19]. Another method to get distinct interface is to use the ⟨100⟩ projection rather than the ⟨110⟩ > projection because the former has four chemically sensitive {002} beams compared with the latter with two of them [44.20]. The interfaces of epitaxial systems can also be well studied by the high-resolution high-angle annular dark-field scanning TEM (HAADF-STEM) technique, which is extremely sensitive to the atomic number of the atoms in the high-resolution image (Sect. 44.1.4).

Polarity

Liliental-Weber et al. studied the atomic structure of Mg-rich hexagonal pyramids in GaN film which exhibit opposite polarity from the matrix material [44.21]. By taking a series of HRTEM images in a through-focal sequence, the electron wave exiting the specimen was reconstructed and atomic resolution was achieved. With such a high resolution, the exchange of Ga and N sublattices inside and outside the defect can be unambiguously determined. In this case a special crystal projection needs to be chose, e.g., [11$\bar{2}$0], such that two different atoms with different atomic numbers can be resolved. Figure 44.4 shows the reconstructed images

taken from different regions related to the pyramidal defect. The inversion of the polarity within the defect (BC) compared with the matrix (AB) can be clearly observed.

44.1.3 Diffraction Techniques

As mentioned previously, a parallel electron beam gives rise to diffraction patterns composed of sharp spots. Generally such a diffraction pattern is achieved by selecting a small region using an aperture inserted in the image plane of the objective lens, forming a selected-area diffraction (SAD) pattern. When the electron beam converges to image a small area of the specimen, a CBED pattern comprising diffraction discs forms. Compared with the limited spatial resolution of conventional SAD (usually down to a fraction of a micrometer), CBED gives resolution down to nanometer. Moreover, CBED discs contain more information than simple diffraction spots in SAD. The application of CBED includes phase identification, symmetry determination (point and space group), thickness measurement, strain and lattice parameter measurement, structure factor determination, etc.

Selected-Area Diffraction (SAD)

A SAD pattern can be treated as the magnified image of a planar section through the reciprocal lattice taken normal to the incident beam direction. It can be used to:

1. Identify the orientation of the specimen
2. Find the proper *g* vector for defect analysis
3. Identify the structure of defects such as precipitates, twins, stacking faults, etc.
4. Determine the orientation relationship between phases
5. Determine the long-range order parameters.

The accuracy of analysis of a diffraction pattern depends upon the accuracy of measurement. Some factors of importance are:

1. The shape factor of the features, which determines the shape of the intensity distribution about the rel-points
2. Instrument alignment and beam divergence
3. Specimen perfection
4. Curvature of the reflecting sphere and relative orientation of the specimen
5. Double diffraction, giving rise to reflections of zero structure factor.

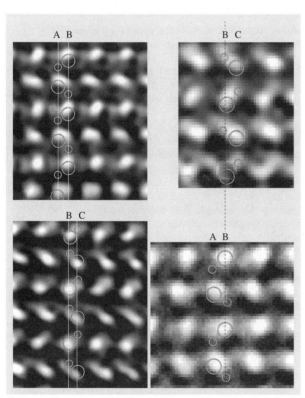

Fig. 44.4 Reconstructed exit wave phase of inversion domains in Mg-rich GaN. Note the change of polarity within the defect (BC) compared with the matrix (AB) (after [44.21], © Elsevier 2005)

Several examples of the application of the SAD technique are described in the following.

Twins in Diamond Cubic and Zincblende Structure Semiconductor Crystals. Twinning is a phenomenon which leads to the existence of domains in a crystal which have orientation relationships usually described by a simple symmetry operation. The most commonly observed type of symmetry operation involves a mirror image of the structure in the twinning plane, although the 180° (or equivalently 60°) rotation of the structure about the normal to the twin plane is also quite common in semiconductors with diamond cubic and zincblende structure. Twins can be produced either during growth or by mechanical deformation. The existence of twins can be discerned using the SAD technique, which is sensitive to the differences in orientation between a *twin* and *matrix*. If the twinned domains lay side by side with the untwined *matrix* regions and the incident beam straddles both regions then a simple superposition of two diffraction patterns will occur, one from the matrix and one from the twin. For example, for a (110)-oriented diamond cubic crystal, a twin might be created by 180° rotation about the ($\bar{1}$11) twin plane normal. This produces twin domains which have ($\bar{1}$10) surface orientation (i.e., the same as the matrix) but which have different in-plane orientation. This is shown in the stereographic projections in Fig. 44.5, where it can be seen that the orientation relationship between the two structures can not only be considered from the point of view of the 180° rotation about the ($\bar{1}$11) twin plane normal but can also be considered either as a mirror across (1$\bar{1}$2) or as a rotation by 70°32′ about the (110) plane normal. The diffraction pattern associated with such side-by-side twins should therefore consist of the (110) matrix patterns with a second *twin* pattern being produced from the first by either a mirror operation across (1$\bar{1}$2) or as a rotation by 70°32′ about the (110) zone axis. However, if the twin and matrix domains overlap one another the diffraction pattern becomes slightly more complicated, with extra spots appearing at 1/3 intervals along the ⟨111⟩ directions. Figure 44.6 shows an example SAD pattern of (110) Ge epifilm with overlapping matrix and twin domains. Extra spots located at $\mathbf{n} \times \mathbf{g}/3$ can be clearly observed. Some of these extra spots can arise from multiple diffraction effects [44.22]. However, the full pattern can be understood by considering the tripling of the periodicity which occurs in this direction when the twin and matrix structures are overlapped [44.23]. Further details regarding analysis of twins can be found in the book by *Edington* [44.24].

Atomic Ordering of Semiconductor Alloys. For alloys containing different kinds of atoms which have attractive interactions, the desire to maximize the number of

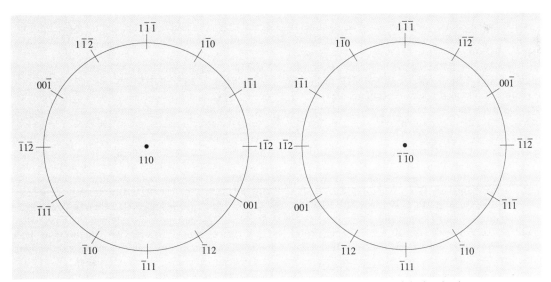

Fig. 44.5 Stereographic projections showing the orientation relationship between the original and twin structures

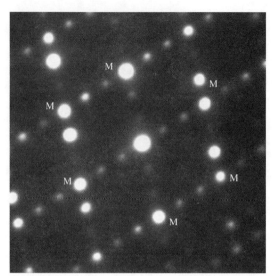

Fig. 44.6 SAD pattern of (110)Ge epifilm with overlapping matrix and twin domains, the main diffraction spots from matrix are indicated by "M"

unlike neighbors can lead to the formation of an ordered structure. Consequently, the usual structure factor rules determining the allowed reflections are relaxed and superstructure reflections in between the original Bragg reflections can be introduced. The intensity of such reflections is related to the difference between the atomic scattering factors of the atoms involved, as opposed to the intensity of fundamental reflections, which is related to their sum. A comprehensive review is given by *Marcinkowski* [44.25]. Studies of such long-range order structure can be carried out using the SAD technique to analyze the superstructure reflections. TEM specimens in various tilted orientations are required to obtain SAD patterns along different directions [44.26, 27]. The extra diffraction spots are indexed and compared with those of the standard ordered structure in order to determine the long-range order. With the help of computer simulation, the structural ordering can be determined using smaller mutual tilts [44.28]. In various III–V alloy semiconductors such as AlGaAs, In(Al)GaP, InGaAsP, InAlAs, GaAsSb, GaAsP, and InAsSb, several types of ordered structures such as CuAu, CuPt, chalcopyrite, and famatinite, have been observed. In crystals grown on (001) substrates, observed ordered structure is mostly of CuPt type in which a doubling of the periodicity of the column III or V atoms occurs on the {111} plane [44.26].

Conventional Convergent–Beam Electron Diffraction (CBED)

When the incident electron waves impinging on the specimen comprise a convergent cone of rays, this generates a continuous range of Ewald spheres in reciprocal space. This causes higher-order Laue zones (HOLZs) to become visible in diffraction patterns, such that three-dimensional diffraction information is exhibited on the diffraction pattern. The fact that such information can be obtained from areas as small as a few nanometers in diameter is a major advantage of CBED. The diffraction discs in a CBED pattern are created exactly from the same area, and each disc contains the intensities determined by the orientation of the incident beam with respect to the (hkl) lattice plane. Several examples of the application of CBED are presented in the following sections.

Thickness Determination. CBED is a convenient method to determine the TEM specimen thickness. In this technique, the sample is tilted to form a two-beam condition. The intensity fringes (K-M fringes) in the diffracted discs due to various s are related to the sample thickness by $s_i^2/n_k^2 + 1/(\xi_g^2 n_k^2) = 1/t^2$ (where s_i is the diffraction deviation for the i-th fringe, n_k is an integer, ξ_g is the extinction length for the reflection, and t is the thickness of the sample). By choosing a proper n_k to plot a straight line of s_i^2/n_k^2 versus $1/n_k^2$, the thickness t can be determined by the intercept of the line [44.29]. For such measurement, the region of the foil selected should be flat without distortion and the beam must be focused at the plane of the specimen.

Polarity. There are several TEM methods for polarity determination: CBED, TEM-EDS, HRTEM, bend contour analysis, and EELS. Their relative capabilities to determine polarity depend strongly upon the species of crystal and the morphology of specimens to be examined. Since the CBED pattern is sensitive to both defects sample thickness in the illuminated area of the specimen, the use of this method is sometimes limited. *Mitate* et al. studied the polarity of GaN, ZnO, AlN and GaAs with both TEM-EDS (energy-dispersive x-ray spectroscopy) and the conventional CBED methods [44.30]. The results show that the CBED method is useful for ZnO and GaN, while the TEM-EDS method must be used for AlN and GaAs and can also be used for GaN and ZnO. It is clearly indicated that the TEM-EDS method is of use for the case where the difference in atomic scattering factor between the two constituent elements is small, while the CBED method is useful for the case where the difference is large. In the presence of foil bending, it may be difficult to perform CBED experiments because any displacement of the electron beam on the sample is accompanied by a change of the local crystal orientation. In this case, the polarity can be determined from bend contours.

Sphalerite Structure. *Taftø* and *Spence* [44.31, 32] proposed a method of determining the deviations from centrosymmetry by taking advantage of the strong coherent multiple scattering normally present in CBED. In this technique, the TEM foil is tilted so that both the 200 reflection and two high odd-index reflections are simultaneously excited. When illuminating with a convergent electron beam, the effect of the dynamical interaction of these two weak reflections with the diffracted 200 beam is shown as cross-like contrast which intersects with the broad 200 Bragg line. Either bright or dark contrast suggests that the interference between the two weak reflections with the 200 reflection is either constructive of destructive and thus the type of the reflection hkl

can then be revealed. An example image showing the detailed features of the $\bar{2}00$ and 200 diffraction discs for a thickness of 100 nm is shown in Fig. 44.7a,b. Figure 44.7c shows the two double-scattering paths which contribute to the 200 reflection.

Wurtzite Structure. For polarity determination of wurtzite crystals such as GaN, conventional CBED is generally carried out along the $\langle 01\bar{1}0 \rangle$ axis. The intensity distributions within the $+g$ and $-g$ diffraction discs for polar directions such as [0002] and [000$\bar{2}$] are different and can be used for determination of atom distributions within the unit cell. To ensure the accurate determination of the polarity, the rotation angle and the 180° ambiguity between the image and the diffraction pattern need to be taken into account. Computer simulation of the intensity distribution taking into account the sample thickness, Debye–Waller factor, and absorption coefficient is necessary to verify the results. A good match between the experimental and simulated pattern is required for at least two different sample thicknesses to ensure that the interpretation is correct. Figure 44.8 shows a TEM image of an inversion domain boundary in an ammonothermally grown bulk GaN crystal [44.33]. The CBED patterns taken on either side of the boundary are shown as insets. An inversion

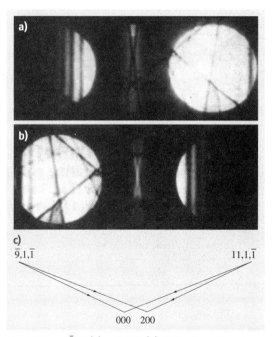

Fig. 44.7a–c $\bar{2}00$ (**a**) and 200 (**b**) diffraction discs showing the destructive and constructive interaction between the high-order odd reflections and $\bar{2}00/200$ reflection; (**c**) two double-scattering paths which contribute to the 200 reflection

Fig. 44.8 TEM image showing a inversion domain boundary in a GaN bulk crystal; *insets* show the inversion of the intensity distribution in 0002 diffraction discs on either side of the boundary

of the intensity distribution in the 200 reflections on either side of the boundary can be clearly observed.

Lattice Parameter, Strain, and Composition. CBED has become a well-established technique to study lattice strain and composition in crystals with high spatial resolution and high accuracy [44.34, 35]. The high-order Laue zone (HOLZ) lines that arise from very high-order reflections are very sensitive to changes in lattice parameters. The geometry of HOLZ lines in the bright-field disc of a CBED pattern is also sensitive to the accelerating voltage. Hence, if the accelerating voltage of the incident electrons is maintained constant, changes in the angular position of the HOLZ lines can be directly correlated to variations in lattice parameter. The concept of effective accelerating voltage E_e is introduced in the pattern simulation to compensate for systematic shifts of HOLZ lines caused by dynamical effects [44.34]. The lattice parameters can be determined by matching the observed HOLZ pattern with computer simulations based on the kinematical theory of diffraction. Typically, the lattice parameter information can be obtained from a very small region defined by the incident probe size. The technique enables measurement of lattice parameter changes with a precision of approximately ±0.0002 nm in many cases. *Rozeveld* and *Howe* conducted a detailed analysis of the lattice parameter error and developed a procedure to optimize the lattice parameter precision [44.36]. This method can be used to determine the six lattice parameters (a, b, c, α, β, and γ) from a single CBED pattern. *Zipprich* et al. [44.37] studied the strain and composition of Si/SiGe multilayer systems by analyzing the rocking curves in CBED patterns and comparing the experimental results with kinematical two-beam calculations. The resulting parameters are then further refined by dynamical simulations. The best accuracy for strain determination in this study is achieved by using perfectly flat specimen regions, recorded with the smallest possible beam diameter that yields sufficient signal intensity. The accuracy of this measurement can be greatly enhanced by application of the recently developed chromatic corrector for the modern TEM.

Strain determination methods using cross-sectional specimens, e.g., in HRTEM or in the measurement of the shift of Bragg lines in CBED patterns, are generally influenced by surface relaxation (i.e., *thin-foil effects*). *Jacob* et al. [44.38] studied the strained GaAs/In$_x$Ga$_{1-x}$As/GaAs substrate structure with both finite-element calculation and experimental TEM imaging, showing that the surface strain relaxation of cross-sectional TEM samples is significant regardless of the TEM foil thickness. Therefore, it is preferable to make strain measurements on plan-view samples.

Large-Angle Convergent-Beam Electron Diffraction (LACBED)

Large-angle convergent-beam electron diffraction (LACBED) is a relatively newly established TEM technique, first introduced by *Tanaka* et al. to overcome the limitations of conventional CBED that occur when the beam convergence becomes larger than the Bragg angle [44.39]. In the LACBED technique, the convergent electron beam is brought to a focus and the specimen is moved either above or below the object plane. Since the sample is defocused, a larger illuminated area forms shadow images in the diffraction discs. The transmitted beam is selected by means of the selected-area aperture and the disc pattern is made of deficiency lines. These Bragg lines are superimposed with the shadow image in the transmitted disc, i.e., information about reciprocal space (the Bragg lines) and direct space (the shadow image) are simultaneously present. In this way, a LACBED pattern can be considered as an image-diffraction mapping technique. There is no rotation between the shadow image and the diffraction pattern if the specimen is below the object plane, while there is a 180° rotation if the specimen is above. To a good approximation, the spatial resolution in the shadow image is given by the minimum probe size. Some examples of the application of LACBED are presented in the following sections.

Dislocation Analysis. LACBED was originally applied to the characterization of dislocation Burgers vectors by *Cherns* and *Preston* [44.40]. Since then, LACBED has evolved into a technique to investigate strain fields in materials. It can be used to unambiguously determine the magnitude and the sign of a dislocation Burgers vector and has been applied to the characterization of partial dislocations, stair-rod dislocations, grain boundary dislocations, etc. It is also an essential technique to study point defects, planar defects, and other crystalline phenomena [44.41].

It was found that splitting of HOLZ lines and Kikuchi lines occur when a convergent electron beam is brought close to a dislocation due to its long-range displacement, providing $\mathbf{g} \cdot \mathbf{b} \neq 0$ [44.42]. However since the displacement field of a dislocation varies from place to place, features in a CBED pattern are sensitive to the exact location of the probe, which is difficult to determine experimentally. This difficulty is overcome by the use of LACBED, in which the specimen is defocused and a larger area including a segment of dislocation is illuminated. The displacement and the multiplicity of the splitting of a LACBED Bragg line when it intersects a dislocation may be simply related to the sign and magnitude of the dislocation under a wide range of diffracting conditions. Simple rules discovered by *Cherns* and *Preston* [44.43] can be summarized as follows, and a corresponding schematic is shown in Fig. 44.9:

1. The number (n) of interfringes present at a splitting is given by $n = |\boldsymbol{g} \cdot \boldsymbol{b}|$.
2. The sign of $\boldsymbol{g} \cdot \boldsymbol{b}$ is given by the characteristic twisting of the Bragg line: when $\boldsymbol{g} \cdot \boldsymbol{b} > 0$, then at $x > 0$, the contour is bent to the $s < 0$ side, and at $x < 0$, the contour is bent to the $s > 0$ side. On the other hand, when $\boldsymbol{g} \cdot \boldsymbol{b} < 0$, then the contour is bent in the opposite sense (s indicates the deviation vector, while x indicates the horizontal distance from the dislocation line).

The method requires the observation of at least three splittings in order to solve the set of three linear equations: $\boldsymbol{g}_1 \cdot \boldsymbol{b} = n_1$, $\boldsymbol{g}_2 \cdot \boldsymbol{b} = n_2$, and $\boldsymbol{g}_3 \cdot \boldsymbol{b} = n_3$.

Partial dislocations with small Burgers vectors, such as stair-rod dislocations with $\boldsymbol{b} = 1/6\langle 110 \rangle$, are notoriously difficult to analyze directly in image mode. *Cherns* et al. successfully studied Shockley and Frank partial dislocations as well as stair-rod dislocations in Si, GaAs, and CdTe with LACBED [44.44].

Stacking Faults. LACBED can be employed to extract information on the type and the magnitude of the displacement associated with an inclined stacking fault [44.41, 45–47]. It also has the advantage of being applicable to in-plane stacking faults, in contrast to conventional imaging [44.41]. When the inclination of the fault is known, the LACBED technique can directly determine the fault type. The asymmetry of the LACBED pattern due to the fault indicates the sign of the fault. Also, with a more detailed comparison of theory and experiment, an estimate of the phase shift can be determined. In the case of {111} intrinsic or extrinsic stacking faults present in face-centered cubic (fcc) structures, the phase shift α could be 0 or $\pm 2\pi/3$. As a result, the rocking curves are modified and the CBED lines with $\alpha = \pm 2\pi/3$ are split into a main line and a subsidiary one, the subsidiary line being on one side or on the other depending upon the sign of α and s. The best effect is observed when the incident beam is

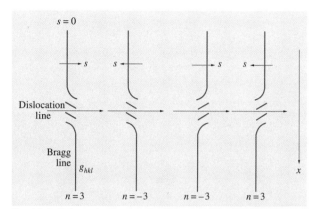

Fig. 44.9 Schematic showing the Cherns and Preston rules

located in the center of the stacking fault. From such a pattern, it is possible to identify $\boldsymbol{R}$ [44.48].

Strain Measurement. The LACBED method enables examination of multilayers in plan view since rocking curves over a sufficient angle can be observed and allow for the analysis of the contributions from the component layers [44.41]. In the case of strained heterostructures, a shift of the ZOLZ or HOLZ lines in LACBED discs has been observed due to the relative misfit strain between epilayer and substrate. The strain in an epitaxial system may cause relative rotation of the individual planes in the epilayer and/or the substrate. Such a rotation can be observed as a splitting of the Bragg lines of medium- or high-index reflections. For large rotations, the peak splitting approximates to the relative angle of rotation resulting from the misfit stresses in bicrystals and multilayers. Misfit strains down to 0.1% can be measured by LACBED [44.41]. *Hovsepian* et al. studied the composition of GeSi quantum dots embedded in two Si layers with the LACBED technique [44.49]. The asymmetry of rocking curves of inclined planes (inclined to 001 growth direction) is caused by the phase shift $\boldsymbol{g} \cdot \boldsymbol{R}$, where $\boldsymbol{R}$ is total normal displacement across the layer. By comparing to the two-beam kinematically calculated rocking curves, the composition of the Ge quantum dots can be determined.

44.1.4 STEM, EELS, and EFTEM in Microanalysis

Often combined with conventional TEM, scanning transmission electron microscopy (STEM), electron energy-loss spectroscopy (EELS), and energy-filtered

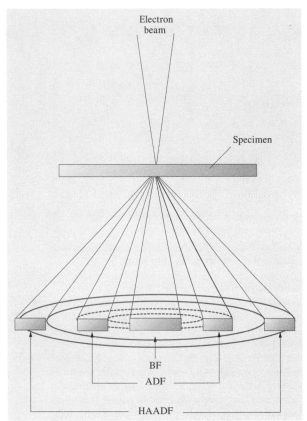

Fig. 44.10 Schematics showing different STEM detector arrangement

transmission electron microscopy (EFTEM) have found unique applications in a wide range of applications in materials analysis. In many research areas, such as the semiconductor industry, STEM, EELS, and EFTEM have become essential tools to assist process development and defect analysis.

STEM Imaging

STEM has been used extensively in conjunction with conventional TEM, especially due to the advance in refining electron beam spot size by the availability of field-emission guns and lens aberration correctors [44.51, 52].

STEM image formation is achieved through scan synchronization of the electron beam and a television (TV) monitor with input signals from an electron detector. The electron detector is placed conjugate to the back focal plane of a microscope. Depending on the collection angle of the detector (Fig. 44.10), three major STEM methods are defined: bright field (BF), annular dark field (ADF), and high-angle annular dark field (HAADF). The BF detector is normally a circular disk which collects the transmitted electron beam and low-angle scattered electrons. Both ADF and HAADF detectors are annular disks with a hole in the middle, enabling a limited part of the scattered electrons to be collected for the STEM signal. Figure 44.11 shows examples of the same feature (TiN and $TiSi_x$ crystal growth when forming a contact in a semiconductor device) viewed by three different STEM methods. With proper selection of camera lengths or by using an objective lens aperture, BF STEM gives essentially the same image as BF TEM. HAADF collects only high-angle (e.g., > 40 mrad) scattered electrons. The contrast was attributed to Rutherford scattering [44.53, 54], so the local intensity is proportional to $\sum n_i Z_i^2$, where Z is the atomic number and n_i is the number of elements of ele-

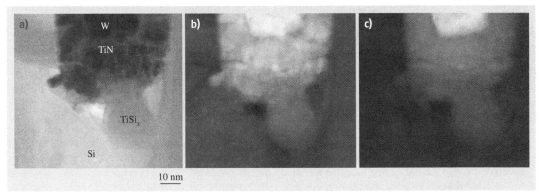

Fig. 44.11a–c Example STEM images (after [44.50]): (**a**) BF STEM, (**b**) ADF STEM, and (**c**) HAADF STEM

ment i present in an atom column. This simple approach gives a good explanation as to why the HAADF STEM is Z-contrast. Recent studies have found that HAADF STEM is better modeled by thermal diffuse scattering [44.55–57]. Using the thermal diffuse scattering model, the Z-dependent contrast HAADF STEM can be quantitatively analyzed. Because of its Z-dependency, HAADF STEM is also called Z-contrast STEM.

Positioned in the medium collection-angle range between BF and HAADF, the ADF collector collects diffracted electron beams as well as certain low-angle thermally scattered electrons. Thus the ADF image has mixed Z and diffraction contrasts, as shown in Fig. 44.11 [44.50]. Instrumentally, the ADF and HAADF often share the same detector, and the collection angle is normally controlled by camera length. Although different STEM methods have been used in different situations, the HAADF STEM has proven to be a unique and powerful tool by offering incoherent and Z-contrast imaging.

EELS Elemental Analysis

Recent advancements in instruments and software have made EELS a practical tool in various microanalysis applications. In combination with HAADF STEM imaging, EELS analysis offers rich information with regards to chemistry, chemical bonding, and thickness with high spatial resolution. When electrons pass through a specimen, some experience inelastic scattering. A magnetic spectrometer (either post-column or in column [44.58]) spreads the inelastically scattered electrons based on their energies. By plotting the number of electrons versus energy loss, we obtain an electron energy-loss spectrum.

The features of an EELS spectrum are illustrated in Fig. 44.12. The horizontal axis is the energy loss. However, it is often labeled *energy*, instead of *energy loss* ΔE, for simplicity. There are three general regions in a complete EELS spectrum: the zero-loss peak (ZLP), the low-loss region, and the core-loss edge. The zero-loss peak is formed by electrons that are not scattered

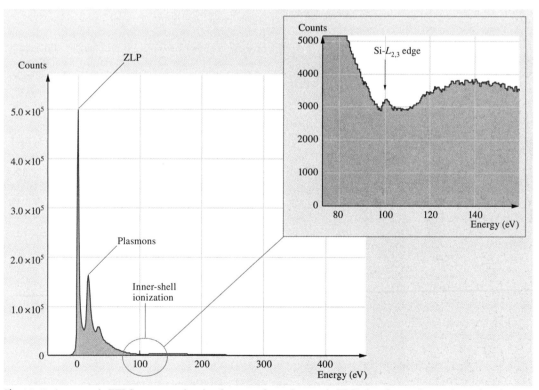

Fig. 44.12 An example EELS spectrum showing features of zero-loss peak (ZLP), low-loss region, and core-loss features (after [44.50])

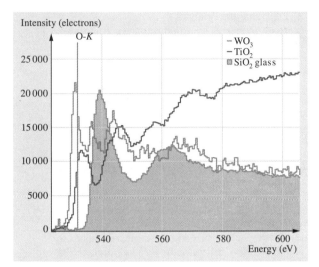

Fig. 44.13 Comparison of oxygen near edge fine structures from WO_3, TiO_2, and SiO_2 ◀

or scattered elastically in the specimen with negligible energy loss. A low-loss region is the energy loss range up to $\approx 50\,eV$. The signal in this region often includes plasmon excitation, due to collective excitation of valence electrons [44.58]. The core-loss edge in the spectrum occurs when an incident electron transfers sufficient energy to an orbital electron to move it to higher energy levels. The atom is said to be ionized. Thus, this process is also called inner-shell ionization loss. The character of the inner-shell energy loss is *edge*. The onset of the edge corresponds to the minimum energy for the ionization (the energy difference between the original energy level and the lowest available target level).

Important information often contained in EELS is the chemical bonding information. A bonding change can result in a variation of energy-loss near-edge structure (ELNES). Some examples of variation of near-edge fine structures are shown in Fig. 44.13, which compares the O-K edge (energy loss of K-shell electrons) for WO_3, TiO_2, and SiO_2 [44.50]. In these comparisons, we see pronounced differences in the shape of the O-K edge due to the difference of W–O, Ti–O and Si–O bonding. This valuable information is often of interest in microanalysis and helps to identify chemical information in certain ambiguous situations [44.60]. EELS and EDS are often complementary to each other in their efficiencies and accuracies of detecting certain elements [44.58, 61]. However, EELS generally offers higher spatial resolution and less signal overlapping than EDS. In STEM mode, the spatial resolution of EELS is essentially that of STEM. Nanoscale chemical analysis can be routinely obtained with a modern STEM/EELS system.

EFTEM

By combining an electron energy-loss spectrometer and imaging-forming lenses, it is possible to form an image with electrons of a selected energy. This technique is known as energy-filtered transmission electron

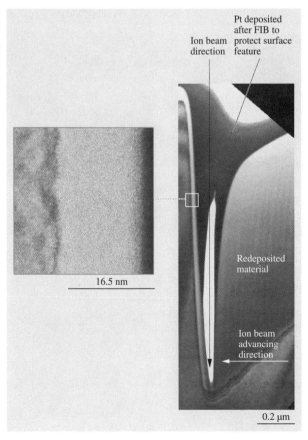

Fig. 44.14 A cross-sectional view of Si surface prepared by FIB using 30 keV Ga^+. Note the amorphous layer caused by ion-beam damage. The milled surface is not parallel to the beam direction because of the radial intensity distribution of the ion beam (after [44.59]) ◀

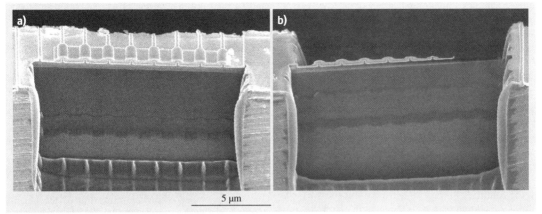

Fig. 44.15a,b Sample preparation for TEM using an FIB/SEM system. (**a**) The front side of the sample where the milling can be stopped at a desired location with monitoring by SEM. (**b**) Back-side view of the finished sample (after [44.50])

microscopy. EFTEM adds a new dimension to conventional TEM and its advantages include:

1. Cleaner imaging by the removal of inelastic scattered electrons (imaging using zero-loss energy window)
2. Element-specific mapping
3. Removal or suppression of diffraction contrast
4. Contrast enhancement.

Some examples from the application of EFTEM can be found in Sect. 44.2.3.

44.1.5 FIB for TEM Sample Preparation

The success of TEM analysis largely relies on the quality of the TEM sample. The development of the focused ion beam (FIB) system [44.62, 63] has greatly enhanced TEM sample preparation capabilities [44.59, 64]. In a FIB system, a focused ion beam (normally Ga^+) is used to micromachine the target material by ion beam sputtering. The emitted secondary electrons or ions can be used for in situ imaging. Direct ion-beam imaging makes the location control of FIB milling possible. Due to the capability of monitoring FIB milling process, combination of SEM and FIB has become common and necessary for TEM sample preparation.

A cross-sectional view of a FIB-milled surface in Si is shown in Fig. 44.14 [44.65]. Figure 44.15 shows two SEM pictures of TEM sample preparation in a FIB/SEM system. The benefits offered by FIB in TEM sample preparation include:

1. The ability to be location specific with an accuracy of a few nanometers
2. The absence of mechanical damage to the sample
3. A similar cutting rate for different materials (because the milling front is controlled mostly by focused beam movement as opposed to sputtering rate)
4. The ability to isolate a small sample from a bulk material with a lift-out probe.

An undesirable artifact from FIB is the ion-beam damage to the sample prepared, as shown in Fig. 44.14. For many materials, such as crystalline Si, the structural damage can introduce certain problems with imaging and analysis, especially for thin samples. Low-energy ion-beam milling, either within FIB or with a conventional ion miller, can generally improve sample quality by reducing the thickness of the damaged layer [44.59].

44.2 Selected Examples of Application of TEM to Semiconductor Systems

In the following, TEM studies of defects, strain, interface, etc. of two kinds of semiconductor systems will be detailed as examples: conventional heteroepitaxial systems and large-mismatch heteroepitaxial systems. Examples for the application of STEM, EELS, and EFTEM in semiconductor devices are given in Sect. 44.2.3.

44.2.1 Studies of Conventional Heteroepitaxial Semiconductor Systems

In low-mismatch systems, the theories of *van der Merwe* [44.66] and *Matthews* [44.67] describing the accommodation of lattice misfit are well established and have been experimentally verified. According to these theories, the epigrowth is assumed to follow a two-dimensional, layer-by-layer mode. When the thickness of the epifilm is lower than a critical thickness h_c, the film is pseudomorphic with the substrate and the mismatch is accommodated by elastic strain. At the critical thickness h_c, misfit dislocations form to accommodate the strain plastically as well as elastically. The formation of the misfit dislocations can occur in two ways:

1. Threading dislocations replicated from the substrate into the film are forced, under the influence of the mismatch stress, to glide leaving a trailing interfacial misfit segment connecting the original substrate threading segment and the mobile threading segment.
2. Dislocation half-loops nucleate at the surface of the film and glide toward the film–substrate interface, leaving a misfit segment and two threading segments (the surface nucleation will likely occur at some heterogeneity such as a step, surface sites where contamination is present, and possibly at the valleys of surface undulations caused by morphological instabilities) [44.68, 69].

This ultimately results in an interface composed of large coherent regions separated by rows of misfit dislocations. If the substrate is assumed to be rigid and the elastic strain is built up only on the film side, the lattice misfit is composed of two parts [44.67]

$$f_0 = \varepsilon_{\text{pl}} + \varepsilon_{\text{el}},$$

where ε_{pl} is the strain accommodated by dislocations and ε_{el} is the strain accommodated elastically, i.e.,

$$\varepsilon_{\text{el}} = \frac{(\langle a_0 \rangle - a_0)}{a_0},$$

$$\varepsilon_{\text{pl}} = \frac{b_f}{D_f},$$

where $\langle a_0 \rangle$ is the average lattice parameter of the grown film, b_f is the Burgers vector of the misfit dislocations, and D_f is the average spacing between two misfit dislocations.

If the total misfit is relieved by plastic strain, the spacing between misfit dislocations should be $D = b_f / f_0$. However, this is difficult to achieve because the nucleation and glide of dislocations needs to overcome an energy barrier and its driving force presumes a sufficient amount of residual strain.

For metal films, the experimentally determined critical thickness agrees reasonably well with the value predicted using the equilibrium theory of van der Merwe and Matthews. For semiconductor films with diamond and zincblende structure, however, experimental observations revealed much larger values of the critical thickness and a slower relaxation of the elastic strain than would be expected from equilibrium calculations [44.70].

Misfit Dislocations

As aforementioned, epilayers which are mismatched with respect to the substrate due to either different structure or composition can be relaxed or partially relaxed by the formation of misfit dislocations with edge character once the epilayer exceeds a critical thickness. In most common cases of heteroepitaxial structures with diamond or zincblende-type structures with $\langle 001 \rangle$ orientation, the mismatch is accommodated by an orthogonal network of misfit dislocations along two perpendicular in-plane $\langle 110 \rangle$ directions. Figure 44.16 shows a plan-view TEM image taken along the [001] direction from a $\text{Si}/\text{Si}_{0.85}\text{Ge}_{0.15}$ interface. Arrowed is a characteristic fingerprint of the Hagen–Strunk mechanism [44.71]. It has been always observed that the density of misfits along these two directions are different (i.e., not equally spaced). As the dislocation spacing approaches that needed to relax the layer fully the asymmetry between the two orthogonal arrays becomes less marked. This has been attributed to the differing mobilities of α and β dislocations which are more significant at lower stresses [44.72]. For small-mismatch systems, 60° misfit dislocations are dominant, although occasionally pure edge misfit dislocations can be observed in low-misfit systems; for example, *Fitzgerald* et al. [44.73] reported observation of such dislocations at the $\text{In}_x\text{Ga}_{1-x}\text{As}/\text{GaAs}$ ($x \approx 0.12$, mismatch $= 0.085\%$) interface and attributed the generation of such sessile dislocations to the reaction of two glissile 60° dislocations with Burgers vectors in the same $\{111\}$ glide plane. For higher-mismatch systems (misfit > 1.5–2%), the interfacial misfit dislocations are predominantly of pure edge character with Burgers vectors $\pm a/2[1\bar{1}0]$ [44.74]. The explanation is the three-dimensional growth due to the high strain which occurs when the epitaxial deposit is very thin and the edge dislocations climb to the interface at the edge of the island. Figure 44.17 shows a HRTEM image taken from Ge/Si interface, showing that all three observed misfit dislocations are of 90° pure edge character.

For pure edge misfit dislocations, the reflections of {220} can be easily applied to satisfy both the $\boldsymbol{g}\cdot\boldsymbol{b}=0$ and $\boldsymbol{g}\cdot(\boldsymbol{b}^{\wedge}\boldsymbol{u})=0$ criteria and make them invisible. However, the conventional Burgers vector analysis based on the simple $\boldsymbol{g}\cdot\boldsymbol{b}=0$ invisibility criterion was not successful for the case of 60° misfit dislocations due to the strong residual contrast. Precise Burgers vector determination is difficult in (001) plan-view specimens because the reflections which give $\boldsymbol{g}\cdot\boldsymbol{b}=0$ and $\boldsymbol{g}\cdot(\boldsymbol{b}^{\wedge}\boldsymbol{u})=0$ ($\boldsymbol{g}=422$, etc.) require high tilt angle. In practice, it is generally assumed that, if misfit dislocations lying along [110] and [1$\bar{1}$0] cannot be made invisible with 220 and 2$\bar{2}$0 reflections, then they are neither pure screw nor pure edge and therefore are likely to be 60°. Moreover, it is typical for 60° dislocations to show residual contrast in ⟨400⟩ diffraction. In addition, 60° dislocations show good contrast when the ⟨220⟩$\boldsymbol{g}$ vector is parallel to $\boldsymbol{u}$ and stronger contrast when the ⟨220⟩$\boldsymbol{g}$ vector and $\boldsymbol{u}$ are perpendicular. However, precise Burgers vector determination of 60° dislocations is still a problem. Nevertheless, several methods to analyze 60° misfit dislocations have been reported and are presented in the following.

Fig. 44.16 Plan-view image taken along the [001] growth direction showing the orthogonal misfit dislocation network at the strained Si/Si$_{0.85}$Ge$_{0.15}$ interface; the *arrow* indicates a characteristic fingerprint of the Hagen–Strunk mechanism

One of the techniques of analyzing 60° dislocations is the $(\boldsymbol{g}\cdot\boldsymbol{b})s$ criteria that has been utilized by *Stach* [44.75]. This is based on dynamical theory calculations of the electron scattering in the vicinity of dislocations, through which it can be shown that the position of the dislocation image with respect to its actual

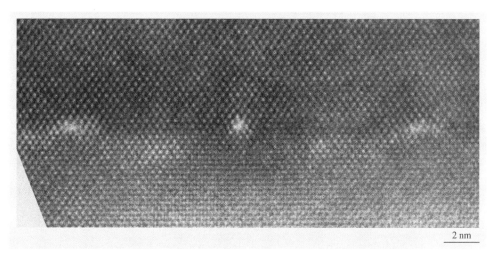

Fig. 44.17 HRTEM image showing the dominant misfit dislocations of pure edge character at the Ge/Si interface

core position varies with changes in deviation parameter as given by the quantity $(\boldsymbol{g}\cdot\boldsymbol{b})s$. Therefore, when $\boldsymbol{g}\cdot\boldsymbol{b}=0$, the position of the dislocation core will not vary with changes of s. This means that, even when effective invisibility is not observed according to the usual $\boldsymbol{g}\cdot\boldsymbol{b}=0$ condition, it is possible to determine the value of $\boldsymbol{b}$ by tracking the position of the dislocation image as s is changed. This is done most efficiently using high magnification on the video screen on the TEM. In Stach's study, two-beam bright-field and dark-field images were recorded using 400, 040, 311, and $13\bar{1}$ diffractions and the position of the dislocation image was observed as the deviation parameter was varied from $s \ll 0$ to $s \gg 0$.

In another approach, bend contour analysis is used to investigate the 60° misfit dislocations. This technique was originally explored by *Bollmann* [44.76] and was recently revisited for large-area analysis of misfit arrays [44.77]. Basically, in a two-beam bright- or dark-field image, a characteristic splitting occurs when a dislocation crosses a bend contour with the number of splitting fringes given by $n = \boldsymbol{g}\cdot\boldsymbol{b}$. By evaluating the splitting of at least three bend contours from different reflections, the complete Burgers vector can be determined. In spite of the advantages that LACBED has (for example, controllable deviation error) for Burgers vector analysis, the bend contour technique is highly suitable for the analysis of misfit dislocations in small-mismatch semiconductor epistructures (e.g., SiGe film grown on Si). Spiecker studied a large number of misfit dislocations simultaneously in SiGe/Si heterostructure plan-view samples by analyzing the splitting and displacement of the bend contour contrast that occurred as they were crossed by misfit dislocations.

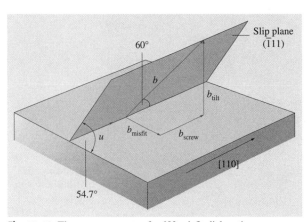

Fig. 44.18 Three components of a 60° misfit dislocations

For Burgers vector analysis, it is convenient to dissociate the 60° dislocation into three components, e.g., $a/2[101] = a/4[1\bar{1}0] + a/4[110] + a/2[001] = \boldsymbol{b}_{\text{misfit}} + \boldsymbol{b}_{\text{screw}} + \boldsymbol{b}_{\text{tilt}}$, as shown in Fig. 44.18. In this particular study, a dislocation with line direction [110] was analyzed. The sign and the magnitude of the screw component is determined by the splitting and twisting of the $\bar{4}40$ band contour while the sign of $\boldsymbol{b}_{\text{misfit}}$ is determined by the displacement of the $4\bar{4}0$ bend contour. To determine the tilt component, the bend contours of higher-order Laue zone reflections $\bar{3}31$ and $\bar{3}33$ were used. The main advantage of this method is its applicability to simultaneous determination of the Burgers vectors of a large number of dislocations distributed over large sample regions.

Another technique capable of providing information of misfit dislocation Burgers vectors was reported by *Dixon* and *Goodhew* [44.78]. This method used detailed analysis of the interactions between orthogonal arrays of misfit dislocations to approximately estimate the distribution of possible Burgers vectors. Three types of interaction were considered:

1. Those where the dislocations have the same Burgers vectors (parallel or antiparallel), which leads to the formation of two L-shaped segments
2. Those where the Burgers vectors are oriented at 60° to each other, which leads to the formation of linking $1/2\langle 110\rangle$ dislocation
3. Those where the Burgers vectors are perpendicular to each other, which are predicted to have no interaction.

For the case of an epitaxial interface where interfacial dislocations of 60° character are predominant, with edge character being also present in reduced number (screw character dislocations being absent) and assuming that the five possible Burgers vectors are present in equal numbers for each set of dislocations, there will be 25 types of interaction. By analyzing all the possibilities, it was found that 4/25 (16%) of intersections would be L-shaped, 16/25 (64%) would have links, and 5/25 (20%) would not react. For the case where no pure edge dislocations were found at the interface, only 16 reactions are possible, of which 4/16 (25%) intersections should be L-shaped, 8/16 (50%) would have links, and 4/16 (25%) would not have reacted. The linking segments formed at intersections are very short and assumed to be undistinguishable from the unreacted intersections. The reported observation showed that 18% of the intersections were L-shaped, implying that the interface probably contained dislocations of all four 60° types together with some pure edge dislocations.

Lattice imaging can, in principle, be used to distinguish between edge and 60° misfit dislocations. Both types of dislocation have the same Burgers vector magnitude ($a/2\langle 110 \rangle$) but for edge-type dislocations the Burgers vector is oriented parallel to the interface while the 60°-type dislocations form a 45° angle with the interface. Therefore on [110] lattice images, the edge dislocations should show two terminating {111} planes at the core, while the 60° type should show only one terminating plane.

Critical Thickness

Extensive studies have been carried out by TEM to elucidate the practical critical thickness of the epifilms. Four general models have been identified defining critical thickness:

1. The Matthews and Blakeslee model [44.79] for the turnover of a single threading dislocation during the growth of a single epilayer
2. The Matthews and Blakeslee model for the turn over of a single threading dislocation in a multilayer structure which was subsequently pulled out into a half-loop by the internal stress in the layer
3. The Miles and McGill model [44.80], basically a modified and more sophisticated version of (1) and (2)
4. The People and Bean model [44.81], which concerns the critical thickness that allows the nucleation of fresh dislocation line.

Dixon et al. [44.82] performed a systematic study of the critical thickness in the $In_xGa_{1-x}As/GaAs$ ($x < 0.25$) epistructure. In their study, two critical thicknesses are identified: threading dislocation are turned over, forming misfits at a thickness predicted by the Matthews and Blakeslee model; and, at larger thicknesses, fresh dislocations are nucleated at the interface and the critical thickness fits well with the predictions of the People and Bean model. Dixon et al. also concluded that some misfit dislocations were present which did not act to relieve misfit strain.

Stacking Faults and Partial Dislocations

The dissociation of 60° misfit dislocations and the existence of stacking faults are widely observed in conventional small-mismatch epistructures. However, the geometry and the formation mechanism of the stacking faults have been the subject of extended research.

Theoretically [44.83], the order in which the partials can nucleate is determined by the atomic configuration on the {111} glide planes. Under conditions of tensile stress, the first partial to nucleate is the 90° Shockley partial followed by the 30° partial. This order is reversed if the stress field is compressive. The resolved shear stress on the {111} slip plane is, in both cases, in the same direction as the Burgers vector of the 90° partial. This means that the force exerted by the stress field on the 90° partial is twice as large as the force on the 30° Shockley partial. The consequence of this is that, in a tensile stress field, the 90° partial nucleates first, forming a stacking fault, since it experiences the largest force. On the other hand, the 30° partial, which would annihilate the stacking fault, feels a weaker force so that it may not nucleate until later in the growth process. As a result, stacking faults are very often observed extending from the interface to the film surface. In contrast, if the stress field is compressive, the 30° partial begins the nucleation process. Since this partial experiences a smaller force, a higher nucleation barrier results. Once the 30° partial is formed it will again be trailed by the 90° partial that is driven both by a higher force from the stress field and from a force associated with the stacking fault. This means that the extent of dissociation will be very small in a compressive field. An undissociated 60° perfect dislocation can then easily cross-slip, i.e., change form one {111} glide plane to another. Such effects are observable if we compare misfit dislocation grids of a film under tensile and compressive strain; in the film under tensile strain, the misfit segments form straight lines and a number of stacking faults can be observed, whereas in the film under compressive strain the misfit dislocations exhibit higher curvature, evidence for cross-slip having occurred, and little evidence for the existence of stacking faults.

A variety of observations on different heterosystems have been reported. *Kimura* et al. [44.84] studied the development of defects in strained Si/SiGe heterostructures of supercritical thickness. TEM observations showed that, when the strain energy is increasing in the Si film, 60° misfit dislocations at the interface dissociate into Shockley partials with the 30° partial being located in the Si film and the 90° partial being located in the SiGe layer. The observed stacking faults are believed to form by the 30° partial gliding out to the surface of the Si film. *Marshall* et al. [44.85] reported that the predominant configuration of 60° misfit dislocations in low-misfit SiGe/Si films ($< 15\%$Ge or 0.6% misfit) involves dissociation at the interface, with the stacking fault extending into the substrate. *Zou* and *Cock-*

ayne [44.86] studied the equilibrium geometries of the dissociated misfit dislocations in single-semiconductor heterostructures. They predicted that for, a tensile-strained layer grown on a (001) substrate, the 30° partial is located in the strained layer and the 90° partial is located in the substrate. Their experimental observation in low-strained [001] $In_{0.1}Ga_{0.9}As/GaAs$ single heterostructures [44.87] showed that the dissociation of misfit dislocation is dominant in the structure, with the 90° partial being located above the strained interface and the 30° partial in the buffer layer. *Hirashita* et al. [44.88] reported the observation of stacking faults extending from a strained Si surface to a strained Si/SiGe interface, which are accompanied by 90° Shockley partials at the interface. These defects are increasingly formed in the strained Si layers on SGOI substrates while the strained layer thickness increases. It was concluded that the operative mechanism involved the generation of 90° partial dislocations at the Si surface which propagated on {111} planes towards the SiGe interface, relaxing the cumulative tensile strain in strained Si layers. *Fitzgerald* et al. [44.89] reported mechanisms for the reaction of partials in lattice-mismatched $In_xGa_{1-x}As/GaAs$ heterostructures. In films which are under compression, the trailing partial which is closest to the epilayer surface is the 90° partial, while the leading partial is the 30° partial. Therefore, the two leading 30° partials can form a stair-rod dislocation ($\boldsymbol{b} = a/6\langle 1\bar{1}0\rangle$), leaving two edge partials ($\boldsymbol{b} = a/6\langle 1\bar{1}2\rangle$). In a tensile epilayer, the leading partials are edge partials, and the 30° partials are closer to the surface. In this case, the leading 90° partials would form an edge dislocation with $\boldsymbol{b} = a/3\langle 1\bar{1}0\rangle$.

Figure 44.19a is a cross-sectional image of two stacking faults formed at a strained $Si/Si_{0.5}Ge_{0.5}$ interface. The reaction between stacking faults can be clearly observed. A stair-rod dislocation is formed at the junction and between two faults and leads to the annihilation of all the faults. The squared area is enlarged and shown in Fig. 44.19b. The stacking sequence can be clearly seen as *CBACBCBA*... as the stacking fault is crossed.

Graded Buffer and Insertion of Strained Layers

Compositionally graded buffers have been successfully employed in lattice-mismatched epitaxy to incorporate high-quality relaxed layers onto conventional semiconductor substrates [44.90, 91]. A typical example is the use of SiGe graded buffers as virtual substrates for the production of high-mobility complementary metal oxide semiconductor (CMOS) structures and for III–V integration on Si. Some of the major requirements for such application include: low threading dislocation density (TDD), low surface roughness, and high degree of relaxation of the 4% lattice mismatch between Si and Ge. The compositionally graded structure provides mul-

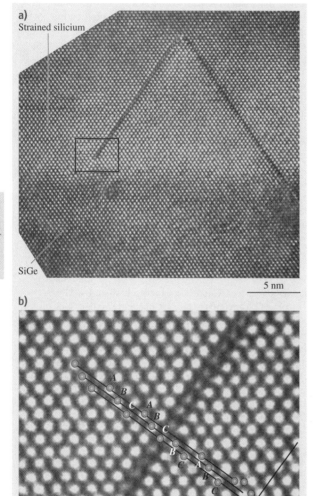

Fig. 44.19a,b A cross-sectional HRTEM image showing the stacking faults and their reaction at a strained $Si/Si_{0.5}Ge_{0.5}$ interface (**a**) and a magnified image of the squared stacking fault (**b**)

tiple low-mismatch interfaces, preventing dislocation nucleation and facilitating the glide of existing threading dislocations at each interface to relieve mismatch strain. Such grading results in relaxed cap layers with threading dislocation densities in the 10^5–10^6 cm^{-2} range, whereas direct growth of uniform composition relaxed layer with high lattice mismatch results in threading dislocation densities in the 10^8–10^9 cm^{-2} range.

Another technique of filtering threading dislocations by the insertion of superlattice was reported by *Blakeslee* [44.92]. The study suggested that, by properly designing superlattice structures, multistrained interfaces are provided for threading dislocations to lie on. Such structure also reduces the harmful interactions between dislocations. The essential ingredients of the proper design include:

1. The superlattice thickness should be considerably larger than the equilibrium critical thickness in order to provide the necessary excess stress to move the dislocation.
2. The strain gradients should be as gentle as possible everywhere except in the superlattice itself.

Compared with the step-graded structure, the superlattice structure confines the dislocations better. However, in this study, the relaxation of the film was not discussed.

Park et al. [44.93] studied the effects on strain relaxation and threading dislocation density of the insertion of thin layers, strained in tension or compression, into compositionally graded SiGe. Figure 44.20 shows cross-sectional images of three samples for comparison. Figure 44.20a shows the sample grown as the control condition, having no inserted strained layers, consisting of a 2 μm linearly graded SiGe buffer where the Ge content increased from 0 to 20% at a grading rate of 10 %/μm followed by a 0.5 μm-thick $Si_{0.8}Ge_{0.2}$ cap layer. The structure exhibits a quite uniform distribution of dislocations throughout the graded buffer. Figure 44.20b shows the case of two inserted layers of pure Si, strained in tension, the bottom layer being inserted at the 6% Ge location and the top layer being inserted at the 12% Ge location. Figure 44.20c shows the insertion of two SiGe layers strained in compression, the bottom layer with 12% Ge being inserted at the 6% Ge location and the top layer with 24% Ge being inserted at the 12% Ge location. Accumulation of dislocations was found at the bottom of the layers strained in compression and at the top of the layers strained in ten-

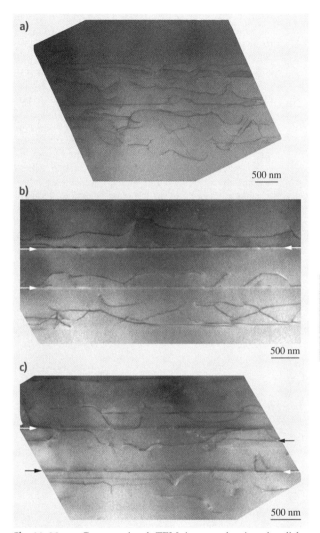

Fig. 44.20a–c Cross-sectional TEM images showing the dislocation behavior in the compositionally graded $Si_{0.8}Ge_{0.2}$ buffer without/with insertion of tensile/compressed strained layers (**a**) reference sample without insertion of layers; (**b**) insertion of two tensile layers: top layer (**c**) insertion of compression layer. Note: *arrows* indicate inserted layers

sion. This can be easily understood based on the sense of the dislocations. Compared with the sample with inserted compressive layers, the sample with tensile layers exhibits heavier accumulation of dislocations. Moreover, in the samples with inserted layers strained in tension, dislocations are pinned more heavily at the top inserted tensile layer (0% Ge at 12% Ge) than at the lower inserted tensile layer (0% Ge at 6% Ge), whereas in the sample with layers strained in compression, dislocations were pinned more heavily at the lower inserted compressive layer (12% Ge at 6% Ge) than at the top inserted compressive layer (24% Ge at 12% Ge). The inserted strained layers in this study were always 20 nm thick, which is below the equilibrium critical thickness. Results showed that the relaxation in the cap layer is largely improved without significant increase of threading dislocations in the samples with inserted layers in compression compared with the control sample. Misfit dislocations are observed in the cap layer in both types of sample, in contrast to the control sample, indicating the disturbance of relaxation due to the insertion of stressed layers.

Observation of Dislocations in Aspect Ratio Trapping Epigrowth

Aspect ratio trapping (ART) is an epitaxial technique involving selective growth in patterned openings bounded by substantially vertical dielectric side-walls, enabling dislocations to be trapped if the aspect ratio (h/w) of the opening is sufficiently large. Recent research reported by *Park* et al. [44.94] showed that ART could be effective for Ge grown on Si in trenches up to 400 nm wide and of arbitrary length. Figure 44.21 shows a plan-view TEM image recorded from such a sample that was thinned from the substrate side down to a thickness of ≈ 200 nm. Both the Si substrate and the first ≈ 300 nm of epifilm were removed, leaving only a defect-free Ge layer; this demonstrates the efficacy of the ART technique in eliminating threading dislocations in Ge films grown on Si substrates. *Bai* et al. [44.95] carried out analysis of the mechanisms by which dislocation elimination is achieved. Detailed TEM studies reveal that facets, when formed early on in the growth process, play a dominant role in determining the configurations of threading dislocations in the films. These dislocations are shown to behave as *growth dislocations*; during growth they are oriented approximately along the facet normal, and so are deflected out from the central regions of the trenches. This suggests a strategy of facet engineering by which the efficacy of threading dislocation trapping might be further improved. TEM images in Fig. 44.22 show the redirection of dislocations close to the normal of the encountered facets. The thin SiGe marker layers of approximately 10–15% Si content are periodically inserted to delineate the growth front.

44.2.2 TEM Studies of Large-Mismatch Heteroepitaxial Systems

Wurtzite polytypes of AlN, GaN, and InN and their alloys are suitable for numerous device applications such as short-wavelength light sources or detectors, and high-power and high-frequency devices. Due to the difficulty in obtaining GaN substrates, current GaN devices are fabricated on epitaxially grown films on various substrates. Most research on GaN epitaxy in the last two decades has been concentrated on growth on sapphire or SiC substrates. Sapphire is currently the most commonly used substrate for GaN epigrowth due to its relatively low cost. The large mismatch between GaN and sapphire (16%) leads to a very high density of interfacial dislocations. SiC is another candidate for the substrate which has smaller mismatch (3.4%). Si attracts attention as a substrate for GaN epigrowth since the first molecular-beam epitaxy (MBE) grown on GaN light-emitting diode (LED) on Si was demonstrated in 1998. The lattice mismatch between GaN and Si is -17%, yielding biaxial tensile stress in the GaN/Si interface. Owing to the high mismatch in lattice parameter and thermal properties between GaN and nonnative substrates, a high density of struc-

Fig. 44.21 Plan-view (along the [001] growth direction) TEM image of the top layer (≈ 200 nm from the film surface) of a Ge film grown with ART technique

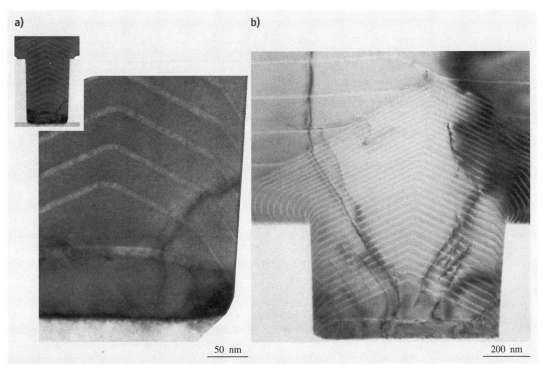

Fig. 44.22a,b Cross-sectional TEM images (viewed along the [110] trench direction) showing the redirection of threading dislocations under the influence of faceting (**a**) dislocation redirection in a trench of ≈ 300 nm wide (the linear white contrast features correspond to the inserted SiGe growth marker layers), where the *inset* shows the overview of the Ge film grown in the trench (the film was overgrown beyond the side wall); (**b**) dislocation redirection in a trench ≈ 800 nm wide

tural defects exists in GaN films. The large in-plane mismatch induces a high density of interfacial dislocations ($\approx 10^{13}$ cm^{-2}). Due to the three-dimensional (3-D) growth at the initial stage, a high density of threading dislocations ($\approx 10^8$–10^{10} cm^{-2}) form in the GaN film to accommodate the twist between neighboring islands. The variations on the surface of substrates cause planar defects such as inversion domain boundaries (IDBs), stacking mismatch boundaries (SMBs), prismatic stacking faults (PSFs) or basal plane stacking faults (BSFs). Extensive research has been carried out to reduce the threading dislocation density in GaN epilayers. The employment of a low-temperature nucleation buffer layer before subsequent GaN growth was found to drastically improve epitaxial quality. Other strategies to improve the crystalline quality of III-nitride epifilms, such as lateral epitaxial overgrowth, vicinal surface epitaxy, insertion of low-temperature layers, use of porous substrates, etc., have been widely studied.

Interface

Unlike epitaxial systems with sphalerite structures (e.g., Si$_x$Ge$_{1-x}$/Si), large-mismatch systems such as GaN/SiC, GaN/sapphire, AlN/sapphire, etc., cannot occur by the formation and glide of dislocation half-loops to the interface producing misfit dislocation segments due to the absence of an effective slip system. As such the mechanisms of epilayer relaxation in these systems are far from understood. Furthermore, the stress caused by island coalescence and thermal processing as well as the high density of defects introduce further complexity into the relaxation mechanism. Some of the basic understanding of this topic can be summarized as follows.

Coincidence Lattice (Pseudosemicoherent Interfaces).
In large-mismatch systems which have mismatch parameters larger than ≈ 4–5%, the interface is incoherent, as shown in Fig. 44.23a, and there is no continuity

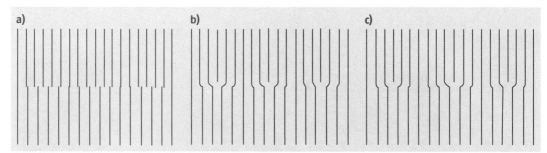

Fig. 44.23a-c Models for a large-mismatch interface: (**a**) perfect coincidence lattice with a lattice plane ratio of $m/n = 6/5$; (**b**) coincidence lattice with coherence relaxations within the unit cells; (**c**) as in (**b**), but with a slight deviation of $1/80$, forming a new coincidence unit with $19/16 = 6/5 - 1/80$

between the lattice planes on the two sides of the interface [44.96]. However, the bonding between films and substrates in such systems suggest the appearance of coherency along some planes on the two sides of the interface separated by *geometrical misfit dislocations* (or *mismatch dislocations*), which separate these *pseudosemicoherent planes* (Fig. 44.23b). Perfect coincidence sites between the epilayer lattice a_e and substrate lattice a_s would occur when $a_s/a_e = m/n$, where m and n are positive integers. If $m = n + 1$, there is one extra lattice plane in each unit cell of the coincidence site lattice, i.e., a geometrical misfit dislocation is generated. The character of these dislocations depends on the symmetry of the coincidence lattice. In this sense, the Burgers vector of such a geometrical dislocation must not necessarily be an invariant vector as it is in the bulk lattices.

In contrast to the low-mismatch case, in which misfit dislocations are generally produced by lattice dislocations that may have Burgers vector not necessarily parallel to the interface, in large-mismatch systems the mismatch dislocations exist right from the start of film deposition and often have an in-plane Burgers vector.

Another difference between misfit dislocations in low-mismatch system and mismatch dislocations in high-mismatch system is that the latter lack long-range strain fields [44.97].

Near-Coincidence Lattice. An alternative model of generation of mismatch dislocations is that the deviation can also be accommodated by another type of secondary defect, as shown schematically in Fig. 44.23c [44.98]. Such secondary defects interrupt the periodicity of the original coincidence lattice and form a new coincidence unit.

Reported Models. Zheleva et al. suggested a domain-matching model for large-mismatch epitaxial systems [44.99]. In this model, a domain is defined by the minimum number of lattice planes that gives a value of unity for the difference between m (the number of epifilm lattice planes) and n (the number of substrate lattice planes). The residual domain strain can be calculated in both in-plane directions. Sun et al. suggested a concept of extended atomic distance mismatch (EADM) [44.100]. EADM $= (Id - I'd')/(I'd')$ (where I and I' are integers, d and d' are atomic distances of the epilayer and substrate, respectively. I and I' are determined in the following way: $d : d' \sim I : I'$, where $I : I'$ is the smallest irreducible integral ratio for $d : d'$. The difference between I and I' is one that introduces a periodic edge-type dislocation.)

TEM Observations of Large-Mismatch Interfaces. Kehagias et al. studied misfit relaxation at the AlN/Al$_2$O$_3$(0001) interface in both plan-view and cross-sectional geometry [44.101]. From moiré fringes shown in plan-view TEM images, a general case of a matching ratio of AlN:Al$_2$O$_3$ equal to $8:9$ was observed, which is confirmed in the cross-sectional geometry. However, occasionally, AlN:Al$_2$O$_3$ ratios of $6:7$ and $9:10$ were also observed. Threading dislocation densities were also measured using the moiré fringes.

HRTEM performed on cross-sectional samples is generally used to observe interfaces and defects in the mismatched systems. In wurtzite structures, since the dominant slip system is $\langle 11\bar{2}0 \rangle\{0001\}$, the $\langle 11\bar{2}0 \rangle$ projection is generally used for investigation, with dislocation cores being viewed *end-on*, i.e., imaged along the line direction of the dislocation. This configuration is based on the presumption that the misfit disloca-

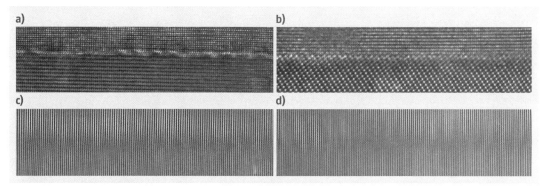

Fig. 44.24a–d HRTEM images taken at the AlN/sapphire interface: (**a**) along zone axis $[11\bar{2}0]_{AlN}$; (**b**) along zone axis $[1\bar{1}00]_{AlN}$; (**c**) reconstructed image from masked FFT of (**a**) showing the extra half-planes in the substrate; (**d**) reconstructed image from masked FFT of (**b**) showing the extra half-planes in the substrate

tions are of 60° mixed type with line direction along $\langle 11\bar{2}0 \rangle$ and Burger vector along another equivalent $\langle 11\bar{2}0 \rangle$ direction. However, studies of AlN film grown on sapphire substrate show extra half-planes along both $\langle 11\bar{2}0 \rangle_{AlN}$ and $\langle 1\bar{1}00 \rangle_{AlN}$ directions, as illustrated in Fig. 44.24 [44.13]. Further studies are required to reach any conclusive analysis.

Defects in AlN/GaN Films Originating from SiC Substrate Steps

One of the strategies that has been explored to reduce the defects in GaN epifilms grown on nonnative substrates is the utilization of vicinal, offcut substrates. Offcut 6H-SiC substrates have been recently shown to reduce the stress level inside the films through the combined effects of mutual tilt between the epilayer and substrate, which helps to relax out-of-plane mismatch, and by the generation of geometric partial misfit dislocations (GPMDs) which serve both to relax in-plane mismatch and to accommodate stacking differences between the epilayer and substrate at some proportion of the steps at the substrate–film interface [44.102]. One of the critical factors in offcut epitaxy is the effect of the surface steps present on substrates on the quality of

Fig. 44.25 HRTEM image taken at zone axis $[11\bar{2}0]$ showing a PSF forming at a I_1-type step

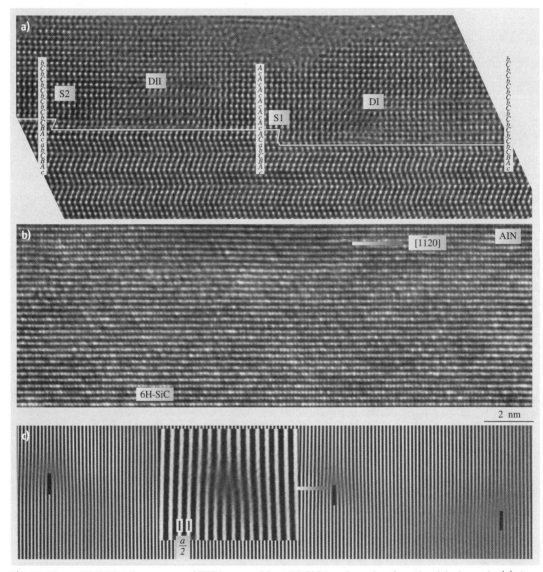

Fig. 44.26a–c HRTEM and reconstructed FFT images of the AlN/SiC interface taken from the vicinal sample: (**a**) along zone axis [11$\bar{2}$0]; (**b**) along zone axis [10$\bar{1}$0]; (**c**) reconstructed FFT image with $g = 1\bar{2}10$ from (**b**)

subsequently grown epifilms. *Vermaut* et al. [44.103] investigated the dislocation character of various types of steps at the 2H-AlN/6H-SiC interface and defined three types of interface steps: I_1-type steps with dislocation character of $1/3\langle 10\bar{1}0\rangle + 1/2 c_{AlN}$, I_2-type steps with dislocation character of $1/3\langle 10\bar{1}0\rangle + c_{AlN}$, and E-type steps.

Prismatic Stacking Faults (PSFs) Originating at I_1 Type Steps. Figure 44.25 shows a HRTEM image taken at an I_1 step region [44.104] along zone axis [11$\bar{2}$0] (60° inclined to the offcut direction). The AlN/SiC interface is delineated by a white line separating SiC on the lower side from AlN on the upper side. The stacking sequences revealed in such HRTEM images can be un-

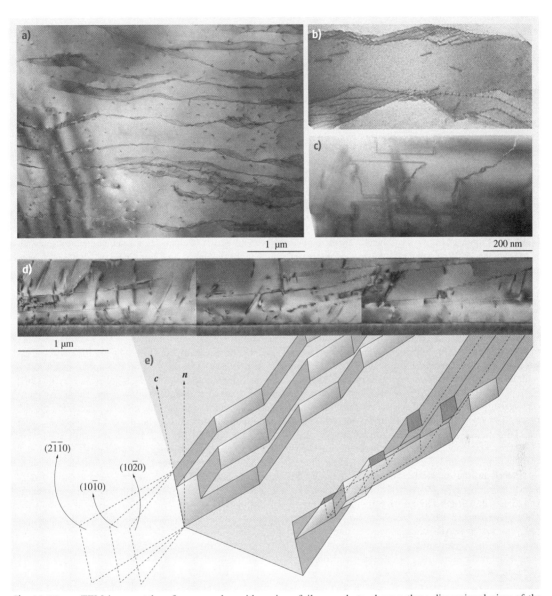

Fig. 44.27a–e TEM images taken from samples with various foil normals to show a three-dimensional view of the planar boundaries. (**a,b**) (0001) plan view of the defects with different magnification; (**c**) view along [$\bar{1}2\bar{1}0$]; (**d**) view along [$10\bar{1}0$]; (**e**) schematic representation of the *boat*-shaped walls created by the intersecting faults. The *c*-axis of the substrate is indicated along with the surface normal n

derstood with the aid of a notation system based on that introduced by *Pirouz* and *Yang* [44.105]. It can be seen that, in the vicinity of the two-bilayer-step riser, the first two AlN bilayers tend to duplicate the stacking positions exhibited to that step riser, i.e., *cA*. However, further out on the terrace, that stacking sequence changes to *Ba* (most probably under the influence of the stacking sequence under the terrace). The relationship between

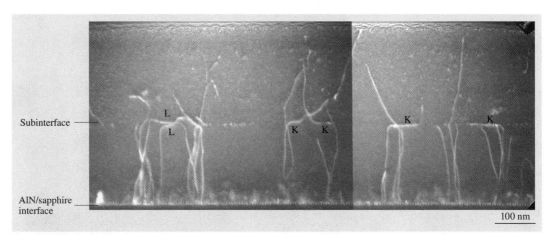

Fig. 44.28 A montage of two weak-beam dark-field images taken with $g = 0002$, showing the subinterface corresponding to when the V/III ratio and both fluxes were increased. Large kinks (indicated by K) or dipole half-loops (indicated by L) are formed at this subinterface

these two stacking sequences suggest a displacement vector of $1/6\langle 2\bar{2}03\rangle$, which can be accommodated by a stacking mismatch boundary (SMB). However, observations reveal that the defects formed at such steps are predominantly PSFs, which add an additional lattice shift and have a displacement vector of $1/2\langle 1\bar{1}01\rangle$. This might be due to the lower energy of PSFs than SMBs.

Geometrical Partial Misfit Dislocations (GPMDs) Formed at I_2-Type Steps. Figure 44.26a shows a HRTEM image taken from an offcut sample along $[11\bar{2}0]$ zone axis (inclined to the offcut direction), showing two steps, S1 and S2 (in the same sense due to the offcut), at the substrate surface. Both steps S1 and S2 are of I_2 type [44.106] with dislocation vector $1/3\langle 10\bar{1}0\rangle$. DI and DII are two regions exhibiting lattice distortion. It can be seen that, in the vicinity of S2, the first two AlN bilayers tend to duplicate the stacking positions exhibited by that step riser (on the higher side of the step), i.e., bC. However, further out on the terrace, the stacking sequence changes to cA through the distorted region DII (distorted region DI has similar features). Thus, it is clear that the distortions are confined close to the interface, and the stacking sequences above the distorted regions continuously overspan the step without interruption. Figure 44.26b shows an HRTEM image taken from the same sample along zone axis $[10\bar{1}0]$ (perpendicular to the offcut direction), and Fig. 44.26c is the corresponding reconstructed fast Fourier transform (FFT) image. Three dislocations are revealed with the extra half-planes on the substrate side. Analysis shows that the Burgers vector has no out-of-plane component and its projection onto the $(10\bar{1}0)$ plane is $1/6\langle 11\bar{2}0\rangle$. The localized dislocation core and well-defined extra half-plane imply that the dislocation line is along $[10\bar{1}0]$. These misfit dislocations can only be 30° complete dislocations (with Burgers vectors of $\pm 1/3[\bar{2}110]$ or $\pm 1/3[11\bar{2}0]$) or 60° partial dislocations. Since the former is energetically unreasonable, the misfit dislocations (MDs) should be 60° partials. These *unpaired* partials, with line direction $\langle 10\bar{1}0\rangle$ and Burgers vector $1/3\langle 1\bar{1}00\rangle$, are suggested as geometrical partial misfit dislocations (GPMDs) that are formed at I_2 steps to accommodate both the lattice mismatch and stacking sequence mismatch simultaneously [44.102]. The distorted regions (DI and DII) shown in Fig. 44.26a may be due to the strain field associated with dislocations of this kind.

Intersecting Stacking Fault Structures in GaN/AlN/SiC Epitaxy

Bai et al. [44.104] conducted a systematic study of intersecting planar boundary structures observed in GaN epifilms grown on vicinal 6H-SiC substrates. These structures are shown to comprise stacking faults that fold back and forth from the basal plane. The prismatic stacking faults, with fault vector $1/2\langle 10\bar{1}1\rangle$, nucleate at steps on the substrate surface as a consequence of the different stacking sequences exposed on either side of the step. Once nucleated, PSFs intersecting the vertical step risers in the AlN buffer and eventually in the GaN film are replicated during the predominantly step-

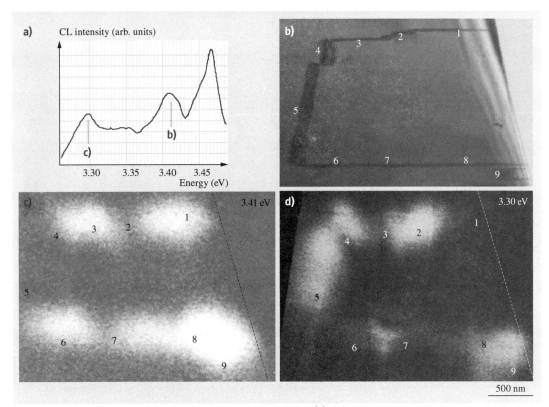

Fig. 44.29a–d TEM-CL observation of prismatic stacking faults. (**a**) A low-temperature CL spectrum acquired over a large area revealing three emission peaks. The luminescence intensity is plotted on a logarithmic scale. (**b**) Cross-section TEM image of the region of interest showing the presence of stacking faults. CL images from the same region corresponding to emission energies of (**c**) 3.41 eV and (**d**) 3.30 eV (after [44.107], © AIP 2006)

flow growth and propagate into the growing crystal. As a consequence of the different growth rates experienced on either side of the intersection of a PSF with a vertical step riser, the PSF may be redirected onto an equivalent {11$\bar{2}$0} plane, leaving an I_1 BSF between the bottom of the redirected section of the PSF and the top of that portion of the original PSF which was below the terrace. This leads to the formation of folded PSF/BSF fault structures which exhibit various configurations. Such folded stacking fault configurations form walls which enclose domains of different stacking sequence. Figure 44.27 shows a series of TEM images recorded with foil normals parallel to [0001], [1$\bar{2}$10], and [10$\bar{1}$0]. These images reveal the existence of folded planar defect structures which thread diagonally through the epifilm and intersect the sample surface [44.104]. These complex configurations typically adopt *boat-like* shapes, as shown schematically in Fig. 44.27e. An unfolded or folded BSF forms the bottom of the boat while other folded BSFs/PSFs form the two side walls of the *boat*. These folded boundary configurations act as walls separating domains with different stacking sequence. Detailed contrast analysis confirms that domain walls consist of intersecting PSFs on equivalent (11$\bar{2}$0) and (2$\bar{1}\bar{1}$0) planes and I_1 BSFs on (0001).

Dislocation Redirection in AlN/Sapphire Epilayer Driven by Growth Mode Modification

Bai et al. [44.108] reported TEM observation of redirection of threading dislocations in AlN epilayers grown on sapphire substrate. The threading dislocations experience redirection of their line orientation which is found to coincide with imposed increases in both of

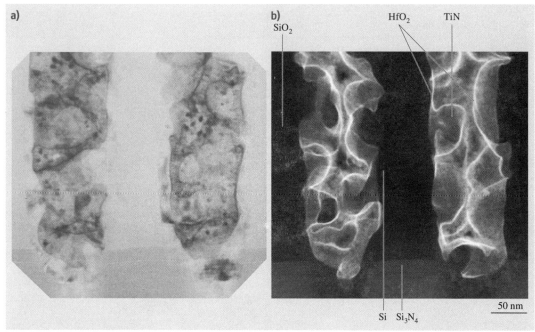

Fig. 44.30a,b The lower portion of two semiconductor memory capacitor cells. (**a**) Bright-field TEM image; (**b**) Z-contrast STEM image (after [44.50])

V/III ratio and overall flux rate leading to the formation of an internal subinterface delineated by the changes in dislocation orientation. Threading dislocations either experience large kinks and then redirect into threading orientation or form dipole half-loops via annihilation of redirected threading segments of opposite sign, with the latter leading to a significant dislocation density reduction. These phenomena can be accounted for by a transition of growth mode. At the point where the growth mode changes from atomic step-flow to two-dimensional (2-D) layer-by-layer growth, macrosteps sweep over dislocation outcrops, forcing the creation of large kinks and/or dipole half-loops as proposed by *Klapper* [44.109]. Figure 44.28 shows a weak-beam dark-field image taken with $g = 0002$, clearly portraying two phenomena:

1. Where a large kink forms in the dislocation line at this subinterface and the threading dislocations (TD) reorients back into approximately [0001] direction.
2. Where two TDs are redirected towards each other forming a dipole half-loop.

The reorientation of the TDs upon increase of V/III ratio and flux magnitudes can be understood from the point of view of a transition in growth mode at this juncture of growth.

TEM–CL Observation of Prismatic Stacking Faults

Use of a TEM equipped with a scanning attachment and coupled with a cathodluminescence (CL) light collector/spectrometer system offers the possibility of correlating microstructural information (diffraction contrast) with spatially resolved spectroscopy. This method is attractive for use with optoelectronic materials. *Mei* et al. [44.107] and *Liu* et al. [44.110] established a direct correlation between stacking faults in *a*-plane GaN epilayers and luminescence peaks in the 3.29–3.41 eV range. Combined TEM-CL allows the structural features of stacking faults to be determined by diffraction contrast, with the optical emission characteristics being observed by highly spatially resolved monochromatic cathodluminescence in the exact same regions. Figure 44.29a shows a CL spectrum taken from the observing feature. There are two distinct peaks at 3.41 and 3.30 eV, besides the dominant bulk GaN donor-bound exciton peak at 3.47 eV. Diffraction-contrast TEM analysis was performed in the same region where the localized emission was observed. By correlating the CL images in Fig. 44.29c,d with the TEM image in Fig. 44.29b, the location of the 3.41 eV emission can be identified as basal-plane stacking faults, while the emission at 3.30 eV is associated with prismatic stacking faults.

44.2.3 Application of STEM, EELS, and EFTEM

Z-Contrast STEM

A TEM and a HAADF STEM image of similar sections of two semiconductor memory capacitor cells are shown in Fig. 44.30. In the TEM image, the HfO_2 dielectric layer cannot be seen clearly because of the interference of diffraction contrast from TiN and Si. In comparison, the Z-contrast STEM gives a clean image, emphasizing materials differences. The continuity and thickness variation of the HfO_2 layer can be better studied in the Z-contrast STEM.

Figure 44.31 shows a HAADF STEM image of nano WN_x particles deposited on 100 nm-thick SiO_2 film. Due to the overlapping SiO_2, the visibility and shape definition of particles are poor in conventional TEM, as shown by the HREM imaging (inset of Fig. 44.31). Here the Z-contrast STEM shows its advantage by giving high definition of nanoparticles and their low-density shells.

EELS Application in Microanalysis

EELS Study of Mn Diffusion. Figure 44.32 illustrates EELS elemental profiles across a magnetoresistive random-access memory (MRAM) stack (Ta/MnIr/NiFe/Al_2O_3/NiFe/Ta) as measured along the scan line [44.60]. From the Mn profile in Fig. 44.32b, diffusion of Mn through the FeNi layer into the Al_2O_3 layer due to thermal process is observed. In transition metals, L_3 and L_2 white lines ($2p^{3/2} \rightarrow 3d^{3/2}3d^{5/2}$, $2p^{1/2} \rightarrow 3d^{3/2}$) are observed in the energy-loss spectra [44.111]. The comparison of the Mn $L_{2,3}$ edges from the MnIr layer and NiFe/Al_2O_3 interface (Fig. 44.33) shows a significant difference in the fine structure, especially in the intensity ratio of $L_3 : L_2$ [44.60]. Studies have shown that the relative intensities of L_3 and L_2 of Mn are highly sensitive to the 3d occupancy and thus the valence state [44.112, 113]. The intensity ratio of white lines $I(L_3)/I(L_2)$ can be calculated from the spectra using the double-step background-fitting procedure with the step at the

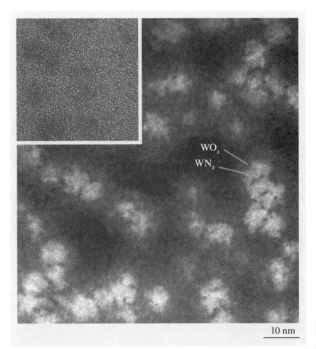

Fig. 44.31 Z-contrast STEM image of WN_x nanoparticles on 100 nm thick amorphous SiO_2 film. The particles are surrounded by low-density shells, possibly WO_x. The *inset* shows a HREM image of the particle (after [44.50])

peak [44.111, 112]. Figure 44.34 plots $I(L_3)/I(L_2)$ versus the Mn valence state for the reference data [44.112] and for the data from this MRAM study. Deduced from the correlation to the reference data, the valence state of Mn in Al_2O_3 is found to be around +2.2. Thus, Mn diffused into the Al_2O_3 layer and accumulated there in an oxidized state [44.60].

EELS Fine Edge to Study Interface Material. Figure 44.35 shows a HAADF STEM image of Al/WN_x interface with an EELS elemental line scan [44.50]. A nitrogen-rich interface is observed in the line scan profile. The interface is also observable in the STEM image. From the elemental profile, it is difficult to identify the phase of the interface material. The EELS fine edge structure can be strongly affected by chemical bonding variations. This fine structure allows us to probe the information beyond elemental identification. Figure 44.36 shows a comparison of Al-$L_{2,3}$ edge for the spectra from interface location and standard samples. The major difference in the fine edge structures of

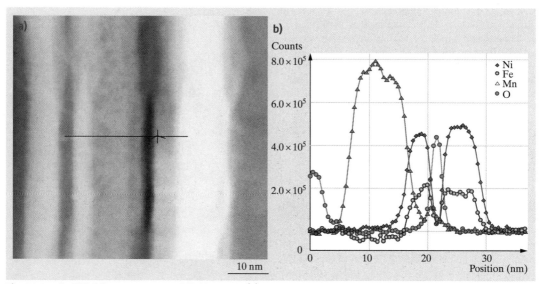

Fig. 44.32a,b EELS line scan across a MRAM stack. (a) STEM image showing the features and the scan line location. (b) Elemental profiles along the scan line as detected by EELS

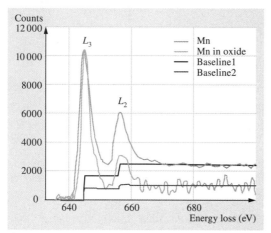

Fig. 44.33 Comparison of manganese $L_{2,3}$ edge spectra from Mn layer and from Al_2O_3 layer. The baselines are used for calculating the $L_3 : L_2$ ratio

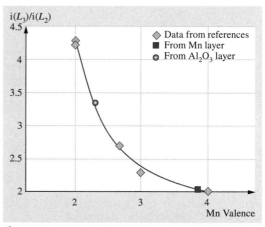

Fig. 44.34 A plot of white line intensity ratio $I(L_3)/I(L_2)$ versus the valence state of Mn (after [44.60])

Al and AlN is indicated by the three arrows (a, b, and c). Here we observe that the spectrum from interface is very similar to that of AlN, except for the small shoulder indicated by arrow a. In comparison with that of Al metal, we find this extra shoulder is most likely caused by overlapping Al metal. This analysis is further supported by comparison of EELS in the low-loss region (Fig. 44.37). In Fig. 44.37, we can see the distinguishing features for Al and AlN. The low-loss spectrum from the interface again resembles the summation of Al and AlN spectra. Thus, by combining the information from EELS line scan (elemental identification) and the fine edge structure comparison, we conclude that the interface layer between Al/WN_x is AlN_x. The existence of Al metal at the interface is mostly likely caused by overlapping.

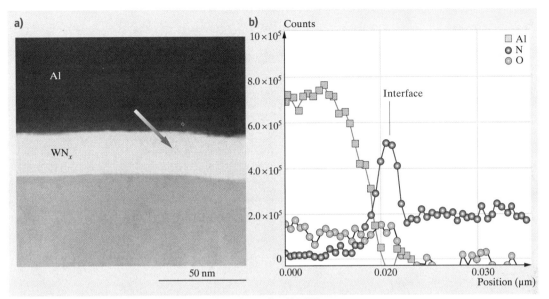

Fig. 44.35 EELS line scan across an Al/WN$_x$ interface (after [44.50])

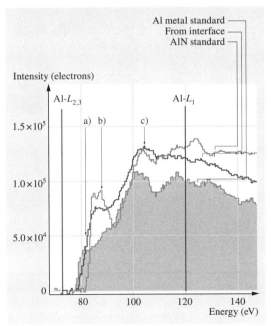

Fig. 44.36 Comparison of Al-$L_{2,3}$ edges for interface in the study and standard sample (after [44.50])

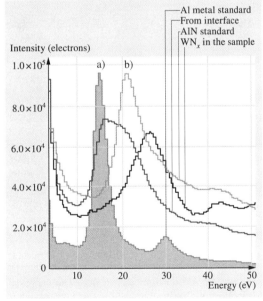

Fig. 44.37 Comparison of low-loss features for the locations in the study and the standard samples (after [44.50])

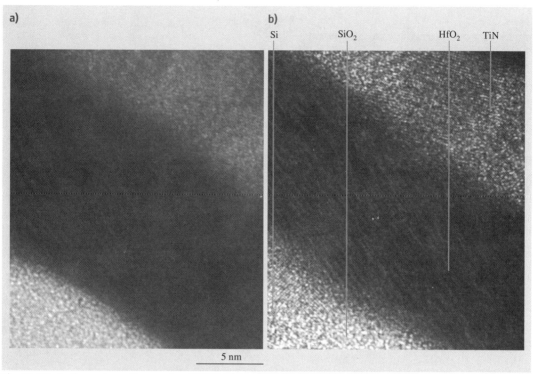

Fig. 44.38a,b Comparison of conventional HREM (**a**) and energy-filtered HREM (**b**)

EFTEM Applications

EFTEM to Image Thick Sample. Figure 44.38 shows is a conventional HREM image of a crystalline HfO_2 layer in contact with Si and TiN. Because the sample is not sufficiently thin, part of the blurring of the HREM image is due to inelastic electrons. Figure 44.38b shows an energy-filtered (zero-loss) image of the same feature. The imaging improvement by energy filtering is obvious through the lattice images of the TiN, HfO_2, and Si regions.

EFTEM to Map Elemental Distributions. For many elements, EFTEM enables fast and distinctive elemental mapping. Figure 44.39 shows a series of elemental maps of a defect in a semiconductor device. The chemical distribution is clearly revealed. Compared with other elemental mapping techniques such as EDS elemental mapping, EFTEM elemental mapping can achieve much higher spatial resolution.

EFTEM to Enhance Contrast. EFTEM can be used to enhance contrast in situations when TEM would show low contrast. An example is given in Fig. 44.40. As indicated by the arrow, the Si low-loss image (using the Si plasma peak at about 15 eV) clearly reveals Si particles where the bright-field TEM image shows low or no contrast.

EFTEM to Reduce Diffraction Contrast. EFTEM can often be used as an alternative method to Z-contrast STEM to reduce diffraction contrast. This can be critical in situations such as electron tomography of crystalline materials where orientation-dependent contrast is to be avoided [44.114]. Most EFTEM, except zero-loss imaging, offers a certain reduction of diffraction contrast. The diffraction contrast reduction can be seen in both Figs. 44.38 and 44.39. EFTEM offers a wide variety of energy-filtering controls. Some EFTEM methods, such as jump-ratio imaging [44.114]

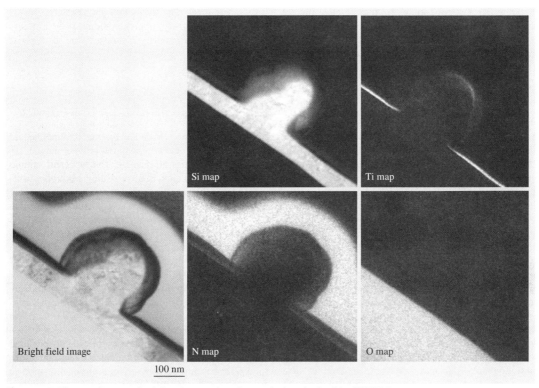

Fig. 44.39 Elemental maps of a defect in a semiconductor device obtained by EFTEM

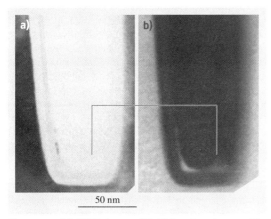

Fig. 44.40 (a) TEM image. (b) Si low-loss image. The *arrow* indicates equivalent area in the two images

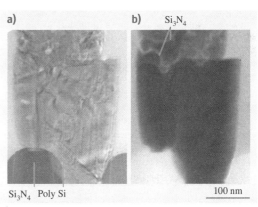

Fig. 44.41a,b The result of diffraction contrast reduction by EFTEM. (a) TEM bright-field image; (b) energy-filtered 74 eV low-loss (energy window width = 10 eV, centered at 74 eV) image (after [44.50])

(dividing the post-edge image by the pre-edge image) and certain low-loss-region imaging [44.115], are more effective in reducing diffraction contrast. Figure 44.41 shows a comparison of a TEM and the 74 eV low-loss image. The prominent diffraction contrast in polycrystalline Si is effectively suppressed in the 74 eV low-loss image, where the majority contrast is due to material difference.

44.3 Concluding Remarks: Current Application Status and Development

As an efficient technique to explore the microstructure of various materials, TEM has been extensively applied in both the research and industrial fields. Through observation and analysis of structural features and defects, it plays a key role in the evaluation of materials quality, the investigation of defect formation mechanisms, and the correlation between processing and microstructural properties.

Conventional TEM techniques such as diffraction-contrast imaging, selected-area diffraction, convergent-beam electron diffraction, and large-angle convergent-beam electron diffraction are extensively used to characterize defects in both bulk and epitaxially grown crystals. Examples of such application include the study of threading dislocation density reduction in wurtzite-type epifilms, defects produced by composition variations in semiconductor alloys, and strain relaxation and dislocation behavior in strained Si/SiGe episystems. LACBED has the advantage of being able to characterize most types of crystal defect (point defects, perfect and partial dislocations, stacking faults, antiphase boundaries, and grain boundaries). In addition to the aforementioned basic functions, many techniques, such as EDS, EELS, STEM, EFTEM, holography, and tomography, have been incorporated into the TEM system. The modern TEM system is a high-resolution probe with extensive capabilities for exploration of the internal structure of various materials.

In recent years, TEM has become even more important, as the structural dimensions in many research and industrial applications are rapidly approaching the nanometer scale. The high-resolution TEM/STEM with various analytical functions has become an indispensable tool for nanoscale defect analyses. Take the semiconductor industry as an example; the trend for ever-shrinking device dimension has demanded that defect analyses be localized to the nanometer scale. This has led to increased use of the TEM as an integral part of process development and failure analysis. Examples include distributions and density of crystalline defects at the thin-film metallization–silicon substrate interface, defect formation and involvement during dopant implantation and thermal treatments, and measurement of local strains in the critical channel region of fully processed devices. The powerful sample preparation technique, FIB, providing accurate specimen thickness control over large area as well as being highly site specific, is especially useful in nanoscale devices. In addition to its importance in imaging nanoscale microstructures, TEM is also the only technique that can produce high count rates in diffraction patterns from individual nanotubes or other nanostructures. Capable of revealing small lattice-parameter changes, CBED strain analysis finds its application in the semiconductor industry in analyzing strain variations within device substrates. Furthermore, with the advances of nanotechnology, electron tomography has found an increasingly important role in the physical sciences. For materials with complex nanoscale structures, electron tomography offers a promising solution to overcome the difficulties caused by sample thickness limitations.

The progress of materials analysis by TEM largely relies on the advances of instrumentation. Various technologies have been applied to improve the spatial resolution of TEM/STEM imaging and the energy resolution of EELS. Recent progress includes lens aberration correctors to improve TEM and STEM resolution and monochromators for reducing energy spread of the electron beam. With the help of these new technologies, atomic-scale defect analysis will become more precise and convenient. The use of aberration correctors allows the relaxation of the tight spacing around the specimen. One benefit is naturally the easier accommodation of in situ apparatus within the TEM. The in situ experiments of interest to the readers may include in situ TEM observation of defects under the influence of mechanical and/or electrical stress. Meanwhile, the quality of electron microdiffraction has been greatly improved with the availability of energy-filtered imaging, field-emission guns, liquid-helium-cooled sample stages, charge-coupled device (CCD) cameras, monochromators, aberration correctors, and energy-loss spectrometers with parallel detection, together with much faster computers. It has now become possible to measure low-order structure factors from crystals of known structure using the quantitative convergent-beam diffraction method with accuracy equal to or better than that of the x-ray Pendellösung method. The well-developed analytical capabilities (EDS, EELS) of TEM provide compositional information with spatial resolution in the nanometer and subnanometer range. Combined with structural information from high-resolution phase-contrast imaging, such capabilities make TEM an indispensable tool for the development and application of new materials in the fields of semiconductors, ultrafine-grain materials or thin films. Energy-filtering TEM (EFTEM) has proven to be the key tool for nano-analytical applications, since it uses the rich information provided by the energy-loss spectrum in a spatially resolved manner with short acquisition times.

As with any analytical tools, TEM has its own limitations. The most notable ones are as follows:

1. TEM observation is very localized. It is not a suitable way for large-scale sampling.
2. Due to dynamical scattering, quantitative diffraction analysis can be difficult compared with x-ray and neutron diffraction.
3. TEM sample preparation is destructive.
4. There are possible surface effects, which become increasingly important with reduced sample thickness.
5. Electron-beam damage can be significant in some situations.
6. Structural defects can be generated or changed by sample preparation and electron-beam irradiation.

New developments in TEM instruments have enabled major improvements on two fronts: the ability to see ever smaller things and the ability to see the *unseeable*. Various contrasts with TEM/STEM have enabled various structural, defect, and chemical analysis with resolution as high as 0.1 nm. With improvements on both theoretical and instrumental fronts, we expect new advances in the field of transmission electron microscopy. At the same time, we may say that TEM itself has matured to be a premier research tool in materials analysis.

References

44.1 J.W. Edington: *Practical Electron Microscopy in Materials Science* (Van Nostrand Reinhold, New York 1976) pp. 113–116

44.2 J.W. Edington: *Practical Electron Microscopy in Materials Science* (Van Nostrand Reinhold, New York 1976) p. 63

44.3 J.W. Edington: *Practical Electron Microscopy in Materials Science* (Van Nostrand Reinhold, New York 1976) pp. 109–145

44.4 D.J.H. Cockayne, I.L.F. Ray, M.J. Whelan: Investigation of dislocation strain fields using weak beams, Philos. Mag. **20**, 1265–1270 (1969)

44.5 J.W. Edington: *Practical Electron Microscopy in Materials Science* (Van Nostrand Reinhold, New York 1976) pp. 145–149

44.6 D.B. Holt: Polarity reversal and symmetry in semicondcuting compounds with the sphalerite and wurtzite structures, J. Mater. Sci. **19**(2), 439–446 (1984)

44.7 R. Serneels, M. Snykers, P. Delavignette, R. Gevers, S. Amelinckx: Friedel's law in electron diffraction as applied to the study of domain structures in non-centrosymmetrical crystals, Phys. Status Solidi (b) **58**, 277–292 (1973)

44.8 L.T. Romano, J.E. Northrup, M.A. O'Keefe: Inversion domains in GaN grown on sapphire, Appl. Phys. Lett. **69**(16), 2394–2396 (1996)

44.9 J. Jasinski, Z. Liliental-Weber, H. Lu, W.J. Schaff: V-shaped inversion domains in InN grown on c-plane sapphire, Appl. Phys. Lett. **85**(2), 233–235 (2004)

44.10 G. Thomas, M.J. Goringe: *Transmission Electron Microscopy of Materials* (Wiley, New York 1979) pp. 36–40

44.11 W.L. Bell, G. Thomas: *Electron Microscopy and Structure of Materials* (Univ. of California Press, Berkeley 1972) p. 23

44.12 J.W. Edington: *Practical Electron Microscopy in Materials Science* (Van Nostrand Reinhold, New York 1976) pp. 87–88

44.13 J. Bai: Studies of defects and strain relaxation in III-nitride epifilms. Ph.D. Thesis (State University of New York at Stony Brook, Stony Brook 2006)

44.14 M. Dudley: Lecture notes for ESM512, State University of New York at Stony Brook (2002)

44.15 C. Kisielowski, B. Freitag, X. Xu, S.P. Beckman, D.C. Chrzan: Sub-angstrom imaging of dislocation core structures: how well are experiments comparable with theory, Philos. Mag. **86**(29–31), 4575–4588 (2006)

44.16 A.H. Buist, A. van den Bos, M.A.O. Miedema: Optimal experimental design for exit wave reconstruction from focal series in TEM, Ultramicroscopy **64**, 137–152 (1996)

44.17 J.M. Gibson: High resolution electron microscopy of interfaces between epitaxial thin films and semiconductors, Ultramicroscopy **14**, 1–10 (1984)

44.18 X. Xu, S.P. Beckman, P. Specht, E.R. Weber, D.C. Chrzan, R.P. Erni, I. Arslan, N. Browning, A. Bleloch, C. Kisielowski: Distortion and segregation in a dislocation core region at atomic resolution, Phys. Rev. Lett. **95**, 145501 (2005)

44.19 B.C. de Cooman, N.-H. Cho, Z. Elgat, C.B. Carter: HREM of compound semiconductors, Ultramicroscopy **18**, 305–312 (1985)

44.20 D.J. Smith, Z.G. Li, P. Lu, M.R. McCartney, S.-C. Tsen: Characterization of thin films, interfaces and surfaces by high-resolution electron microscopy, Ultramicroscopy **37**, 169–179 (1991)

44.21 Z. Liliental-Weber, T. Tomaszewicz, D. Zakharov, M.A. O'Keefe: Defects in p-doped bulk GaN crystals grown with Ga polarity, J. Cryst. Growth **281**, 125–134 (2005)

44.22 H. Bender, A. Veirman, J. Landuyt, S. Amelinckx: HREM investigation of twinning in very high dose phosphorus ion-implanted silicon, Appl. Phys. A **39**, 83–90 (1986)

44.23 X.J. Wu, F.H. Li, H. Hashimoto: TEM study on overlapped twins in GaAs crystal, Philos. Mag. B **63**, 931–939 (1991)

44.24 J.W. Edington: *Practical Electron Microscopy in Materials Science* (Van Nostrand Reinhold, New York 1976) pp. 73–81

44.25 M.J. Marcinkowski: *Electron Microscopy and Strength of Crystals* (Univ. of California Press, Berkeley 1971) p. 333

44.26 O. Ueda, Y. Nakata, T. Fujii: Study on microstructure of ordered InGaAs crystals grown on substrates by transmission electron microscopy, Appl. Phys. Lett. **58**(7), 705–707 (1991)

44.27 T.-Y. Seong, A. G.Norman, G.R. Booker, A.G. Cullis: Atomic ordering and domain structures in metal organic chemical vapor deposition grown InGaAs (001) layers, J. Appl. Phys. **75**(12), 7852–7865 (1994)

44.28 N. Amir, K. Cohen, S. Stolyarova, A. Chack, R. Beserman, R. Weil, Y. Nemirovsky: Long-range order in CdZnTe epilayers, J. Phys. D Appl. Phys. **33**, L9–L12 (2000)

44.29 D.B. Williams, C.B. Carter: *Transmission Electron Microscopy* (Plenum, New York 1996) pp. 321–323

44.30 T. Mitate, Y. Sonoda, N. Kuwano: Polarity determination of wurtzite and zincblende structures by TEM, Phys. Status Solidi (a) **192**(2), 383–388 (2002)

44.31 J. Taftø, J.C.H. Spence: A simple method for the determination of structure-factor phase relationships and crystal polarity using electron diffraction, J. Appl. Crystallogr. **15**, 60–64 (1982)

44.32 K. Ishizuka, J. Taftø: Quantitative analysis of CBED to determine polarity and ionicity of ZnS-type crystals, Acta Cryst. B **40**, 332–337 (1984)

44.33 J. Bai, M. Dudley, B. Raghothamachar, P. Gouma, B.J. Skromme, L. Chen, P.J. Hartlieb, E. Michaels, J.W. Kolis: Correlated structural and optical characterization of ammonothermally grown bulk GaN, Appl. Phys. Lett. **84**(17), 3289–3291 (2004)

44.34 Y. Tomokiyo, S. Matsumura, T. Okuyama, T. Yasunaga, N. Kuwano, K. Oki: Dynamical diffraction effect on HOLZ-pattern geometry in Si-Ge alloys and determination of local lattice parameter, Ultramicroscopy **54**(2-4), 276–285 (1994)

44.35 A. Hovsepian, D. Cherns, W. Jäger: Analysis of ultrathin Ge layers in Si by large angle convergent beam electron diffraction, Philos. Mag. A **79**(6), 1395–1410 (1999)

44.36 S.J. Rozeveld, J.M. Howe: Determination of multiple lattice parameters from convergent-beam electron diffraction pattern, Ultramicroscopy **50**(1), 41–56 (1993)

44.37 J. Zipprich, T. Fuller, F. Banhart, O.G. Schmidt, K. Eberl: The quantitative characterization of SiGe layers by analyzing rocking profiles in CBED patterns, J. Microsc. **194**(1), 12–20 (1999)

44.38 D. Jacob, Y. Androussi, T. Benabbas, P. Francois, A. Lefebvre: Surface relaxation of strained semiconductor heterostructures revealed by finite-element calculations and transmission electron microscopy, Philos. Mag. A **78**(4), 879–891 (1998)

44.39 M. Tanaka, R. Saito, K. Ueno, Y. Harada: Large-angle convergent-beam electron diffraction, J. Electron Microsc. **29**(4), 408–412 (1980)

44.40 D. Cherns, A.R. Preston: Convergent beam diffraction studies of crystal defects, Proc. 11th Int. Congr. Electron Microsc., Kyoto, Vol.1, ed. by T. Imura, S. Marusa, T. Suzuki (The Japanese Society of Electron Microscopy, Tokyo 1986) pp. 207–208

44.41 D. Cherns, A.R. Preston: Convergent beam diffraction studies of interfaces, defects, and multilayers, J. Electron Microsc. Tech. **13**, 111–122 (1989)

44.42 R.W. Carpenter, J.C.H. Spence: Three-dimensional strain-field information in convergent-beam electron diffraction patterns, Acta Crystallogr. A **38**, 55–61 (1982)

44.43 D. Cherns, A.R. Preston: Convergent beam diffraction studies of crystal defects, Proc. 11th Int. Congr. Electron Microsc., Kyoto, Vol.1, ed. by T. Imura, S. Marusa, T. Suzuki (The Japanese Society of Electron Microscopy, Tokyo 1986) p. 721

44.44 D. Cherns, J.-P. Morniroli: Analysis of partial and stair-rod dislocations by large angle convergent baem electron diffraction, Ultramicroscopy **53**(2), 167–180 (1994)

44.45 K.K. Fung: Convergent-beam electron diffraction study of transverse stacking faults and dislocations, Ultramicroscopy **17**, 81–86 (1985)

44.46 C.T. Chou, L.J. Zhao, T. Ko: Higher-order Laue zone effects of stacking-faulted crystals, Philos. Mag. A **59**(6), 1221–1243 (1989)

44.47 D.E. Jesson, J.W. Steeds: Higher-order Laue zone diffraction from crystals containing transverse stacking faults, Philos. Mag. A **61**, 385–415 (1990)

44.48　J.P. Morniroli: CBED and LACBED characterization of crystal defects, J. Microsc. **223**(3), 240–245 (2006)

44.49　A. Hovsepian, D. Cherns, W. Jäger: Analysis of ultra-thin Ge layers in Si by large angle convergent beam electron diffraction, Philos. Mag. A **79**(6), 1395–1410 (1999)

44.50　S.X. Wang: EELS fine edge structure and quantification analyses, Internal report of Micron Technology (2005)

44.51　O.L. Krivanek, P.D. Nellist, N. Dellby, M.F. Murfitt, Z. Szilagyi: Toward sub-0.5 Å electron beams, Ultramicroscopy **96**, 229–237 (2003)

44.52　D.A. Blom, L.F. Allard, S. Mishina, M.A. O'Keefe: Early results from an aberration-corrected JEOL 2200FS STEM/TEM at Oak Ridge National Laboratory, Microsc. Microanal. **12**, 483–491 (2006)

44.53　S.J. Pennycook, L.A. Boatner: Chemically sensitive structure-imaging with a scanning transmission electron microscope, Nature **336**, 565–567 (1988)

44.54　S.J. Pennycook, J. Narayan: Direct imaging of dopant distributions in silicon by scanning transmission electron microscopy, Appl. Phys. Lett. **45**, 385–387 (1984)

44.55　D.E. Jesson, S.J. Pennycook: Incoherent imaging of crystals using thermally scattered electrons, Proc. R. Soc. Lond. Ser. A **449**, 273–393 (1995)

44.56　S.J. Pennycook, D.E. Jesson: High-resolution incoherent imaging of crystals, Phys. Rev. Lett. **64**, 938–941 (1990)

44.57　P. Rez: Scattering cross sections in electron microscopy and analysis, Microsc. Microanal. **7**, 356–362 (2001)

44.58　R.F. Egerton: *Electron Energy-Loss in the Electron Microscope* (Plenum, New York 1986)

44.59　L.A. Giannuzzi, J.L. Drown, S.R. Brown, R.B. Irwin, F.A. Stevie: Focused ion beam milling and micromanipulation lift-out for site specific cross-section TEM specimen preparation, Mater. Res. Soc. Symp. Proc. **480**, 19–27 (1997)

44.60　S.X. Wang, M.M. Kowalewski: TEM and PEELS study of Mn diffusion in an MRAM structure, Microsc. Microanal. **9**(Suppl. 2), 496–497 (2003)

44.61　C.C. Ahn, O.L. Krivanek: *EELS Atlas* (Gatan Inc./Arizona State Univ., Warrendal/Tempe 1983)

44.62　J. Orloff, L.W. Swanson: Optical column design with liquid metal ion sources, J. Vac. Sci. Technol. **19**, 1149–1152 (1981)

44.63　T. Ishitani, T. Ohnishi, Y. Madokoro, Y. Kawanami: Focused-ion-beam "cutter" and "attacher" for micromachining and device transplantation, J. Vac. Sci. Technol. B **9**, 2633–2637 (1991)

44.64　P. Gasser, U.E. Klotz, F.A. Khalid, O. Beffort: Site-specific specimen preparation by focused ion beam milling for transmission electron microscopy of metal matrix composites, Microsc. Microanal. **10**, 311–316 (2004)

44.65　S.X. Wang: TEM study of surface damage and profile of a FIB-prepared Si sample, Microsc. Microanal. **10**(Suppl. 2), 1158–1159 (2004)

44.66　J.H. van der Merwe: Strains in crystalline overgrowths, Philos. Mag. **7**(80), 1433–1434 (1962)

44.67　J.W. Matthews (Ed.): *Epitaxial Growth* (Academic, New York 1975) p. 559, Part B

44.68　D.E. Jesson, S.J. Pennycook, J.-M. Baribeau, D.C. Houghton: Direct imaging of surface cusp evolution during strained-layer epitaxy and implications for strain relaxation, Phys. Rev. Lett. **71**, 1744–1747 (1993)

44.69　D.D. Perovic, G.C. Weatherly, J.-M. Baribeau, D.C. Houghton: Heterogeneous nucleation sources in molecular beam epitaxy-grown Ge_xSi_{1-x}/Si strained layer superlattices, Thin Solid Films **183**(1/2), 141–156 (1989)

44.70　P.M.J. Marée, J.C. Barbour, J.F. van der Veen, K.L. Kavanagh, C.W.T. Bulle-Lieuwma, M.P.A. Viegers: Generation of misfit dislocations in semiconductors, J. Appl. Phys. **62**(11), 4413–4420 (1987)

44.71　W. Hagen, H. Strunk: New type of source generating misfit dislocations, Appl. Phys. **17**(1), 85–87 (1978)

44.72　K.R. Breen, P.N. Uppal, J.S. Ahearn: Interface dislocation structures in $In_xGa_{1-x}As$/GaAs mismatched epitaxy, J. Vac. Sci. Technol. B **7**, 758–763 (1989)

44.73　E.A. Fitzgerald, D.G. Ast, P.D. Kirchner, G.D. Pettit, J.M. Woodall: Structure and recombination in InGaAs/GaAs heterostructures, J. Appl. Phys. **63**(3), 693–703 (1988)

44.74　E.A. Fitzgerald: Dislocations in strained-layer epitaxy-theory, experiment, and applications, Mater. Sci. Rep. **7**(3), 91 (1991)

44.75　E.A. Stach, R. Hull, R.M. Tromp, F.M. Ross, M.C. Reuter, J.C. Bean: In-situ transmission electron microscopy studies of the interaction between dislocations in strained SiGe/Si(001) heterostructures, Philos. Mag. A **80**(9), 2159–2200 (2000)

44.76　W. Bollmann: Size and sign of the Burgers vector from transmission micrographs, Philos. Mag. **13**(125), 935–944 (1966)

44.77　E. Spiecker, W. Jäger: Quantitative large-area analysis of misfit dislocation arrays by bend contour contrast evaluation. In: *Microscopy of Semiconducting Materials*, Inst. Phys. Conf. Ser., Vol. 180, ed. by A.G. Cullis, P.A. Midgley (Institute of Physics, London 2003) pp. 259–264

44.78　R.H. Dixon, P.J. Goodhew: On the origin of misfit dislocations in InGaAs/GaAs strained layers, J. Appl. Phys. **68**(7), 3163–3168 (1990)

44.79　J.W. Matthews, A.E. Blakeslee: Defects in epitaxial multilayers: I. Misfit dislocations, J. Cryst. Growth **27**, 118–125 (1974)

44.80　R.H. Miles, T.C. McGill: Structural perfection in poorly lattice matched heterostructures, J. Vac. Sci. Technol. B **7**(4), 753–757 (1989)

44.81 R. People, J.C. Bean: Calculation of critical layer thickness versus lattice mismatch for Ge_xSi_{1-x}/Si strained-layer heterostructures, Appl. Phys. Lett. **47**(3), 322–324 (1985)

44.82 R.H. Dixon, P.J. Goodhew: On the origin of misfit dislocations in InGaAs/GaAs strained layers, J. Appl. Phys. **68**(7), 3163–3168 (1990)

44.83 P.M.J. Marée, J.C. Barbour, J.F. van der Veen, K.L. Kavanagh, C.W.T. Buile-Lieuwrna, M.P.A. Viegers: Generation of misfit dislocations in semiconductors, J. Appl. Phys. **62**(11), 4413–4420 (1987)

44.84 Y. Kimura, N. Sugii, S. Kimura, K. Inui, W. Hirasawa: Generation of misfit dislocations and stacking faults in supercritical thickness strained-Si/SiGe heterostructures, Appl. Phys. Lett. **88**, 031912–031914 (2006)

44.85 A.F. Marshall, D.B. Aubertine, W.D. Nix, P.C. McIntyre: Misfit dislocation dissociation and Lomer formation in low mismatch SiGe/Si heterostructures, J. Mater. Res. **20**(2), 447–455 (2005)

44.86 J. Zou, D.J.H. Cockayne: Theoretical consideration of equilibrium dissociation geometries of 60° misfit dislocations in single semiconductor heterostructures, J. Appl. Phys. **77**(6), 2448–2453 (1995)

44.87 J. Zou, D.J.H. Cockayne: Equilibrium dissociation configuration of misfit dislocations in low strained $In_{0.1}Ga_{0.9}As/GaAs$ single heterostructures, Appl. Phys. Lett. **63**(16), 2222–2224 (1993)

44.88 N. Hirashita, N. Sugiyama, E. Toyoda, S.-I. Takagi: Strain relaxation processes in strained-Si layer on SiGe-on-insulator substrates, Thin Solid Films **508**, 112–116 (2006)

44.89 E.A. Fitzgerald, D.G. Ast, P.D. Kirchner, G.D. Pettit, J.M. Woodall: Structure and recombination in InGaAs/GaAs heterostructures, J. Appl. Phys. **63**(3), 693–703 (1988)

44.90 Y.H. Xie, E.A. Fitzgerald, P.J. Silverman, A.R. Kortan, B.E. Weir: Fabrication of relaxed GeSi buffer layers on Si(100) with low threading dislocation density, Mater. Sci. Eng. B **14**, 332–335 (1992)

44.91 C.W. Leitz, M.T. Currie, A.Y. Kim, J. Lai, E. Robbins, E.A. Fitzgerald, M.T. Bulsara: Dislocation glide and blocking kinetics in compositionally graded SiGe/Si, J. Appl. Phys. **90**(4), 2730–2736 (2001)

44.92 A.E. Blakeslee: The use of superlattices to block the propagation of dislocations in semiconductors, Mater. Res. Soc. Symp. Proc. **148**, 217–227 (1989)

44.93 J.S. Park, M. Curtin, J. Bai, S. Bengtson, M. Carroll, A. Lochtefeld: Thin strained layers inserted in compositionally graded SiGe buffers and their effects on strain relaxation and dislocation, J. Appl. Phys. **101**, 053501 (2007)

44.94 J.S. Park, J. Bai, M. Curtin, B. Adekore, M. Carroll, A. Loctefeld: Defect reduction of selective Ge epitaxy in trenches on Si(001) substrates using aspect ratio trapping, Appl. Phys. Lett. **90**, 052113 (2007)

44.95 J. Bai, J.S. Park, Z. Cheng, M. Curtin, B. Adekore, M. Carroll, A. Lochtefeld, M. Dudley: Study of the defect elimination mechanisms in aspect ratio trapping Ge growth, Appl. Phys. Lett. **90**(10), 101902 (2007)

44.96 Y. Ikuhara, P. Pirouz: High resolution transmission electron microscopy studies of metal/ceramics interfaces, Microsc. Res. Tech. **40**(3), 206–241 (1998)

44.97 A. Trampert: Private communication (2005)

44.98 A. Trampert, K.H. Ploog: Heteroepitaxy of large-misfit systems: Role of coincidence lattice, Cryst. Res. Technol. **35**(6/7), 793–806 (2000)

44.99 T. Zheleva, K. Jagannadham, J. Narayan: Epitaxial growth in large-lattice-mismatch systems, J. Appl. Phys. **75**(2), 860–871 (1994)

44.100 C.J. Sun, P. Kung, A. Saxler, H. Ohsato, K. Haritos, M. Razeghi: A crystallographic model of (00.1) aluminum nitride epitaxial thin film growth on (00.1) sapphire substrate, J. Appl. Phys. **75**(8), 3964–3967 (1994)

44.101 T. Kehagias, P. Komninou, G. Nouet, P. Ruterna, T. Karakostas: Misfit relaxation of the AlN/Al_2O_3 (0001) interface, Phys. Rev. B **64**, 195329 (2001)

44.102 X.R. Huang, J. Bai, M. Dudley, B. Wagner, R.F. Davis, Y. Zhu: Step-controlled strain relaxation in the vicinal surface epitaxy of nitrides, Phys. Rev. Lett. **95**, 086101 (2005)

44.103 P. Vermaut, P. Ruterana, G. Nouet, H. Morkoç: Structural defects due to interface steps and polytypism in III-V semiconducting materials: A case study using high-resolution electron microscopy of the 2H-AlN/6H-SiC interface, Philos. Mag. A **75**(1), 239–259 (1997)

44.104 J. Bai, X. Huang, M. Dudley, B. Wagner, R.F. Davis, L. Wu, E. Sutter, Y. Zhu, B.J. Skromme: Intersecting basal plane and prismatic stacking fault structures and their formation mechanisms in GaN, J. Appl. Phys. **98**(6), 063510 (2005)

44.105 P. Pirouz, J.W. Yang: Polytypic transformations in SiC: the role of TEM, Ultramicroscopy **51**(1-4), 189–214 (1993)

44.106 J. Bai, X. Huang, M. Dudley: High-resolution TEM observation of AlN grown on on-axis and off-cut SiC substrates, Mater. Sci. Semicond. Process. **9**, 180–183 (2006)

44.107 J. Mei, S. Srinivasan, R. Liu, F.A. Ponce, Y. Narukawa, T. Mukai: Prismatic stacking faults in epitaxially laterally overgrown GaN, Appl. Phys. Lett. **88**, 141912 (2006)

44.108 J. Bai, M. Dudley, W. Sun, H. Wang, M. Khan: Reduction of threading dislocation densities in AlN/sapphire epilayers driven by growth mode modification, Appl. Phys. Lett. **88**(5), 051903 (2006)

44.109 H. Klapper: Generation and propagation of dislocations during crystal growth, Mater. Chem. Phys. **66**, 101–109 (2000)

44.110 R. Liu, A. Bell, F.A. Ponce, C.Q. Chen, J.W. Yang, M.A. Khan: Luminescence from stacking faults in gallium nitride, Appl. Phys. Lett. **86**, 021908 (2005)

44.111 D.H. Pearson, C.C. Ahn, B. Fultz: Phys. Rev. B **47**, 8471–8478 (1993)

44.112 Z.L. Wang, J.S. Yin, Y.D. Jiang, J. Zhang: Appl. Phys. Lett. **70**, 3362–3364 (1997)

44.113 J.L. Mansot, P. Leone, P. Euzen, P. Palvadeau: Microsc. Microanal. Microstruct. **5**, 79–90 (1994)

44.114 P.A. Midgley, M. Wayland: 3-D electron microscopy in the physical sciences: the development of Z-contrast and EFTEM tomography, Ultramicroscopy **96**, 413–431 (2003)

44.115 L. Tsung, D. Matheson, C. Skelton, R. Turner, J. Ringnalda: Energy contrast from Si low loss at 74 eV for semiconductor devices, Microsc. Microanal. **9**(Suppl. 2), 490–491 (2003)

1520

45. Electron Paramagnetic Resonance Characterization of Point Defects

Mary E. Zvanut

Electron paramagnetic resonance (EPR) spectroscopy identifies, counts, and monitors point defects in a wide variety of materials. Unfortunately, this powerful tool has faded from the literature in recent years. The present trend away from fundamental studies and towards technological challenges, and the need for fast diagnostic tools for use during and after materials growth has weakened the popularity of magnetic resonance tools. While admittedly the use of EPR in industrial laboratories for routine materials characterization is limited, EPR spectroscopy can be, and has been, successfully used to provide reams of information directly relevant to technologically significant materials.

The interpretation of EPR spectra involves an understanding of basic quantum mechanics and a reasonable investment of time. Once a defect is identified, however, the spectra may be used as a fingerprint that can be used in additional studies addressing the chemical kinetics, charge transport, and electronic energies of the defect and surrounding lattice. Numerous examples are provided in this chapter. In addition, the fundamental information extracted from EPR analysis should not be forgotten. Perhaps knowing the distribution of spin states about the core of a defect will not expedite the production of material X for use as device Y, but it may provide the seed of knowledge with which to build the 21st century's technological revolution. We must remember that the basic understanding of semiconductors developed in the middle of the last century spawned the solid-state transistor, which unquestionably produced the computer revolution in the latter half of the 20th century.

This chapter will acquaint the reader with the fundamental methods used to interpret EPR data and summarize many different experiments which illustrate the applicability of the technique to important materials issues.

45.1	Electronic Paramagnetic Resonance	1522
45.2	EPR Analysis	1524
	45.2.1 Zeeman Effect	1524
	45.2.2 Nuclear Hyperfine Interaction	1526
	45.2.3 Interactions Involving More than One Electron	1529
	45.2.4 Total Number of Spins	1533
45.3	Scope of EPR Technique	1534
	45.3.1 Defects in a Thin Film on a Substrate	1534
	45.3.2 Defects at an Interface	1535
	45.3.3 Defects at Surfaces	1536
	45.3.4 Nondilute Systems	1537
45.4	Supplementary Instrumentation and Supportive Techniques	1538
	45.4.1 Photo-EPR	1539
	45.4.2 Correlation with Electrically Detected Trapping Centers and Defect Levels	1541
	45.4.3 Heat Treatment and EPR	1543
45.5	Summary and Final Thoughts	1545
References		1546

Electron paramagnetic resonance (EPR), or electron spin resonance (ESR), provides fundamental chemical and structural information about a point defect. For example, by detecting the unpaired (paramagnetic) electron on the nitrogen donor in SiC, EPR results show that nitrogen substitutes for carbon and that the electron–nuclear hyperfine interaction differs by a factor of ten for the different symmetry sites in the 4H- and 6H-SiC polytypes. The total number of uncompensated nitrogen donors may also be determined. Note that EPR senses specific charge states of the defect. For instance, in the case of nitrogen in SiC only neutral nitrogen with the donor electron *on* the nucleus is detected. Also, it is important to realize that EPR probes the ground state of a defect; thereby providing information about the *as-grown*, unperturbed defect. Illumination or heat may be used during a measurement to alter the charge state or produce the excited state, providing additional information. Ultimately, the electron wavefunction overlap with the core nucleus and surrounding neighbors may be determined. While this last bit of information may not be pertinent to most immediate applications, taken together the data paint a thorough picture of a specific center, enabling additional knowledge to be gleaned from the same defect in similar materials as well as other defects in the same host.

Having read other chapters in this Handbook, the reader may wonder how EPR compares with other techniques. EPR does *not* provide information about the lattice symmetry like x-ray diffraction or the overall chemical composition of the material as can be obtained from energy-dispersive x-ray analysis. Most importantly, the concentration of defects is not directly obtained from an EPR spectrum. The total number of a specific defect can be estimated, sometimes within a factor of two, but the distribution of centers is not sensed by magnetic resonance. Therefore, any concentration measurement must be inferred from etching studies or measurements of different sized samples. Finally, many readers may be familiar with nuclear magnetic resonance (NMR). Although the physical principle of NMR and EPR are the same, implementation of the techniques is entirely different, as a reader familiar with NMR will realize when they read this chapter. Unlike NMR, EPR has never been successfully adapted for scanning, as has NMR, where the technique has become the basis of magnetic resonance imaging. Whereas many of the methods discussed in this text provide the electronic or chemical identification of a defect, EPR can provide a complete picture – after great investment of time and analysis. Thus, while EPR provides a great deal of information, it cannot be used as a substitute for conventional characterization methods used routinely on a large volume of material. Rather, the technique provides many important details about point defects that are not detectable by other means.

As with any technique, correlation with other experimental tools augments the information provided. The use of optical excitation, for example, sometimes enables one to extract defect levels (ionization energies) or probe the excited state of a defect. In addition, comparison of EPR spectra with optical absorption studies has enabled the identification of many optical absorption bands and the calculation of optical cross sections. Electrical measurements in tandem with EPR studies have also produced important scientific and technological information about trapping centers in electronic devices. In principle, comparison with secondary-ion mass spectroscopy (SIMS) or glow-discharge mass spectroscopy (GDMS) data could provide the fraction of impurities in a particular charge state or located at a specific lattice site.

Before closing this Introduction, the reader should be made aware that EPR measurements were first recorded in 1945, and have permeated the literature for the past 60 years, covering chemistry, biology, physics, and engineering journals. In addition to the defects in materials discussed here, the technique is widely used to study free radicals in solution and biological species. There is active work detecting the many complicated structures responsible for diseases, as well as the more benign moieties responsible for life. Within the framework of materials, there are studies in nanomaterials, amorphous solids, and interfaces, some of which will be mentioned in this chapter. Many texts and articles are available that cover the application of EPR to the multitude of fields mentioned [45.1–6].

The chapter will begin with a review of the EPR instrumentation, briefly explaining the significant parts of the spectrometer and detection system. This will be followed by a section outlining the heart of the analysis using examples to illustrate the many features extracted from EPR spectra. The next two sections will cover the scope of the technique as it applies to defects in solids and the typical correlation experiments involving EPR. Finally, the last section will summarize the material presented and outline the different types of magnetic resonance techniques amenable to the study of defects in materials.

45.1 Electronic Paramagnetic Resonance

Electron paramagnetic resonance is based on the absorption of energy between spin states induced by the presence of an applied magnetic field [45.1, 7, 8]. In order to appreciate the technical aspects of the exper-

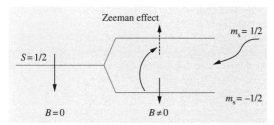

Fig. 45.1 Energy levels for a spin-$\frac{1}{2}$ electron in zero magnetic field (*left*) and nonzero applied field (*right*). The *vertical arrows* represent the spin of the electron; m_s is the magnetic spin quantum number

iment, one must understand the following. If there were no internal magnetic fields affecting the defect and if the defect contains a single unpaired electron (like that of a donor in a semiconductor), the absorbed energy is

$$h\upsilon = \Delta E = \mu_b g B, \qquad (45.1)$$

where h is Planck's constant, υ is the frequency of the absorbed radiation, μ_b is the Bohr magneton, and B is the applied magnetic field. The term $\mu_b g B$ is the energy difference between the spin states shown in Fig. 45.1. We leave discussion of g until the next section. The point here is that, if you place your sample in a magnetic field while illuminating with different frequency radiation, you should be able to adjust the incident photon energy until absorption is detected, similar to an optical absorption measurement. However, because $g \approx 2$ and convenient magnetic fields are on the order of Tesla, υ is typically in the microwave region. Unfortunately, microwaves are not conveniently manipulated like optical photons; thus the actual situation requires placing a sample in a fixed microwave field and applying a series of magnetic fields. One then searches for the field that produces microwave absorption. The energy levels for the single unpaired electron in an applied magnetic field are shown schematically in Fig. 45.1 and a typical spectrum illustrating detection of the EPR signal is shown in Fig. 45.2.

The instrumentation for EPR was developed in the 1950s, and little has changed except for the addition of more sophisticated detection circuitry and data-acquisition electronics. Figure 45.3 illustrates the basic experimental setup. The spectrometer consists of the klystron or Gunn diode used to produce the microwave radiation, microwave bridge for setting the desired power, waveguides which transmit the radiation to the sample, and cavity resonator in which a particular mode of the microwave is stored. As stated above,

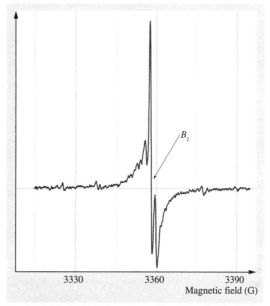

Fig. 45.2 Typical EPR spectrum illustrating the derivative line shape and *zero-crossing* B_z where the g value is calculated. The spectrum was obtained from 4H-SiC at 4 K with the magnetic field parallel to the c-axis of the crystal. The *dotted line* represents zero intensity

wavelengths in the microwave region are not easily varied, so a fixed frequency of 10 GHz is commonly used. The dimensions of the cavity establish a specific mode of the microwave frequency, in a similar way that a specific length of string determines the possible modes of a mechanical wave. The microwave energy stored in the standing wave is ultimately the energy absorbed by the sample when the resonance condition (45.1) is fulfilled. In order to produce an EPR signal, two more features must be understood: coupling of the microwaves in the

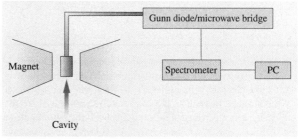

Fig. 45.3 Schematic diagram of EPR instrumentation; see text for description of the purpose for each component

cavity to the detection circuitry and the generation of the applied magnetic field. The former employs standard microwave electronics, as is discussed in detail in many texts [45.7]. The latter simply uses a standard magnet power supply and electromagnet that communicate with the spectrometer to set the magnetic field amplitude and ramp rate.

The instrumentation described above leads to the detection of microwave absorption at defects. However, the reader familiar with spectroscopy will recognize that the spectrum in Fig. 45.2 is not a simple absorption; rather it reflects the derivative of the absorption. In order to detect the small amount of paramagnetic defects typical of most materials, a type of phase-sensitive detection is employed. In practice, an oscillating magnetic field is superimposed on the ramped field and the change in the absorption is detected. This AC detection method produces the derivative spectrum shown in Fig. 45.2.

45.2 EPR Analysis

The theory of EPR rests on the concept of the *spin Hamiltonian*, which sums all of the energy sources affecting the electron dipole moment at the defect. The most obvious contribution comes from the applied magnetic field, but there are many others, some of which can be much larger than the effect of the applied field. The only ones considered here are the nuclear magnetic field and the spin–spin interactions. Below, each effect is treated individually and is accompanied by examples reflecting the type of interaction described.

45.2.1 Zeeman Effect

If there were no internal fields affecting the defect, the absorbed energy may be described by the interaction of the applied magnetic field B with the electron spin at the defect. The appropriate Hamiltonian is the Zeeman term

$$H = \mu_b S \cdot \mathbf{g} \cdot B,\qquad(45.2)$$

where μ_b is the Bohr magneton, S is the total spin of the electron, and B is the applied magnetic field. The **g**-tensor is related to the proportionality factor between the quantized electron magnetic dipole moment and total angular momentum. In EPR **g** takes on a significant role as will be seen later. Assuming simple spin-$\frac{1}{2}$ wavefunctions, $\left|\frac{1}{2},\frac{1}{2}\right\rangle, \left|\frac{1}{2},-\frac{1}{2}\right\rangle$, the energy solutions are

$$E = \mu_b g B m_s,\qquad(45.3)$$

where m_s is the z-component of the spin angular momentum $\pm\frac{1}{2}$. (The "bra-ket" notation is a standard method for denoting the wavefunction of the electron where here we use only the spin part: $|s, m_s\rangle$ where s is the electron spin and m_s is the magnetic spin quantum number. In general, the spin wavefunction for a paramagnetic electron at a defect may be a linear combination of s, m_s states.) The energy difference between the two m_s states, $\mu_b g B \Delta m_s$, is $\mu_b g B$ as seen in (45.1). The two spin levels are illustrated in Fig. 45.1 for a fixed value of magnetic field. Experimentally, the g-value is extracted from the magnetic field at which the intensity of the spectrum crosses zero intensity, as is indicated in Fig. 45.2 by B_z. In the simple case of a single electron free from the influence of any other magnetic fields, the g-value is the Lande free electron value. In a crystal, g is shifted by an amount that depends on the local environment of the defect. The shift is generally caused by a small amount of angular momentum that is not included in (45.2), but enters the theory as a perturbation and is incorporated into the g-value [45.1, 7–9]. Thus, each defect has a characteristic **g**-tensor determined by

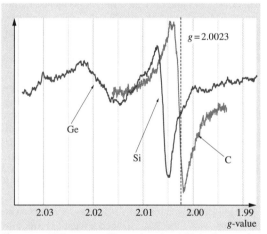

Fig. 45.4 EPR spectra illustrating the shift in the g-value as the mass of the central nucleus increases. Data was obtained from SiGe (*brown*) and SiC (*grey*) at 4 K. The elemental labeling (Ge, Si and C) indicates the central nuclei of the defects represented by the signal highlighted with an *arrow*. The *vertical dashed line* represents the g-value of a free electron. Note that the *x*-axis decreases to the right

its surroundings. Figure 45.4 illustrates the effect of the nearest neighbors on the g shift for a simple *dangling bond* center on a carbon atom in SiC (spectrum labeled "C"), silicon atom (Si) in SiGe and germanium (Ge) atom in SiGe (black spectrum). Here the x-axis is interpreted in terms of g through (45.1) with ΔE as the microwave energy; the free electron g-value, indicated by the dashed line, is used to approximate g for the defects. The EPR signal shifts to lower magnetic field (higher g) as the atomic number of the atom increases. Although by no means a hard rule, the shift is typical of this simple type of defect and reflects the change in the spin–orbit coupling parameter with increasing atomic number [45.9].

In the Zeeman term of (45.2), **g** is a tensor that incorporates the angular dependence of the interaction between the magnetic field and electron spin angular momentum. Whereas the hyperfine tensor discussed next provides chemical information about the point defect, the **g**-tensor provides structural details because it reflects the defect symmetry. The tensor is obtained from the g-value measured at each orientation of the sample with respect to the incident magnetic field. The interpretation of (45.2) in terms of the symmetry of g and the method used to extract this information is described in many texts. Suffice it to say that in general g may be written as

$$g^2 = g_X^2 \cos^2 \theta_X + g_Y^2 \cos^2 \theta_Y + g_Z^2 \cos^2 \theta_Z, \quad (45.4)$$

where θ_i is the angle between the i-th ($i = X, Y,$ and Z) axes of the defect and the applied magnetic field. It is important to realize that $X, Y,$ and Z are *not* necessarily the x-, y-, and z-axes of the crystal structure. Furthermore, the orientation of the defect axes with respect to, for example, the horizontal distance between the poles of the magnet, is not generally known. The procedure for extracting $X, Y,$ and Z from angular measurements with respect to the known crystal axes is thoroughly described in [45.1].

To illustrate, consider a defect in a hexagonal material where measurements are made in a plane containing the c-axis. Measuring B_z for a series of spectra, calculating g, and plotting against the angle between B and the c-axis can produce the data shown in Fig. 45.5. The filled squares represent the g-values of the Mg-related acceptor signal in GaN and the solid line is a fit to the equation

$$g^2 = g_\parallel^2 \cos^2 \theta + g_\perp^2 \sin^2 \theta \quad (45.5)$$

where θ is the angle between the magnetic field and the principle axis of the defect. Note that (45.5) is a special case of (45.4) where $g_X = g_Y = g_\perp$ and $g_Z = g_\parallel$.

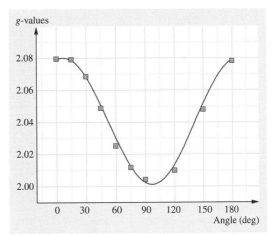

Fig. 45.5 Plot of the g-values calculated from EPR spectra obtained with the magnetic field in the plane of the c-axis and measured at selected angles with respect to the c-axis of a GaN film. The data (*filled squares*) were obtained from the Mg-related acceptor signal at 4 K. The fit to (45.5) is illustrated by the *solid line*

In the case of the Mg acceptor, θ_i is the same as the laboratory angle measured between B and the c-axis because the principle axis turns out to be the c-axis. A fit of (45.5) to the data in Fig. 45.5 shows that $g_\parallel = 2.096$ and $g_\perp = 2.008$. Although the specific values of $g_\parallel$ and $g_\perp$ provide some information, the main conclusion obtained from these data is the determination of the symmetry of the center. A defect exhibiting the angular dependence of (45.5) is said to have axial symmetry. In this case, the magnesium-related acceptor has axial symmetry about the c-axis. Confirmation of the axial symmetry requires rotation about two additional crystal axes. Specifically, rotation about the axis of symmetry, c-axis, should reveal spectra that do not depend on rotation angle. Unfortunately, this has not yet been possible for GaN because most samples are films grown in a predetermined orientation. Figure 45.6 shows measurements of the g-shift for the boron acceptor in bulk 6H-SiC, where results obtained from three rotation planes are plotted [45.10]. The Greek letters η, ξ, and ζ designate the high-symmetry cubic directions $[1\bar{1}0]$, $[112]$, and $[\bar{1}\bar{1}0]$, respectively. The sets of lines in each orientation plane represent data from the three different symmetry sites in 6H-SiC and the four possible bonding directions. Note that in the last panel, when the sample is rotated about the c-axis, one set of data forms a horizontal line. These results were obtained from the

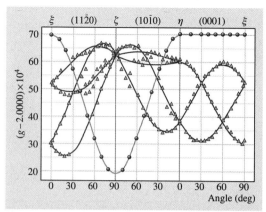

Fig. 45.6 The shift of the g-value from the free electron value when an EPR spectrum is measured in the rotation planes indicated above each panel in the graph: (11$\bar{2}$0), (10$\bar{1}$0), and (0001). The *vertical lines* indicate the crystal directions in the cubic system: η, [1$\bar{1}$0]; ξ, [112]; ζ, [$\bar{1}\bar{1}$0]. Data were obtained from the shallow boron acceptor in 6H-SiC at 4 K (after [45.10], © IOP 1998)

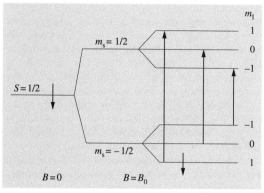

Fig. 45.7 Energy-level diagram from a spin-$\frac{1}{2}$ electron: in zero magnetic field (*leftmost*); applied field in absence of any other magnetic fields (*middle*); in the presence of a nuclear magnetic field (*rightmost*). m_I is the magnetic spin quantum number for the nucleus. The *downward arrow* represents the electron spin; the *upward arrows* represent allowed transitions

hexagonal site where the defect axis is oriented along the c-axis of the crystal. When the data for this site is followed into the other two panels, the angular pattern illustrated in Fig. 45.5 is revealed, as expected for a site with axial symmetry about the c-axis.

At this point, it is not at all clear how the chemical identity of a defect is determined. One might guess that there are calculations relating the **g**-tensor to specific types of point defects. Unfortunately, the **g**-tensor is difficult to calculate accurately because the wavefunction must be known over the entire crystal space, a situation that is difficult to achieve using even the most powerful computers. However, the story does not end with the Zeeman term. The two remaining terms to be discussed in this chapter, the nuclear hyperfine and the fine-structure terms, provide a great deal more information about the chemical, structural, and electronic state of the defect.

45.2.2 Nuclear Hyperfine Interaction

When nuclei of nonzero spin are sufficiently close to a paramagnetic defect, an additional magnetic field must be considered, that of the nuclear magnetic moment. The term is written as

$$H = \sum_j S \cdot A \cdot I_j , \qquad (45.6)$$

where **A** is the hyperfine tensor and **I** is the spin of the j-th nucleus surrounding the defect. This local nuclear magnetic field splits the electronic levels shown in Fig. 45.1, so that the situation becomes that shown in Fig. 45.7. Here m_I is the nuclear magnetic quantum number analogous to m_s. The size of the separation between the m_I levels A depends on the strength of the interaction. The nuclear hyperfine interaction may be different for the same defect in different materials as well as for different centers in the same material. However, I is a property of a given nucleus, so that it can be used to distinguish different defects, practically independent of the host. Specifically, the number of nuclear hyperfine lines originating from transitions between the different levels in Fig. 45.7 is proportional to I, thereby giving us the first clue into the chemical identity of the center. Since the selection rule allows only transitions with $\Delta m_s = \pm 1$ and $\Delta m_I = 0$ the Zeeman EPR line will divide into $2I + 1$ lines as shown by the upward arrows in Fig. 45.7. All the hyperfine lines will have the same intensity, and they will be spaced equally from the Zeeman EPR line. Because the intensity of a transition depends on the total number of defects causing the absorption, the isotopic abundance of the nonzero spin nuclei and the number of like nuclei determine the ratio of hyperfine line intensity to the intensity of the total spectrum. (Intensity here refers to the total integrated intensity of the hyperfine lines and the spectrum.)

Two examples are discussed to provide an understanding of the nuclear hyperfine portion of EPR spectra: the shallow nitrogen donor in SiC and the positively charged carbon vacancy in SiC. The former illustrates a 100% abundant nuclear spin entity and the latter is a case where the nuclear spin is much less than 100%. Although both examples are defects in SiC, the situations are quite general. The only effect of the environment is the strength and symmetry of the interaction, neither of which is critical to the basic understanding of the hyperfine term.

Figure 45.8 shows the characteristic EPR fingerprint of isolated N atoms in SiC: three evenly spaced lines of equal intensity. (Nitrogen has nuclear spin 1 and is 100% abundant.) The vertical arrows point to the three hyperfine lines arising from the interaction between the magnetic field and the nitrogen nuclei situated at the cubic sites in 4H-SiC. The large line marked with an arrow is a distorted spectrum of nitrogen on the hexagonal site in SiC. This site exhibits the three hyperfine lines as shown for the cubic site when different EPR parameters are used. Semiclassically, one could picture the origin of the three lines as follows: the Zeeman energy at which the paramagnetic electron absorbs the incident radiation E_z is shifted by the interaction between the hyperfine interaction between the electron and magnetic nucleus. Since almost all nitrogen atoms have a nuclear spin of 1, the Zeeman energy at each nitrogen has equal probability of being lowered ($m_I = -1$), increased ($m_I = +1$) or unaffected ($m_I = 0$), depending on the relative orientations of the electron and nuclear magnetic dipole moments. Thus, three lines of equal intensity are produced: one representing $E_z - E_{hf}$, one $E_z + E_{hf}$, and one E_z, where E_{hf} is the hyperfine interaction energy. The magnetic field separation between the lines is proportional to E_{hf}. The hyperfine parameter A is proportional to this energy, and is often quoted in terms of magnetic field units T through the conversion A/gB, where g is the g-value for the spectrum and B_z is the magnetic field separation between the adjacent lines. The g-value is obtained from the average B_z of the outer two lines, or in this case, B_z of the central line. Exhaustive analysis of the nitrogen spectrum may be found in numerous papers [45.11, 12].

From the above, it should be apparent that observation of the nuclear hyperfine lines is the key to determining the chemical origin of an EPR center. Not all impurities have isotopes with 100% abundant nonzero nuclear spin. For example, only 4.5% of all Si atoms (^{29}Si) and a mere 1.1% of carbon atoms (^{13}C) are spin $\frac{1}{2}$. Therefore, any intrinsic defect in SiC has a very low probability of being situated near a nucleus of nonzero spin. In these cases, the spectrum consists of a strong Zeeman line due to defects involving spin-zero Si and C nuclei, and pairs of equally spaced smaller lines due to the very few defects involving spin-$\frac{1}{2}$ Si and C nuclei. Figure 45.9 shows an EPR measurement of V_c^+. The satellite lines A and B arise from the spin-$\frac{1}{2}$ nearest-neighbor Si nuclei. The ratio of the relative integrated intensity of set A to the intensity of the entire spectrum is about 5% and that for set B is 15%. The outer set is attributed to those centers for which the spin-$\frac{1}{2}$ nucleus is located along the c-axis; the inner set to those in which any one of the three remaining Si neighbors is spin $\frac{1}{2}$. The darkest circle in the sketch in the upper right corner of the figure represents the single *unique* axial silicon (A lines), while the three lighter circles represent the other nearest neighbors (B lines). In summary, detection of two sets of hyperfine lines with intensity equivalent to interactions with four Si atoms distinguished by two different energies paints a picture of a carbon vacancy slightly distorted along the c-axis. The positive charge states is determined primarily from theoretical calculations as is discussed later. A complete analysis of this center along with that of the V_c^+ located on the other symmetry site of 4H-SiC is discussed in [45.13].

Hyperfine lines of low-atomic-abundance nuclear spins are often difficult to detect, particularly if the number of defects is also small. This is often the case for Si- and C-based semiconductors, where defect densities are below 10^{16} cm^{-3} and the abundance of the nonzero

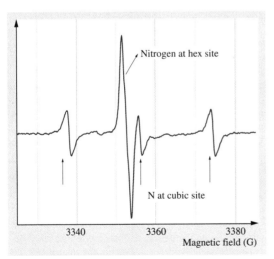

Fig. 45.8 EPR spectrum of nitrogen in 6H-SiC obtained at 30 K with the magnetic field parallel to the c-axis

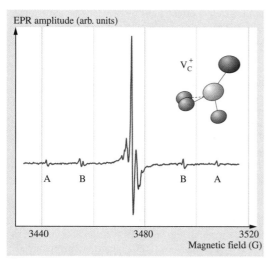

Fig. 45.9 EPR spectrum of the positively charged carbon vacancy in 4H-SiC obtained at 4 K with the magnetic field parallel to the c-axis. The *lines* marked A and B are the hyperfine lines due to neighboring Si atoms. The *model* in the *upper-right corner* shows the four Si neighbors of the carbon vacancy, where the *brown circles* produce the B *lines* and the *dark gray circle* produces the A *lines*. The *light gray circle* represents the carbon vacancy

spin isotopes is less than a few percent. In these cases, isotopic enrichment allows for enhanced hyperfine detection. For instance, recently isotopic enrichment was used to enhance the hyperfine of Se impurities in Si and intrinsic defects in SiGe alloys [45.14, 15].

The angular dependence of the hyperfine interaction allows one to map out the local spin density, or effectively determine the wavefunction of the paramagnetic electron. To understand this one has to appreciate that the type of spatial wavefunction will determine the relative directions of the electron–nuclear dipole coupling. Any orbital with $L > 0$ will be directionally dependent and therefore produce an angular-dependent hyperfine interaction. For example, a p_z-shell electron will have lobes along a specific axis, so that, if one measured an EPR signal along this axis, the hyperfine separation would be different from that measured along any other direction. Perhaps the more illuminating example is to consider $L = 0$, or an s-shell electron. Here, no angular dependence would be expected and the hyperfine parameter A should be isotropic, as is the case of phosphorus in Si. In many materials, a paramagnetic electron is a hybrid orbital containing contributions from s-, p-, and sometimes higher L-states. The separation between the hyperfine lines of nitrogen in Fig. 45.8 exhibit a small angular dependence. Contributions from s-like and p-like orbitals are deconvolved so that the percentage of the wavefunction that is s-like and p-like are determined. For the nearly isotropic nitrogen donor spectrum in 4H-SiC on the cubic site, the wavefunction at the impurity is found to be almost entirely s-like with less than 1% p-character [45.12]. The calculation of the amount of s- and p-character from angular-dependent hyperfine is straightforward and may be found in many texts [45.1]. Although not directly applicable to routine materials characterization, this type of information is extremely helpful to theorists calculating the strength of the hyperfine interaction because it provides a realistic starting function for determination of the hyperfine energies. It is these energies that refine the picture of the defect, particularly those involving nuclear spins common to many elements.

The nearest neighbors may not be the only nuclei contributing to the EPR spectra of a specific defect. When more distance neighbors are sensed, they often show up as sets of satellite lines more closely spaced than those of the nearest neighbors. For example, the pair of lines adjacent to the central line of the V_c^+ spectrum in Fig. 45.9 represents contributions from next-nearest-neighbor carbon atoms. Their intensity reflects the fact that any one of the 12 next nearest neighbors may be a ^{13}C nucleus. It should be pointed out that the analysis can eventually provide the probability that the paramagnetic electron resides on any one of the neighbors, effectively mapping out the spin density in the vicinity of the defect.

Theoretical Calculations of Hyperfine

As with any spectroscopic technique the experimental results may be compared with theory to extract additional information about the defect. In some cases, comparison with theory is the only means to interpret the EPR data in terms of a specific defect because the nuclear hyperfine is not detectable or is ambiguous. Luckily, unlike the **g**-tensor, the **A**-tensor is sensitive to at most the second or third nearest neighbors so that accurate calculations are feasible. Hyperfine calculations are particularly powerful tools to determine the defect structure because the strength of the nuclear spin–electron spin interaction is sensitive to the orientation and charge states of the environment. The entire **A**-tensor for different types of defects may often be predicted from density functional theory and the local spin-density approximation. The results produce an enormous amount of points that can be compared with experimental data, thus reinforcing the interpretation of the data in terms a specific defect structure. For example, the **A**-tensor for the positively charged carbon vacancy discussed above was calculated for the defect located at different symmetry sites of the two different polytypes of SiC (4H and 6H), as well as for different charge states [45.16]. The **A**-tensor was also extracted from the complete angular dependence of the EPR spectrum. Comparison of the two results showed that centers known as EI5 and EI6 are V_c^+ located at the hexagonal and cubic sites, respectively [45.13]. Similar comparison between theory and experiment revealed the spectra for HEI1 to be due to the negatively charged carbon vacancy [45.17]. Of course, not all defects in all materials are amenable to reasonable calculation. A large impurity atom and a low degree of symmetry can overwhelm the computational power of even the most modern computers. Nevertheless, comparison of EPR spectra to theoretical calculations of the nuclear hyperfine tensor has enabled the description of countless defects in innumerable types of materials.

To summarize, identification of nuclear hyperfine lines in EPR spectra is critical to determining the chemical origin of the center. The lines have the following characteristics:

1. Nearly equal intensity
2. Nearly equal separation and/or separated equally from the Zeeman line
3. The number of lines is $2I + 1$.

Also, all the lines should exhibit the same dependence on microwave power because they represent the same physical entity. Once the nuclear spin (I) is determined from the spectra, the possible types of nuclei contributing to the spectrum may be determined. Knowledge of the material composition and growth conditions often refines the type of nuclei expected to be involved in the defect. The angular dependence provides the basic components of the wavefunction, which can be used in theoretical calculations to determine the strength of the hyperfine interaction for different defect structures and charge states. Finally, comparison of the theoretically calculated **A**-tensor with the angular-dependent experimental data provides a reasonably definitive picture of the defect.

This section has emphasized the importance of the nuclear spin in detection of defects by magnetic resonance. However, too much of a good thing can create problems. In particular, difficulties arise when the crystal host is composed of atoms with 100% abundant nuclear spin. Depending on the strength of the interaction of the defect with host, the presence of many different sources of hyperfine interaction can lead to a series of barely resolved lines or even produce one broad EPR signal with all the powerful hyperfine information buried in its breadth. This is thought to be the cause of the limited information extracted from spectra in GaN because Ga has two isotopes, both with nuclear spin $\frac{3}{2}$, and nitrogen has one nearly 100% $I = 1$ isotope. The single broad line assigned to the Mg-related acceptor may be affected by unresolved hyperfine. The reader should be cautioned that the phrase *depending on the interaction with the host* is critical here. For instance the characteristic line pattern for Fe^{3+} and Mn^{2+}, transition metals that typically interact minimally with the host, are easily observed in GaN crystals [45.18, 19].

45.2.3 Interactions Involving More than One Electron

The above discussion suggests that theoretical calculations, which can predict the hyperfine interaction energy, can also be used to distinguish between the different charge states of a defect. However, in some cases, the charge state can be inferred from the experimental spectrum itself. This occurs if the number of electrons at a defect couple to a total spin greater than $\frac{1}{2}$. From the rules of adding spin angular momentum, it is known that n electrons can yield a total spin between 0 and $\frac{n}{2}$. In fact, Hund's rules tell us that the high spin is favored as the ground state; thus, all multiple electron defects should be paramagnetic. Of course Hund's rules do not strictly apply to a center surrounded by the many perturbing fields in a crystal lattice. Nevertheless, in some situations an EPR spectrum may best be described using a Hamiltonian of spin great than $\frac{1}{2}$. When this occurs, the term that must be included in the analysis is

$$H = \mathbf{S} \cdot \mathbf{D} \cdot \mathbf{S}, \quad (45.7)$$

where **D** is the fine-structure term, present only when $S > \frac{1}{2}$. Different physical situations can necessitate the use of this term including spin–orbit interaction and dipole–dipole coupling between different electrons. Here we will not be concerned with the origin of the term, but highlight two situations where the quadrupole term is used: an excited state of an $S = 0$ center and transition-metal impurities.

The EPR spectra discussed thus far represent the ground state of a defect. With the addition of optical illumination, one can populate the higher energy levels. This is particularly useful when the ground state of the center is an $S = 0$ EPR inactive state. Often, detection

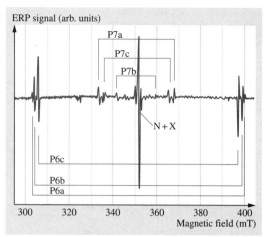

Fig. 45.10 EPR spectrum of P6 and P7 centers obtained from neutron-irradiated 6H-SiC with the magnetic field oriented parallel to the c-axis. The *sharp lines* in the center are due to nitrogen and an unidentified defect (after [45.20], © APS 2001)

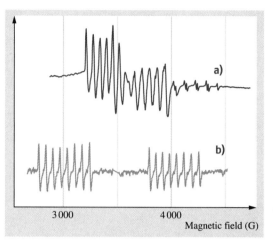

Fig. 45.11a,b EPR spectra of vanadium in 4H-SiC: (a) V^{4+} measured at 4 K with the magnetic field 10° from c-axis; (b) V^{3+} measured at 30 K with the magnetic field perpendicular to the c-axis

of an excited state may be verified by the fact that, after removal of the light source, the spectrum immediately returns to its pre-illumination condition due to the inherently short lifetime of the excited state. The P6 and P7 centers in n-type 6H-SiC irradiated with neutrons provide an example [45.20]. Figure 45.10 shows the EPR spectrum obtained from neutron-irradiated 6H-SiC during illumination with white light from a halogen lamp. The two sets of paired lines highlighted by brackets represent two similar defects (P6 and P7) on the three different symmetry sites of 6H-SiC (a, b, c). Pertinent to this discussion is that each line of the pairs reflects EPR transitions between spin states $m_s = -1$ to $m_s = 0$ and $m_s = 0$ to $m_s = 1$ of the $S = 1$ center. These lines are often referred to as fine structure. P6 and P7 are thought to represent different orientations of the same defect, although the exact model for the defects remains under debate [45.20, 21]. Note that recent data indicate that the centers are observed in heavily N-doped material without illumination, implying that $S = 1$ is the ground state. Whatever the case, the presence of the paired lines identifies the center as $S = 1$.

The transition element vanadium provides an example of how both the type of impurity and its charge state are determined directly from observation of the spectrum. Figure 45.11 shows two spectra obtained from 4H-SiC: a multiplet of nearly equal intensity lines adjacent to a second set of lines with much lower intensity (Fig. 45.11a) and a pair of octets (Fig. 45.11b). According to the theory presented in the last section, the set of eight equally spaced lines in Fig. 45.11b indicates a 100% abundant $I = \frac{7}{2}$ nucleus. Checking the tables and considering typical unintentional impurities in SiC, it is concluded that the spectrum (Fig. 45.11b) arises from a vanadium atom on the cubic site [45.22]. A similar pair of octets, reflecting the hexagonal site, is found beyond the magnetic field range shown. Unfortunately, the spectrum in Fig. 45.11a is more complicated; suffice it to say that studies have shown that this spectrum also arises from a vanadium impurity, where the multiplet with high intensity arises from the cubic site and the set of low-intensity lines originate from the hexagonal site [45.23]. The presence of a single set of lines for each symmetry site in spectrum Fig. 45.11a and double set in Fig. 45.11b suggests that the former is an $S = \frac{1}{2}$ center while the latter is $S = 1$. V^{4+} in SiC has one unpaired electron, which would produce spin $\frac{1}{2}$, and V^{3+} has two, which could couple to spin 1. Thus, by simply examining the EPR line pattern, the impurity and its charge states are immediately determined.

How a high-spin defect produces a set of EPR lines at different magnetic fields is not immediately obvious from anything discussed thus far. Simply redrawing the $s = \frac{1}{2}$ energy diagram of Fig. 45.1 for $s = 1$ produces the levels shown in Fig. 45.12a, where the horizontal axis now represents a varying magnetic field and the verti-

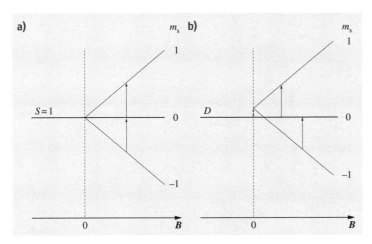

Fig. 45.12a,b Energy levels of a spin-1 electron system in a varying magnetic field: (a) in the absence of any zero-field splitting; (b) in the presence of a perturbing field causing a splitting of the spin states in zero magnetic field. The *vertical arrows* represent the fixed quanta of microwave energy available for the transitions

cal axis is energy. The equal sized arrows represent the fixed quantum of microwave energy available for the transitions. It is apparent that the transitions between the different m_s states occur at the same magnetic field. What then, produces the separated pattern of lines observed for high-spin defects? Basically, any perturbing field that removes the degeneracy of the $m_s = \pm 1$ and $m_s = 0$ states at zero magnetic field will yield noncoincident resonance absorptions. Figure 45.12b illustrates the resulting energy levels where the separation D on the vertical axis is referred to generally as the zero-field splitting. Once the states are separated at zero magnetic field, the $m_s = -1$ to $m_s = 0$ and $m_s = 0$ to $m_s = 1$ transitions no longer occur at the same magnetic field; thus, separate EPR lines will appear at each transition. In general $2S$ EPR resonances will occur. While several types of interactions can separate the energy of the degenerate spin states, the most common is the anisotropic magnetic dipole–dipole interaction. The calculation required to demonstrate the effect of the zero-field splitting is straightforward and is shown in many texts [45.1, 8].

The fine-structure lines produced by high-spin centers have different characteristics than those of hyperfine lines discussed at the end of the previous section. For example, the intensity of each EPR line is not the same. For any $S > 1$ center, the intensities of resonance lines from the various m_s transitions exhibit different, but predictable, variations. Figure 45.13a shows this for Fe^{3+} in $SrTiO_3$ measured with the magnetic field oriented along the c-axis of the sample. The five lines highlighted arise from the five transitions of the $S = \frac{5}{2}$ center. The remaining line represents Cr^{3+}. The relative integrated intensities of the Fe^{3+} lines, $5:8:9:8:5$, follow from the transition-matrix element between the $\frac{5}{2}$ to $\frac{3}{2}$, $\frac{3}{2}$ to $\frac{1}{2}$, $\frac{1}{2}$ to $-\frac{1}{2}$, $-\frac{1}{2}$ to $-\frac{3}{2}$, and $-\frac{3}{2}$ to $-\frac{5}{2}$ spin states

$$g\mu_B B_x \left\langle \frac{5}{2}, m_s^i \left| S_x \right| \frac{5}{2}, m_s^f \right\rangle , \quad (45.8)$$

where B_x is the microwave magnetic field perpendicular to the applied field, m_s^i and m_s^f represents the magnetic spin quantum number for the initial and final states,

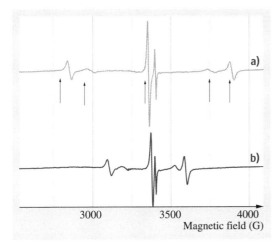

Fig. 45.13a,b EPR spectrum of Fe^{3+} in $SrTiO_3$ obtained at room temperature. The *arrows* point to the spin transitions that arise from Fe^{3+}; the remaining EPR line is due to Cr^{3+}. Data were obtained with the magnetic field (a) at $0°$ and (b) at $30°$ with respect to the (100) direction

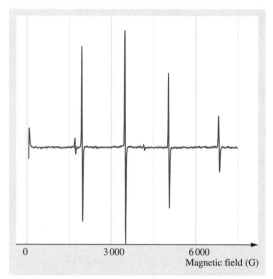

Fig. 45.14 EPR spectrum of Fe^{3+} in a GaN crystal obtained at 4 K

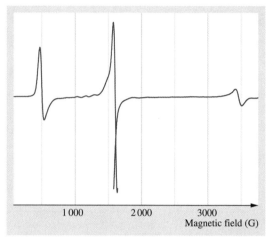

Fig. 45.15 EPR spectrum of Fe^{3+} in a LiNbO$_3$ crystal obtained perpendicular to the c-axis at room temperature

respectively, and S_x is the spin angular momentum operator that ultimately raises the electron from the lower state to the upper state. See [45.1, appendix C] for details. This intensity pattern would apply to any $S = \frac{5}{2}$ center with the simple $|s, m_s\rangle$ wavefunctions. Similar patterns may be predicted for other $S > \frac{1}{2}$ centers. For example, Fig. 45.14 shows that the line pattern for the $S = \frac{5}{2}$ impurity Fe^{3+} in GaN also consists of five lines of unequal intensity. The sharpness of the lines in GaN compared with those in SrTiO$_3$ most likely reflects a lower density of extended defects or less strain. The line width also explains why the relative amplitudes are different in the two samples; however, it is not obvious why certain transitions should be broadened more than others.

Not all situations involving high-spin defects are as straightforward as the two mentioned here. Sometimes the crystal field, the electric field generated by the ions or ligands surrounding the defect, dominant spin–spin and spin–orbit interactions ultimately producing wavefunctions which are linear combination of the simple spin states, $|s, m_s\rangle$. Crystal-field effects are exhaustively discussed in many texts [45.24]. For EPR, the only point is that the ground-state orbital wavefunction is determined by the field strength, thereby establishing the type of spin wavefunction appropriate for a particular defect. When the crystal field is much greater than other interactions, an EPR spectrum greatly different from the two iron spectra in Figs. 45.13 and 45.14 is produced. Figure 45.15, showing Fe^{3+} in LiNbO$_3$, illustrates the point. Here, the trigonal crystal-field effect is of the same order of magnitude as the Zeeman effect, yielding an EPR signal significantly different from the five-line pattern seen previously. Additional interaction terms must be added to the Hamiltonian in order to unravel the meaning of the spectrum [45.25, 26].

Like the Zeeman and hyperfine terms, the fine-structure term can produce angular dependence. In some cases, a simple shifting of lines occurs as seen in Fig. 45.13b. However, often spectra dominated by D exhibit angular-dependent intensities. This occurs when the appropriate wavefunctions are linear combinations of the $|s, m_s\rangle$ states. Because the crystal field is directional, different wavefunctions are produced at different orientations of the sample with respect to the applied B field. The angular-dependent wavefunctions then lead to angular-dependent transition-matrix elements (45.8) and, therefore, varying spectral intensities. Note that high magnetic fields and high frequencies may be employed to avoid complications due to the large crystal field. However, using microwave frequencies larger than 10 GHz is tedious, involving different types of waveguides, cavities, and microwave bridges than those used for the lower frequencies. Often, it is easier to deconvolve the complicated experimental data using straightforward calculations than to execute an EPR measurement at sufficiently high frequency as to avoid the crystal-field effects.

To summarize, we have considered three energy contributions to a point defect in the presence of an applied magnetic field: the electronic Zeeman term (45.2), the nuclear hyperfine interaction (45.6), and the fine-structure term (45.7). In general, all three effects may be present, so that the appropriate Hamiltonian to begin analysis of a spectrum is

$$H = \mu_b S \cdot g \cdot B + \sum_j S \cdot A \cdot I + S \cdot D \cdot S. \quad (45.9)$$

For defects with $S > 1$, additional terms may be added depending on the nature of the defect and its surroundings. Because many of the defects typically encountered involve only two or more of the terms above, no discussion of the additional terms is presented here. Equation (45.9) is not the most *user-friendly* equation ever presented to the average reader. For those with peripheral interest in EPR, understanding how the number and separation of EPR lines are used to determine defect structure should be sufficient to appreciate the power of the technique. More interested readers will find the quantum-mechanical calculation techniques required for complete analysis in the many references referred to throughout this section, specifically [45.1, 8].

45.2.4 Total Number of Spins

In addition to a physical description, EPR data may provide the total number of centers of a specific defect. This is accomplished by comparing the spectrum of an unknown quantity of a defect with that obtained from a known quantity. Significantly, the EPR signal from the standard need not arise from the same defect or even a different defect in the same material. This convenience is afforded by the fact that the spin-flip probability is usually independent of the local environment. Common calibration standards include 2,2-diphenyl-1-picrylhydrazyl (DPPH) and the phosphorus signal in powdered, heavily doped, n-type Si. In principle, the comparison can lead to an absolute number of spins with an accuracy of 50% with sensitivity as low as 10^{10} spins [45.1]. However, the reader should be warned that the sensitivity depends strongly on line width and number of lines. The number quoted is based on a single resonance with 1 G line width. For a signal of 10 G line width composed of five lines, the minimum detectable spins increases by a factor of at least 50. Other factors such as temperature and microwave saturation also limit sensitivity. Such issues are discussed in [45.1, appendix E].

A few items to remember regarding the absolute spin measurement are:

1. Only the paramagnetic state of any defect is being measured. For instance, the number of acceptor impurities may be calculated. However, the result is limited to acceptors that have captured the hole in the valence band. For this reason, EPR of acceptors (and donors) is often performed at low temperature to more closely reflect the total number. Compensation may further reduce the number of EPR-active acceptor sites. The amount of the impurity calculated from the acceptor signal will not include aggregates, complexes or other forms of the impurity; however, different EPR signals may be related to these entities and measured separately.
2. Only the total number of centers is determined. Additional experiments are needed to find the spatial distribution of defects. When concentration is given in an EPR study, the centers are assumed to be uniformly distributed throughout the material unless otherwise stated.
3. Defects arising from complex wavefunctions, such as high-spin centers in a strong crystal field, will exhibit intensities that are dependent on orientation. In these cases, one must first determine the various transition rates before the number of defects may be calculated.

Traditionally, the absolute number of spins is not the focus of an EPR study. This is partially because of the many caveats discussed above, as well as the difficulty of generating an accurate standard. One powerful aspect of the spectroscopy that is commonly employed, however, is determination of the variation in number of defects by measuring the relative amplitude of EPR signals. For a single EPR resonance that does not change shape during the course of a study, the amplitudes of a signal may be used to indicate varying defect densities. This is the approach used in many of the experiments discussed below. For the types of centers mentioned in item 3, however, careful alignment of the samples is required between measurements so that the amplitude changes truly reflect the number of spins and not angular-dependent transition probabilities.

45.3 Scope of EPR Technique

Having presented the basis for interpreting EPR spectra, without question the most challenging feature of EPR for most readers, the remainder of the chapter focuses on the power of EPR in terms of the types of defects detectable and typical correlation studies. All of the examples in the section above are simple point defects, either an intrinsic defect or single-atom impurity. Many other forms of defects are detected including substitutional, interstitial, antisite, vacancies, vacancy pairs, antisite–vacancy pairs, and impurity–vacancy pairs. The only requirement is that the defect be paramagnetic in the as-grown material or be able to be made paramagnetic with an external perturbation.

EPR was first used on bulk crystals, and indeed this is where the full power of the technique is realized. However, with the ongoing push towards miniaturization and increasing desire for a *chip-based* world, films less than one micrometer thick and particles with less than 100 nm diameter are typically encountered. While this has been the case in electronics for more than a generation, miniaturization of optical and even microwave devices is increasingly popular. Indeed, one of the most recent initiatives involves growing films using crystals with well-known microwave or magnetic properties. Ultimately, these will be deposited onto a full wafers with future integration into Si electronics as the goal.

The utility of applying EPR to films is not obvious because, although the technique is sensitive to as few as 10^{11} centers, for traditional defect analysis these defects must be isolated. A simple calculation shows that a micrometer-thick film with 10^{11} centers uniformly distributed yields 10^{15} cm^{-3} defects. While this is not an unrealistic number for the types of films of interest today, the calculation represents the most hopeful situation: the minimum spin detection limit and the thickest films of interest. Nevertheless, the example does show that studying films is not out of the question. Indeed many successful experiments are reported in the literature. Most take the advantage of stacking many film–substrate samples so that the signal intensity may be maximized. Careful alignment is required in these cases so that the crystallinity of the samples is not compromised. The examples below illustrate several different types of film–substrate studies:

1. Intrinsic defect in the *bulk* of a micrometer-thick film
2. Defect at a crystalline substrate–amorphous film interface
3. Near-surface impurities on a polycrystalline film.

45.3.1 Defects in a Thin Film on a Substrate

The first situation addressed is the study of a simple point defect in a film, a donor in GaN. The only difference between the film and bulk experiments in this case is the preparatory steps for the measurement. Usually, several film–substrate samples are stacked together to increase the total amount of GaN being studied. Furthermore, GaN is typically grown on sapphire or SiC. In either case, the substrate must be carefully studied to distinguish the substrate EPR signals from those of the film. The microwave absorption utilized in EPR detection completely penetrates most semiconductors and insulators, so that the technique senses the substrate and film equally. Luckily, the well-known EPR signatures of defects in sapphire are highly anisotropic, so that their contribution to the spectrum may often be minimized by prudent orientation of the sample with respect to the magnetic field. The only defect to be avoided in n-type SiC substrates is the nitrogen donor, which is easily resolved from the donor in GaN.

Most of the information about the donor EPR signal in GaN is contained within a work by *Carlos* and coworkers [45.27]. No hyperfine could be detected in the spectrum, leaving the chemical origin of the center uncertain, but several EPR characteristics suggest that the resonance represents an electron in a donor band. Work in our laboratory shows that the spectrum is found only in n-type samples, and that the signal intensity increases with increasing donor density for $n = 1 \times 10^{14} - 1 \times 10^{17}$ cm^{-3} [45.28]. Carlos and coworkers concentrated on the spectroscopic characteristics of the EPR signal to demonstrate the donor assignment. Their measurements suggest that the *g*-value is typical of a donor electron and the angular dependence reflects that of the hexagonal lattice. Furthermore, the line width indicates that the paramagnetic electron is not *attached* to the donor atom, and the temperature dependence of the line width, shown in Fig. 45.16, eliminates a conduction-band electron. The different symbols represent data obtained from various thickness samples as indicated on the figure. The main point here is that all sets of data exhibit a decreasing line width until 20 K, followed by an increasing line width. The former region is

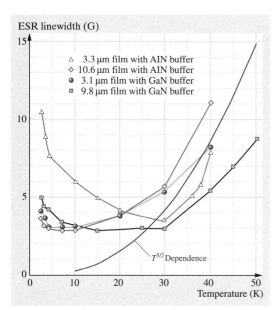

Fig. 45.16 Line width of the donor signal in GaN measured at selected temperatures. The *smooth brown line* is a $T^{5/2}$ fit to all of the data; *other lines* merely connect data points (after [45.27], © APS 1993)

thought to be due to motional effects and the latter due to coupling to acoustic phonons, neither of which should occur if conduction-band electrons were involved.

Although there are several other EPR studies of point defects in crystalline thin films, the small signal size inherent to the low sample volume limits the number of intensive investigations. Homoepitaxial films are highly unlikely to produce meaningful results about the film because the layer of interest cannot be separated spectroscopically from the bulk substrate. Consequently, many film studies were performed on amorphous material, the most common of which was SiO_2 films on Si substrates. Although many studies focused on oxide films irradiated by γ-rays or higher energy, several studies addressed the intrinsic defects in the oxide films [45.29–35]. In some cases, the defects, specifically an oxygen vacancy known as an E′ center, was successfully correlated with many of the electrical trapping effects in metal–oxide–semiconductor field-effect transistor (MOSFET) devices [45.34, 35]. Other studies of film–substrate systems in which the point defect resides in the *bulk* of the film include diamond and MgO [45.36–38].

45.3.2 Defects at an Interface

The second type of center addressed in this section is one located at an interface, specifically the Si–SiO_2 interface. In Si devices, paramagnetic defects at the Si–oxide interface are known to be directly related to electrically active trapping sites that alter device performance [45.39, 40]. For this reason, much EPR work has concentrated on a dangling bond defect located at the semiconductor–oxide interface [45.39–44]. Several different types are found in pure silicon-based interfaces, each involving an unpaired electron on a Si atom located on the semiconductor side of the interface [45.40, 42]. Centers with a Ge dangling bond and C dangling bond are seen in SiGe–oxide and SiC–oxide interfaces, respectively [45.41, 43, 44]. The dangling-bond-like defects are referred to collectively as P_b centers.

In general, EPR cannot selectively detect centers at surfaces and interfaces because the microwave radiation penetrates the entire semiconductor substrate. However, for a perfectly flat surface, the angular dependence of the EPR resonance may provide enough information to deduce the surface nature of the center, as was done for the Si P_b center located at the interface between a (111) Si substrate and amorphous SiO_2 layer [45.42]. To understand the difference between interfacial and bulk angular dependence one must reconsider the discussion of the **g**-tensor presented earlier. **g** is a tensor because the absorbed energy depends on the orientation of the applied magnetic field and a preferred direction of the dipole moment. However, in a crystal a specific defect may be located at one of several different symmetry-related sites. For example, the simple dangling bond in bulk Si may be directed in any one of four (111) bonding directions, all of which may make a different angle with respect to the applied field depending on the orientation of the sample (Fig. 45.17). Since the defect may be any of the four (111) bonding directions of tetrahedral Si, the EPR spectrum should reveal four lines when the magnetic field is oriented at a general angle with respect to the surface normal of the sample. For special orientations, such as B rotated in a (111) plane,

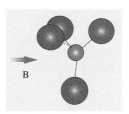

Fig. 45.17 Schematic model of a tetrahedrally coordinated atom (*small circle*) with the magnetic field directed at an arbitrary angle with respect to any one of the [111] directed bonds. The dangling bond could be any one of the four bonds with the *large circle* removed

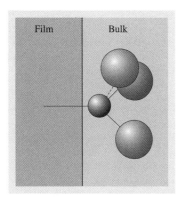

Fig. 45.18 Schematic model of a tetrahedrally coordinated atom at an interface, where one of the [111]-directed bonds points perpendicular to the interfacial plane. The *vertical solid line* marks the ideal interface plane between the bulk and the substrate

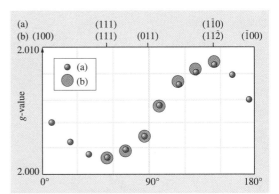

Fig. 45.19 g-values of the P_b center measured in oxidized Si with the magnetic field in the plane (a) (*small circles*) and plane (b) (*large circles*) (after [45.42], © AIP 1979)

the bond perpendicular to the plane should produce one isotropic line and the remaining three dangling bonds would generate one anisotropic line three times larger than the isotropic one.

Now perhaps the reader can see how an interface defect would be different. At an interface (or surface) not all of the four directions would be identical. The easiest case to imagine is the (111) surface, where one of the four bonds is perpendicular to the plane of the interface along a now unique (111) direction (Fig. 45.18). The other three possible bonds would be pointing into the *bulk* of the Si substrate, and if paramagnetic, would produce a different **g**-tensor than the interfacial defect. Since silicon is a perfected material, generally the back bonds are unbroken and do not contribute an EPR signal. Therefore, if the magnetic field were in the plane of the interface and perpendicular to the dangling bond, one would expect to see only a single isotropic EPR line due to the dangling bond at the interface. Unfortunately rotation in the plane of the interface is not realistic, but Poindexter and coworkers performed measurements with B in the $(11\bar{2})$ and $(1\bar{1}0)$ planes, revealing the expected angular dependence. Figure 45.19 shows the g-value versus angle with respect to the (100) direction for a P_b center in thermally oxidized (111) Si. The appearance of only a single g-value at each angle indicates that only one EPR line is observed at each angle. The coincidence of the unfilled and filled circles, which represent data obtained from two different planes of rotation, imply that the angular dependence of the spectrum is identical for the two different planes. Both observations are expected for a dangling bond between the (111) surface of Si and the overlying oxide film.

The relationship of the P_b center to Si was confirmed by observation of the nuclear hyperfine interaction. Detecting the hyperfine in this case is a heroic experiment considering the low atomic abundance of ^{29}Si and small total number of defects at an interface. Nevertheless, *Brower* stacked 35 oxidized Si wafers together to observe the hyperfine lines, confirming that the nucleus associated with the dangling bond is Si [45.45]

As one might expect, other types of interfaces may be studied with EPR. However, to the author's knowledge, the only reports in the literature focus on the semiconductor–oxide system, specifically SiC–SiO$_2$ and SiGe–SiO$_2$ [45.41–44]. In both cases oxidized porous material or oxygen-implanted substrates were used, and the interfacial nature was confirmed through etching studies. The planar interfaces necessary for the angular-dependent studies are not achievable at this time for these types of materials. In principle, many other types of interfaces could be examined. The limitation, however, is always preparing the samples in such a way as to maximize the amount of interface in the EPR cavity. Multilayer heterostructures should make ideal samples for study, but the author is not aware of any attempts to date.

45.3.3 Defects at Surfaces

A surface may be thought of as a special case of an interface, one in which one side of the interface is the ambient atmosphere. In principle, the Si P_b center should be observable on bare Si. And indeed, one *should* be able to detect the center if one could do an EPR measurement in vacuum. Si oxidizes readily in air at room temperature, thus an unoxidized surface is virtually unobtainable under the conditions required for an EPR study. Nevertheless, many surface defects are reported in the literature, but most relate to the roughened surface caused by cutting or polishing [45.46]. Such centers exhibit an isotropic g-value and are often passivated in the presence of hydrogen [45.47].

For single crystals, one may differentiate between bulk and surface defects in much the same way as was discussed for bulk and interface. However, for polycrystalline materials the situation is somewhat different, as is illustrated in the following example. Cr-doped polycrystalline $SrTiO_3$ films, 1700 nm and 350 nm thick, were grown on sapphire substrates [45.48]. The powder-pattern spectrum typical of a polycrystalline material was resolved into two separate spectra: one with a g-value of 1.977, typical of Cr^{3+} in $SrTiO_3$, and a second center with $g = 1.974$. The authors show that the latter is consistent with a Cr^{3+} impurity located near the surface. As discussed earlier, the symmetry of a surface defect is inevitably lower than that of a bulk defect and angular-dependent measurements may be used to exploit the difference. However, the random nature of the polycrystallites requires a different data analysis than that of a pure crystal. Calculations by *Deigen* and *Glinchuck* show that the g-value for a surface defect should be shifted from that found in the bulk by an amount proportional to the angle between the applied field and the surface normal [45.49]. In the Cr-doped films, the EPR signal represented by $g = 1.974$ shifts as the sample is rotated in the magnetic field. The angular dependence of this portion of the signal agrees with the theory predicted for a surface center. Also, the authors point out that the ratio of the 1.974 signal intensity to that for the 1.977 signal is larger in the 350 nm films than in the 1700 nm ones, as expected for a surface center in a thinner sample.

The trapping of impurities within the *bulk* of sub-μm-sized particle is a well-known difficulty, and several studies have employed EPR to distinguish *bulk* and surface impurities in these nanoparticles. The Mn^{2+} impurity in ZnS provides just one example [45.50]. EPR of nanoparticles have also been used to address the relationship between the defects and ferromagnetic or ferroelectric behavior [45.51]. Si surface centers, not surprisingly, are often addressed in EPR studies of Si nanodimensional materials [45.52, 53].

45.3.4 Nondilute Systems

This chapter, as well as much of the EPR literature, focuses on low concentrations of isolated point defects separated by at least several atomic units. The interpretation of spectra requires a significantly different approach when the defect–defect distance gets smaller. More specifically, new terms such as the exchange interaction enter the Hamiltonian if nearby spins on separate defects begin to interact. *Ferher* et al. demonstrated this for P-doped Si [45.54] using samples with two different donor concentrations, 1×10^{17} and 4×10^{17} cm^{-3}. The spectra reveal several pairs of EPR lines between the hyperfine lines from the isolated phosphorus atoms. Feher demonstrated that the number of pairs increased as the concentration of phosphorus increased, suggesting that at sufficiently high density the spacing between some of the dopants is suitable for electron–electron exchange. The work was extended by *Maekawa* and *Kinoshita* studying Si doped with 10^{16}–10^{19} cm^{-3} phosphorus atoms [45.55]. Temperature-dependent measurements confirmed the role of the exchange interaction suggested by Feher for the most lightly doped samples and revealed the presence of electron-hopping and impurity-band conduction at the highest temperatures.

Bencini and *Gatteschi* discuss the role of the exchange interaction in EPR spectra for a variety of different circumstances ranging from transition-metal dopants to protein-based systems [45.56]. All cases including exchange and superexchange are discussed. The emphasis of these types of studies is distinctively different from that discussed above for the nondilute systems. For example, spin–spin correlation as well as spin–spin and spin–lattice relaxation times, are emphasized. Nuclear hyperfine may often be lost in the typically large line widths of exchange-dominated systems. Thus, the interaction of the defects with the local environment are the focus of the study, rather than the detailed atomic structure of a specific defect. Finally, the case of a nondilute system of a ferromagnetic material should be pointed out, in which the collection of spins creates a magnetic field without the application of an applied field. The magnetic resonance of such a system, referred to as ferromagnetic resonance, requires entirely different analysis from the paramagnetic resonance discussed in this chapter. The reader should consult [45.57,58] for information on ferromagnetic resonance.

Much of the literature of nondilute systems is concerned with one- and two-dimensional systems of spins, reminiscent of dangling bonds at an extended defect. Unfortunately, sensitivity may present a limitation. In the case of strongly coupled spins, the minimum detectable number of spins would be severely crippled by the line width typical of dilute systems. Specifics are difficult to estimate, but most nondilute systems produce line widths hundreds of G wide, thereby decreasing EPR sensitivity from the typical 10^{11} spins to 10^{14} spins. Consider a material with 10^8 cm^{-2} line defects each 1 μm long, consisting of 1×10^3 defects. This yields 1×10^{11} defects in each cm^2 of material, an amount on the edge of detectability for even a 1 G line width. However, in emerging materials where the concentration of extended defects may greatly exceed 10^8 cm^2, EPR detection may not be unrealistic.

45.4 Supplementary Instrumentation and Supportive Techniques

As with any technique, correlation studies using data obtained from other techniques or the incorporation of additional instrumentation enhances the amount of information gained from the study. Many different types of techniques and layers of instrumentation have expanded the capabilities of EPR over the years. First, one example using additional instrumentation will be presented: photo-EPR. Then several correlation studies will be discussed, including electrical measurements and thermal annealing. Additional examples involving expanded instrumentation are briefly reviewed in the final section of the chapter.

In this section the term *defect level* is used extensively, so the reader must fully understand what it means. A defect level is similar to an ionization energy in that it represents the difference in energy between two charge states. In the case of ionization, however, the final state is represented by an electron infinitely far away from the ion. For a defect in a crystal, the electron is located in the conduction band after removal from the defect. The defect level is the energy necessary to remove an electron from a defect and place it in the conduction band. Similarly, a defect level may be viewed as the energy required to excite an electron from the valence band to the defect. The level is represented schematically as shown in Fig. 45.20. The lines labeled E_v and E_c are the valence- and conduction-band edges, respectively. The line labeled $X^{-/0}$ is the defect level, representing the energy required to change the charge state of defect X from negative to neutral. The inverse, the energy required to change defect X from neutral to negative, is the same level. Since a defect level is always quoted with respect to a band edge, it is typically written as $E_c - E_x$ or $E_v + E_x$. From the discussion, it should be clear that a defect level has meaning only with respect to specific charge states. That is, the statement *the defect level of boron in* SiC is ambiguous. One should say *the defect level of boron from the neutral to negative charge state*, or $B^{-/0}$. However, for typical acceptor and donor impurities the charge states are well known and often omitted. For less common impurities the complete statement is imperative to avoid confusion.

Several different descriptive terms are used in conjunction with a defect level. *Shallow* and *deep* are used to distinguish the energy difference between the level and a band. The former generally refers a defect level that may be depopulated at room temperature; the latter implies any level sufficiently far from a band such that it is stable at room temperature. Recently a third descriptor, *mid-gap level*, has been introduced. This term applies to a level that is close to the center of the bandgap of a wide-bandgap semiconductor. Operationally, it generally refers to a level more than about 1 eV from either band edge, and one that is not easily detected by conventional thermal techniques such as deep-level transient spectroscopy.

Some defects have more than one level in the bandgap. The term *amphoteric* is used if the levels are $X^{0/+}$ and $X^{-/0}$, where the former is referred to as a donor level and the latter as an acceptor level. Confusion arises when one talks about transition-metal defects because the original work was performed on ionic insulators rather than semiconductors. Thus, the ionic notation is used. For example, for vanadium with five outer electrons in tetrahedrally bonded SiC, the neutral state is referred to as V^{4+} because four of the five outer electrons are transferred to the four *positive* C atoms surrounding the site. The $V^{4+/5+}$ level is referred to as the donor level because the vanadium becomes neutral (V^{4+}) after releasing an electron to the conduction band. Similar V^{3+} is negative and $V^{3+/4+}$ is the acceptor level.

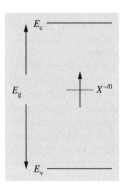

Fig. 45.20 Schematic energy diagram illustrating the concept of a defect level $X^{-/0}$ (see text)

Another confusion that may arise is the difference between a defect level measured using a standard thermal technique such as deep-level transient spectroscopy (DLTS) and an optical method, such as optical absorption. The former detects transitions between ground states, whereas in the latter measurement the final state may be an excited state of the optically induced charge state. In other words, using thermal methods one sense only zero-phonon transitions, whereas optical excitation induces the Franck–Condon (vertical) transition. By definition, a defect level is a thermally determined entity. The energy difference between the zero-phonon and Franck–Condon levels may be interpreted in terms of a structural relaxation of the defect, as is discussed by *Godlewski* et al. [45.59–61].

45.4.1 Photo-EPR

As mentioned earlier, optical illumination may excite the EPR active state of a defect in two different ways. Diamagnetic ($S = 0$) centers may be excited to their triplet ($S = 1$) state where the $\Delta m_s = 1$ transitions within the $S = 1$ manifold may be seen. An example involving the P6/P7 centers was presented. In this section, photo-EPR involving a change in charge state is discussed. Radiation such as near-infrared, visible or ultraviolet may ionize a defect by removing an electron to the conduction band or by exciting an electron from the valence band. This not only enables detection of previously nonparamagnetic centers, but may lead to determination of the electrical level of the defect [45.59–61]. For instance, the transition of vanadium from the 4+ to 5+ charge state $V^{4+/5+}$ was found to be 1.6 eV above the valence-band edge E_v using photo-EPR [45.22]. Several groups have presented studies for the defect level of the carbon vacancy V_c [45.62–64], and *Son* and coworkers used photo-EPR to address the levels of several different defects in SiC [45.65].

The technique consists of measuring the resonance signal during illumination with sub-bandgap light, usually from a 100 W-lamp monochromator system. Two different approaches may be used: steady-state and time-dependent photo-EPR. Almost all of the studies involve the former because the latter involves extensive time and is feasible only for defects with sufficiently long relaxation times. The typical time for an EPR scan, tens of seconds, limits the temporal resolution and the type of defects studied. Nevertheless, *Godlewski* and coworkers applied time-dependent EPR to impurities in ZnS and GaAs [45.59, 60]. Experimentally, the intensity of an EPR signal is monitored as a function of time while illuminating the sample with a fixed photon energy, and the time dependence is fitted to one of many equations depending on the types of transitions involved. Cases involving interaction between acceptors, donors, and conduction/valence bands are thoroughly covered in a review article by *Godlewski* [45.61].

To understand the analysis, consider the simplest case of a single transition involving defect ionization of an electron to the conduction band. The time dependence of the signal intensity recorded at a specific wavelength should follow a first-order kinetic process

$$\Delta n(t) \propto 1 - \exp\left(\frac{-t}{\tau}\right), \qquad (45.10)$$

where Δn represents the normalized change in defect concentration during illumination and τ is the time constant for the process. If the incident light intensity I can be measured, the cross section σ can be calculated from $\frac{1}{\tau} = I\sigma$. The intensity of light inside the sample may be estimated if all of the optical properties of the material are known. Typically samples are transparent to the illumination wavelengths used in photo-EPR, so the incident intensity serves as a good approximation to the intensity of light at the defect. Ideally, a plot of the cross section versus excitation wavelength produces a curve with a threshold reflecting the defect level of the center. Figure 45.21 illustrate some of these ideas, where the time evolution of the V_c^+ signal in 4H-SiC is shown for excitation energies of 1.73 eV (Fig. 45.21a) and 2.3 eV (Fig. 45.21b). The data represent two processes: one for ionization of an electron from V_c^0 to the conduction band (Fig. 45.21b) and the second for excitation of an electron from the valence band to V_c^+ (Fig. 45.21a). The good fit to a single exponential (solid line) supports a model based on a simple transition involving only one defect. When measurements are made with sufficiently low light intensity such that the number of V_c^+ defects generated by the light is proportional to I, the cross sections are shown to be on the order of 10^{-15} and 10^{-16} cm^2 for ionization to the conduction band and excitation from the valence band, respectively [45.64]. Unfortunately, the very long time constants typical of these samples precluded a detailed study of the energy dependence of the cross section. However, the two energies measured at the shortest time constants, 2.3 eV and 1.78 eV, may be used as estimates for the transitions energy from the defect to E_c and from E_v to the defect. The fact that they sum to a quantity larger than the bandgap of 4H-SiC (3.26 eV) suggests that some of the energy is consumed by relaxation of the defect upon capture or release of an electron. The value for

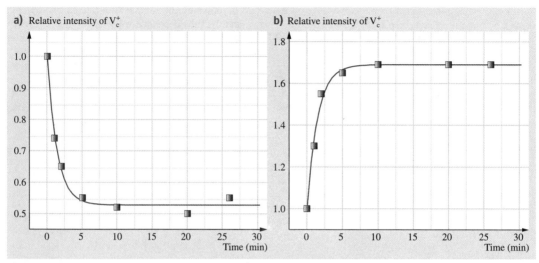

Fig. 45.21a,b Time-dependent photo-EPR data of V_c^+ measured in high-purity semi-insulating 4H-SiC. The photon energy used was (**a**) 1.78 and (**b**) 2.3 eV. The *solid lines* are exponential fits to the data

the relaxation energy extracted from the data, 0.8 eV, is consistent with that predicted by theory, supporting the simple interpretation of the results [45.66]. Analysis of steady-state photo-EPR data yields a similar result for the relaxation energy [45.66].

A more thorough application of time-dependent photo-EPR is presented by *Godlewski* [45.61]. Here, defect relaxation as well as phonon coupling and a model of the purely electronic cross section are included in the energy dependence of the photo-EPR optical cross section used to fit the data. Thermalization measurements are included along with the optically induced spectra to study the influence of defects with levels located very close to a band edge. The defect levels and relaxation energy are determined for $Cr^{+/++}$ in ZnS and GaAs. These values, which compare favorably with those obtained from optical absorption studies, provide validity to the photo-EPR approach.

Although powerful, time-dependent photo-EPR requires that the transition times be long enough to be detected by the relatively slow EPR measurements and short enough to be measured within the lifetime of an experimentalist. The steady-state photo-EPR method provides a simpler approach to determining defect levels; however, the technique cannot distinguish the influence of multiple transitions and is therefore limited to pure materials in which a single dominant defect prevails. The term *steady state* photo-EPR refers to measuring the EPR signal after a fixed period of illumination at a selected wavelength. Results obtained on high-purity SiC are shown in Fig. 45.22, for V_c^+ (brown circles), nitrogen (triangles), and boron (gray circles). In general, the threshold at 1.5 eV in the V_c^+ data is interpreted as the transition of the electron from E_v to the defect and the one at 1.7 eV is thought to represent the transition from the defect to the conduction band. The thresh-

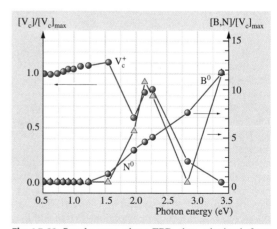

Fig. 45.22 Steady-state photo-EPR data obtained from high-purity semi-insulating 4H-SiC. The *left vertical axis* reflects the relative concentration of V_c^+ (*brown circles*); the *right*, boron (*gray circles*) or nitrogen (*triangles*)

olds are equivalent to transition energies obtained using thermal excitation in that they typically represent zero-phonon transitions. As such, the difference between each threshold value and the energy obtained from the time-dependent data of Fig. 45.21 is the difference between the Franck–Condon and zero-phonon transitions. In other words, the difference is a measure of the relaxation energy. However, as Fig. 45.22 shows, several other photoactive centers are detected, complicating this interpretation. Thus, while interpretation of the data shown in Fig. 45.22 provides a working model, many additional studies including comparison with other techniques need to be performed to completely understand the transitions.

45.4.2 Correlation with Electrically Detected Trapping Centers and Defect Levels

The importance of crystalline semiconductors to the electronics industry has spurred an overwhelming number of studies correlating charge-trapping centers and defect levels to impurities and intrinsic defects. For example, instabilities in metal–oxide–semiconductor field-effect transistors (MOSFETs) initiated numerous experimental programs linking charge-trapping centers in the thermally grown gate oxides with well-known EPR defects in quartz and glass [45.5, 6]. In crystalline materials, the electrical levels in semiconductors detected by deep-level transient spectroscopy (DLTS) and temperature-dependent Hall measurements (TDH) have been attributed to EPR-detected defects with similar thermal properties [45.68, 69]. Also, the interface defects discussed above were widely studied in terms of their electrically detected counterparts. Below, a few studies that employ electrical–EPR correlation measurements are summarized.

Before reading the examples, one must appreciate the limitations of comparisons between EPR and most electrical techniques. The same sample is seldom used for both measurements because the area required for an electrical method such as capacitance–voltage (C–V) or DLTS is often an order of magnitude less that that required for EPR. Besides, a metal contact is often necessary for electrical measurement, while a thick flat metal surface suitable for good electrical contact severely cripples the sensitivity of EPR. Several variations have been devised to avoid these difficulties, the most successful of which is described below, but in most cases separate pieces are used for the two different types of measurements.

Acceptor Activation and Passivation

In the bulk of semiconductors, many of the correlation studies focus on the identification of acceptors or donors and the mechanism by which they are passivated by hydrogen. The dominant GaN acceptor, magnesium, provides one example. Using samples with magnesium concentration in the range 10^{18}–10^{19} cm^{-3} Glaser and coworkers studied a broad, axially symmetric signal found only in Mg-doped GaN. Additional work by others showed that the passivation of holes and the EPR defect followed the same trends [45.70, 71]. Activation by heat treatments in an inert environment produced similar results for the EPR signal and holes also [45.72]. Figure 45.23 shows a comparison of the resistivity measured by *Nakamura* and coworkers (gray squares) and the intensity of the Mg-related EPR signal (brown circles) observed in our laboratory during consecutive annealing treatments in dry N_2 [45.67]. In this comparison, not only are the samples different physical pieces, but they were grown by different groups using somewhat different methods. Nakamura's samples were grown by a metalorganic chemical vapor deposition (MOCVD) technique at 1035 °C on a GaN buffer layer using a sapphire substrate, while our samples were grown by organometallic CVD on an AlN buffer layer using an n-type 4H-SiC substrate. Nevertheless, the comparison is revealing: the resistivity decreases as the EPR signal intensity increases. This behavior is consistent with the release of hydrogen from an acceptor.

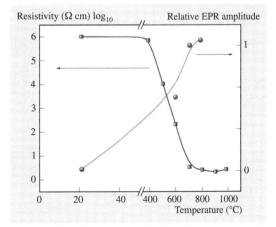

Fig. 45.23 Relative amount of Mg-related acceptors in GaN (*brown circles*) and resistivity (*gray squares*) after sequential annealing treatments in N_2 (after [45.67], © Phys. Soc. Japan 1992)

Similar studies of samples grown by molecular-beam epitaxy and heat treatments in hydrogen are all consistent with the interpretation of acceptor activation by release of hydrogen [45.71]. Although the need to activate the CVD-grown GaN acceptor with a postgrowth N_2 anneal was well accepted by the time of the EPR/annealing study, the spectroscopy confirms that the entity directly involved in the activation process is the acceptor. Observation of the behavior in only Mg-doped GaN suggests the relationship with magnesium, and infrared (IR) studies have indicated that the acceptor is a Mg–N complex, consistent with the broad EPR signal [45.71, 73, 74]. Other studies of acceptor and donor passivation are described by *Gendron* and coworkers, and separately *Gerardi* et al., studying N-doped (n-type) and Al-doped (p-type) SiC [45.75, 76].

Deep Levels: DLTS

Naturally, there is a desire to know the physical entity responsible for the deep levels in semiconductors. In the early days of Si, DLTS and EPR were often performed in the same laboratory in an effort to understand the nature of the deep levels. The C_iC_s (interstitial C–substitutional C pair) in Si provides one of the most complete examples linking a deep level to an EPR center [45.77, 78]. In this case the metastable properties of the center were monitored by several techniques, EPR, DLTS, photoluminescence, and optically detected magnetic resonance, allowing for strong correlations to be developed. The details are found in [45.78]. More recently, the concentration of the Z1/Z2 DLTS signal in 4H-SiC was shown to correlate with an EPR signal known as SI-5, a high-spin center thought to be a divacancy or V_C–C_{Si} pair [45.68]. Figure 45.24 illustrates the comparison between the concentration of the SI-5 EPR signal and Z1/Z2. In addition to the one-to-one correspondence in the number of centers, it was pointed out that illumination was required to observe SI-5, consistent with the negative U character attributed to the DLTS signal. Once the model for SI-5 is confirmed, the DLTS/EPR study will provide a rare assignment of a physical entity in SiC to a deep level detected by DLTS.

Compensating Defects

The importance of semi-insulating semiconductor materials to the formation of high-power electronics has spurred investigations into the deep level, or more appropriately mid-gap level, responsible for compensation. The levels are detected with several techniques including optical DLTS, temperature-dependent Hall measurements, and photo-EPR. A comparison of the latter two techniques has led to the assignment of vanadium and several different intrinsic defects as compensating centers in SiC [45.69]. However, the conclusions should be accepted with a great deal of caution. First, electrical measurements of these high-resistivity materials are crippled by the difficulty of making ohmic contacts. Also, temperature-dependent Hall measurements are often limited to resistivity measurements, from which the carrier density can be obtained only with an assumed model for the temperature dependence of the mobility. Finally, in some cases defect levels are extracted from steady-state photo-EPR, the limitations of which were outlined earlier. Nevertheless, the EPR and electrical studies of these compensating centers have generated a great deal of information on which to further advance the growth and characterization of SiC.

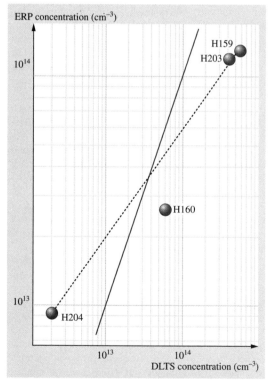

Fig. 45.24 Concentration of the P6, P7 centers versus concentration of the Z1/Z2 defect measured in the different samples indicated on the graph. The *solid line* represents a 1 : 1 correlation; the *dashed line* is a straight line fit (after [45.68])

With more time, the true nature of the compensating defect(s) may be revealed.

Interface Defects and Defect Level

One last study that deserves mention is the association of electrically detected states at the interface between Si and SiO_2 with the Si P_b center [45.39]. Unique among all the investigations described thus far, the interface state–P_b center correlation was performed in situ. That is, the capacitance–voltage (C–V) measurements necessary to detect the interface states were measured on the EPR samples in the microwave cavity. In this way, sample-to-sample uncertainty was eliminated. The experiment requires special sample preparation because metal layers necessary for electrical contact severally reduce EPR sensitivity. The investigators soldered wires to a thick, $0.0028\,cm^2$ contact pad that was positioned adjacent a $0.53\,cm^2$ thin (50 nm) contact. As a varying electrical bias was applied, the small dot was used for the C–V measurement and the large area was used for EPR. The data of Figs. 45.25 and 45.26 show the interface state density and relative change in the number of P_b centers measured at selected values of the applied bias, respectively. The x-axes of both figures were derived from the applied gate voltage using standard analysis, and the y-axis of Fig. 45.26 was extracted from the derivative of the EPR signal intensity as a function of bandgap energy [45.39, 79]. The similarity of the two data sets clearly shows that the dangling bonds at the Si–SiO_2 interface are responsible for D_{it} located 0.3 and 0.8 eV above E_v. Many other electrical measurements and annealing studies have indirectly reaffirmed this conclusion.

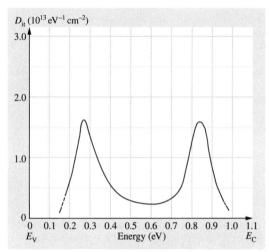

Fig. 45.25 Interface-state density peaks obtained from an oxidized Si sample measured in situ during an EPR measurement (after [45.39], © AIP 1984)

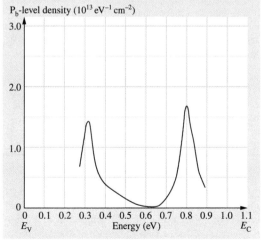

Fig. 45.26 Concentration of P_b centers obtained from EPR measurements made on the same sample as in Fig. 45.25 (after [45.39], © AIP 1984)

45.4.3 Heat Treatment and EPR

Monitoring changes in the EPR signal during heat treatment may determine many interesting properties of defects. The changes may be simply intensity changes as the defect is gradually passivated by an external species introduced during the anneal or a complete transformation of the spectrum from one type of center to another. The heat treatment may be performed either in situ from 4 to 400 K or ex situ up to any temperature desired. Keeping the sample in the cavity is a tremendous advantage during an annealing study because it avoids uncertainties due to sample alignment; however, a typical cavity can withstand heat only slightly above room temperature. Special EPR cavities are available that can be used up to $1000\,°C$; however the cavities are extremely expensive and found mostly at user facilities. At the opposite extreme, temperatures approaching millikelvins are achievable in an EPR system, but not on a regular basis in a typical laboratory. Luckily, many centers are observable between the temperatures easily accessible in most facilities, 4–400 K.

The temperature range of ex situ annealing studies is limited only by the thermal properties of the sample and furnace availability. Since the sample must be remounted in the cavity after each anneal, proper sample alignment is critical, but this is easily accomplished by thoughtful design of a sample holder. Another consideration in the annealing study is possible changes in the cavity quality factor Q, a measurement of the system sensitivity. The sample partially determines Q, so that any changes caused by heat treatments may change the sensitivity of the measurement and thus perturb the results. The most obvious concern is activation of shallow impurities that contribute to conductivity, one factor that affects the sensitivity of the cavity directly. To avoid confusion a standard can be mounted with the sample on the holder. The standard should contain a well-established isotropic signal with narrow line width and a g-value well separated from that of the sample. When EPR measurements are performed below the freeze-out temperature for carriers produced from shallow impurities, the sensitivity changes during the annealing study are less of a concern; nevertheless, it is good practice to include the standard during any experiment in which EPR intensities will be compared.

Another factor to consider when doing annealing studies is that the amplitude of an EPR signal is inversely proportional to the measurement temperature. This is a concern whenever spectra that are to be compared are measured at different temperatures. In this situation, one must account for the temperature dependence of the EPR signal by normalizing all spectra to a chosen temperature before analysis. The temperature dependence of an EPR signal, often referred to as the Curie law, originates from the statistical difference between spin populations in the initial and final energy levels producing the EPR transition. Boltzmann statistics states that this population difference is exponentially dependent on the temperature and the energy difference between the spin-up and spin-down states. The temperature dependence may often be approximated as $\frac{1}{T}$ because $k_B T$ is typically greater than the energy difference between the spin states. For measurements below 4 K, however, the exact expression should be used [45.1].

In situ EPR annealing studies have revealed information about the chemical kinetics among different charge states of defects as well as the mechanisms of charge transport in materials [45.54, 55, 75, 80]. The work of *Merkle* and *Maier* on the association of metal impurities with oxygen vacancies in SrTiO$_3$ provides an example where chemical kinetics are thoroughly analyzed using samples with different starting concentrations of impurities [45.80]. By monitoring the evolution of Fe^{3+} and Mn^{5+} with temperature, the authors extracted the reaction enthalpies and entropies for the association of the metal impurities with oxygen vacancies. The EPR signals for Fe^{3+}, Mn^{5+} and the oxygen-vacancy-related species, Fe^{3+}V$_0$ and Mn^{5+}V$_0$, were measured between room temperature and 170 K. Data analysis suggests that almost all the oxygen vacancies in Fe-doped material are bound to the iron impurities at 300 K, but the situation with the Mn is not as clear. In general, the results have interesting implications for those growing nominally pure titanate films. If the films are free of metal contaminates, will the chronic oxygen-deficiency problem associated with complex oxides be minimized or even eliminated? Such a question will remain unanswered until a sufficient quantity of films of repeatable quality is available for thorough materials characterization, including EPR.

Ex situ EPR/annealing studies range from simply noting the temperature at which an EPR signal intensity changes dramatically to a complete kinetic analysis of the chemical reactions involved in the thermal annealing process. The former do not need to be discussed here, but the latter is addressed below using the Si P$_b$ center as an example.

A very thorough study of hydrogen release from a passivated P$_b$ center is provided by *Brower* and *Myer* [45.81, 82]. Using (111) Si wafers oxidized to maximize the number of P$_b$ centers, Brower measured the intensity of the EPR signal after isothermal vacuum heat treatments over a temperature range of 500 and 595 °C. The low signal-to-noise ratio prevented a more extensive temperature range. The data were found to fit a first-order kinetic equation that includes the temporal profile of the furnace and temperature-dependent rate constant. According to the model, the number of P$_b$ centers remaining after a anneal at temperature T is

$$P_{b\text{-calc}} = N_0 \left(1 - \exp\left\{ -k_{d0} \int_{T\text{profile}} \exp\left[-\frac{E_d}{k_B T}(t) \right] dt \right\} \right),$$

(45.11)

where k_{d0} is the first-order rate constant, E_d is the activation energy for dissociation, and N_0 is the maximum number of P$_b$ centers measured by EPR. The analysis proceeded as follows. First, k_{d0} was obtained from a fit to the time evolution of the P$_b$ center at a fixed temperature. The rate constants obtained from a range of temperatures were then fitted to a first-order kinetic equation to obtain E_d. The k_{d0} and E_d values were then used in (45.11) to determine $P_{b\text{-calc}}$. Figure 45.27 shows agreement between the values calculated us-

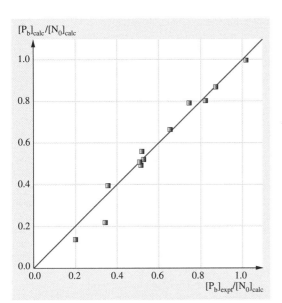

Fig. 45.27 Fraction of P_b centers calculated using (45.11) versus fractional P_b density measured after hydrogenation and subsequent vacuum annealing. The *straight line* represents a 1 : 1 correspondence (after [45.82])

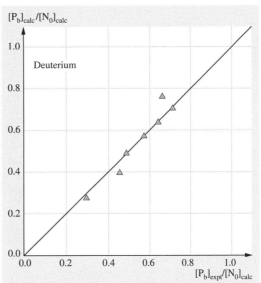

Fig. 45.28 Fraction of P_b centers calculated using (45.11) versus fractional P_b density measured after deuterium heat treatment and subsequent vacuum annealing. The *straight line* represents a 1 : 1 correspondence (after [45.82])

ing (45.11) and the experimental data points obtained between 500 and 595 °C. Figure 45.28 shows that similarly good agreement was obtained when D_2 instead of H_2 was released from the interface by the vacuum annealing process. The deuterium study provided support for the involvement of hydrogen in the annealing process in that the ratio of the k_{d0} values extracted from the H_2 and D_2 data agreed with that expected from the different vibrational frequencies for hydrogen and deuterium. Complete analysis of the data yields values of $k_{d0} = 1.2 \times 10^{12}\,\text{s}^{-1}$ and $E_d = 2.56 \pm 0.06\,\text{eV}$ for the dissociation process $P_bH \rightarrow P_b + H$. Earlier studies of the passivation process $P_b + H \rightarrow P_bH$ had yielded $E_d = 1.66 \pm 0.06\,\text{eV}$. Brower used both sets of values to describe the dissociation of H_2 in semiconductors. Similar studies involving passivation of donors and acceptors in semiconductors exist in the literature [45.71, 75, 76]. Most are not as thorough as the one described here, but all provide insight into the chemical entities involved in the annealing treatments.

45.5 Summary and Final Thoughts

Electron paramagnetic resonance spectroscopy ultimately determines the spin wavefunction by probing the Δm_s transitions at paramagnetic defects. After intensive measurement and analysis, which often includes a comparison with theoretical calculations, a picture of a defect emerges including the identity of the nucleus (nuclei), symmetry, charge state, and absolute number of defects. The experimental procedure may be summarized as follows. A spectrum is obtained by placing a sample in a bath of 10 GHz microwaves situated between the poles of an electromagnet. The magnetic field is ramped to remove the spin degeneracy of the ground state. When the energy difference between the m_s levels equals that of the microwave energy, the sample absorbs the microwaves and this absorption is detected by the external circuitry. The spectrum reflects the derivative of the absorption because the ramped magnetic field is modulated. Phase-sensitive detection is used to monitor the microwave intensity returned from the sample.

Once a spectrum is obtained, it can be interpreted in terms of the Zeeman, nuclear hyperfine, and nuclear fine-structure terms as shown in (45.9). The g-value is determined from the first term and zero crossing of the EPR signal, and the angular dependence of g establishes the symmetry of the center. The second term relates directly to the nuclei that are involved in the defect. The number of hyperfine lines and their intensity is determined by the spin and isotopic abundance of the nucleus. This is often enough to identify the chemical elements forming the defect. Together with the angular dependence of the separation between the hyperfine lines, data may be compared with theory to establish the defect structure and charge state. Additional information is obtained from the fine-structure term, which reflects the spin multiplicity and can sometimes directly determine the charge state of the center. Although not every defect provides all this information, many may be thoroughly described by their electron paramagnetic resonance spectrum, as is illustrated many times throughout this chapter.

Throughout the past five decades, multitudes of intensive investigations have identified intrinsic defects and impurities in semiconductors and insulators. Often, a spectrum measured from a newly obtained sample is easily identified after a reasonable literature search. With the type of defect known, additional experiments may be performed to determine other parameters such as defect level, transport mechanisms or dissociation energy. Examples of such studies are discussed throughout this chapter.

Finally, the reader should be made aware of the many sophisticated versions of magnetic resonance that are not described in this chapter. The most powerful for determining defect structure is electron nuclear double resonance (ENDOR). Developed by Feher in the early 1950s, the technique allows for direct detection of the nucleus by monitoring the nuclear spin flips during an EPR experiment. The technique avoids the ambiguity often encountered when measuring hyperfine line intensities by monitoring the nuclear resonance directly. Because the sensitivity of ENDOR is lower than that of EPR, films are rarely measured. On the other hand, a variation of magnetic resonance that monitors the change in luminescence when sweeping through the resonance field is ideal for studying small samples. Optically detected magnetic resonance (ODMR) combines the spectroscopic selectivity of EPR with the high sensitivity of photoluminescence. ODMR uses the applied magnetic field to flip the electron spin, so that previously spin-forbidden electronic transitions become allowed or allowed transitions become spin-forbidden. Basically, one excites the sample at a luminescence peak and ramps the magnetic field through resonance, causing the peak to increase or decrease depending on the exact nature of the transition. A second technique that is amenable to thin films is electrically detected magnetic resonance (EDMR). An electron–hole recombination current is monitored while the magnetic field is ramped through resonance. When the spins flip, the current changes because the electron–hole recombination transition is controlled by spin selection rules. Although most of the spectroscopic information is lost in the generally broadened signals, the high sensitivity of EDMR allows for studies of interfaces in a transistor-type structure. Thus, the technique is invaluable if direct comparison with electronic device characteristics is desired. Additional variations on EPR exist, including measurements in the time domain. However such pulsed-EPR techniques are seldom applied to crystals and are not typically classified as *materials characterization* techniques because of their inherent complexity and the fundamental nature of the information extracted. All of the magnetic resonance methods mentioned in this closing comment are outlined in [45.1], where further details may be obtained.

References

45.1 J.A. Weil, J.R. Bolton, J.E. Wertz: *Electron Paramagnetic Resonance* (Wiley, New York 1994)

45.2 C.P. Poole Jr., C.P. Poole, F.J. Owens: *Introduction to Nanotechnology* (Wiley, Hoboken 2003), Chap. 3

45.3 C. More, V. Belle, M. Asso, A. Fournel, G. Roger, B. Guigliarelli, P. Bertrand: EPR spectroscopy: a powerful technique for the structural and functional investigation of metalloproteins, Biospectroscopy **5**, S3–S18 (1999)

45.4 H.M. Swartz, N. Khan, J. Buckey, R. Corni, L. Gould, O. Grinberg, A. Hartford, H. Hopf, H. Hou, E. Hug, A. Iwasaki, P. Lesniewski, I. Salikhov, T. Walczak: Clinical applications of EPR: Overview and perspectives, NMR Biomed. **17**, 335 (2004)

45.5 D.L. Griscom: Electron spin resonance, Glass Sci. Technol. **4B**, 151–160 (1990)

45.6 D.L. Griscom: Defect structure of glasses, J. Non-Cryst. Solids **73**, 51 (1985)

45.7 C.P. Poole: *Electron Spin Resonance: A Comprehensive Treatise on Experimental Technique* (Interscience, New York 1967)

45.8 W. Gordy, W. West (Ed.): *Techniques of Chemistry*, Vol. 15 (Wiley, New York 1980)

45.9 M.H.L. Pryce: A modified perturbation procedure for a problem in paramagnetism, Proc. Phys. Soc. A **63**, 25 (1950)

45.10 S. Greulich-Weber, F. Feege, K.N. Kalabukhova, S.N. Lukin, J.-M. Spaeth, F.J. Adrian: EPR and ENDOR investigations of B acceptors in 3C-, 4H- and 6H-silicon carbide, Semicond. Sci. Technol. **13**, 59 (1998)

45.11 H.H. Woodbury, G.W. Ludwig: Electron spin resonance studies in SiC, Phys. Rev. **124**, 1083 (1961)

45.12 N.T. Son, E. Janzen, J. Isoya, S. Yamasaki: Hyperfine interaction of the nitrogen donor in 4H-SiC, Phys. Rev. B **70**, 193207 (2004)

45.13 T. Umeda, J. Isoya, N. Morishita, T. Ohshima, T. Kamiya: EPR identification of two types of carbon vacancies in 4H-SiC, Phys. Rev. B **69**, 121201 (2004)

45.14 P.T. Huy, C.A.J. Ammerlaan, T. Gregorkiewicz, D.T. Dong: Hydrogen passivation of the selenium double donor in silicon: A study by magnetic resonance, Phys. Rev. B **61**, 7448 (2000)

45.15 H.J. von Bardeleben, M. Schoisswohl, J.L. Cantin: Electron paramagnetic resonance study of defects in oxidized and nitrided porous Si and $Si_{1-x}Ge_x$, Colloids Surf. A Physicochem. Eng. Asp. **115**, 277 (1996)

45.16 M. Bockstedte, M. Heid, O. Pankratov: Signature of instrinsic defects in SiC: Ab initio calculations of hyperfine tensors, Phys. Rev. B **67**, 193102 (2003)

45.17 T. Umeda, Y. Ishitsuka, J. Isoya, N.T. Son, N. Morishita, T. Ohshima, H. Itoh: EPR and theoretical studies of negatively charged carbon vacancy in 4H-SiC, Phys. Rev. B **71**, 193202 (2005)

45.18 K. Maier, M. Kunzer, U. Kaufmann, J. Schneider, B. Monemar, I. Akasaki, H. Amano: Iron acceptors in gallium nitride (GaN), Mater. Sci. Forum **143–147**, 93–98 (1994)

45.19 P.G. Baranov, I.V. Illyin, E.N. Mokhov, A.D. Roenkov: Identification of manganese trace impurity in GaN crystals by electron paramagnetic resonance, Semicond. Sci. Technol. **11**, 1843 (1996)

45.20 T. Lingner, S. Greulich-Weber, J.-M. Spaeth: Structure of the silicon vacancy in 6H-SiC after annealing identified as the carbon vacancy-carbon antisite pair, Phys. Rev. B **64**, 245212 (2001)

45.21 N.T. Son, P. Carlsson, J. ul Hassan, E. Janzen, T. Umeda, J. Isoya, A. Gali, M. Bockstedte: Divacancy in 4H-SiC, Phys. Rev. Lett. **96**, 055501 (2006)

45.22 K. Maier, J. Schneider, W. Wilkening, S. Leibenzeder, R. Stein: Electron spin resonance studies of transition metal deep level impurities in SiC, Mater. Sci. Eng. B **11**, 27–30 (1992)

45.23 K. Maier, H.D. Müller, J. Schneider: Transition metals in silicon carbide (SiC): Vanadium and titanium, Mater. Sci. Forum **81–87**, 1183–1194 (1992)

45.24 A. Abragam, B. Bleaney: *Electron Paramagnetic Resonance of Transition Ions* (Dover, New York 1986)

45.25 F. Mehran, B.A. Scott: Electron paramagnetic resonance of $LiNbO_3:Fe^{3+}$, Solid State Commun. **11**, 15 (1972)

45.26 H.H. Towner, Y.M. Kim, H.S. Story: EPR studies of crystal field parameters in $LiNbO_3:Fe^{3+}$, J. Chem. Phys. **56**, 3676 (1972)

45.27 W.E. Carlos, F.A. Freitas Jr., M. Asif Khan, D.T. Olson, J.N. Kuznia: Electron-spin-resonance studies of donors in wurtzite GaN, Phy. Rev. B **48**, 17878 (1993)

45.28 Haiyan Wang: unpublished data (2003)

45.29 A. Stesmans, K. Vanheusden: Generation of delocalized E'_δ defects in buried Si oxide by hole injection, J. Appl. Phys. **76**, 1681 (1994)

45.30 A. Stesmans, F. Scheerlinck: Natural intrinsic EX center in thermal SiO_2 on Si:^{17}O hyperfine interacton, Phys. Rev. B **50**, 5204 (1994)

45.31 W.L. Warren, P.M. Lenahan: Electron spin resonance study of high field stressing in metal-oxide-silicon device oxides, Appl. Phys. Lett. **49**, 1296 (1986)

45.32 M.E. Zvanut, T.L. Chen, R.E. Stahlbush, E.S. Steigerwalt, G.A. Brown: Generation of thermally induced defects in buried SiO_2 films, J. Appl. Phys. **77**, 4329 (1995)

45.33 M.E. Zvanut, F.J. Feigl, J.D. Zook: A defect relaxation model for bias instabilities in metal-oxide-semiconductor capacitors, J. Appl. Phys. **64**, 2221 (1988)

45.34 P.M. Lenahan, P.V. Dressendorfer: Hole traps and trivalent silicon centers in metal/oxide/silicon devices, J. Appl. Phys. **55**, 3495 (1984)

45.35 W.L. Warren, E.H. Poindexter, M. Offenberg, W. Müller-Warmuth: Paramagnetic point defects in amorphous silicon dioxide and amorphous silicon nitride thin films I. a-SiO_2, J. Electrochem. Soc. **139**, 872 (1992)

45.36 N. Mizuochi, M. Ogura, H. Watanabe, J. Isoya, H. Okuchi, S. Yamasaki: EPR study of hydrogen-related defects in boron-doped p-type CVD homoepitaxial diamond films, Diam. Relat. Mater. **13**, 2096 (2004)

45.37 M.E. Zvanut, W.E. Carlos, J.A. Freitas, K.D. Jamison, R.P. Hellmer: An identification of phosphorous in diamond thin films using electron paramagnetic resonance spectroscopy, Appl. Phys. Lett. **65**, 2287 (1994)

45.38 M. Sterrer, E. Fischbach, T. Risse, H.J. Freund: Geometric characterization of a singly charged oxygen vacancy on a single crystalline MgO (001) film by electron paramagnetic resonance spectroscopy, Phys. Rev. Lett. **94**, 186101 (2005)

45.39 E.H. Poindexter, G.J. Gerardi, M.-E. Rueckel, P.F. Caplan, N.M. Johnson, D.K. Biegelsen: Elec-

45.40 E.H. Poindexter, P.J. Caplan: Interface states and electron spin resonance centers in thermally oxidized (111) and (100) silicon wafers, J. Appl. Phys. **52**, 679 (1981)

tronic traps and P_b centers at the Si/SiO$_2$ interface: band-gap energy distribution, J. Appl. Phys. **56**, 2844 (1984)

45.41 J.L. Cantin, H.J. von Bardeleben, Y. Shishkin, Y. Ke, R.P. Devaty, W.J. Choyke: Identification of the carbon dangling bond center at the 4H-SiC/SiO$_2$ interface by an EPR study of oxidized porous SiC, Phys. Rev. Lett. **92**, 015502 (2004)

45.42 P.J. Caplan, E.H. Poindexter, B.E. Deal, R.R. Razouk: ESR centers, interface states, and oxide fixed charge in thermally oxidized silicon wafers, J. Appl. Phys. **50**, 5847 (1979)

45.43 J.L. Cantin, M. Schoisswohl, H.J. von Bardeleben, V. Morazzani, J.J. Ganem, I. Trimaille: EPR study of the defects in porous Si/SiO$_x$N$_y$ and Si$_{0.80}$Ge$_{0.20}$/SiGeO$_2$. In: *The Physics and Chemistry of SiO$_2$ and Si/SiO$_2$ Interfaces*, Proc., Vol.96-1, ed. by H.Z. Massoud, E.H. Poindexter, C.R. Helms (The Electrochemical Society, Pennington 1996) p. 28

45.44 M.E. Zvanut, W.E. Carlos, M.E. Twigg, R.E. Stahlbush, D.J. Godbey: Interfacial point defects in heavily implanted silicon germanium alloys, J. Vac. Sci. Technol. B **10**, 2026 (1992)

45.45 K.L. Brower: ^{29}Si hyperfine structure of unpaired spins at the Si/SiO$_2$ interface, Appl. Phys. Lett. **43**, 1111 (1983)

45.46 P.J. Macfarlane, M.E. Zvanut: Characterization of paramagnetic defect centers in three polytypes of dry heat treated, oxidized SiC, J. Appl. Phys. **88**, 4122 (2000)

45.47 P.J. Macfarlane, M.E. Zvanut: Reduction and creation of paramagnetic centers on surfaces of three different polytypes of SiC, J. Vac. Sci. Technol. B **17**, 1627 (1999)

45.48 M.D. Glinchuk, I.P. Bykov, A.M. Slipenyuk, V.V. Laguta, L. Jastrabik: ESR study of impurities in strontium titanate films, Phys. Solid State **43**, 841 (2001)

45.49 M.R. Deigen, M.D. Glinchuk: Theory of local electronic states on the surface of a non-metallic crystal, Surf. Sci. **3**, 243 (1965)

45.50 P.A.G. Beermann, B.R. McGarvey, B.O. Skadtchenko, S. Muralidharan, R.C.W. Sung: Cationic substitution sites in Mn^{2+}-doped ZnS naoparticles, J. Nanopart. Res. **8**, 235 (2006)

45.51 D. Pan, G. Xu, L. Lv, Y. Yong, X. Wang, J. Wan, G. Wang, Y. Sui: Observation and manipulation of paramagnetic oxygen vacancies in co-doped TiO$_2$ nanocrystals, Appl. Phys. Lett. **89**, 082510 (2006)

45.52 A.V. Brodovoi, S.G. Bunchuk, V.V. Polropivny, V.V. Skorokhod: Magnetic properties of nanoporous Si powder, Int. J. Nanotechnol. **3**, 57 (2006)

45.53 R.P. Wang: Defects in Si nanowires, Appl. Phys. Lett. **88**, 142104 (2006)

45.54 G. Feher, R.C. Fletcher, E.A. Gere: Exchange effects in spin resonance of impurity atoms in silicon, Phys. Rev. **100**, 1784 (1955)

45.55 S. Maekawa, N. Kinoshita: Electron spin resonance in phosphorous doped silicon at low temperatures, J. Phys. Soc. Jpn. **20**, 1447 (1965)

45.56 A. Bencini, D. Gatteschi: *EPR of Exchange Coupled Systems* (Springer, Berlin, Heidelberg 1990)

45.57 J.H. Van Vleck: Concerning the theory of ferromagnetic resonance absorption, Phys. Rev. **78**, 266 (1950)

45.58 C.E. Patton: Microwave resonance and relaxation. In: *Magnetic Oxides*, ed. by D.J. Craik (Wiley, London 1975), Chap. 10

45.59 M. Godlewski: Photoelectron paramagnetic resonance studies of ionization transitions of chromium impurities in ZnS and GaAs, J. Appl. Phys. **56**, 2901 (1984)

45.60 M. Godlewski, Z. Wilamowski, M. Kaminska, W.E. Lamb, B.C. Cavenett: Photo-EPR and ODMR investgations of radiative processes in ZnS:Cr,Sc, J. Phys. C Solid State Phys. **14**, 2835 (1981)

45.61 M. Godlewski: On the application of the photo-EPR technique to the studies of photoionization, DAP recombination, and non-radiative recombination processes, Phys. Status Solidi (a) **90**, 11 (1985)

45.62 N.T. Son, B. Magnusson, E. Janzen: Photoexcitation-electron-paramagnetic-resonance studies of the carbon vacancy in 4H-SiC, Appl. Phys. Lett. **81**, 3945–3947 (2003)

45.63 M.E. Zvanut, V.V. Konovalov: The level position of a deep intrinsic defect in 4H-SiC studied by photoinduced electron paramagnetic resonance, Appl. Phys. Lett. **80**, 410 (2002)

45.64 M.E. Zvanut, H. Wang, W. Lee, W.C. Mitchel, W.D. Mitchell: Deep level point defects in semi-insulating SiC. In: *Silicon Carbide and Related Materials*, ed. by R. Devaty, D. Larkin, S. Saddow (Trans Tech, Switzerland 2006) p. 517, or Mater. Sci. Forum **527–529**, 517 (2006)

45.65 N.T. Son, P. Carlsson, B. Magnusson, E. Janzen: Characterization of semi-insulating SiC, Mater. Res. Soc. Symp. Proc. **911**, 0911-B06-03 (2006)

45.66 H. Wang: Investigation of defect energy levels in SI 4H-SiC using EPR and photo-EPR. Ph.D. Thesis (Univ. of Alabama, Birmingham 2006)

45.67 S. Nakamura, T. Mukai, M. Senoh, N. Iwasa: Thermal annealing effects on P-type Mg-doped GaN films, Jpn. J. Appl. Phys. **31**, L139 (1992)

45.68 N.Y. Garces, W.E. Carlos, E.R. Glaser, S.W. Huh, H.J. Chung, S. Nigam, A.Y. Polyakov, M. Skowronski: Relationship between the EPR SI-5 signal and the 0.65 eV electron trap in 4H- and 6H-

SiC polytypes, Mater. Sci. Forum **527–529**, 547–550 (2006)

45.69 M.E. Zvanut, W. Lee, W.C. Mitchel, W.D. Mitchell, G. Landis: The acceptor level for vanadium in 4H and 6H SiC, Physica B **376/377**, 346 (2006)

45.70 M. Palczewska, B. Suchanek, R. Dwilinski, K. Pakula, A. Wagner, M. Kaminska: Paramagnetic defects in GaN, MRS Internet J. Nitride Semicond. Res. **3**, 45 (1998)

45.71 D.M. Matlock, M.E. Zvanut, H. Wang, J.R. DiMaio, R.F. Davis, J.E. Van Nostrand, R.L. Henry, D. Koleske, A. Wickenden: The effects of oxygen, nitrogen, and hydrogen annealing on Mg acceptors in GaN as monitored by electron paramagnetic resonance spectroscopy, J. Electron. Mater. **34**, 34 (2005)

45.72 M.E. Zvanut, D.M. Matlock, R.L. Henry, D. Koleske, A. Wickenden: Thermal activation of Mg-doped GaN as monitored by electron paramagnetic resonance spectroscopy, J. Appl. Phys. **95**, 1884–1887 (2004)

45.73 J. Neugebauer, C.G. Van de Walle: Hydrogen in GaN: Novel aspects of a common impurity, Phys. Rev. Lett. **75**, 4452 (1995)

45.74 V.J.B. Torres, S. Oberg, R. Jones: Theoretical studies of hydrogen passivated substitutional magnesium acceptor in wurzite GaN, MRS Internet J. Nitride Semicond. Res. **2**, 35 (1997)

45.75 G.J. Gerardi, E.H. Poindexter, D.J. Keeble: Paramagnetic centers and dopant excitation in crystalline silicon carbide, Appl. Spectrosc. **50**, 1428 (1996)

45.76 F. Gendron, L.M. Porter, C. Porte, E. Bringuier: Hydrogen passivation of donors and acceptors in SiC, Appl. Phys. Lett. **67**, 1253 (1995)

45.77 G.D. Watkins: Defect metastability and bistability, Mater. Sci. Forum **38–41**, 39 (1989)

45.78 L.W. Song, X.D. Zhan, B.W. Benson, G.D. Watkins: Bistable defect in silicon: The interstitial-carbon–substitutional-carbon pair, Phys. Rev. Lett. **60**, 460 (1988)

45.79 E.H. Nicollian, J.R. Brews: *MOS (Metal Oxide Semiconductor) Physics and Technology* (Wiley, New York 1982)

45.80 R. Merkle, J. Maier: Defect association in acceptor-doped $SrTiO_3$: Case study for $Fe'_{Ti}V_O$ and $Mn_{Ti}V_O$, Phys. Chem. Chem. Phys. **5**, 2297 (2003)

45.81 K.L. Brower, S.M. Myers: Chemical kinetics of hydrogen and (111) $Si-SiO_2$ interface defects, Appl. Phys. Lett. **57**, 162 (1990)

45.82 K.L. Brower: Dissociate kinetics of hydrogen-passivated (111) $Si-SiO_2$ interface defects, Phys. Rev. B **42**, 3444 (1990)

1550

46. Defect Characterization in Semiconductors with Positron Annihilation Spectroscopy

Filip Tuomisto

Positron annihilation spectroscopy is an experimental technique that allows the selective detection of vacancy defects in semiconductors, providing a means to both identify and quantify them. This chapter gives an introduction to the principles of the positron annihilation techniques and then discusses the physics of some interesting observations on vacancy defects related to growth and doping of semiconductors. Illustrative examples are selected from studies performed in silicon, III-nitrides, and ZnO.

A short overview of positron annihilation spectroscopy is given in Sect. 46.1. The identification of vacancies and their charge states is described in Sect. 46.2; this section also discusses how ion-type acceptors can be detected due to the positrons' shallow Rydberg states around negative ions. The role of vacancies in the electrical deactivation of dopants is discussed in Sect. 46.3, and investigations of the effects of growth conditions on the formation of vacancy defects are reviewed in Sect. 46.4. Section 46.5 gives a brief summary.

46.1 Positron Annihilation Spectroscopy 1552
    46.1.1 Positron Implantation and Diffusion in Solids 1552
    46.1.2 Positron States and Annihilation Characteristics... 1553
    46.1.3 Positron Trapping at Point Defects 1556
    46.1.4 Experimental Techniques 1557

46.2 Identification of Point Defects and Their Charge States 1560
    46.2.1 Vacancies in Si: Impurity Decoration 1560
    46.2.2 Vacancies in ZnO: Sublattice and Charge State 1562
    46.2.3 Negative Ions as Shallow Positron Traps in GaN.. 1564

46.3 Defects, Doping, and Electrical Compensation 1565
    46.3.1 Formation of Vacancy–Donor Complexes in Highly n-Type Si 1566
    46.3.2 Vacancies as Dominant Compensating Centers in n-Type GaN 1568

46.4 Point Defects and Growth Conditions 1569
    46.4.1 Growth Stoichiometry: GaN Versus InN 1570
    46.4.2 GaN: Effects of Growth Polarity 1572
    46.4.3 Bulk Growth of ZnO 1573

46.5 Summary 1576

References 1576

Many techniques are applied to identify defects in semiconductors on the atomic scale. The role of the positron annihilation method is due to its ability to selectively detect vacancy-type defects. This is based on two special properties of the positron: it has a positive charge and it annihilates with electrons. An energetic positron which has penetrated into a solid rapidly loses its energy and then lives a few hundred picoseconds in thermal equilibrium with the environment. During its thermal motion the positron interacts with defects, which may lead to trapping into a localized state. Thus the final positron annihilation with an electron can happen from various states.

Energy and momentum are conserved in the annihilation process, where two photons of about 511 keV are emitted in opposite directions. These photons carry information on the state of the annihilated positron. The positron lifetime is inversely proportional to the electron density encountered by the positron. The momentum of the annihilated electron causes an angular deviation from the 180° straight angle between the two 511 keV photons and creates a Doppler shift in their energy. Thus the observation of positron annihilation radiation gives experimental information on the electronic and defect structures of solids.

The sensitivity of positron annihilation spectroscopy to vacancy-type defects is easy to understand. The free positron in a crystal lattice feels strong repulsion from the positive ion cores. An open-volume defect such as a vacant lattice site is therefore an attractive center where the positron gets trapped. The reduced electron density at the vacant site increases the positron lifetime. In addition, the missing valence and core electrons cause substantial changes in the momentum distribution of the annihilated electrons. Two positron techniques have been efficiently used in defect studies in semiconductors, namely the positron lifetime and the Doppler broadening of the 511 keV line. There are three main advantages of positron annihilation spectroscopy, which can be listed as follows. First, the identification of vacancy-type defects is straightforward. Second, the technique is strongly supported by theory, since the annihilation characteristics can be calculated from first principles. Finally, positron annihilation can be applied to bulk crystals and thin layers of any electrical conduction type.

The experimental and theoretical bases of the positron annihilation spectroscopy of vacancies in metals and alloys were developed in the 1970s. Its applications started gradually to widen to semiconductors in the beginning of the 1980s. At that time the low-energy positron beam was also developed and opened an avenue for defect studies of epitaxial layers and surface regions. The positron annihilation method has had a significant impact on defect spectroscopy in solids by introducing an experimental technique for the unambiguous identification of vacancies. Native vacancies have been observed at high concentrations in many compound semiconductors, and their role in doping and compensation can now be quantitatively discussed.

In addition to vacancy defects, positrons may become confined to interesting regions of low-dimensional structures in semiconductors such as quantum wells, heterointerfaces and quantum dots due to favorable affinity or internal electric fields. The annihilation radiation carries information on the details of the electronic and atomic structures and the chemical composition of the annihilation site. We will, however, in this chapter concentrate on studies of vacancy defects which combined with, e.g., electrical measurements provide quantitative information on electrical compensation.

The aim of this chapter is twofold. We first want to introduce the principles of the positron annihilation techniques and then to discuss the physics of some interesting observations on vacancy defects related to growth and doping of semiconductors. For the sake of coherence, the illustrative examples are selected from the studies performed by the positron group of the Helsinki University of Technology in Si, III-nitrides, and ZnO. For full information on all the published works on positron annihilation in semiconductors, we refer to earlier review articles [46.1–4], books and book chapters [46.5–10], the proceedings of the International Conference on Positron Annihilation (ICPA), and other references therein.

46.1 Positron Annihilation Spectroscopy

In this section we review the principles of positron annihilation spectroscopy and describe the experimental techniques. Thermalized positrons in lattices behave like free electrons and holes. Analogously, positrons have shallow hydrogenic states at negative ions such as acceptor impurities. Furthermore, vacancies and other centers with open volume act as deep traps for positrons. These defects can be detected experimentally by measuring either the positron lifetime or the momentum density of the annihilating positron–electron pairs (Doppler broadening of the annihilation radiation). For the sake of clarity, we will concentrate on the measurements of those two quantities, as these two methods are the most used in defect studies in semiconductors. Descriptions of other techniques can be found in, e.g., [46.9].

46.1.1 Positron Implantation and Diffusion in Solids

The basic principle of a positron experiment is shown in Fig. 46.1. Positrons are easily obtained from radioactive (β^+) isotopes such as ^{22}Na (other possible isotopes are, e.g., ^{58}Co, ^{64}Cu, and ^{68}Ge). The most commonly used isotope is ^{22}Na, where the positron emission is accompanied by a 1.27 MeV photon. This photon is used in positron lifetime experiments as the time signal of the positron emission from the source. The stopping profile of positrons from the β^+ emission is exponential [46.6, 11]. For the ^{22}Na source, where the positron energy distribution extends to $E_{max} = 0.54$ MeV, the positron mean stopping depth is about 110 μm in Si and 40 μm in GaN. The positrons emitted directly from a radioactive source thus probe the bulk of a solid.

Low-energy positrons are needed for studying thin overlayers and near-surface regions. Positrons from the β^+ emission are first slowed down and thermalized in a moderator. This is usually a thin film placed in front of the positron source and made of a material (e.g., Cu or W) that has a negative affinity for positrons. Thermalized positrons close to the moderator surface are emitted into the vacuum with energy of the order of 1 eV and a beam is formed using electric and magnetic fields. The positron beam is accelerated to a tunable energy of 0–40 keV, giving the possibility to control the positron stopping depth in the sample. The typical intensity of a positron beam created in this way is 10^4–10^6 e$^+$s^{-1}. Another way of producing a positron beam is through the electron–positron pair production process. This, however, requires a remarkably larger facility (e.g., a nuclear reactor) and hence not many such beams exist in spite of the advantage of obtaining a beam intensity several orders of magnitude higher than that of a conventional beam.

For monoenergetic positrons, the stopping profile can be described by a derivative of a Gaussian function, i.e., a Makhov profile, with a mean stopping depth [46.2, 11]

$$\bar{x} = 0.886 x_0 = A E^n \text{ (keV)}, \quad (46.1)$$

where E is the positron energy, $A = (4/\rho)$μg/cm^2; $n \approx 1.6$, and ρ is the density of the material in g/cm^3. The mean stopping depth varies with energy from 1 nm up to a few microns. A 20 keV energy corresponds to 2 μm in Si and 0.8 μm in GaN. The width of the stopping profile is rather broad and hence the positron energy must be carefully chosen so that, e.g., the signal from an overlayer is not contaminated by that from the substrate or the surface.

In a solid, the fast positron rapidly loses its energy through ionization and core electron excitation. Finally, the positron momentum distribution relaxes to a Maxwell–Boltzmann distribution through electron–hole excitation and phonon emission. The thermalization time at 300 K is 1–3 ps, i.e., much less than a typical positron lifetime of 200 ps [46.12, 13]. The

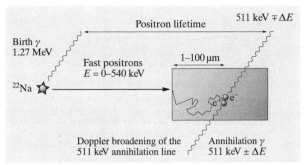

Fig. 46.1 Schematic figure of a positron experiment, where the positrons are implanted into a sample from a ^{22}Na source. The positron lifetime is determined as the time difference between the 511 keV annihilation photons and the 1.27 MeV photon emitted together with the positron from ^{22}Na. The Doppler shift ΔE results from the momentum of the annihilating electron–positron pairs

positron behaves thus as a fully thermalized particle in semiconductors.

The transport of thermalized positrons in solids can be described by diffusion theory. The positron diffusion coefficient has been measured in several semiconductors by implanting low-energy positrons at various depths and observing the fraction which diffuses back to the entrance surface [46.14–16]. The diffusion coefficient D_+ at 300 K is in the range 1.5–3 cm^2 s^{-1}. The total diffusion length during the finite positron lifetime τ is

$$L_+ = (6 D_+ \tau)^{1/2} \approx 5000 \text{ Å}. \quad (46.2)$$

If defects are present, the positron may get trapped before annihilation and this naturally reduces the effective diffusion length. On the other hand, the presence of an electric field, due to, e.g., charging of the sample surface, increases the effective diffusion length.

46.1.2 Positron States and Annihilation Characteristics

Positron Wavefunction and Lifetime, and the Momentum Distribution of the Annihilation Radiation

After implantation and thermalization positrons in semiconductors behave like free carriers (i.e., the positron state is a Bloch state in a defect-free lattice). Various positron states yield specific annihilation characteristics that can be experimentally observed in positron lifetime and Doppler broadening experiments. The positron wavefunction can be calculated from the one-particle Schrödinger equation [46.3]

$$-\frac{\hbar^2}{2m}\nabla^2\Psi_+(r) + V(r)\Psi_+(r) = E_+\Psi_+(r), \quad (46.3)$$

where the positron potential consists of two parts

$$V(r) = V_{\text{Coul}}(r) + V_{\text{corr}}(r). \quad (46.4)$$

The first term is the electrostatic Coulomb potential and the second term takes into account the electron–positron correlation effects. Many practical schemes exist for solving the positron state Ψ_+ from the Schrödinger equation [46.3, 17].

A positron state can be characterized experimentally by measuring the positron lifetime and the momentum distribution of the annihilation radiation. These quantities can also be calculated once the corresponding electronic structure of the solid system is known. The positron annihilation rate λ, the inverse of the positron lifetime τ, is proportional to the overlap of the electron and positron densities

$$1/\tau = \lambda = \pi r_0^2 c \int d\mathbf{r}\, |\Psi_+(r)|^2 n(r)\gamma[n(r)], \quad (46.5)$$

where r_0 is the classical electron radius, c is the velocity of light, $n(r)$ is the electron density, and $\gamma(n)$ is the enhancement factor that accounts for the pile-up of the electron density at the positron beyond the (average) density $n(r)$ [46.3]. The momentum distribution $\rho(p)$ of the annihilation radiation is a nonlocal quantity and requires knowledge of all the electron wavefunctions Ψ_i overlapping with the positron. It can be written in the form

$$\rho(\mathbf{p}) = \frac{\pi r_0 c}{V}\sum_i \left|\int d\mathbf{r}\, e^{-i\mathbf{p}\mathbf{r}}\Psi_+(r)\Psi_i(r)\sqrt{\gamma_i(r)}\right|^2, \quad (46.6)$$

where V is the normalization volume and $\gamma_i(r)$ may depend only on i or on r. A Doppler broadening experiment measures the longitudinal momentum distribution along the direction of the emitted 511 keV photons, defined here as the z-axis

$$\rho(p_L) = \int_{-\infty}^{\infty}\int_{-\infty}^{\infty} dp_x\, dp_y\, \rho(\mathbf{p}). \quad (46.7)$$

It should be noted that the momentum distribution $\rho(\mathbf{p})$ of the annihilation radiation is mainly that of the annihilating electrons *seen by the positron*, because the momentum of the thermalized positron is negligible.

Fig. 46.2 Delocalized positron density in a perfect ZnO lattice according to theoretical calculations. The *c*-axis of the wurtzite structure is in the *vertical direction* in the figure plane. The positions of the Zn and O atoms are marked with *larger* and *smaller thick open circles*, respectively. The contour spacing is 1/7 of the maximum value, the darkest line denoting the highest value

As an example, the calculated positron density in a perfect ZnO wurtzite lattice is shown in Fig. 46.2. For details of the calculations see [46.18]. The positron is delocalized in a Bloch state with $k_+ = 0$. Due to the Coulomb repulsion by positive ion cores, the positron wavefunction has its maximum at the interstitial space between the atoms. The positron energy band $E_+(k)$ is parabolic and free-particle-like with an effective mass of $m^* \approx 1.1\, m_0$ [46.3].

Deep Positron States at Vacancy Defects

In analogy to free carriers, also the positron has localized states at lattice imperfections. At vacancy-type defects where ions are missing, the repulsion sensed by the positron is lowered and the positron feels these kinds of defects as potential wells. As a result, localized positron states at open-volume defects are formed. The positron ground state at a vacancy-type defect is generally deep: the binding energy is about 1 eV or more [46.3]. Figure 46.3 shows the calculated density of the localized positron at unrelaxed Zn and O vacancies in ZnO. The positron wavefunction is confined in the open volume formed at the vacancy. The localiza-

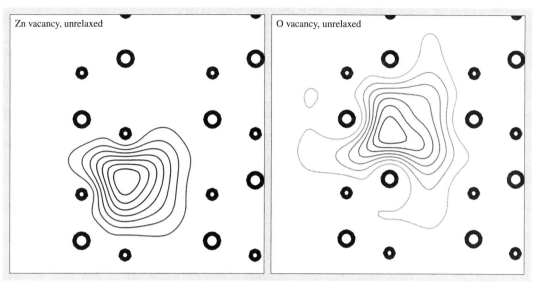

Fig. 46.3 Localized positron density in a perfect ZnO lattice according to theoretical calculations. The c-axis of the wurtzite structure is in the *vertical direction* in the figure plane. The positions of the Zn and O atoms are marked with *larger* and *smaller thick open circles*, respectively. The contour spacing is $1/7$ of the maximum value, the *darkest line* denoting the highest value

tion is clearly stronger in the case of the Zn vacancy because the open volume of V_{Zn} is much larger than that of V_O.

In a vacancy defect the electron density is locally reduced. This is reflected in the positron lifetimes, which are longer than in the defect-free lattice. For example, the calculated lifetimes in the unrelaxed Zn and O vacancies are 45 ps and 3 ps longer than in the perfect lattice. The longer positron lifetime at V_{Zn} is due to the larger open volume compared with that of V_O. The positron lifetime measurement is thus a probe of vacancy defects in materials. Positron annihilation at a vacancy-type defect leads also to changes in the momentum distribution $\rho(p)$ probed by the Doppler broadening experiment. The momentum distribution arising from valence electron annihilation becomes narrower due to a lower electron density. In addition, the localized positron at a vacancy has a reduced overlap with ion cores, leading to a considerable decrease in annihilation with high-momentum core electrons. In our model case of ZnO, where the dominant contribution to the high-momentum part of the distribution comes from the Zn 3d electrons, the changes in the momentum distribution are more pronounced when positrons are trapped at Zn vacancies.

The comparison of the measured positron lifetimes and Doppler broadening spectra with the theoretically calculated data for specific defects provides a very efficient tool for identification of the observed vacancy defects. Several ab initio approaches have been studied in recent years [46.17, and the references therein]. The agreement between theory and experiment is excellent in terms of differences or ratios between the data for defects and the perfect lattice. Also quantitative agreement has been obtained in many materials, but all the theoretical schemes applied so far seem to have problems with the treatment of 3d electrons, either under- or overestimating their contribution to the positron annihilation data. For example, in the case of ZnO, the calculated lifetimes in the perfect lattice range from about 140 to 180 ps [46.19, 20], depending on the scheme, while the experimentally determined bulk lifetime is 170 ps [46.20, 21]. On the other hand, the calculated differences in the lifetimes and the ratios of the momentum distributions between the defect and bulk states agree very well with experiments [46.17, 20, 22].

Shallow Positron States at Negative Ions

A negatively charged impurity atom or an intrinsic point defect can bind positrons at shallow states even if these defects do not contain open volume [46.23, 24]. Being a positive particle, the positron can be localized at the hydrogenic (Rydberg) state of the Coulomb field around a negatively charged center. The situation is analogous to the binding of an electron to a shallow donor atom. The positron binding energy at the negative ion can be estimated from the simple effective-mass theory

$$E_{\text{ion},n} = \frac{13.6\,\text{eV}}{\varepsilon^2}\left(\frac{m^*}{m_0}\right)\frac{Z^2}{n^2} \approx 10\text{--}100\,\text{meV}, \tag{46.8}$$

where ε is the dielectric constant, m^* is the effective mass of the positron, Z is the charge of the negative ion, and n is the quantum number. With $m^* \approx m_0$, $Z = 1\text{--}3$, and $n = 1\text{--}4$, (46.8) yields typically $E_{\text{ion}} = 10\text{--}100\,\text{meV}$, indicating that positrons are thermally desorbed from the Rydberg states at 100–300 K.

The hydrogenic positron state around a negative ion has a typical extension of 10–100 Å and thus positrons probe the same electron density as in the defect-free lattice. As a consequence, the annihilation characteristics (positron lifetime, positron–electron momentum distribution) are not different from those in the lattice. Although the negative ions cannot be identified with the experimental parameters, information on their concentration can be obtained in the positron lifetime and Doppler broadening experiments when they compete with vacancies in positron trapping [46.23, 24].

46.1.3 Positron Trapping at Point Defects

Positron Trapping Rate and Trapping Coefficient

The positron transition from a free Bloch state to a localized state at a defect is called positron trapping. The trapping is analogous to carrier capture. However, it must be fast enough to compete with annihilation. The positron trapping rate κ onto a defect D is proportional to the defect concentration c_D

$$\kappa_D = \mu_D c_D. \tag{46.9}$$

The trapping coefficient μ_D depends on the defect and the host lattice. Since the positron binding energy at vacancies is typically $> 1\,\text{eV}$, the thermal escape (detrapping) of positrons from the vacancies can usually be neglected. Due to the Coulomb repulsion, the trapping coefficient at positively charged vacancies is so small that the trapping does not occur during the short positron lifetime of a few hundred picoseconds [46.25]. Therefore, the positron technique does not detect vacancies or other defects in their positive charge states. The trapping coefficient at neutral vacancies is typically $\mu_D \approx 10^{14}\text{--}10^{15}\,\text{s}^{-1}$ independently of temperature [46.25–27]. This value means that neutral vacancies are observed when their concentration is $\geq 10^{16}\,\text{cm}^{-3}$.

The positron trapping coefficient at negative vacancies is typically $\mu_D \approx 10^{15}\text{--}10^{16}\,\text{s}^{-1}$ at 300 K temperature [46.25–27]. The sensitivity to detect negative vacancies is thus $\geq 10^{15}\,\text{cm}^{-3}$. The experimental fingerprint of a negative vacancy is the increase of μ_D with decreasing temperature [46.26, 27]. The $T^{-1/2}$ dependence of μ_D is simply due to the increase of the amplitude of the free positron Coulomb wave in the presence of a negative defect as the thermal velocity of the positron decreases [46.25]. The temperature dependence of μ_D allows to experimentally distinguish negative vacancy defects from neutral ones.

The positron trapping coefficient μ_{ion} at the hydrogenic states around negative ions is of the same order of magnitude as that at negative vacancies [46.24, 28]. Furthermore, the trapping coefficient exhibits a similar $T^{-1/2}$ temperature dependence. Unlike in the case of vacancy defects, the thermal escape of positrons from the negative ions plays a crucial role at usual experimental temperatures. The principle of detailed balance yields the following equation for the detrapping rate δ_{ion} from the hydrogenic state [46.3]

$$\delta_{\text{ion}} = \mu_{\text{ion}}\left(\frac{2\pi m^* k_B T}{h^2}\right)^{3/2} \exp\left(-\frac{E_{\text{ion}}}{k_B T}\right). \tag{46.10}$$

Typically ion concentrations above $10^{16}\,\text{cm}^{-3}$ influence positron annihilation at low temperatures ($T < 100\,\text{K}$), but the ions are not observed at high temperatures ($T > 300\,\text{K}$), where the detrapping rate (46.10) is large.

Kinetic Trapping Model

In practice the positron annihilation data is analyzed in terms of kinetic rate equations describing the positron transitions between the free Bloch states and localized states at defects [46.8–10]. Very often the experimental data show the presence of two defects, one of which is a vacancy and the other is a negative ion. The probability of a positron to be in the free state is $n_B(t)$, to be trapped at vacancies is $n_V(t)$, and to be trapped at ions is $n_{\text{ion}}(t)$. We can write the rate equations as

$$\frac{dn_B}{dt} = -(\lambda_B + \kappa_V + \kappa_{\text{ion}})n_B + \delta_{\text{ion}} n_{\text{ion}}, \tag{46.11}$$

$$\frac{dn_V}{dt} = \kappa_V n_B - \lambda_V n_V, \quad (46.12)$$

$$\frac{dn_{ion}}{dt} = \kappa_{ion} n_B - (\lambda_{ion} + \delta_{ion}) n_{ion}, \quad (46.13)$$

where λ, κ, and δ refer to the corresponding annihilation, trapping, and detrapping rates.

Assuming that the positron at $t = 0$ is in the free Bloch state, (46.11–46.13) can be solved and the probability of a positron to be alive at time t is obtained as

$$n(t) = n_B(t) + n_V(t) + n_{ion}(t) = \sum_{i=1}^{3} I_i \exp(-\lambda_i t), \quad (46.14)$$

indicating that the lifetime spectrum $-dn(t)/dt$ has three exponential components. The fractions of positron annihilations at various states are

$$\eta_B = \int_0^\infty dt\, \lambda_B n_B(t) = 1 - \eta_{ion} - \eta_V, \quad (46.15)$$

$$\eta_V = \int_0^\infty dt\, \lambda_V n_V(t) = \frac{\kappa_V}{\lambda_B + \kappa_V + \frac{\kappa_{ion}}{1+\delta_{ion}/\lambda_{ion}}}, \quad (46.16)$$

$$\eta_{ion} = \int_0^\infty dt\, \lambda_{ion} n_{ion}(t)$$
$$= \frac{\kappa_{ion}}{(1+\delta_{ion}/\lambda_{ion})\left(\lambda_B + \kappa_V + \frac{\kappa_{ion}}{1+\delta_{ion}/\lambda_{ion}}\right)}. \quad (46.17)$$

These equations are useful because they can be related with the experimental average lifetime τ_{ave} (the center of mass of the lifetime spectrum), the positron–electron momentum distribution $\rho(p_L)$, and the shape parameters S and W of the Doppler-broadened annihilation line (representing annihilations with low-momentum valence electrons and high-momentum core electrons, respectively) as follows

$$\tau_{ave} = \eta_B \tau_B + \eta_{ion} \tau_{ion} + \eta_V \tau_V, \quad (46.18)$$

$$\rho(p_L) = \eta_B \rho_B(p_L) + \eta_{ion} \rho_{ion}(p_L) + \eta_V \rho_V(p_L), \quad (46.19)$$

$$S = \eta_B S_B + \eta_{ion} S_{ion} + \eta_V S_V, \quad (46.20)$$

$$W = \eta_B W_B + \eta_{ion} W_{ion} + \eta_V W_V. \quad (46.21)$$

Equations (46.15–46.21) allow the experimental determination of the trapping rates κ_V and κ_{ion}, and consequently the defect concentrations can be obtained from (46.12). Furthermore, these equations enable the combination of positron lifetime and Doppler broadening results, and various correlations between τ_{ave}, $\rho(p_L)$, S, and W can be studied.

At high temperatures all positrons escape from the hydrogenic state of the negative ions and no annihilations take place at them. In this case the lifetime spectrum has two components

$$\tau_1^{-1} = \tau_B^{-1} + \kappa_V, \quad (46.22)$$

$$\tau_2 = \tau_V, \quad (46.23)$$

$$I_2 = 1 - I_1 = \frac{\kappa_V}{\kappa_V + \lambda_B - \lambda_D}. \quad (46.24)$$

The first lifetime τ_1 represents the effective lifetime in the lattice in the presence of positron trapping at vacancies. Since $\kappa_V > 0$ and $I_2 > 0$, τ_1 is less than τ_B. The second lifetime component τ_2 characterizes positrons trapped at vacancies, and it can be directly used to identify the open volume of the vacancy defect. When $\eta_{ion} = 0$ and $\delta_{ion}/\lambda_{ion} \gg 1$ the determination of the positron trapping rate and vacancy concentration is straightforward using (46.15–46.21)

$$\kappa_V = \mu_V c_V = \lambda_B \frac{\tau_{ave} - \tau_B}{\tau_V - \tau_{ave}}$$
$$= \lambda_B \frac{S - S_B}{S_V - S} = \lambda_B \frac{W - W_B}{W_V - W}. \quad (46.25)$$

Notice that in this case τ_{ave}, S, and W depend linearly on each other. The linearity of experimental points in (τ_{ave}, S), (τ_{ave}, W), and (S, W) plots thus provides evidence that positrons annihilate from two distinguishable states, indicating that they are trapped at only a single type of vacancy defect in the samples.

46.1.4 Experimental Techniques

Positron Lifetime Spectroscopy

Positron lifetime spectroscopy is a powerful technique in defect studies, because the various positron states appear as different exponential decay components. The number of positron states, and their annihilation rates and relative intensities can be determined. In a positron lifetime measurement, one needs to detect the start and stop signals corresponding to the positron entrance and annihilation times in the sample, respectively (Fig. 46.1). A suitable start signal is the 1.27 MeV photon that accompanies the positron emission from the ^{22}Na isotope. The 511 keV annihilation photon serves as the stop signal. The positron source is prepared by sealing about 10 μCi (about 10^5–10^6 Bq) of radioactive isotope between two thin foils. The source is then sandwiched between two identical pieces (e.g., $5 \times 5 \times 0.5$ mm^3) of the sample material. This technique is standard for bulk crystal studies. Pulsed positron beams have been constructed for lifetime spectroscopy in thin layers [46.29, 30], but so far they have not been used much in defect studies.

The conventional lifetime spectrometer consists of start and stop detectors, each of them made by coupling a fast scintillator to a photomultiplier. The timing pulses are obtained by differential constant-fraction discrimination. The time delays between the start and stop signals are converted into amplitude pulses, the heights of which are stored in a multichannel analyzer. Thanks to the development of fast analog-to-digital converters (ADCs), digital data readout techniques have recently become viable [46.31–34]. This allows direct digitization of the detector pulses and performance of the timing and energy windowing with software instead of the conventional analog electronics, simplifying the measurement setup significantly. About 10^6 lifetime events are recorded in 1 h. The experimental spectrum represents the probability of positron annihilation at time t and it consists of exponential decay components

$$-\frac{dn(t)}{dt} = \sum_i I_i \lambda_i \exp(-\lambda_i t) , \quad (46.26)$$

where $n(t)$ is the probability for the positron to be alive at time t. The decay constants $\lambda_i = 1/\tau_i$ are called the annihilation rates and they are the inverses of the positron lifetimes τ_i. Each positron lifetime has intensity I_i. In practice the ideal spectrum of (46.26) is convoluted by a Gaussian resolution function which has a width of 200–250 ps (full-width at half-maximum, FWHM). About 5–10% of positrons annihilate in the source material and proper *source corrections* must be made. Due to the finite time resolution, annihilations in the source materials, and random background, typically only 1–3 lifetime components can be resolved in the analysis of the experimental spectra. The separation of two lifetimes is successful only if the ratio λ_1/λ_2 is ≥ 1.3–1.5.

Figure 46.4 shows positron lifetime spectra recorded in as-grown and electron-irradiated high-quality ZnO bulk crystals [46.21, 35]. Positrons enter the sample and thermalize at time $t = 0$. The vertical axis of Fig. 46.4 gives the number of annihilations at a time interval of 25 ps. In the as-grown sample the positron lifetime spectrum has a single component of $170 \pm$

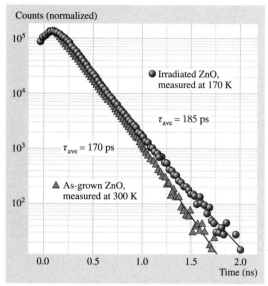

Fig. 46.4 Examples of positron lifetime spectra in as-grown and electron-irradiated high-quality ZnO samples. A constant background and annihilations in the source materials have been subtracted from the spectra, which consist of 2×10^6 recorded annihilation events. The *solid lines* are fits of the sum of exponential components convoluted with the resolution function of the spectrometer. The data in the as-grown sample was recorded at 300 K and has only a single component of 170 ± 1 ps. The spectrum in the electron-irradiated crystal was recorded at 170 K and can be decomposed into two components of $\tau_1 = 155 \pm 5$ ps, $\tau_2 = 230 \pm 10$ ps, and $I_2 = 38 \pm 5\%$

1 ps at 300 K corresponding to positron annihilations in the defect-free lattice. The electron-irradiated sample has two lifetime components, the longer of which ($\tau_2 = 230$ ps) is due to positrons annihilating when trapped at irradiation-induced Zn vacancies. For more discussion see Sect. 46.2.

The experimental results are often presented in terms of the average positron lifetime τ_{ave}, defined as

$$\tau_{\text{ave}} = \int_0^\infty dt\, t \left(-\frac{dn}{dt}\right) = \int_0^\infty dt\, n(t) = \sum_i I_i \tau_i . \quad (46.27)$$

The average lifetime is a statistically accurate parameter, because it is equal to the center of mass of the experimental lifetime spectrum. Hence it can be cor-

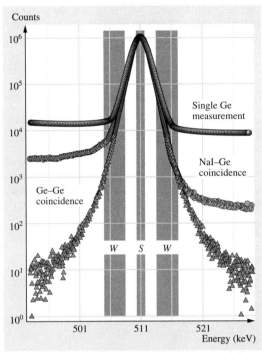

Fig. 46.5 Doppler broadening spectra obtained in different data collection modes. The definition of the shape parameters S and W is shown

rectly calculated from the intensity and lifetime values even if the decomposition only represents a good fit to the experimental data without any physical meaning. For example, the positron average lifetimes in the two spectra of Fig. 46.4 are 170 ps (as-grown ZnO) and 185 ps (electron-irradiated ZnO). The difference is very significant because changes below 1 ps in the average lifetime can be reliably observed in experiments.

Doppler Broadening Spectroscopy

Doppler broadening spectroscopy is often applied especially in the low-energy positron beam experiments, where lifetime spectroscopy is usually very difficult due to the missing start signal. The motion of the annihilating electron–positron pair causes a Doppler shift in the annihilation radiation (Fig. 46.1)

$$\Delta E_\gamma = \frac{1}{2} c p_L, \quad (46.28)$$

where p_L is the longitudinal momentum component of the pair in the direction of the annihilation photon emission. This causes the broadening of the 511 keV annihilation line (Fig. 46.5). The shape of the 511 keV peak thus gives the one-dimensional momentum distribution $\rho(p_L)$ of the annihilating electron–positron pairs. A Doppler shift of 1 keV corresponds to a momentum value of $p_L = 3.91 \times 10^3 \, m_0 c$ (≈ 0.54 a.u.).

The Doppler broadening can be experimentally measured using a Ge gamma detector with a good energy resolution (Fig. 46.5). For measurements of bulk samples, the same source–sample sandwich is used as in the lifetime experiments. For layer studies, the positron beam hits the sample and the Doppler broadening is often monitored as a function of the beam energy. The typical resolution of a detector is around 1–1.5 keV at 500 keV. This is considerable compared with the total width of 2–3 keV of the annihilation peak, meaning that the experimental line shape is strongly influenced by the detector resolution. Therefore, various shape parameters are used to characterize the 511 keV line. Their definitions are shown in Fig. 46.5 as well.

The low-electron-momentum parameter S is defined as the ratio of the counts in the central region of the annihilation line to the total number of the counts in the line. In the same way, the high-electron-momentum parameter W is the fraction of the counts in the wing regions of the line. Due to their low momenta, mainly valence electrons contribute to the region of the S parameter. On the other hand, only core electrons have momentum values high enough to contribute to the W parameter. Therefore, S and W are called the valence and core annihilation parameters, respectively. The S parameter is integrated from the Gaussian part (given by the momentum distribution of the unbound or only weakly bound valence electrons) so that it includes roughly 50% of the total counts in the peak. The lower limit of the W parameter window is chosen so that the dominant contribution to that part of the spectrum comes from the exponential tails (linear in the semi-log plot) of the core electron distributions. In order to have as good statistics as possible, the upper limit is set as high as reasonable from the data scatter point of view. The proper choice of the lower limit of the W parameter window depends on the studied material.

The high-momentum part of the Doppler broadening spectrum arises from annihilations with core electrons which contain information on the chemical identity of the atoms. Thus detailed investigation of core electron annihilation can reveal the nature of the atoms in the regions where positrons annihilate. In order to study the high-momentum part in detail, the experimental background needs to be reduced. A second gamma detector is placed opposite to the Ge detector and the only events that are accepted are those for which both 511 keV photons are detected [46.36, 37]. Depending on the type of the second detector, electron momenta even up to $p \approx 60 \times 10^{-3} m_0 c$ (≈ 8 a.u.) can be measured with the coincidence detection of the Doppler broadening.

46.2 Identification of Point Defects and Their Charge States

The annihilation characteristics of trapped positrons serve as fingerprints in defect identification. The positron lifetime at a defect is a basic quantity for two reasons: it reflects the open volume of the defect and it can be predicted by theoretical calculations. However, the lifetime experiment alone is not enough for the direct identification of the sublattice of the vacancy in compound semiconductors or to determine whether the vacancy is isolated or complexed with impurity atoms, as it is insensitive to the chemical surroundings of the defect. Doppler broadening experiments provide information on the momentum distribution of the annihilating electrons. By the coincidence technique one can reveal the core electron momentum distribution that carries information about the type of atoms in the region of annihilation. In the case of a vacancy, the positron wavefunction is localized and overlaps predominantly with the core electrons of the neighboring atoms. Therefore, vacancies on different sublattices can be distinguished and impurities associated with vacancies may be identified. Finally, by varying the sample temperature during the experiments, one can distinguish between neutral and negative charge states of the vacancies due to the different thermal behavior of the positron trapping, and detect negatively charged non-open volume defects.

46.2.1 Vacancies in Si: Impurity Decoration

To illustrate how the positron lifetime together with Doppler broadening experiments can be used to identify vacancy defects, we review here results obtained in highly n-type Si doped with phosphorus (P) and arsenic (As). Electron irradiation with energies of the order of 1–2 MeV is a convenient experimental approach to produce a controlled concentration of vacancies. By changing the fluence one can vary the fraction of positrons annihilating at vacancies. Electron irradiation at 2 MeV creates vacancies and interstitials as primary defects, both of which are mobile in Si at 300 K [46.39]. Hence the vacancies produced in the irradiation disappear, e.g., by recombination with interstitials or by diffusion to the surface, or become stabilized by impurities or intrinsic defects, such as other vacancies. As both the vacancies and interstitials are mobile, the recombination and divacancy formation processes are not very efficient; in experiments about 1% of the primarily created vacancies have been observed to form divacancies when competing stabilizing defects, such as oxygen or dopant atoms (e.g., P or As), are not present [46.40].

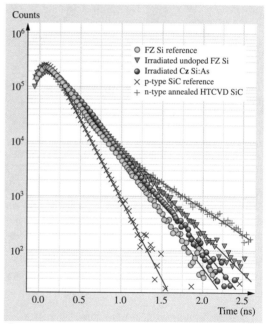

Fig. 46.6 Positron lifetime spectra in as-grown and 2 MeV electron-irradiated Si samples. Positrons annihilate in the as-grown sample with a single lifetime of 220 ps corresponding to delocalized positrons in the lattice. In the irradiated samples the experiments reveal vacancies with positron lifetimes of 250 ps (V–As pair in CZ Si:As sample doped with $[\mathrm{As}] = 1020\,\mathrm{cm}^{-3}$) and 300 ps (divacancy in undoped FZ Si sample) [46.38]. To illustrate the sensitivity of the lifetime measurement, data measured in a p-type SiC reference sample (single lifetime component of 150 ps) and a high-temperature-annealed n-type SiC sample (two components, of which the higher is 450 ps) are shown [46.38]

Figure 46.6 shows the positron lifetime spectra measured in unirradiated float-zone (FZ)-refined Si, electron-irradiated As-doped Czochralski (CZ)-grown Si ([As] = 10^{20} cm^{-3}), and electron-irradiated FZ Si samples [46.38]. To illustrate the sensitivity of the lifetime measurement, data measured in a p-type SiC reference sample (single lifetime component of 150 ps) and a high-temperature-annealed n-type SiC sample (two components, of which the higher is 450 ps) are shown [46.41]. The unirradiated samples have only a single positron lifetime component of about 220 ps, which is practically constant as a function of temperature [46.40]. This behavior shows that no vacancies are observed by positrons and all annihilations take place at the delocalized state in the bulk lattice, with the lifetime $\tau_B = 220$ ps. The presence of vacancy defects in the electron-irradiated FZ Si samples is evident in Fig. 46.6. The lifetime spectrum has two components, the longer of which, $\tau_2 = 300 \pm 5$ ps, corresponds to positrons trapped at vacancy defects. The lifetime of $\tau_2 = 300 \pm 5$ ps is significantly larger than expected for a monovacancy but it is equal to the calculated lifetime for annihilation at divacancies [46.42]. Vacancy defects are clearly present in the electron-irradiated As-doped Si samples as well (Fig. 46.6). In this case the spectrum has only a single component due to the high concentration of vacancies, causing saturation trapping of positrons. This occurs when the vacancy concentration exceeds 10^{18} cm^{-3}, consistent with the expected introduction rate in electron-irradiated heavily n-type doped Si [46.26, 27, 39]. The lifetime component is $\tau_V = 250 \pm 5$ ps, a lifetime characteristic of a single vacancy according to theoretical calculations [46.42]. The lifetime component $\tau_V = 250 \pm 5$ ps is detected in as-grown As-doped Si ([As] = 10^{20} cm^{-3}) as well (not shown in the figure) [46.43], but the vacancy concentration is clearly smaller than in the electron-irradiated samples. The average lifetimes are $\tau_{ave} = 232$ ps and $\tau_{ave} = 250$ ps in the as-grown and irradiated samples, respectively [46.38, 41].

In order to identify the monovacancies in detail, Doppler broadening experiments using the two-detector coincidence technique have been performed. The lifetime results from above can be used to determine the fraction of positrons annihilating at vacancies $\eta = (\tau_{ave} - \tau_B)/(\tau_V - \tau_B)$ (Sect. 46.1). Since the momentum distribution in the lattice $\rho_B(p)$ can be measured in the reference sample, the distributions $\rho_V(p)$ at vacancies can be decomposed from the measured spectrum $\rho(p)$. They are shown in Fig. 46.7 for the monovacancies observed in as-grown Si([As] = 10^{20} cm^{-3})

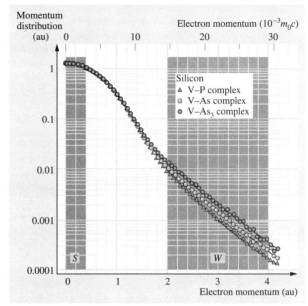

Fig. 46.7 The positron–electron momentum distribution at the various vacancy–impurity pairs, identified in electron-irradiated Si ([P] = 10^{20} cm^{-3}) (*full circles*) and in as-grown (*full circles*) and irradiated (*open triangles*) Si([As] = 10^{20} cm^{-3}). The results of theoretical calculations are shown by the *solid curves*

as well as in irradiated Si([As] = 10^{20} cm^{-3}) and Si([P] = 10^{20} cm^{-3}).

The momentum distributions $\rho_V(p)$ at vacancies indicate large differences at higher momenta ($p > 1.8$ a.u.), where annihilation with core electrons is the most important contribution (Fig. 46.7). Since the core electron momentum distribution is a specific characteristic of a given atom, the differences between the spectra indicate different atomic environments of the vacancy in each of the three cases. Because in both Si ($Z = 14$) and P ($Z = 15$) the 2p electrons constitute the outermost core electron shell, the core electron momentum distributions of these elements are very similar. The crucial difference in the core electron structures of Si, P, and As is the presence of 3d electrons in As. The overlap of positrons with the As 3d electrons is much stronger than with the more localized Si or P 2p electrons. The large intensity of the core electron momentum distribution is thus a clear sign of As atoms surrounding the vacancy.

The 2 MeV electron irradiation creates vacancies and interstitials as primary defects, both of which are mobile at 300 K. In heavily n-type Si the donor atom may capture the vacancy and form a vacancy–impurity pair [46.39]. The monovacancy detected in heavily P-doped Si is thus the V–P pair. Similarly, it is natural to associate the electron irradiation-induced vacancy in Si([As] $= 10^{20}$ cm^{-3}) with a V–As pair. The influence of As next to the vacancy is clearly visible as the enhanced intensity in the high-momentum region (Fig. 46.7). An even stronger signal from As is seen in the as-grown Si([As] $= 10^{20}$ cm^{-3}). A linear extrapolation of the intensity of the distribution shows that the native complex is V–As$_3$, i.e., the vacancy is surrounded by *three* As atoms.

The identifications are confirmed by theoretical results [46.42, 43], which are in very good agreement at both low and high momenta. The theory reproduces the linear increase of the intensity of the core electron momentum distribution with increasing number of As atoms surrounding the vacancy. For the V–As$_3$ complex the agreement with the experimental result is excellent (Fig. 46.7), whereas the intensities calculated for V–As$_2$ and V–As$_4$ are much too small or large, respectively. In the valence electron momentum range, the calculated curves for V–As and V–As$_3$ also fit very well with the experiment. To conclude, the theoretical calculations strongly support the experimental defect identifications that (i) vacancies complexed with a single donor impurities are detected in electron-irradiated P- and As-doped Si, and (ii) the native defect in Si([As] $= 10^{20}$ cm^{-3}) is a vacancy surrounded by three As atoms.

46.2.2 Vacancies in ZnO: Sublattice and Charge State

The elemental sensitivity of the Doppler broadening spectrum demonstrated above can be helpful in the identification of the sublattice of a vacancy in a binary compound. Under certain conditions, vacancies complexed with impurities can be distinguished from isolated vacancies also in compound semiconductors [46.44]. In materials such as GaN and ZnO, where the group III–II element is significantly heavier (and larger) than the rather light group V–VI element, already the positron lifetime experiments often give conclusive identification of the group III–II vacancies, and the differences in the Doppler spectra are pronounced. On the other hand, it is not evident that the group V–VI vacancies are detected at all. In materials such as GaAs or ZnSe, in which the two elements are quite similar, vacancies on both sublattices are more likely to trap positrons, but the differences between the vacancy-specific parameters can be quite subtle [46.8].

The lifetime in the defect-free ZnO lattice has been measured to be about 170 ps. After 2 MeV electron irradiation at room temperature a longer lifetime component $\tau_2 = 230 \pm 10$ ps is detected in the measured spectrum (Fig. 46.4). In addition, the average lifetime increases to $\tau_{ave} = 178$ ps at room temperature and up to 185 ps at lower temperatures. The longer lifetime component can be directly associated with Zn vacancies based on comparison to theoretical calculations that predict a difference of 60 ps between the Zn vacancy and the bulk lifetimes, when lattice relaxations around the vacancy are taken into account [46.20]. On the other hand, the positron lifetime in the O vacancy, even with a strong outward relaxation, would be at most 20–25 ps longer than in the bulk [46.35]. In order to put the identification on an even firmer basis, the Doppler broadening results obtained in the irradiated ZnO samples can be compared with the theoretical ones. Figure 46.8 represents both the experimental and theoretically calculated ratios of the Zn-vacancy-specific momentum distribution to that of the defect-free lattice. The experimental data for the Zn vacancy are extracted from the spectrum measured at 170 K with the help of the annihilation fractions obtained from the lifetime measurements at the same temperature in the same way as in the case of the Si vacancies in Sect. 46.2.1. The data calculated for the O vacancy are similar to those in the defect-free lattice and are not shown in the figure. The agreement between the theoretically calculated and experimentally determined ratio curves is excellent, strengthening

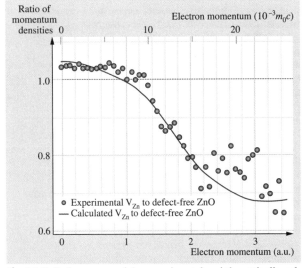

Fig. 46.8 Ratios curves of the experimental and theoretically calculated momentum densities specific to the Zn vacancy in ZnO

the identification of the irradiation-induced vacancy detected in ZnO as the Zn vacancy.

Figure 46.9 shows the positron lifetime measured as a function of temperature in as-grown and electron-irradiated high-quality (EaglePicher) n-type ZnO samples [46.35]. At 300–500 K the average positron lifetime in the as-grown sample is constant or very slightly increasing due to thermal expansion of the crystal lattice. It provides the lifetime of the positron in the delocalized state in the ZnO lattice, $\tau_B = 170$ ps at 300 K. The increase in the average positron lifetime with decreasing temperature at 10–300 K is a clear indication of the presence of negatively charged vacancies, the positron trapping coefficient of which increases with decreasing temperature (Sect. 46.1). A longer lifetime component of $\tau_2 = 265 \pm 25$ ps could be separated from the lifetime spectrum at 10 K, indicating that the vacancy in question is the Zn vacancy. The large uncertainty in τ_2 is due to the fact that the increase in τ_{ave} is very small, only 3 ps (the vacancy signal is weak due to a low concentration). The same increase in the average positron lifetime with decreasing temperature at 10–300 K is observed in the data from the irradiated sample (Fig. 46.9). Here τ_{ave} is clearly above the bulk value τ_B and thus the decomposition of the lifetime spectra could be performed with much greater accuracy. The longer lifetime component is presented in the upper part of Fig. 46.9, and its average value is that specific to the Zn vacancy as discussed above, $\tau_2 = 230 \pm 10$ ps, within experimental accuracy the same as in the as-grown sample. The decrease in the average positron lifetime with decreasing temperature in the irradiated samples below 200 K is due to negative non-open volume defects (negative ions) acting as shallow traps for positrons and competing with the vacancies in trapping of positrons. Their effects are discussed in more detail in the next section.

The Doppler broadening parameters S and W, defined in Sect. 46.1, measured simultaneously with the positron lifetime fall on a line plotted against each other with temperature as the running parameter, which typically indicates the presence of only two distinguishable positron states (bulk and vacancy). The negative-ion-type defects (shallow traps) do not cause deviations from the straight line, since the annihilation parameters of positrons trapped at these defects coincide with those of the bulk. However, as can be seen in Fig. 46.10, the points measured at 90–190 K fall off the straight line determined by the annihilations in the defect-free lattice and the Zn vacancy in the (S, τ_{ave}) and (W, τ_{ave}) plots. This implies that a third positron state can be distinguished in the lifetime versus Doppler parameter data, although the Doppler data alone are linear. In order to cause a deviation from the straight line, the localization to this defect needs to be strong, implying that the defect has a distinguishable open volume. The open volume of this defect cannot be very large, since the independence of temperature of the longer lifetime component (Fig. 46.9) shows no evidence of possible mixing of several lifetime components. Hence, the lifetime specific to this defect needs to be sufficiently far from τ_2 (and closer to τ_1), below 200 ps. In addition, in order to produce the deviation from the straight line seen in Fig. 46.10, the defect-specific lifetime needs to be above τ_{ave} over the whole temperature range, i.e., above 185 ps. One additional aspect of the third type of defects is evident from the data. The fraction of positrons annihilating as trapped at this defect is vanishing at room temperature and clearly

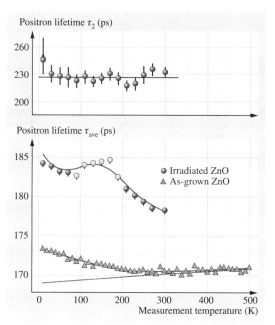

Fig. 46.9 The positron lifetime parameters of as-grown and electron-irradiated ZnO samples plotted as functions of measurement temperature. The data markers are drawn as open in the temperature range 90–190 K, where the effect of the O vacancies is the most visible. The *dashed line* shows the fitted bulk lifetime, where the temperature dependence is due to thermal expansion of the lattice. The *solid curves* represent the fitting of the temperature-dependent trapping model to the data [46.35]

smaller than the annihilation fractions at the Zn vacancies and negative-ion-type defects below 90 K, but larger at the intermediate temperature 90–190 K. This implies that the enhancement of positron trapping with decreasing temperature is larger at this defect at temperatures 190–300 K, but saturates around 150 K, where the Zn vacancies and the negative-ion-type defects become more important. This indicates that the third type of defect is neutral, and the temperature dependence of positron trapping observed at temperatures close to room temperature is due to either thermal escape from the defect or a change in the charge state (from neutral to positive) of the defect. Based on these considerations

and the lifetime value of 190–200 ps, a prominent candidate for this defect is the O vacancy that has a donor nature and would naturally have a smaller open volume than the Zn vacancy.

46.2.3 Negative Ions as Shallow Positron Traps in GaN

In addition to vacancy defects, negatively charged impurities and intrinsic defects with no open volume (called collectively negative ions), can trap positrons at shallow hydrogen-like states, as explained in Sect. 46.1. They can only be detected when they compete with vacancies in the trapping of positrons, and due to the weak localization of the positron at these defects they cannot be identified. However, their concentration can be estimated and compared with those (obtained by, e.g., secondary-ion mass spectrometry, SIMS) of the known impurities. Figure 46.11 shows a typical example of the temperature-dependent positron lifetime data when negative ions compete with vacancies in positron trapping. The data are measured in thick high nitrogen pressure (HNP)-grown bulk GaN crystals, with both the N and Ga polarity faces facing the positron source, and in thick homoepitaxial GaN layers grown with hydride vapor-phase epitaxy (HVPE) grown on both polarity faces of the bulk crystals [46.45]. All these samples are n-type.

The average positron lifetime is clearly above the bulk lifetime $\tau_B = 160 \pm 1$ ps in the HNP GaN and N-polar HVPE GaN samples (Fig. 46.11), indicating that positrons are trapped at vacancies. The lifetime spectra recorded at 300–500 K in those samples can be decomposed into two components. The positrons trapped at vacancies annihilate with the longer lifetime $\tau_V = \tau_2 = 235 \pm 10$ ps, characteristic of the Ga vacancy [46.44, 46, 47] that is negatively charged in n-type and semi-insulating GaN [46.44, 47, 48]. The average positron lifetime measured in the Ga-polar HVPE GaN samples coincides with the bulk lifetime $\tau_B = 160 \pm 1$ ps, indicating that the vacancy concentration in those samples is below the detection limit of about 10^{15} cm^{-3}.

At low temperatures the average positron lifetime in the HNP and N-polar HVPE GaN samples decreases and the lifetime at the Ga vacancy τ_V remains constant (Fig. 46.11). This behavior indicates that the fraction $\eta_V = (\tau_{ave} - \tau_B)/(\tau_V - \tau_B)$ of positrons annihilating at vacancies decreases. Since the positron trapping at negative Ga vacancies should be enhanced at low temperatures (Sect. 46.1), the decrease of η_V is due to other defects which compete with Ga vacancies as positron

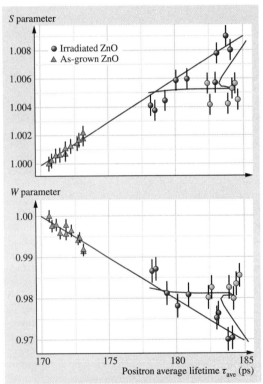

Fig. 46.10 The S and W parameters measured in as-grown and electron-irradiated ZnO samples plotted as functions of τ_{ave} with temperature as the running parameter. The data markers are drawn as open in the temperature range 90–190 K, where the effect of the O vacancies is the most visible. The *solid lines* connect the parameters of the bulk lattice to those (not shown) specific to the Zn vacancy. The *solid curves* are obtained from the fitting of the temperature-dependent trapping model [46.35]

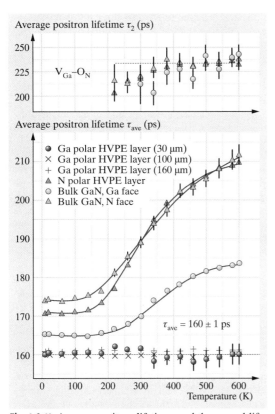

Fig. 46.11 Average positron lifetimes and the second lifetime components extracted from the lifetime spectra measured in HNP and HVPE GaN samples. The *solid curves* represent the fits of the temperature-dependent trapping model. The *solid lines* in the upper and lower parts of the figure show the average values of τ_{ave} and τ_2 in regions of no temperature dependence [46.45]

traps. Negative ions are able to bind positrons at shallow (< 0.1 eV) hydrogenic states in their attractive Coulomb field (Sect. 46.1). Since they possess no open volume, the lifetime of positrons trapped at them is the same

as in the defect-free lattice, $\tau_{\text{ion}} = \tau_B = 160 \pm 1$ ps. The average lifetime increases above 150 K when positrons start to escape from the ions and a larger fraction of them annihilates at vacancies.

The temperature dependence of the average lifetime can be modeled with the kinetic trapping equations introduced in Sect. 46.1. The positron trapping coefficients at negative Ga vacancies μ_V and negative ions μ_{ion} vary as $T^{-1/2}$ as a function of temperature [46.3,8]. The positron escape rate from the ions can be expressed as $\delta(T) \propto \mu_{\text{ion}} T^{-3/2} \exp(-E_{\text{ion}}/k_B T)$, where E_{ion} is the positron binding energy at the Rydberg state of the ions (46.10). The fractions of annihilations at Ga vacancies η_V and at negative ions η_{ion} are given in (46.16–46.17) and they depend on the concentrations $c_V = \kappa_V/\mu_V$ and $c_{\text{ion}} = \kappa_{\text{ion}}/\mu_{\text{ion}}$ of Ga vacancies and negative ions (46.9), respectively, as well as on the detrapping rate $\delta_{\text{ion}}(T)$ (46.10). We take the conventional value $\mu_V = 2 \times 10^{15}$ s^{-1} for the positron trapping coefficient at 300 K [46.8, 9]. Inserting the annihilation fractions η_B, η_{ion}, and η_V from (46.15–46.17) into the equation for the average lifetime $\tau_{\text{ave}} = \eta_B \tau_B + \eta_{\text{ion}} \tau_{\text{ion}} + \eta_V \tau_V$ (46.18), the resulting formula can be fitted to the experimental data of Fig. 46.11 with c_V, c_{ion}, μ_{ion}, and E_{ion} as adjustable parameters. As indicated by the solid lines in Fig. 46.11, the fits reproduce well the experimental data with the positron binding energy of $E_{\text{ion}} = 60 \pm 10$ meV and trapping coefficient $\mu_{\text{ion}} = (7 \pm 4) \times 10^{16} (T/K)^{-0.5}$. These values are close to those obtained previously in, e.g., GaAs and GaN [46.8, 28, 46].

Even though the negative ions cannot be identified based on the positron data alone, their concentrations can be determined and compared with acceptor impurity concentrations measured with SIMS. In as-grown GaN samples, the negative ion concentrations obtained from positron experiments correlate with the concentrations of magnesium (Mg) impurities in both HVPE- and HNP-grown samples [46.45, 46]. This indicates that, as expected, the Mg impurities act as compensating defects in n-type GaN.

46.3 Defects, Doping, and Electrical Compensation

The concentrations of the defects detected by positrons can be estimated from their respective annihilation fractions. This can be done in a straightforward manner even when the exact defect structure such as the decoration by impurities of a vacancy remains unresolved. The experimental annihilation fractions depend on the trapping rates to the different defects through (46.16–46.17), and the trapping rate κ_D to a specific defect is directly proportional to the defect concentration c_D (46.9). The proportionality constant is called the trapping coefficient μ_D (Sect. 46.1). The values and behavior as a function of temperature of the trapping coefficients depend on the type and charge state of the defect, and have been estimated both experimentally and by theoretical calculations [46.25–27]. Even though the absolute magnitude of the trapping coefficient may be off by a factor of two or three from the physically proper value, differences in vacancy concentrations in a given material can be detected with the high accuracy of the lifetime experiment: changes as low as 1 ps can be reliably detected in the average positron lifetime of about 200 ps. In this section studies of vacancy–donor complexes in Si and GaN are presented.

46.3.1 Formation of Vacancy–Donor Complexes in Highly n-Type Si

Doping levels up to 10^{20} cm^{-3} are used in current device technologies. In n-type doping of Si with arsenic, however, fundamental material problems start to appear when the impurity concentration increases above $\approx 3 \times 10^{20}$ cm^{-3} [46.50, 51]. The concentration of the free carriers (electrons) does not increase linearly with the doping concentration, indicating that inactive impurity clusters or compensating defects are formed. Furthermore, the diffusion coefficient of As starts to increase rapidly at [As] > 3×10^{20} cm^{-3}, demonstrating that new migration mechanisms become dominant [46.52].

Both the electrical deactivation of dopants and the enhanced As diffusion have often been attributed to the formation of vacancy–impurity complexes [46.51]. According to theoretical calculations, vacancies surrounded by several As atoms (V–As$_n$), $n > 2$) have negative formation energies, suggesting that these complexes are abundantly present at any doping level [46.53, 54]. The formation of these defects is however limited by kinetic processes such as the migration of As. The calculations predict that also the V–As$_2$ complex is mobile at relatively low temperatures, enabling the formation of higher-order V–As$_n$ complexes [46.54, 55]. As shown in Sect. 46.2, the dominant structure of vacancy complexes has been identified as V–As$_3$ in Czochralski (CZ) Si doped up to [As] = 10^{20} cm^{-3} [46.43, 56].

In order to verify the formation mechanism of the V–As$_3$ complexes in highly As-doped Si, electron-irradiated Si ([As] = 10^{20} cm^{-3}) samples, where V–As pairs were observed as dominant vacancy complexes defects, were subjected to thermal annealing experiments. Figure 46.12 shows the behavior of the average positron lifetime measured at room temperature as a function of the annealing temperature [46.49, 57]. For comparison, data obtained in P-doped Si ([P] = 10^{20} cm^{-3}) samples subjected to similar irradiation and annealing are also shown in the figure. In Si ([As] = 10^{20} cm^{-3}) the lifetime is around 242 ps up to 1100 K, indicating the presence of monovacancies. The peak at 500 K is due to the formation and annealing of divacancy defects [46.49, 57]. In the P-doped sample the average lifetime is initially higher due to a slightly higher concentration of divacancies, and decreases dramatically already after the annealing at 700 K, indicating that the vacancy concentration decreases at a lower temperature than in As-doped Si.

Coincidence Doppler broadening measurements were performed simultaneously with the lifetime experiments. As explained in Sect. 46.2.1, the vacancy defects observed after irradiation in As-doped Si are the V–As pairs (Fig. 46.7). After annealing at 600 K, the intensity of the high-momentum part (2–4 a.u.) has increased, and the defect in question can be identified as the V–As$_2$ complex by comparison with theoretical cal-

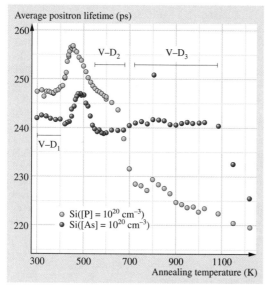

Fig. 46.12 Average positron lifetime measured at room temperature as a function of annealing temperature for the electron-irradiated Si samples. The annealings were performed isochronally (30 min at each temperature) in vacuum [46.49]. The regions where each of the vacancy–donor complex (V–D$_n$) is the dominant defect are also shown

culations. After annealing at 775 K, the intensity of the high-momentum part further increases up to the level of the V–As$_3$ complex (Fig. 46.7). The measurement after annealing at 1000 K gives the same Doppler-broadened spectrum, indicating that the defects are still the V–As$_3$ complexes formed around 700 K. After 1100 K the average lifetime (Fig. 46.12) starts to decrease, indicating that the vacancy defects are annealing away.

The average lifetime data in Fig. 46.12 suggests that similar dissociation and complexing of vacancy–donor pairs occurs in P-doped Si as well. It is worth noticing that the divacancy-related peak in the average lifetime is at a slightly lower temperature and the peak is slightly higher, demonstrating the fact that the V–P pairs formed in the irradiation are a slightly less stable than the V–As pairs. This, together with the initially higher average lifetime, shows that the formation of V$_2$ is more efficient in P-doped Si. In both the As- and P-doped samples no divacancies are observed (based on the separation of the lifetime components) after the annealing at 600 K. In P-doped Si, the high momentum part of the Doppler-broadened spectrum cannot be used in the identification of the complexes with different numbers of P atoms due to the similarity of the core electron distributions of P and Si (Sect. 46.2.1). On the other hand, the distribution in the valence region (momentum values below 0.5 a.u.) after the 600 K annealing is clearly broader than in the as-irradiated sample. This effect is observed in the case of As doping as well and is reproduced by theory. The broadening indicates the increased presence of positive P ions next to the vacancy causing increased valence electron density, leading to increased electron momentum. The valence region broadens further after 775 K annealing. Hence a similar conversion of V–P to V–P$_2$ and finally V–P$_3$ is observed in P-doped Si as for the V–As$_n$ complexes in As-doped Si. However, the dissociations seem to be more dominant in P-doped Si, since only a small fraction of the V–P pairs are finally converted to V–P$_3$ by annealing. The difference may be related to the lower binding energy of the P-decorated vacancy complex as manifested by the lower annealing temperature of the V–P pairs compared with that of the V–As pairs [46.57].

In addition to particle irradiation, vacancies are formed in thermal equilibrium at relevant concentrations in highly doped Si when the sample temperature exceeds 650 K [46.58,59]. The vacancies are formed directly next to the dopant atoms, which reduces the high formation energy of the isolated vacancy of about 3 eV in highly n-type Si by the Coulomb binding energy and the ionization energy down to about 1 eV [46.58]. The vacancy–donor pairs diffuse rapidly at these temperatures and finally form complexes with three donor atoms either already at the annealing temperature or during cooling down. These results have been obtained by both isochronal and isothermal annealing experiments, and positron measurements both at room temperature in between the annealings and in situ during the annealings. An interesting feature is that the detection of V–As$_3$ complexes at high temperatures is difficult due to positrons escaping from them above 500 K, indicating that the binding energy is about 0.25 eV [46.59], significantly lower than that typically observed (at least about 1 eV) for vacancy defects in semiconductors, as for example, in V–P$_3$.

The vacancy concentrations can be estimated in a straightforward manner when the vacancy-specific and bulk annihilation parameters are known. Equation (46.25) provides the relationship between the trapping rate and the experimental parameters. The vacancy concentration is proportional to the trapping coefficient, and can be obtained, e.g., from the Doppler data as

$$[V] = \frac{N_{at}}{\mu_V \tau_b} \frac{W - W_B}{W_V - W} , \qquad (46.29)$$

where N_{at} is the atomic density. In the following, we have used the positron trapping coefficient $\mu_V = 10^{15}\,\text{s}^{-1}$ for the Si vacancies and $\tau_B = 218\,\text{ps}$ for the positron lifetime in defect-free Si.

The concentration of the V–As$_3$ complexes in Czochralski-grown Si([As] = 10^{20} cm^{-3}) is only 0.1% of the As concentration [46.43], and the material does not show substantial electrical deactivation. In addition, molecular-beam epitaxy (MBE) can be applied to achieve metastable n-type doping with Sb that becomes compensated only at 10^{21} cm^{-3} [46.60]. In order to illustrate the importance of vacancy defects in the compensation, selected electrical and positron results obtained with a slow positron beam in MBE-grown Si thin layers with [As] > 10^{20} cm^{-3} and [Sb] = 2.7×10^{19}–3.7×10^{21} cm^{-3} [46.61, 62] are shown in Table 46.1. Here only the Doppler broadening spectra have been measured and the identification of the vacancy–donor complexes is based on the characteristic data acquired in the experiments on the electron-irradiated samples presented above.

The results show that the as-grown As-doped MBE Si layers are efficiently compensated, but the doping can be activated through rapid thermal annealing (RTA), a routinely used technique in device processing for this purpose. The total vacancy concentrations in these sam-

Table 46.1 Donor concentrations and electrical activities of highly As- and Sb-doped MBE Si samples. The total vacancy concentrations have been determined by combining electrical and positron experiments [46.61, 62]

Donor concentration (cm^{-3})	Description	Electrical activity (%)	Total [V] (cm^{-3})
[As] = 1.5×10^{20}	As-grown (720 K)	20	4×10^{19}
[As] = 1.5×10^{20}	RTA 1170 K	98	1×10^{18}
[As] = 3.5×10^{20}	As-grown (720 K)	2	1×10^{20}
[As] = 3.5×10^{20}	RTA 1170 K	85	1.5×10^{19}
[Sb] = 2.7×10^{19}	As-grown (550 K)	90	$\leq 10^{19}$
[Sb] = 5.9×10^{20}	As-grown (550 K)	70	5×10^{19}
[Sb] = 9.4×10^{20}	As-grown (550 K)	70	1.5×10^{20}
[Sb] = 3.7×10^{21}	As-grown (550 K)	6	9×10^{20}

ples, which are high enough to be dominant in the compensation, are dramatically reduced in the RTA treatments. The vacancies in question are dominantly the V–As$_3$ complexes, but also more complicated defect structures such as V$_2$–As$_5$ give a nonvanishing contribution [46.61].

The highly Sb-doped MBE Si layers grown at low temperature, where the V–Sb$_2$ complex is stable [46.57], contain defects such as V–Sb$_2$ and V$_2$–Sb$_2$ instead of V–Sb$_3$ [46.62]. It is clearly seen from the total vacancy concentrations that they play an important role in the compensation of the Sb-doped layers as well. However, the electrical activity of the as-grown samples is far better than those grown at higher temperatures (and doped with As). This can be explained by the migration processes described above, i.e., the V–Sb$_2$ do not diffuse and form V–Sb$_3$ complexes, enabling a larger fraction of the dopants to be isolated and active.

46.3.2 Vacancies as Dominant Compensating Centers in n-Type GaN

Gallium nitride (GaN) is an important wide-band-gap semiconductor for optoelectronic and electronic applications. It can be grown by several methods, each of which have partially different impurity and defect characteristics. The oxygen and silicon impurities dope GaN to an n-type semiconductor. The conductivity of n-type GaN is partly compensated by Ga vacancies [46.46]. It is now possible to perform more exact and quantitative studies of defects in GaN due to the improved quality of the material over the past few years. In particular, GaN grown by hydride vapor-phase epitaxy (HVPE) has low residual impurity (10^{16} cm^{-3}) and dislocation ($< 10^8$ cm^{-2}) densities. In such a material, the relation between intentional doping and compensating defects can be systematically studied.

The average positron lifetimes measured in free-standing HVPE GaN samples with different levels of intentional doping with oxygen are shown as a function of measurement temperature in Fig. 46.13 [46.44, 63]. Also data from an earlier study of nominally undoped but contaminated by O (concentration in the 10^{17} cm^{-3} range) 40 μm thick HVPE GaN on sapphire is shown for comparison [46.47]. The average lifetime is higher than in the defect-free GaN lattice ($\tau_B = 160$ ps) in all the samples, indicating that positrons are trapped at vacancy defects. The vacancy defect present is related to V$_{Ga}$, since the decomposition reveals the component

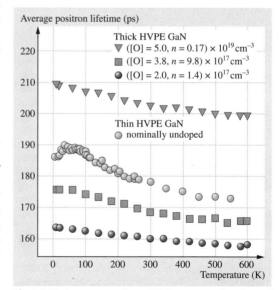

Fig. 46.13 Average positron lifetime as a function of measurement temperature in intentionally oxygen-doped free-standing HVPE GaN samples and one nominally undoped 40 μm thick HVPE GaN/Al$_2$O$_3$ sample

$\tau_2 = 235 \pm 5$ ps. There is a clear increasing trend of the average lifetime with the oxygen and free electron densities, indicating that the Ga vacancy concentration follows that of oxygen.

The increase of the average positron lifetime with decreasing temperature shows that the positron trapping rate increases at low temperatures. The temperature dependence of the average positron lifetime is totally reversible and reproducible, indicating that the concentration of Ga vacancies remains constant. Because the Fermi level is close to the conduction band, the charge states of acceptor-like defects such as V_{Ga} do not change with the measurement temperature. Hence, as explained in Sect. 46.1, this effect is a direct indication that the Ga vacancies are in the negative charge state. This is in excellent agreement with the results of theoretical calculations [46.64] that predict a charge state of 3− for the isolated Ga vacancy and 2− for V_{Ga}–O_N and V_{Ga}–Si_{Ga} complexes in n-type GaN.

The positron data show further that the defects involving Ga vacancies are the *dominant* negatively charged acceptors in the samples. The enhancement of positron trapping at low temperatures would not be observed if other negative centers competed with V_{Ga} as positron traps. For example, negative ions such as Mg_{Ga}^+ localize positrons at hydrogenic states at low temperatures, strongly decreasing the fraction of positron annihilations at vacancy defects and consequently the average positron lifetime (Sect. 46.2.3). The same behavior could be expected for possible neutral or negative charge states of the N vacancy, where the positron lifetime is very close to that of the GaN lattice.

The Ga vacancy concentrations in the GaN samples can be estimated from the positron results using (46.25). They range from 4×10^{15} cm^{-3} in the lowest-doped ([O] $= 2 \times 10^{17}$ cm^{-3}) to about 10^{17} cm^{-3} in the highest-doped sample ([O] $= 2 \times 10^{20}$ cm^{-3}), in correlation with the O concentration. According to electron irradiation studies, the isolated V_{Ga} is mobile already at 600 K [46.65], i.e., at much lower temperatures than applied in the HVPE growth. However, the Ga vacancies bound to defect complexes such as V_{Ga}–O_N have a considerably higher thermal stability [46.45, 65, 66]. In fact, recent detailed studies of the electron momentum density show that the dominant vacancy defect, responsible for the positron lifetime of 235 ps, is a complex of Ga vacancy and oxygen [46.44].

Unlike in the case of vacancy–donor pairs in highly n-type silicon (see previous section), the Ga vacancies are formed as isolated during growth of n-type GaN. They migrate fast at the high growth temperatures that are typically above 1300 K in the case of bulk or quasibulk crystals, and are stabilized (quenched) by donor impurities during cooling down. This is demonstrated by the fact that the V_{Ga}–O_N concentrations are similar in materials grown by HVPE and the high-nitrogen-pressure method when the O concentrations are similar, in spite of the large difference of about 500 K in the growth temperatures [46.45]. In fact, it has been shown that the V_{Ga}–O_N pairs are stable up to about 1300 K [46.66]. On the other hand, the concentration of Ga vacancies in Si-doped n-type GaN is significantly lower than in O-doped n-type GaN with similar free electron concentration due to the lower binding energy of the V_{Ga}–Si_{Ga} pair originating from the larger distance between the individual acceptor (Ga vacancy) and donor (substitutional Si) defects [46.67].

The positron results show that Ga vacancies act as dominant compensating centers in n-type GaN. On the other hand, in p-type GaN, where the formation of Ga vacancies is energetically unfavorable due their acceptor nature, the natural question is whether N vacancies could compensate the doping. The detection of vacancy defects on the N sublattice with positrons is not evident due to the small open volume generated by the missing N atom. Nevertheless evidence of the existence N vacancies complexed with Mg (V_N–Mg_{Ga}) has been obtained with positrons in Mg-doped (p-type) GaN grown by metalorganic chemical vapor deposition (MOCVD) [46.68, 69]. It is worth noting that, even though the vacancy concentrations are similar relative to the doping densities (a few percent at most) in both n- and p-type GaN, the vacancies are dominant compensating centers only in n-type GaN, while in p-type GaN other defects and impurities such as hydrogen play the most important role.

46.4 Point Defects and Growth Conditions

Compound semiconductor thin films can be epitaxially grown by several methods such as metalorganic chemical vapor deposition (MOCVD) or molecular-beam epitaxy (MBE). There are many controllable growth parameters that affect the properties of the overgrown layers, such as the growth rate, stoichiometry, and temperature. In addition, the layer properties may depend on the choice of the substrate material and the orientation of the substrate or the layer. Especially in the case of heteroepitaxy, the layer properties may vary significantly with the distance from the layer/substrate interface. The identities and quantities of both extended and point defects are affected by these parameters. In this section we will also describe how point defects can be studied in thin semiconductor layers by using a variable-energy positron beam.

46.4.1 Growth Stoichiometry: GaN Versus InN

The effect of the growth stoichiometry on the formation of cation vacancies in GaN and InN has been studied by measuring a set of samples grown by MOCVD employing different V/III molar ratios [46.70, 71]. The growth rate as well as the electrical and optical properties of the nitride samples change strongly with the V/III molar ratio [46.72, 73]. All the samples were investigated at room temperature as a function of the positron beam energy E. When positrons are implanted close to the sample surface with $E = 0$–1 keV, the same S parameter of $S = 0.49$ is recorded in all the GaN and InN samples. These values characterize the defects

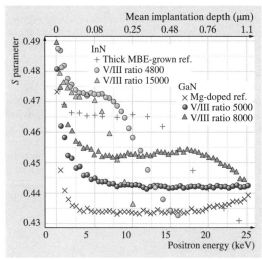

Fig. 46.14 The low-electron-momentum parameter S as a function of the positron implantation energy in three GaN and three InN samples. The *top axis* shows the mean stopping depth corresponding to the positron implantation energy

and chemical nature of the near-surface region of the samples at the depth 0–5 nm. S parameter data from selected GaN and InN samples are shown as a function of positron implantation energy in Fig. 46.14. In the GaN samples the S parameter is constant at 5–15 keV, indicating that all positrons annihilate in the GaN layer (Fig. 46.14), while in the InN samples the region of constant S is different from sample to sample due to the different thicknesses of the layers. The data recorded at the energies where S is constant can be taken as characteristic of the layer. In the case of GaN, the lowest S parameter is obtained in the Mg-doped reference layer [46.46], while for InN the reference value was obtained from a several microns thick layer grown by MBE [46.74]. The values in these samples correspond to positrons annihilating as delocalized particles in the defect-free lattice.

The S parameters in all the measured layers are larger than in the reference samples, as in Fig. 46.14. The increased S parameter indicates that the positron–electron momentum distribution is narrower than in the defect-free reference sample. The narrowing is due to positrons annihilating at vacancy defects, where the electron density is lower and the probability of annihilation with high-momentum core electrons is reduced compared with that of delocalized positrons in the lattice (Sect. 46.1). The increased S parameter is thus a clear sign of vacancy defects present in the measured layers.

The number of different vacancy defects trapping positrons can be investigated through the linearity between the low- and high-electron-momentum parameters S and W. If only a single type of vacancy is present, the W parameter depends linearly on the S parameter when the fraction of positron annihilations at vacancies η_V varies. The plot of the W parameter versus S parameter thus forms a line between the endpoints (S_B, W_B) and (S_V, W_V) corresponding to the defect-free lattice and the total positron trapping at vacancies, respectively. The S and W parameters of all the GaN and InN samples are plotted in Fig. 46.15. All the data points measured in GaN fall on the line connecting the parameters obtained in the Mg-doped GaN reference sample and those determined in earlier studies for the native Ga vacancy [46.46]. Hence Ga vacancies are found in all the GaN samples. Similarly the data points measured in the InN samples fall on the line connecting the parameters of the defect-free InN lattice and those of the In vacancy, also determined in an earlier study [46.74], indicating the presence of In vacancies in these samples. In GaN, the positron trapping fraction η_V and the

S parameter vary from one sample to another due to the different vacancy concentrations, while in InN the data indicates that the In vacancy concentrations are similar in all the samples. It is worth noticing that the data recorded in the InN samples are shifted with respect to that measured in the GaN samples by the differences in the parameter levels in the two materials: the S parameter is higher and W lower in defect-free InN than in defect-free GaN due to the differences in the electronic structures.

The presence of Ga vacancies is expected in n-type undoped GaN due to their low formation energy. The different levels of the S parameter in Figs. 46.14 and 46.15 indicate that the concentration of the Ga vacancies depends on the stoichiometry of growth. On the other hand, in InN the data are similar for all the samples with different V/III ratios, indicating that the In vacancy concentration is independent of the growth stoichiometry.

In order to quantify the concentration of V_{Ga} and V_{In} the S parameter data was analyzed with the positron trapping model. When the cation vacancies are the only defects trapping positrons, their concentration can be determined with the simple formula (46.25)

$$[V_{Ga/In}] = \frac{N_{at}}{\mu_V \tau_b} \frac{S - S_B}{S_V - S} \qquad (46.30)$$

where $\tau_B = 160$ ps (185 ps) is the positron lifetime in the GaN (InN) lattice [46.46, 74], $\mu_V = 3 \times 10^{15}$ s^{-1} is the positron trapping coefficient [46.8], and $N_{at} = 8.775 \times 10^{22}$ cm^{-3} (6.367×10^{22} cm^{-3}) is the atomic density of GaN (InN). For the S parameter at the Ga and In vacancies we take $S_V/S_B = 1.050$ [46.46, 70, 71, 74].

The results in Fig. 46.16 indicate that the concentration of Ga vacancies in GaN is proportional to the stoichiometry of the growth conditions. Rather low $[V_{Ga}] \approx 10^{16}$ cm^{-3} is observed for the sample with the V/III molar ratio of 1000. When the V/III molar ratio becomes 10 000, the V_{Ga} concentration increases by almost three orders of magnitude to $[V_{Ga}] \approx 10^{19}$ cm^{-3}. This behavior shows that empty Ga lattice sites are likely formed in the strongly N-rich environment. In contrast, the concentration of In vacancies in InN remains constant at the level of $[V_{In}] \approx 10^{17}$ cm^{-3} over the whole range of V/III molar ratios from about 3000 to 24 000. The In vacancy concentration in these samples is on the same level as in samples of similar thickness grown by MBE where the growth conditions are much closer to being stoichiometric, suggesting that the In vacancy formation is dominated by thickness-dependent properties such as strain or dislocation

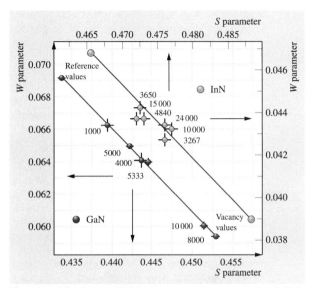

Fig. 46.15 The low- and high-momentum parameters S and W in various GaN and InN samples. The V/III molar ratio of each sample is indicated in the figure. The *straight line* indicates that the same vacancy defect (group III vacancy) is observed in all samples. Note the *different scales* on the *left/right* and *top/bottom* axes

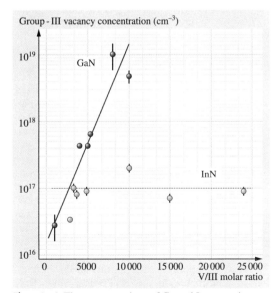

Fig. 46.16 The concentrations of Ga and In vacancies versus the V/III molar ratio in GaN and InN samples grown by MOCVD. The *straight lines* are drawn to emphasize the correlations [46.70, 71]

density. In fact, in MBE-grown InN, the In vacancy concentration drops by several orders of magnitude when the layer thickness increases from a few hundred nanometers to a few microns [46.74].

The difference in the behavior of the cation vacancies in GaN and InN as a function of the V/III molar ratio can be explained by the differences in the calculated vacancy formation energies and the temperatures of the MOCVD growth. In n-type material the calculated formation energy of the Ga vacancy is about 1.3 eV, while it is about 2.6 eV in InN [46.64, 75]. On the other hand, the growth temperature of the GaN samples was around 1000 °C, but only 550 °C in the case of InN samples [46.70, 71]. As the concentrations of the Ga and In vacancies are similar in the samples with low V/III ratios, the formation of the In vacancies must be dictated by other effects, such as strain or presence of dislocation, than the thermal formation (and subsequent stabilization by, e.g., impurities) of an isolated In vacancy in an otherwise perfect lattice. On the other hand, the observed Ga vacancy concentrations are of the same order of magnitude that could be expected from the growth temperature and the calculated formation energy, given that the vacancies (which are mobile already at relatively low temperatures) are stabilized by, e.g., O impurities close to the growth temperature. Hence it is understandable that the stoichiometric conditions affect the final Ga vacancy concentration in GaN more than the In vacancy concentration does in InN.

46.4.2 GaN: Effects of Growth Polarity

The wurtzite structure of GaN introduces effects related to the growth polarity of the layer. In the following, studies of both polar (Ga or N polarity) and nonpolar layers are reviewed [46.45, 47, 76]. The positron lifetime was measured in GaN layers grown by hydride vapor-phase epitaxy (HVPE) on dislocation-free high-pressure (HNP) bulk GaN crystals to thicknesses 30–160 μm [46.45]. Four of the layers were grown on the Ga face and one layer on the N face of the HNP GaN substrate. One of the Ga polar layers (30 μm) was grown in the same run with the N polar layer. Apart from the thickness, the properties of the Ga polar layers were similar to each other.

The average positron lifetimes measured as a function of temperature in the HVPE and HNP GaN samples are shown in Fig. 46.11. As shown in the figure, a second lifetime component of $\tau_2 = 235 \pm 10$ ps could be separated in the lifetime spectra, indicating that positrons annihilate as trapped at Ga vacancy related defects when the average positron lifetime is above the bulk lifetime of 160 ps. On the other hand the decrease of the average positron lifetime with decreasing temperature in the samples (HNP GaN and N-polar HVPE GaN), where τ_{ave} is above τ_B, is a clear indication of the presence of negative ion defects trapping positrons at low temperature to hydrogenic states, where the positron lifetime is equal to τ_B.

Interestingly, the Ga vacancy concentrations (7×10^{17} cm^{-3}) coincide in the N-polar HVPE GaN and the N side of the HNP GaN samples, similarly as the impurity concentrations obtained from secondary-ion mass spectrometry (SIMS) experiments. On the other hand, the difference between the Ga-polar HVPE GaN ([V_{Ga}] < 10^{16} cm^{-3}) and the Ga side of the HNP GaN bulk crystal ([V_{Ga}] = 2×10^{17} cm^{-3}) is significant. These observations support the idea proposed earlier [46.77], namely that the oxygen incorporation (and subsequent Ga vacancy formation) is stronger in the nonpolar directions, in which the N-polar growth mainly proceeds. The difference between the polarities is larger in the HVPE GaN samples than in the HNP bulk GaN crystals. This can be explained by the lower temperature and pressure in HVPE growth, which reduce the oxygen diffusion, and by the presence of more oxygen in the high-pressure growth.

In order to further study the role of growth polarity on the defect incorporation in GaN, a-plane GaN layers (thicknesses 1–25 μm) grown on sapphire were measured with a variable-energy positron beam [46.76]. The S parameter measured in these layers is shown as a function of positron implantation energy in Fig. 46.17. Ga vacancies complexed with oxygen were identified in the layers, and their concentrations are shown as a function of distance from the interface in Fig. 46.18. For comparison, also the data from the c-plane GaN [46.47] are reproduced here. The difference between the polar and nonpolar HVPE GaN layers is clear: the Ga vacancy concentration is constant in the a-plane HVPE GaN, whereas it decreases with increasing distance in c-plane HVPE GaN. SIMS results show that the O concentration is also constant in the HVPE GaN layers, as is the density of extended defects (observed with cross-sectional transmission electron microscopy).

These results give further support for the model based on growth-surface-dependent oxygen incorporation and subsequent Ga vacancy formation. In c-plane heteroepitaxial Ga-polar HVPE GaN, the O concentration profile is determined by the dislocation profile likely due to diffusion from the sapphire substrate. On the other hand, in homoepitaxial c-plane Ga-polar

HVPE GaN, in which the dislocation density is low and the amount of oxygen in the substrate is significantly lower, no vacancies are observed even in the thinnest layers, while in N-polar GaN both the Ga vacancy and O concentrations are high. Hence, as the growth modes are similar in the N-polar and nonpolar GaN, it is natural that the O incorporation from the growth ambient is effective in both, giving rise to a high O concentration and subsequent Ga vacancy concentration, independently of possible extended defects.

46.4.3 Bulk Growth of ZnO

Bulk ZnO crystals can be grown by various methods. In the following, we will compare the positron results [46.21] obtained in ZnO grown by the seeded vapor-phase (VP) [46.78], skull-melt [46.79], hydrothermal (HT) [46.80], and conventional and contactless chemical vapor transport techniques (CVT and CCVT) [46.81, 82]. The VP, skull-melt, CVT, and CCVT materials are all characterized by low (below 10^{17} cm^{-3}) impurity concentrations. In HT ZnO the concentration of the most abundant impurity, lithium, is in the 10^{18} cm^{-3} range in these samples. This is a general property of hydrothermally grown ZnO. All the studied samples were nominally undoped. ZnO grown by the HT method had high resistivity, likely due to Li, while the other materials were all slightly n-type due to residual impurities and/or intrinsic defects.

The average positron lifetimes measured as a function of temperature in all the different bulk ZnO crystals are collected in Fig. 46.19. As seen in the figure, the samples can be roughly divided into two groups, where in one the average lifetime τ_{ave} is very close to the bulk lifetime of $\tau_B = 170$ ps and in the other it is clearly above τ_B, in the 175–185 ps range. The ZnO crystals grown by the VP and skull-melt methods belong to the former and the crystals grown by the HT, CVT, and CCVT methods belong to the latter. As shown in the figure, a second lifetime component of $\tau_2 = 230 \pm 10$ ps (the same in all the samples) could be separated in the lifetime spectra, indicating that positrons annihilate as trapped at Zn vacancy related defects when the average positron lifetime is well above the bulk lifetime of 170 ps.

The ZnO crystals grown by the VP technique were obtained from Eagle–Picher (EP) and ZN-Technologies (ZNT), and the crystals grown by the skull-melt technique from Cermet. As explained in Sect. 46.2.2, the EP material contains Zn vacancies in the doublenegative charge state, evident from the separation of

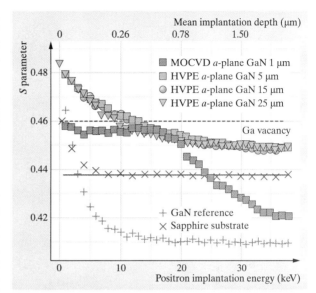

Fig. 46.17 The S parameter in the a-plane GaN layers as a function of positron implantation energy. The *solid* and *dashed lines* show the S parameter in the GaN lattice and at a Ga vacancy, respectively. Also the S parameters measured in the GaN reference and in the sapphire substrate are shown [46.21]

the lifetime components and the increase of the average lifetime with decreasing temperature. The average

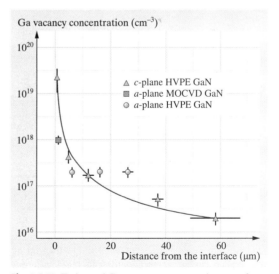

Fig. 46.18 Estimated Ga vacancy concentrations as a function of the distance from the sapphire interface [46.21, 47]

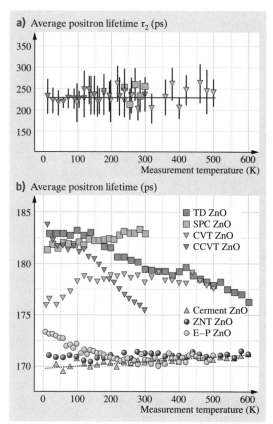

Fig. 46.19a,b The average positron lifetime (**a**) and the higher lifetime components separated from the annihilation spectra (**b**) as a function of measurement temperature in all the measured ZnO samples

small for the lifetimes to be separable from the lifetime spectrum and hence the defects cannot be conclusively identified. Interestingly, the average positron lifetime measured in Cermet ZnO coincides with the lifetime of the ZnO lattice estimated from the EP ZnO data over the whole temperature range of the measurements, and no higher components could be separated from the lifetime spectra at any temperature. This indicates that the concentration of vacancy defects in this material is below the detection limit of positron annihilation spectroscopy, i.e., well below 10^{15} cm^{-3}.

The ZnO crystals grown by the HT method were obtained from Tokyo Denpa (TD) and the Scientific Production Company (SPC). The average positron lifetime is of similar magnitude (180–185 ps) in the TD and SPC ZnO crystals, about 10–15 ps above the lattice ZnO lifetime measured in the EP ZnO, in excellent agreement with earlier reports of positron lifetimes in ZnO crystals grown by the hydrothermal method [46.83, 84]. The main difference between the TD and SPC ZnO crystals is in the behavior of the average positron lifetime with measurement temperature: the average positron lifetime decreases with increasing temperature in TD ZnO, while it increases slightly in SPC ZnO. This indicates that positrons are trapped at negatively charged vacancy defects in TD ZnO, while the data in SPC ZnO suggest that the vacancy defects are neutral and that positrons are also trapped at negatively charged non-open volume defects (negative ions). The negatively charged vacancy defects in TD ZnO are likely to be Zn vacancies.

An average lifetime higher than that in the defect-free ZnO lattice indicates that a fraction of the annihilating positrons must be trapped at vacancy defects. An average lifetime value 10–15 ps different than that in the bulk is typically enough for the separation of lifetimes, as in, e.g., the case of electron-irradiated EP ZnO (Sect. 46.2.2). However, in these hydrothermal samples the separation was possible only in SPC ZnO at 250–300 K, indicating that Zn vacancies are present in the SPC ZnO as well. The problems in the separation of the lifetime components suggest that there is a relatively high intensity of some additional lifetime between the bulk and Zn vacancy lifetimes in the lifetime spectrum. Larger vacancy defects with higher lifetimes than those of the Zn vacancy are not present, as there would then be no problems in the separation. A natural cause for these problems would be the presence of a relatively high concentration of neutral O vacancies (whose specific positron lifetime is $\tau_{V,O} = 195 \pm 15$ ps). In order to create such problems in the separation of the lifetimes, the

lifetime of 173 ps at 20 K corresponds to a concentration of $[V_{Zn}] = 2 \times 10^{15}$ cm^{-3}. The positron lifetime representative of the pure ZnO lattice was fitted to the data obtained above 300 K, and is shown as the dotted line in Fig. 46.19. The slight increase with increasing temperature is due to the thermal expansion of the lattice. The ZNT ZnO is nominally identical to EP ZnO, and the average positron lifetimes in these two materials coincide above 300 K, but at lower temperatures the average lifetime is lower in the ZNT ZnO (about 171 ps at 20 K compared with 173 ps in EP ZnO), but still slightly above the fitted ZnO lattice lifetime. This indicates that a very small but measurable fraction of positrons annihilates at vacancy defects (concentration lower than in EP ZnO, about 10^{15} cm^{-3}), which are presumably Zn vacancies. However, the fraction is too concentration of these O vacancy related defects needs to be in the $[V_O] \cong 10^{17}\,\mathrm{cm^{-3}}$ range. On the other hand, the concentrations of the Zn-vacancy-related defects causing the increase of the average positron lifetime with decreasing temperature in TD ZnO and evident from the separation of the lifetime components in SPC ZnO must be in the low $10^{16}\,\mathrm{cm^{-3}}$ range.

The ZnO samples grown by the CVT and CCVT methods were grown at the Institute of Physics of the Polish Academy of Sciences. The average lifetime is above the bulk lifetime $\tau_B = 170\,\mathrm{ps}$ in both the CVT- and CCVT-grown samples throughout the whole temperature range, indicating the trapping of positrons at vacancy defects. The lifetime spectra measured in both materials could be separated into two components, of which the higher was the V_{Zn}-related lifetime $\tau_2 = 230 \pm 10\,\mathrm{ps}$, as shown in Fig. 46.19. The increase of the average lifetime with decreasing temperature in CCVT ZnO is a clear indication that the Zn vacancies are in the negative charge state. On the other hand, the independence from temperature of the average positron lifetime in CVT-grown ZnO in the range 100–500 K indicates that the observed vacancies are in the neutral charge state. The decrease in the average lifetime below 100 K in the CVT ZnO sample and the flat region at 50–100 K in the CCVT ZnO sample are interpreted as positrons trapping at negative-ion-type defects, which have no open volume and hence produce the annihilation characteristics of the bulk lattice. This is observed only at low temperatures, since the negative-ion-type defects act as shallow traps for positrons, and the escape rate at elevated temperatures is faster than the annihilation rate. The effect of the negative-ion-type defects is small in the CCVT ZnO sample, and hence the Zn vacancies are the dominant negatively charged (acceptor-type) defect. The concentration of the negative Zn vacancies (or related complexes) in CCVT ZnO can be estimated as $[V_{Zn}] \cong 1.5 \times 10^{16}\,\mathrm{cm^{-3}}$. The concentration of the negative-ion-type defects can be estimated to be roughly one order of magnitude lower.

The first lifetime component τ_1 (not shown) is well below the bulk lifetime τ_B in CCVT-grown ZnO, indicating that the one-defect trapping model works. However, in CVT-grown ZnO the first lifetime component coincides with τ_B, indicating the mixing of the bulk component with a defect-specific component with a lifetime close to (but higher than) the bulk lifetime [46.85]. This can be interpreted as positrons trapping at O vacancies. The concentrations of the neutral Zn-vacancy-related complexes in CVT-grown ZnO can be estimated as $[V_{Zn}] \cong 2 \times 10^{16}\,\mathrm{cm^{-3}}$, and the negative ion concentration can be estimated to be of the same order of magnitude. The concentration of the O vacancies in CVT-grown ZnO can be estimated with trapping rate analysis as $[V_O] \cong 10^{17}\,\mathrm{cm^{-3}}$ [46.35, 85].

Interestingly, the Zn vacancy concentrations in the bulk ZnO crystals grown by the chemical vapor transport and hydrothermal methods are very similar, in spite of the growth environment being Zn-rich in the former and O-rich in the latter. Further, the presence of a rather high concentration of O-vacancy-related defects in the ZnO crystals grown by the hydrothermal method is surprising, and is likely to be connected to the high Li concentration in the material. In fact, the possibility that the defects interpreted as O vacancies could instead be some complicated complexes of Zn vacancies with Li and/or H (another light element that is quite abundant in ZnO) cannot be completely ruled out. On the other hand, the O-rich skull-melt and Zn-rich seeded vapor transport methods also produce ZnO crystals that are very much alike from the vacancy point of view. Hence it seems that, in the case of the bulk growth techniques, the formation of the vacancy defects is not greatly affected by the stoichiometry or the partial pressures of the growth environment, but rather by the residual impurities and other intrinsic defects.

An important observation to be made from Fig. 46.19 is that the higher lifetime component separated from the measured lifetime spectra (where it could be performed) is the same in all the samples over the whole measurement temperature range, i.e., $\tau_2 \cong 230\,\mathrm{ps}$. It also coincides with that obtained in electron-irradiated material, hence demonstrating that the Zn vacancies are important defects in ZnO and supporting the determination of their lifetime value. In as-grown ZnO, it is very likely that the Zn vacancies are complexed with either residual impurities or other intrinsic defects, as the irradiation-induced (likely isolated) vacancies have been shown to anneal out from the material already at rather low temperatures of about 500–600 K [46.35]. This applies to the O vacancies as well.

The results obtained for the EP ZnO and CCVT ZnO show that the Zn vacancies act as dominant acceptors in n-type ZnO, similarly to the Ga vacancies in n-type GaN (see Sect. 46.3.2). The situation is more complicated in the ZnO crystals grown by the other techniques due to either too low a concentration of vacancies in general, or too high a concentration of other types of vacancies, such as O vacancies. The Zn vacancies are most likely complexed with donor-type defects, and as they survive the cooling down from the high growth temperatures of about 1000 °C, the stabilizing donors are likely to be located on the O sublattice. In the case of cation vacancies complexed with cation-sublattice-substitutional donor defects, the binding energy is prone to be too low for the vacancies to be effectively stabilized, as in the case of Si donors in GaN [46.67]. As the total open volume of the in-grown Zn vacancies is the same as that of the irradiated Zn vacancies, it is likely that these donor defects are residual impurities, the concentrations of which are of the same order of magnitude as that of the Zn vacancies. On the other hand, the possibility of the Zn vacancies being bound to O vacancies as $V_{Zn}-V_O$ complexes cannot be completely ruled out, as the latter are very difficult to distinguish from the isolated Zn vacancies with positrons. It is important, however, to understand that the O vacancies possibly observed in the hydrothermal- and CVT-grown bulk ZnO crystals are not complexed with the Zn vacancies, as the positrons are sensitive to the total open volume of the defect.

46.5 Summary

Positron annihilation spectroscopy gives microscopic information about vacancy defects in the concentration range $10^{15}-10^{19}$ cm^{-3}. The positron lifetime is the fingerprint of the open volume associated with a defect. It is used to identify mono- and divacancies and larger vacancy clusters. The Doppler broadening of the annihilation radiation measures the momentum distribution of the annihilating electrons. It can be used to identify the nature of the atoms surrounding the vacancy. Consequently, vacancies on different sublattices of a compound semiconductor can be distinguished, and impurities associated with the vacancies can be identified. The charge state of a vacancy defect is determined by the temperature dependence of the positron trapping coefficient. Positron localization into Rydberg states around negative centers yields information about ionic acceptors that have no open volume.

Positron methods can be applied to study vacancies in both bulk crystals and epitaxial layers. The measurements in bulk crystals are straightforward, as (fast) positrons obtained directly from the radioactive source can be used. The studies of epitaxial layers and near-surface regions of bulk crystals require slow positrons. Typically in these cases a monoenergetic positron beam is used, the energy tuning of which allows for depth profiling of the samples in the range from a few nanometers to several microns.

Applications of the technique to Si, GaN, and ZnO have been presented. The Si vacancies complexed with donor impurities have been identified in highly n-type Si and their role as electrically compensating centers has been discussed. Ga vacancies complexed with oxygen are observed as native defects in n-type GaN, while N vacancies complexed with magnesium are detected in p-type GaN. The effects of growth conditions on the formation of group III vacancies have been discussed for GaN and InN. In ZnO, vacancies on both sublattices have been identified and their presence in bulk ZnO crystals grown by various methods is discussed.

References

46.1 R.N. West: Positron studies of condensed matter, Adv. Phys. **22**, 263–383 (1973)
46.2 P.J. Schultz, K.G. Lynn: Interaction of positron beams with surfaces, thin films, and interfaces, Rev. Mod. Phys. **60**, 701–780 (1988)
46.3 M.J. Puska, R.M. Nieminen: Theory of positrons in solids and on solid surfaces, Rev. Mod. Phys. **66**, 841–897 (1994)
46.4 P. Asoka-Kumas, K.G. Lynn, D.O. Welch: Characterization of defects in Si and SiO_2-Si using positrons, J. Appl. Phys. **76**, 4935–4982 (1994)
46.5 P. Hautojärvi (Ed.): *Positrons in Solids*, Topics in Current Physics, Vol. 12 (Springer, Berlin Heidelberg 1979)
46.6 W. Brandt, A. Dupasquier (Eds.): *Positron Solid-State Physics* (North-Holland, Amsterdam 1983)
46.7 A. Dupasquier, A.P. Mills Jr. (Eds.): Positron spectroscopy of solids, Proc. Int. School Phys. Enrico Fermi, CXXV Course (IOS Press, Amsterdam 1995)
46.8 K. Saarinen, P. Hautojärvi, C. Corbel: Positron Annihilation Spectroscopy of Defects in Semiconductors. In: *Identification of Defects in Semiconductors*, Semiconductors and Semimetals, Vol. 51A, ed. by M. Stavola (Academic, New York 1998) pp. 209–285
46.9 R. Krause-Rehberg, H.S. Leipner: *Positron Annihilation in Semiconductors* (Springer, Berlin Heidelberg 1999)

46.10 K. Saarinen: Characterization of native point defects in GaN by positron annihilation spectroscopy. In: *III–V Nitride Semiconductors: Electrical, Structural and Defects Properties*, ed. by M.O. Manasreh (Elsevier, Amsterdam 2000) pp. 109–163

46.11 S. Valkealahti, R.M. Nieminen: Monte Carlo calculations of keV electron and positron slowing down in solids. II, Appl. Phys. A **35**, 51–59 (1984)

46.12 R.M. Nieminen, J. Oliva: Theory of positronium formation and positron emission at metal surfaces, Phys. Rev. B **22**, 2226–2247 (1980)

46.13 K.O. Jensen, A.B. Walker: Positron thermalization and non-thermal trapping in metals, J. Phys.: Condens. Matter **2**, 9757–9776 (1990)

46.14 J. Mäkinen, C. Corbel, P. Hautojärvi, D. Mathiot: Measurement of positron mobility in Si at 30–300 K, Phys. Rev. B **43**, 12114–12117 (1991)

46.15 E. Soininen, J. Mäkinen, D. Beyer, P. Hautojärvi: High-temperature positron diffusion in Si, GaAs, and Ge, Phys. Rev. B **46**, 13104–13118 (1992)

46.16 Y.Y. Shan, P. Asoka-Kumar, K.G. Lynn, S. Fung, C.D. Beling: Field effect on positron diffusion in semi-insulating GaAs, Phys. Rev. B **54**, 1982–1986 (1996)

46.17 I. Makkonen, M. Hakala, J. Puska: Modeling the momentum distributions of annihilating electron–positron pairs in solids, Phys. Rev. B **73**, 035103:1–12 (2006)

46.18 M. Alatalo, B. Barbiellini, M. Hakala, H. Kauppinen, T. Korhonen, M.J. Puska, K. Saarinen, P. Hautojärvi, R.M. Nieminen: Theoretical and experimental study of positron annihilation with core electrons in solids, Phys. Rev. B **54**, 2397–2409 (1996)

46.19 G. Brauer, W. Anwand, W. Skorupa, J. Kuriplach, O. Melikhova, C. Moisson, H. von Weckstern, H. Schmidt, M. Lorenz, M. Grundmann: Defects in virgin and N^+-implanted ZnO single crystals studied by positron annihilation, Hall effect, and deep-level transient spectroscopy, Phys. Rev. B **74**, 045208:1–10 (2006)

46.20 F. Tuomisto, V. Ranki, K. Saarinen, D.C. Look: Evidence of the Zn vacancy acting as a dominant acceptor in *n*-type ZnO, Phys. Rev. Lett. **91**, 205502-1–205502-4 (2003)

46.21 F. Tuomisto, D.C. Look: Vacancy defect distributions in bulk ZnO crystals, Proc. SPIE **6474**, 647413-1–647413-11 (2007)

46.22 I. Makkonen, M.J. Puska: Energetics of positron states trapped at vacancies in solids, Phys. Rev. B **76**, 054119-1–054119-10 (2007)

46.23 K. Saarinen, P. Hautojärvi, A. Vehanen, R. Krause, G. Dlubek: Shallow positron traps in GaAs, Phys. Rev. B **39**, 5287–5296 (1989)

46.24 C. Corbel, F. Pierre, K. Saarinen, P. Hautojärvi, P. Moser: Gallium vacancies and gallium antisites as acceptors in electron-irradiated semi-insulating GaAs, Phys. Rev. B **45**, 3386–3399 (1992)

46.25 M.J. Puska, C. Corbel, R.M. Nieminen: Positron trapping in semiconductors, Phys. Rev. B **41**, 9980–9993 (1990)

46.26 J. Mäkinen, C. Corbel, P. Hautojärvi, P. Moser, F. Pierre: Positron trapping at vacancies in electron-irradiated Si at low temperatures, Phys. Rev. B **39**, 10162–10173 (1989)

46.27 J. Mäkinen, P. Hautojärvi, C. Corbel: Positron annihilation and the charge states of the phosphorus-vacancy pair in silicon, J. Phys.: Condens. Matter **4**, 5137–5155 (1992)

46.28 K. Saarinen, S. Kuisma, J. Mäkinen, P. Hautojärvi, M. Törnqvist, C. Corbel: Introduction of metastable vacancy defects in electron-irradiated semi-insulating GaAs, Phys. Rev. B **51**, 14152–14163 (1995)

46.29 D. Schödlbauer, P. Sperr, G. Kögel, W. Triftshäuser: A pulsing system for low energy positrons, Nucl. Instrum. Methods Phys. Res. B **34**, 258–268 (1988)

46.30 R. Suzuki, Y. Kobayashi, T. Mikado, H. Ohgaki, M. Chiwaki, T. Yamazaki, T. Tomimatsu: Slow positron pulsing system for variable energy positron lifetime spectroscopy, Jpn. J. Appl. Phys. **30**, L532–L534 (1991)

46.31 K. Rytsölä, J. Nissilä, K. Kokkonen, A. Laakso, R. Aavikko, K. Saarinen: Digital measurement of positron lifetime, Appl. Surf. Sci. **194**, 260–263 (2002)

46.32 H. Saito, Y. Nagashima, T. Kurihara, T. Hyodo: A new positron lifetime spectrometer using a fast digital oscilloscope and BaF_2 scintillators, Nucl. Instrum. Methods Phys. Res. A **487**, 612–617 (2002)

46.33 J. Nissilä, K. Rytsölä, R. Aavikko, A. Laakso, K. Saarinen, P. Hautojärvi: Performance analysis of a digital positron lifetime spectrometer, Nucl. Instrum. Methods Phys. Res. A **538**, 778–789 (2005)

46.34 F. Bečvář, J. Čížek, I. Prochazka, J. Janotova: The asset of ultra-fast digitizers for positron-lifetime spectroscopy, Nucl. Instrum. Methods Phys. Res. A **539**, 372–385 (2005)

46.35 F. Tuomisto, K. Saarinen, D.C. Look, G.C. Farlow: Introduction and recovery of point defects in electron-irradiated ZnO, Phys. Rev. B **72**, 085206:1–11 (2005)

46.36 M. Alatalo, H. Kauppinen, K. Saarinen, M.J. Puska, J. Mäkinen, P. Hautojärvi, R.M. Nieminen: Identification of vacancy defects in compound semiconductors by core-electron annihilation: Application to InP, Phys. Rev. B **51**, 4176–4185 (1995)

46.37 P. Asoka-Kumar, M. Alatalo, V.J. Ghosh, A.C. Kruseman, B. Nielsen, K.G. Lynn: Increased Elemental Specificity of Positron Annihilation Spectra, Phys. Rev. Lett. **77**, 2097–2100 (1996)

46.38 K. Saarinen, V. Ranki: Identification of vacancy complexes in Si by positron annihilation, J. Phys.: Condens. Matter **15**, S2791–S2801 (2003)

46.39 G.D. Watkins: The Lattice Vacancy in Silicon. In: *Deep Centers in Semiconductors*, ed. by S.T. Pan-

46.40 H. Kauppinen, C. Corbel, J. Nissilä, K. Saarinen, P. Hautojärvi: Photoionization of the silicon divacancy studied by positron-annihilation spectroscopy, Phys. Rev. B **57**, 12911–12922 (1998)

46.41 R. Aavikko, K. Saarinen, F. Tuomisto, B. Magnusson, N.T. Son, E. Janzén: Clustering of vacancy defects in high-purity semi-insulating SiC, Phys. Rev. B **75**, 085208:1–8 (2007)

46.42 M. Hakala, M.J. Puska, R.M. Nieminen: Momentum distributions of electron–positron pairs annihilating at vacancy clusters in Si, Phys. Rev. B **57**, 7621–7627 (1998)

46.43 K. Saarinen, J. Nissilä, H. Kauppinen, M. Hakala, M.J. Puska, P. Hautojärvi, C. Corbel: Identification of Vacancy-Impurity Complexes in Highly n-Type Si, Phys. Rev. Lett. **82**, 1883–1886 (1999)

46.44 S. Hautakangas, I. Makkonen, V. Ranki, M.J. Puska, K. Saarinen, X. Xu, D.C. Look: Direct evidence of impurity decoration of Ga vacancies in GaN from positron annihilation spectroscopy, Phys. Rev. B **73**, 193301:1–4 (2006)

46.45 F. Tuomisto, K. Saarinen, B. Lucznik, I. Grzegory, H. Teisseyre, T. Suski, S. Porowski, P.R. Hageman, J. Likonen: Effect of growth polarity on vacancy defect and impurity incorporation in dislocation-free GaN, Appl. Phys. Lett. **86**, 031915:1–3 (2005)

46.46 K. Saarinen, T. Laine, S. Kuisma, J. Nissilä, P. Hautojärvi, L. Dobrzynski, J.M. Baranowski, K. Pakula, R. Stępniewski, M. Wojdak, A. Wysmolek, T. Suski, M. Leszczynski, I. Grzegory, S. Porowski: Observation of native Ga vacancies in GaN by positron annihilation, Phys. Rev. Lett. **79**, 3030–3033 (1997)

46.47 J. Oila, J. Kivioja, V. Ranki, K. Saarinen, D.C. Look, R.J. Molnar, S.S. Park, S.K. Lee, J.Y. Han: Ga vacancies as dominant intrinsic acceptors in GaN grown by hydride vapor phase epitaxy, Appl. Phys. Lett. **82**, 3433–3435 (2003)

46.48 D.C. Look, D.C. Reynolds, J.W. Hemsky, J.R. Sizelove, R.L. Jones, R.J. Molnar: Defect donor and acceptor in GaN, Phys. Rev. Lett. **79**, 2273–2276 (1997)

46.49 V. Ranki, J. Nissilä, K. Saarinen: Formation of vacancy-impurity complexes by kinetic processes in highly As-doped Si, Phys. Rev. Lett. **88**, 105506:1–4 (2002)

46.50 A. Lietoila, J.F. Gibbons, T.W. Sigmon: The solid solubility and thermal behavior of metastable concentrations of As in Si, Appl. Phys. Lett. **36**, 765–768 (1980)

46.51 P.M. Fahey, P. Griffin, J.D. Plummer: Point defects and dopant diffusion in silicon, Rev. Mod. Phys. **61**, 289–384 (1989)

46.52 A. Nylandsted Larsen, K. Kyllesbech Larsen, P.E. Andersen, B.G. Svensson: Heavy doping effects in the diffusion of group IV and V impurities in silicon, J. Appl. Phys. **73**, 691–698 (1993)

46.53 K.C. Pandey, A. Erbil, I.G.S. Cargill, R.F. Boehme, D. Vanderbildt: Annealing of heavily arsenic-doped silicon: electrical deactivation and a new defect complex, Phys. Rev. Lett. **61**, 1282–1285 (1988)

46.54 M. Ramamoorthy, S.T. Pantelides: Complex dynamical phenomena in heavily arsenic doped silicon, Phys. Rev. Lett. **76**, 4753–4756 (1996)

46.55 J. Xie, S.P. Chen: Diffusion and clustering in heavily arsenic-doped silicon: discrepancies and explanation, Phys. Rev. Lett. **83**, 1795–1798 (1999)

46.56 D.W. Lawther, U. Myler, P.J. Simpson, P.M. Rousseau, P.B. Griffin, J.D. Plummer: Vacancy generation resulting from electrical deactivation of arsenic, Appl. Phys. Lett. **67**, 3575–3577 (1995)

46.57 V. Ranki, A. Pelli, K. Saarinen: Formation of vacancy-impurity complexes by annealing elementary vacancies introduced by electron irradiation of As-, P-, and Sb-doped Si, Phys. Rev. B **69**, 115205:1–12 (2004)

46.58 V. Ranki, K. Saarinen: Formation of thermal vacancies in highly As- and P-doped Si, Phys. Rev. Lett. **93**, 255502:1–4 (2004)

46.59 K. Kuitunen, K. Saarinen, F. Tuomisto: Positron trapping kinetics in thermally generated vacancy donor complexes in highly As-doped silicon, Phys. Rev. B **75**, 045210:1–5 (2007)

46.60 H.-J. Gossmann, F.C. Unterwald, H.S. Luftman: Doping of Si thin films by low-temperature molecular beam epitaxy, J. Appl. Phys. **73**, 8237–8241 (1993)

46.61 V. Ranki, K. Saarinen: Electrical deactivation by vacancy-impurity complexes in highly As-doped Si, Phys. Rev. B **67**, 041201:1–4 (2003)

46.62 M. Rummukainen, I. Makkonen, V. Ranki, M.J. Puska, K. Saarinen, H.-J.L. Gossmann: Vacancy-impurity complexes in highly Sb-doped Si grown by molecular beam epitaxy, Phys. Rev. Lett. **94**, 165501:1–4 (2005)

46.63 K. Saarinen, S. Hautakangas, F. Tuomisto: Dominant intrinsic acceptors in GaN and ZnO, Phys. Scri. T **126**, 105–109 (2006)

46.64 J. Neugebauer, C.G. Van de Walle: Gallium vacancies and the yellow luminescence in GaN, Appl. Phys. Lett. **69**, 503–505 (1996)

46.65 K. Saarinen, T. Suski, I. Grzegory, D.C. Look: Thermal stability of isolated and complexed Ga vacancies in GaN bulk crystals, Phys. Rev. B **64**, 233201:1–4 (2001)

46.66 F. Tuomisto, K. Saarinen, T. Paskova, B. Monemar, M. Bockowski, T. Suski: Thermal stability of ingrown vacancy defects in GaN grown by hydride vapor phase epitaxy, J. Appl. Phys. **99**, 066105:1–3 (2006)

46.67 J. Oila, V. Ranki, J. Kivioja, K. Saarinen, P. Hautojärvi, J. Likonen, J.M. Baranowski, K. Pakula, T. Suski, M. Leszczynski, I. Grzegory: The influence of dopants and substrate material on the forma-

46.67 tion of Ga vacancies in epitaxial GaN layers, Phys. Rev. B **63**, 045205:1–8 (2001)

46.68 S. Hautakangas, J. Oila, M. Alatalo, K. Saarinen, L. Liszkay, D. Seghier, H.P. Gislason: Vacancy defects as compensating centers in Mg-doped GaN, Phys. Rev. Lett. **90**, 137402:1–4 (2003)

46.69 S. Hautakangas, K. Saarinen, L. Liszkay, J.A. Freitas Jr., R.L. Henry: Role of open volume defects in Mg-doped GaN films studied by positron annihilation spectroscopy, Phys. Rev. B **72**, 165303:1–10 (2005)

46.70 K. Saarinen, P. Seppälä, J. Oila, P. Hautojärvi, C. Corbel, O. Briot, R.L. Aulombard: Gallium vacancies and the growth stoichiometry of GaN studied by positron annihilation spectroscopy, Appl. Phys. Lett. **73**, 3253–3255 (1998)

46.71 A. Pelli, K. Saarinen, F. Tuomisto, S. Ruffenach, O. Briot: Influence of V/III molar ratio on the formation of In vacancies in InN grown by metal-organic vapor-phase epitaxy, Appl. Phys. Lett. **89**, 011911:1–3 (2006)

46.72 O. Briot, J.P. Alexis, S. Sanchez, B. Gil, R.L. Aulombard: Influence of the V/III molar ratio on the structural and electronic properties of MOVPE grown GaN, Solid-State Electron. **41**, 315–317 (1997)

46.73 O. Briot, B. Maleyre, S. Ruffenach: Indium nitride quantum dots grown by metalorganic vapor phase epitaxy, Appl. Phys. Lett. **83**, 2919–2921 (2003)

46.74 J. Oila, A. Kemppinen, A. Laakso, K. Saarinen, W. Egger, L. Liszkay, P. Sperr, H. Lu, W.J. Schaff: Influence of layer thickness on the formation of In vacancies in InN grown by molecular beam epitaxy, Appl. Phys. Lett. **84**, 1486–1488 (2004)

46.75 C. Stampfl, C.G. Van de Walle, D. Vogel, P. Krüger, J. Pollmann: Native defects and impurities in InN: First-principles studies using the local-density approximation and self-interaction and relaxation-corrected pseudopotentials, Phys. Rev. B **61**, 7846–7849 (2000)

46.76 F. Tuomisto, T. Paskova, R. Kröger, S. Figge, D. Hommel, B. Monemar, R. Kersting: Defect distribution in a-plane GaN on Al_2O_3, Appl. Phys. Lett. **90**, 121915:1–3 (2007)

46.77 E. Frayssinet, W. Knap, S. Krukowski, P. Perlin, P. Wisniewski, T. Suski, I. Grzegory, S. Porowski: Evidence of free carrier concentration gradient along the c-axis for undoped GaN single crystals, J. Cryst. Growth **230**, 442–447 (2001)

46.78 D.C. Look, D.C. Reynolds, J.R. Sizelove, R.L. Jones, C.W. Litton, G. Cantwell, W.C. Harsch: Electrical properties of bulk ZnO, Solid State Commun. **105**, 399–401 (1998)

46.79 D.C. Reynolds, C.W. Litton, D.C. Look, J.E. Hoelscher, B. Claflin, T.C. Collins, J. Nause, B. Nemeth: High-quality, melt-grown ZnO single crystals, J. Appl. Phys. **95**, 4802–4805 (2004)

46.80 E. Ohshima, H. Ogino, I. Niikura, K. Maeda, M. Sato, M. Ito, T. Fukuda: Growth of the 2-in-size bulk ZnO single crystals by the hydrothermal method, J. Cryst. Growth **260**, 166–170 (2004)

46.81 A. Mycielski, L. Kowalczyk, A. Szadkowski, B. Chwalisz, A. Wysmołek, R. Stępniewski, J.M. Baranowski, M. Potemski, A. Witowski, R. Jakieła, A. Barcz, B. Witkowska, W. Kaliszek, A. Jędrzejczak, A. Suchocki, E. Łusakowska, E. Kamińska: The chemical vapour transport growth of ZnO single crystals, J. Alloys Compd. **371**, 150–153 (2004)

46.82 K. Grasza, A. Mycielski: Contactless CVT growth of ZnO crystals, Phys. Status Solidi (c) **2**, 1115–1118 (2005)

46.83 S. Brunner, W. Puff, A.G. Balogh, P. Mascher: Characterization of radiation-induced defects in ZnO probed by positron annihilation spectroscopy, Mater. Sci. Forum **363-365**, 141–143 (2001)

46.84 Z.Q. Chen, S. Yamamoto, M. Maekawa, A. Kawasuso, X.L. Yuan, T. Sekiguchi: Postgrowth annealing of defects in ZnO studied by positron annihilation, x-ray diffraction, Rutherford backscattering, cathodoluminescence, and Hall measurements, J. Appl. Phys. **94**, 4807–4812 (2003)

46.85 F. Tuomisto, K. Saarinen, K. Grasza, A. Mycielski: Observation of Zn vacancies in ZnO grown by chemical vapor transport, Phys. Status Solidi (b) **243**, 794–798 (2006)